Simple Tools and Techniques for Enterprise Risk Management

First Edition Book Endorsements

Enterprise Risk Management is a necessary and valuable tool for indentifying, quantifying and mitigating risks across an organization but it is also a significant undertaking in terms of knowledge and application. In these days of fiscal, regulatory and political correctness this book addresses ERM in its broadest sense, providing useful reference and examples. Written in a clear and concise manner, the content should be of tremendous value to anyone involved in risk, audit or corporate governance whether as an analyst or board member.

(Robin Paris, Director, Group Risk, Nestlé)

This book provides an excellent introduction to enterprise risk management set in the context of strong corporate governance. The writing is clear and direct, combining a comprehensive understanding of enterprise risk with a practical and straightforward guide to tools and techniques from strategic to operational level. As a result I have no doubt that it will find its way onto the shelves of the more experienced risk managers.

(Caroline Donaldson, Director, Head of Risk, Network Rail)

Robert Chapman has distilled years of experience and produced a book which is easy to read and full of practical/useful information. Having devised and implemented an enterprise risk management process, I found much of the material instantly recognizable and relevant. My one regret is that this book was not available earlier!

(Matt Smith, Group Risk Manager, Tate & Lyle plc)

This book will be of benefit to all levels of risk practitioner and sets ERM in the context of corporate governance and internal control requirements. It provides a particularly clear description of a risk management process defined by IDEFO diagrams with a useful discussion of internal and external risk factors.

(Andrew Wood, Director, Risk Management, Serco Group plc)

Simple Tools and Techniques for Enterprise Risk Management

Second Edition

Robert J. Chapman PhD

irm
Leading the risk profession

Recommended by the Institute of Risk Management

WILEY
A John Wiley & Sons, Ltd., Publication

This edition first published 2011

Registered Office
John Wiley & Sons Ltd, The Atrium, Southern Gate, Chichester, West Sussex, PO19 8SQ, United Kingdom

For details of our global editorial offices, for customer services and for information about how to apply for permission to reuse the copyright material in this book please see our website at www.wiley.com.

Library of Congress Cataloging-in-Publication Data

Chapman, Robert J.
Simple tools and techniques for enterprise risk management / Robert J. Chapman. – 2nd ed.
p. cm.
ISBN 978-1-119-98997-4 (hbk) – ISBN 978-1-119-99065-9 (ebk) – ISBN 978-1-119-99064-2 (ebk)
1. Risk management. 2. Risk. 3. Uncertainty. 4. Decision making. I. Title.
HD61.C494 2011
658.15′5–dc23

2011042252

ISBN: 978-1-119-98997-4 (hbk) ISBN: 978-1-119-96321-9 (ebk)
ISBN: 978-1-119-99065-9 (ebk) ISBN: 978-1-119-99064-2 (ebk)

A catalogue record for this book is available from the British Library.

Set in 10/12pt Times by Aptara Inc., New Delhi, India
Printed and bound by CPI Group (UK) Ltd, Croydon, CR0 4YY

To Kay, Dominic and Gemma

Contents

List of Figures

Preface to the Second Edition

Since the publication of the first edition in 2006 the landscape of enterprise risk management (ERM) has changed dramatically. Clearly the single most prominent event has been the financial and economic "earthquake", whose epicentre lay in the United States. The "aftershocks" continue to be felt around the globe. I think it is safe to say that never before have governments, regulators, businesses and the public been so preoccupied with risk exposure. Never before has risk management been written about, spoken of or debated with the same intensity. The "man on the street", particularly in Europe and the United States, is now only too acutely aware of the risks to his nation's economy, his employer, his employment and his standard of living. Poor risk management was cited time and time again in the aftermath of the global financial crisis. Clearly, making predictions solely on observations and experience and adopting "bell curve" methods of inference[1] from short time horizons was fundamentally flawed. Risk predictability was found wanting and the ramifications of a lack of forewarning have been devastating. As described in 2009 by Angel Gurría, Secretary-General of the Organisation for Economic Co-operation and Development (OECD), "the current global economic crisis is costing the world trillions of dollars, a protracted recession, millions of lost jobs, a huge loss of confidence in financial markets and a reversal in our efforts to curb global poverty". Bank executives have been pilloried for their risk-seeking behaviour, which at times has been described as reckless. Hector Sants, the chief executive of the UK Financial Services Authority (FSA) at the time of writing, remarked after the crisis: "Remuneration practices – bonuses – have been a symbol; a lightening rod of society's lack of trust in bankers and to address the trust issue this state of affairs has to be recognised and resolved". While a minority of board directors exhibited all of the destructive "d's", from being deceitful, delinquent, devious, dictatorial and dishonest through to disreputable, they kept the media spotlight on board behaviour. Surveys completed by the large accounting firms post the financial crisis lead to the common conclusion that the UK is still not there yet in terms of fully embedding ERM into board behaviour. Clearly ERM (which embraces both corporate governance and ethics) still has a long journey to travel before it is ingrained in the culture of businesses and can be seen to be contributing to business longevity and profitability.

The changes included in this revision reflect world events, national initiatives to address corporate governance failings and the growing importance of project risk management, business ethics, and health and safety management. These last three subjects have been included

[1] Taleb, N. N. (2010) *The Black Swan: The impact of the highly improbable*, Penguin Books, London.

in the business risk taxonomy described in Chapter 9 as additional internal processes, as it is considered they warrant specific attention.

The major differences between the first and second edition are summarised below.

New chapters:

Chapter 4 The Global Financial Crisis 2007–2009, a US Perspective
Chapter 5 Developments in Corporate Governance in Australia and Canada
Chapter 14 Communication and Consultation: Stage 7
Chapter 18 Project Risk Management
Chapter 19 Business Ethics Management
Chapter 20 Health and Safety Management

Significantly modified chapters:

Chapter 2 Developments in Corporate Governance in the UK
Chapter 3 Developments in Corporate Governance in the US
Chapter 9 Risk Identification – Stage 2
Chapter 25 Market Risk
Chapter 26 Social Risk

New appendices:

Appendix 15 Value at Risk, Recommended Reading
Appendix 16 Optimism Bias, Method of Calculation

Reordering of chapters:

Part II, covering the appointment of consultants, has been moved to the end of the book as Part V (Chapters 27–30).

AUDIENCE

Like the first edition, this book is written for a number of audiences: the competent practitioners who may be looking to broaden their approach; board members; non-executive directors who want to become more familiar with the processes and concepts of ERM; company risk directors; members of the Institute and Faculty of Actuaries,[2] the Institute of Risk Management,[3] project risk management practitioners wishing to extend their skills; business analysts; change agents; lecturers; and graduate and undergraduate students. Different parts of the book are aimed at different audiences as described below.

BOOK OVERVIEW

The book is composed of five parts. The target audience is different for each part.

Part I, "Enterprise Risk Management in Context", sets out the impetus behind ERM and describes corporate governance in the UK and overseas. It provides a detailed description of

[2] The Institute and Faculty of Actuaries (the merged body formed in 2010 from the Institute of Actuaries and the Faculty of Actuaries) is the professional body representing actuaries in the United Kingdom. In March 2008, ERM was adopted as one of the six actuarial practice areas, reflecting increased recognition of its importance. A regular newsletter communicates the ongoing work that the profession performs in respect of ERM.

[3] The Institute of Risk Management supported the development of ISO 31000, an international standard for risk management (published 13 November 2009) together with the accompanying standard, ISO 31010 – Risk Assessment Techniques, which followed and the updated Risk Management Vocabulary ISO Guide 73.

the global financial crisis of 2007–2009, the effects of which are still very evident in 2011 in Europe, North America and elsewhere. It explains the relationship between corporate governance, internal control and risk management, and reviews the development of risk management in the private sector. It is aimed at all audiences to set the scene and is particularly focused towards the chief executive, non-executive directors and the board in general.

Part II, "The Risk Management Process", is composed of seven chapters, each of which describes a stage within the overall risk management process. The process stages are based on the stages described within ISO 31000, published in 2009 by the International Organization for Standardization. Part II explains the activities to perform risk management using a standard process definition notation. Process goals, inputs, outputs, mechanisms and controls are fully explained for each stage. Simple tools and techniques are described to accomplish the individual stages. This part is specifically aimed at risk practitioners, chief risk officers, audit committees and business risk managers.

Part III, "Internal Influences – Micro Factors", describes the five sources of risk considered to be controllable (to a degree) by businesses, labelled in this text as financial, operational, technological, project and business ethics. This part is aimed at the audit committee, business risk managers, department heads and risk management practitioners.

Part IV, "External Influences – Macro Factors", describes the six sources of risk considered to be uncontrollable by businesses labelled in this text as economic, environmental, legal, political, market and social. This part is aimed at all audiences, from the chief executive through to the student. These chapters describe the complex world we live in, its changing nature, and those aspects of the environment, in its fullest sense, that may pose threats and upside opportunities to business performance. It is aimed at all those wishing to understand the external influences on businesses today.

Part V, "The Appointment", is composed of four chapters. Chapter 27 describes a consultant selection process on behalf of clients who want to go through a formal auditable process where price is of particular importance. Chapters 28, 29 and 30 describe, from a consultant's perspective, the interview process with a prospective sponsor, the preparation of a proposal and implementation of an assignment post-appointment, respectively. Hence Part V is largely for the benefit of risk practitioners.

HOW TO READ THIS BOOK

Time is precious. How much time do we ever have in any one day to reflect on how we do things and whether there is a better approach? Time between deadlines is commonly short, offering limited opportunity for quiet reflection. Hence this book is purposefully written in such a way that it is hoped that readers can quickly find and focus on the subjects that interest them, rather than having to carry out an extensive search for the instructive guidance they seek. The appropriate approach to reading this book will depend on your exposure to and experience of risk management and where your specific interests lie.

Acknowledgements

FIRST EDITION

In writing this book I owe a debt of gratitude to work colleagues past and present. In particular my thanks go to Peter Doig, Claire Love and Chris Johnson-Newell. My thanks go to Professor Chris Chapman of Southampton University and Dr David Hillson, for their comments and advice. I am grateful to Rachael Wilkie and Chris Swain of John Wiley and Sons Limited, who supported this project. I thank The Financial Times Limited, BBC NewsOnline, The *Observer*, Pearson Education Limited and the Financial Services Agency (FSA), for permission to include extracts from their publications/articles. At the request of the Financial Services Agency (FSA), I advise "use of FSA material does not indicate any endorsement by the FSA of this publication, or the material or views contained within it".

SECOND EDITION

I thank the Chartered Institute of Management Accountants (CIMA), Commonwealth of Australia (Department of the Prime Minister and Cabinet), Financial Services Agency (FSA), Bank of England, HM Treasury, US Federal Reserve, House of Commons, National Audit Office, Home Office, Telegraph Media Group and the UK Institute of Directors for their kind permission to include extracts from their speeches, publications, articles, papers and reports. At the request of the Financial Services Agency, I advise that "use of FSA material does not indicate any endorsement by the FSA of this publication, or the material or views contained within it". In addition at the request of the National Audit Office I advise that "use within this text of National Audit Office (NAO) material does not indicate any endorsement by the NAO of this publication, or the material or views contained within it". In addition, I owe a debt of gratitude to my work colleague Chris Newman for his contribution to the chapter on health and safety.

About the Author

Robert Chapman is currently the Director of Risk Management in the Middle East for AECOM, a publicly traded company on the New York Stock Exchange and listed by *Fortune 500* as one of America's largest companies. Prior to this appointment he was a Director of Risk Management at Hornagold & Hills, Capro Consulting and Osprey Project Management and the Programme Lead for risk management on the HMG joint venture in South Africa, supporting the Paristatal Transnet. He has provided risk management services in Holland, Ireland, South Africa, Qatar, England and the UAE to companies within the pharmaceutical, aviation, marine, rail, broadcast, heritage, water, sport, oil and gas, property development, construction and transportation industries as well to local authorities in the public sector. Dr Chapman has had articles published by *Enterprise Risk* (South Africa), *ExtraProtect* (translated into French and German), *IT Adviser*, *Yorkshire Post*, *Strategic Risk*, *PLC Strategies*, *Project*, the *Architects' Journal* and *PropertyWeek* and refereed papers published by the *Journal of International Project Management* and *Construction Management & Economics*. He was made a Fellow of both the Institute of Risk Management (UK) and the Association for Project Management (UK) for his contribution to the development of the discipline of risk management. Dr Chapman has been recognised by both Transnet in South Africa and the Association for Project Management in the UK as having exceptional risk management skills. He was awarded a PhD in risk management from Reading University in 1998 for research into the impact of changes in personnel on the delivery of objectives for investment projects. Additionally he has completed research on the subject of risk management on behalf of the Architects Registration Council of the United Kingdom (ARCUK). His book entitled *Retaining Design Team Members, a Risk Management Approach* was published by RIBA Enterprises Ltd, London, in 2002 and examines the causes behind employee turnover, the impact it can have and the risk mitigation actions that can be implemented to reduce the likelihood of occurrence. Dr Chapman was a contributory author of the Office of Government Commerce's 2007 publication *Management of Risk, Guidance for Practitioners*, which supports the Prince2 project management methodology. Subsequent to passing the Management of Risk Practitioner exam he became an accredited M_o_R® (Management of Risk) trainer, providing risk management training to a number of diverse companies. Prior to its publication he reviewed and commented upon international risk management standard ISO 31000 on behalf of the British Standards Institute. In addition, he has provided IT risk management guidance to the Chartered Institute of Accountants England and Wales in the form of a risk management handbook.

Part I

Enterprise Risk Management in Context

1
Introduction

A pessimist sees the difficulty in every opportunity; an optimist sees the opportunity in every difficulty.
(Winston Churchill)

Risk management has taken centre stage. It is now the most compelling business issue of our time. Shareholders have repeatedly suffered from erratic business performance. Recent history has shown that risk exposure has not been fully understood and risk management practice has been inadequate. Looking back, while economists have cited many reasons for the Asian financial crisis of 1997–1998, clearly foreign exchange risk was a major contributor. After the New York World Trade Center and Pentagon terrorist attack on 11 September 2001, enterprise risk management was found to be wanting. Business continuity planning had been inadequate. In particular, it was found that greater emphasis needed to be placed on IT disaster recovery, human resource management and communication. After the bankruptcies of Enron in December 2001 and WorldCom in July 2002, inadequate corporate governance and the "soft underbelly" of risk management were exposed, arising primarily from the lack of integrity of financial reporting, a lack of compliance with regulations and operational failures. In late August 2005 Hurricane Katrina struck, reportedly the costliest natural disaster in US history. Oil production, importation and refining were interrupted.[1] Businesses were suddenly exposed to a surge in energy prices, continuity failures and shipping disruption. Costs of production rose and sales fell. More recently, failure to properly understand and manage risk has been cited as the root cause for the global financial crisis of 2007–2010. So severe was this financial tsunami that many economists have described it as the worst financial disaster since the Great Depression of the 1930s. Boards in the financial sector were accused of being greedy, reckless[2] and dysfunctional and in some cases "sheep", falling into the trap of "group think" due to an apparent absence of independent thinking. In addition, there had been a lack of appreciation of risk at both a business and a macro or industry level. Systemic risk in the financial industry had not been recognised, understood or addressed. Regulators on both sides of the Atlantic and the banks themselves failed to recognise the interconnectedness of banks and the potential domino effect of bank failure. If the financial crisis was not excitement enough, the media have had a field day with a number of high-profile and very damaging business ethics failures relating to bribery, insider trading, invasion of privacy and sexual harassment.

[1] As a result of Hurrricane Katrina, at least 20 offshore oil platforms went missing, sunk or adrift.

[2] In an economy where certain businesses are considered "too important to fail" and the taxpayer is called upon to underwrite the risks of banks in the private sector, banks were severely criticised for gambling with taxpayers' money. Banks in the UK had a pivotal role in the global financial crisis and caused economic instability and erosion of national prosperity. The need to nationalise the banks' losses resulted in unemployment particularly in the public sector and left those in employment facing a significant drop in their standard of living. The banks have responded by reinstating extravagant rewards and extraordinary bonuses.

1.1 RISK DIVERSITY

Providing strategic direction for a business means understanding what drives the creation of value and what destroys it. This in turn means that the pursuit of opportunities must entail comprehension of the risks to take and the risks to avoid. Hence, to grow any business entails risk judgement and risk acceptance. A business's ability to prosper in the face of risk, at the same time as responding to unplanned events, good or bad, is a prime indicator of its ability to compete. However, risk exposure continues to grow greater, more complex, diverse and dynamic. This has arisen in no small part from rapid changes in the globalisation of business, speed of communication, the rate of change within markets and technology. Businesses now operate in an entirely different environment compared with just three years ago. Recent experience has shown that as businesses strive for growth, internal risks generated by a business itself can be as large as (or greater than) external risks. The adoption of expansion strategies, such as investment in emerging markets, developing significant new products, acquisition, major organisational restructuring, outsourcing key processes and major capital investment projects can all increase a business's risk exposure.[3]

A review of risk management practices in 14 large global corporations revealed that by the end of the 1990s the range of risks that companies felt they needed to manage had vastly expanded, and was continuing to grow in number (Hunt 2001). There are widespread concerns over e-commerce, which has become accepted and embedded in society with startling speed. According to the Economist Intelligence Unit (2001):

> Many companies perceive a rise in the number and severity of the risks they face. Some industries confront unfamiliar risks stemming from deregulation. Others worry about increasing dependence on business-to-business information systems and just-in-time supply/inventory systems. And everyone is concerned about emerging risks of e-business – from online security to customer privacy.

As a consequence of the diversity of risk, risk management requires a broader approach. This sentiment was echoed by Rod Eddington, former chief executive officer (CEO) of British Airways, who remarked that businesses now require a broader perspective of risk management. He went to say that:

> If you talked to people in the airline industry in the recent past, they very quickly got on to operational risk. Of course, today we think of risk as the whole of business. We think about risk across the full spectrum of the things we do, not just operational things. We think of risk in the context of business risks, whether they are risks around the systems we use, whether they are risks around fuel hedging, whether they're risks around customer service values. If you ask any senior airline person today about risk, I would hope they would move to risk in the true, broader sense of the term. (McCarthy and Flynn 2004)

All stakeholders and regulators are pressing boards of directors to manage risk more comprehensively, rigorously and systematically. Companies that treat risk management as just a compliance issue expose themselves to nursing a damaged balance sheet.

---

[3] Conventional risk management focused on avoiding risks to the business strategy as opposed to managing the risks of the strategy itself, which is where a number of banks have had spectacular failings.

1.2 APPROACH TO RISK MANAGEMENT

This evolving nature of risk and expectations about its management have now put pressure on previous working practices. Historically, within both private and public organisations, risk management has traditionally been segmented and carried out in "silos". This has arisen for a number of reasons such as the way our mind works in problem solving, the structure of business organisations and the evolution of risk management practice. There is clearly the tendency to want to compartmentalise risks into distinct, mutually exclusive categories, and this would appear to be a result of the way we subdivide problems to manage them, the need to allocate tasks within an existing organisational structure and the underlying assumption that the consequences of an unforeseen event will more or less be confined to one given area. In actuality, the fallout from unforeseen events tends to affect multiple business areas and the interrelationships between risks under the categories of operational, financial and technical risk have been overlooked, often with adverse outcomes. Patricia Dunn, former CEO of Barclays Global Investors and former non-executive chairwoman of the board of Hewlett-Packard (HP),[4] has previously identified a failing in approach:

> I think what Boards tend to miss and what management tends to overlook is the need to address risk holistically. They overlook the areas that connect the dots because risk is defined so "atomistically" and we don't have the perspective and the instrument panel that allows us to see risk in a 360 degree way. (McCarthy and Flynn 2004)

Enterprise risk management (ERM) is a response to the sense of inadequacy in using a silo-based approach to manage increasingly interdependent risks. The discipline of ERM, sometimes referred to as strategic business risk management, is seen as a more robust method of managing risk and opportunity and an answer to these business pressures. ERM is designed to improve business performance. While not in its infancy, it is a slowly maturing approach, where risks are managed in a coordinated and integrated way across an entire business. The approach is less to do with any bold breakthrough in thinking, and more to do with the maturing, continuing growth and evolution of the profession of risk management and its application in a structured and disciplined way (McCarthy and Flynn 2004). ERM is about understanding the interdependencies between the risks, how the materialisation of a risk in one business area may increase the impact of risks in another business area. In consequence, it is also about how risk mitigation action can address multiple risks spanning multiple business sectors. It is the illustration of this integrated approach which is the focus of this book.

1.3 BUSINESS GROWTH THROUGH RISK TAKING

Risk is inescapable in business activity. As Peter Drucker explained as far back as the 1970s, economic activity by definition commits present resources to an uncertain future. The one thing that is certain about the future is its uncertainty, its risks. Hence, to take risks is the essence of economic activity. He considers that history has shown that businesses yield greater economic performance only through greater uncertainty – or in other words, through greater risk taking (Drucker 1979).

[4] Hewlett-Packard is referred to in Chapter 19 regarding their unethical behaviour and infringement of privacy.

Nearly all operational tasks and processes are now viewed through the prism of risk (Hunt 2001). Indeed, the term "risk" has become shorthand for any corporate activity. It is thought not possible to "create a business that doesn't take risks" (Boulton *et al.* 2000). The end result of successful strategic direction setting must be capacity to take a greater risk, for this is the only way to improve entrepreneurial performance. However, to extend this capacity, businesses must understand the risks that they take. While in many instances it is futile to try to eliminate risk, and commonly only possible to reduce it, it is essential that the risks taken are the right risks. Businesses must be able to choose rationally among risk-taking courses of action, rather than plunge into uncertainty, on the basis of a hunch, gut feeling, hearsay or experience, no matter how carefully quantified. Quite apart from the arguments for risk management being a good thing in its own right, it is becoming increasingly rare to find an organisation of any size whose stakeholders are not demanding that its management exhibit risk management awareness. This is now a firmly held view supported by the findings of the Economist Intelligence Unit's enterprise risk management survey, referred to earlier. It discovered that 84% of the executives who responded considered that ERM could improve their price/earnings ratio and cost of capital. Organisations that are more risk conscious have for a long time known that actively managing risk and opportunity provides them with a decisive competitive advantage. Taking and managing risk is the essence of business survival and growth.

1.4 RISK AND OPPORTUNITY

There should not be a preoccupation with downside risk. Risk management of both upside risks (opportunities) and downside risks (threats) is at the heart of business growth and wealth creation. Once a board has determined its vision, mission and values, it must set its corporate strategy, its method of delivering the business's vision. Strategy setting is about strategic thinking. Setting the strategy is about directing, showing the way ahead and giving leadership. It is being thoughtful and reflective. Whatever this strategy is, however, the board must decide what opportunities, present and future, it wants to pursue and what risks it is willing to take in developing the opportunities selected. Hence the discipline of risk management should support both the selection and setting of the strategy. However, risk and opportunity management must receive equal attention and it is important for boards to choose the right balance. This is succinctly expressed by the National Audit Office: "a business risk management approach offers the possibility for striking a judicious and systematically argued balance between risk and opportunity in the form of the contradictory pressures for greater entrepreneurialism on the one hand and limitation of downside risks on the other" (National Audit Office 2000). An overemphasis on downside risks and their management can be harmful to any business.

Knight and Petty (2001) stress that risk management is about seeking out the upside risks or opportunities, that getting rid of risk stifles the source of value creation and upside potential. Any behaviour that attempts to escape risk altogether will lead to the least rational decision of all, doing nothing. While risks are important, as all businesses face risk from inception, they are not grounds for inaction but restraints on action. Hence risk management is about controlling risk as far as possible to enable a business to maximise its opportunities. Development of a risk policy should be a creative initiative, exposing exciting opportunities for value growth and innovative handling of risk, not a depressing task, full of reticence, warning and pessimism (Knight and Petty 2001). ERM, then, is about managing both opportunities and risks.

1.5 THE ROLE OF THE BOARD

Even before the global financial crisis, George "Jay" Keyworth, former member of Hewlett-Packard's board, stated that the most important lesson of the last few years is that board members can no longer claim impunity from a lack of knowledge about business risk. The message here is that when something goes wrong, as inevitably it does, board members will be held accountable. The solution is for board members to learn of the potential for adverse events and be sufficiently aware of the sources of risk within the area of business that they are operating in, to be afforded the opportunity to take pre-emptive action (McCarthy and Flynn 2004). The business of risk management is undergoing a fundamental sea change with the discipline of risk management converging at the top of the organisation and being more openly discussed in the same breath as strategy and protection of shareholders. Greater risk taking requires more control. Risk control is viewed as essential to maintaining stability and continuity in the running of businesses. However, in the aftermath of a series of unexpected risk management failures leading to company collapses and other corporate scandals in the UK, investors have expressed concerns about the low level of confidence in financial reporting, board oversight of corporate operations, the safeguards provided by external auditors and the degree of risk management control. These early concerns led to a cry for greater corporate governance, which led to a series of reports on governance and internal control culminating in the Combined Code of Corporate Governance (2003). The incremental development of corporate governance leading up to and beyond the 2003 Code is discussed in Chapter 2. Clearly risk exposure has been growing in an increasingly chaotic and turbulent world, and time has shown that this turbulence has not abated.

The lack of risk management control resides with the board. In 1995 in response to bad press about boards' poor performance and the lack of adequate corporate governance, the Institute of Directors (IoD) published *Standards for the Board*. It proved to be a catalyst for debate on the roles and tasks of a board and on the need to link training and assessed competence with membership of directors' professional bodies. The publication laid out four main objectives for directors. Within the IoD's 2010 factsheet entitled *The role of the board*, apart from one of the objectives being split into two, these objectives remain virtually unchanged as follows:

1. The board must simultaneously be entrepreneurial and drive the business forward while keeping it under prudent control.
2. The board is required to be sufficiently knowledgeable about the workings of the company and answerable for its actions, yet able to stand back from the day-to-day management of the company and retain an objective, longer-term view.
3. The board must be sensitive to the pressure of short-term issues and yet take account of broader, long-term trends.
4. The board must be knowledgeable about "local" issues and yet be aware of potential or actual wider competitive influences.
5. The board is expected to be focused on the commercial needs of the business, while acting responsibly towards its employees, business partners and society as a whole.

The task for boards of course is to ensure the effectiveness of their risk model. With this in mind, here are some action items for the strategic risk management agenda for boards and CEOs to consider:[5]

- Appoint a C-level risk leader empowered not only with the responsibility, but also with the authority to act on all risk management matters.
- Ensure that this leader is independent and can work objectively with the company's external advisers (external audit, legal, etc.) and the governing decision maker and oversight function (the CEO and board).
- Be satisfied as to the adequacy of the depth of current risk analysis actions, from an identification, assessment and mitigation standpoint.
- Be confident that the risk management information that board members receive is accurate, timely, clear and relevant.
- Actively require and participate in regular dialogue with key stakeholders to understand if their objectives have been captured, debated and aligned, are being met and whether stakeholders may derail current initiatives.
- Strive to build a culture where risk management and strategic planning are intertwined.
- Ensure that risk management remains focused on the most serious issues.
- Ensure that risk management is embedded throughout the organisation.

As illustrated in Figure 1.1, risk and opportunity impinge on the four main functions of boards: policy formulation, strategic thinking, supervisory management and accountability. Policy formulation involves setting the culture for the organisation, which should include risk management. Strategic thinking entails selecting markets to pursue and committing resources to those markets on the strength of the risk profile prepared. Supervisory management requires businesses to put in place oversight management and governance processes, including formal risk management. Accountability relates to ensuring that risk mitigation actions have clear owners who are charged with implementing pre-agreed actions to address the risks identified, report changes in risk profiles and engage in ongoing risk management.

1.6 PRIMARY BUSINESS OBJECTIVE (OR GOAL)

The primary objective of a business is to *maximise* the *wealth* of its *shareholders* (owners). In a market economy, the shareholders will provide funds to a business in the expectation that they will receive the maximum possible increase in wealth for the level of risk which must be faced. When evaluating competing investment opportunities, therefore, the shareholders will weigh the returns from each investment against the potential risks involved. The use of the term "wealth" here refers to the market value of the ordinary shares. The market value of the shares will in turn reflect the future returns the shareholders will expect to receive over time from the shares and the level of risk involved. Shareholders are typically not concerned with returns over the short term, but are concerned with achieving the highest possible returns over the long term. Profit maximisation is often suggested as an alternative objective for a business. Profit maximisation is different from wealth maximisation. Profit maximisation is usually seen as a short-term objective, whereas wealth maximisation is a long-term objective.

---

[5] These recommendations were made in the first edition of this text published in 2006, prior to the global financial crisis and the Walker Review of 2009 described in Chapter 2.

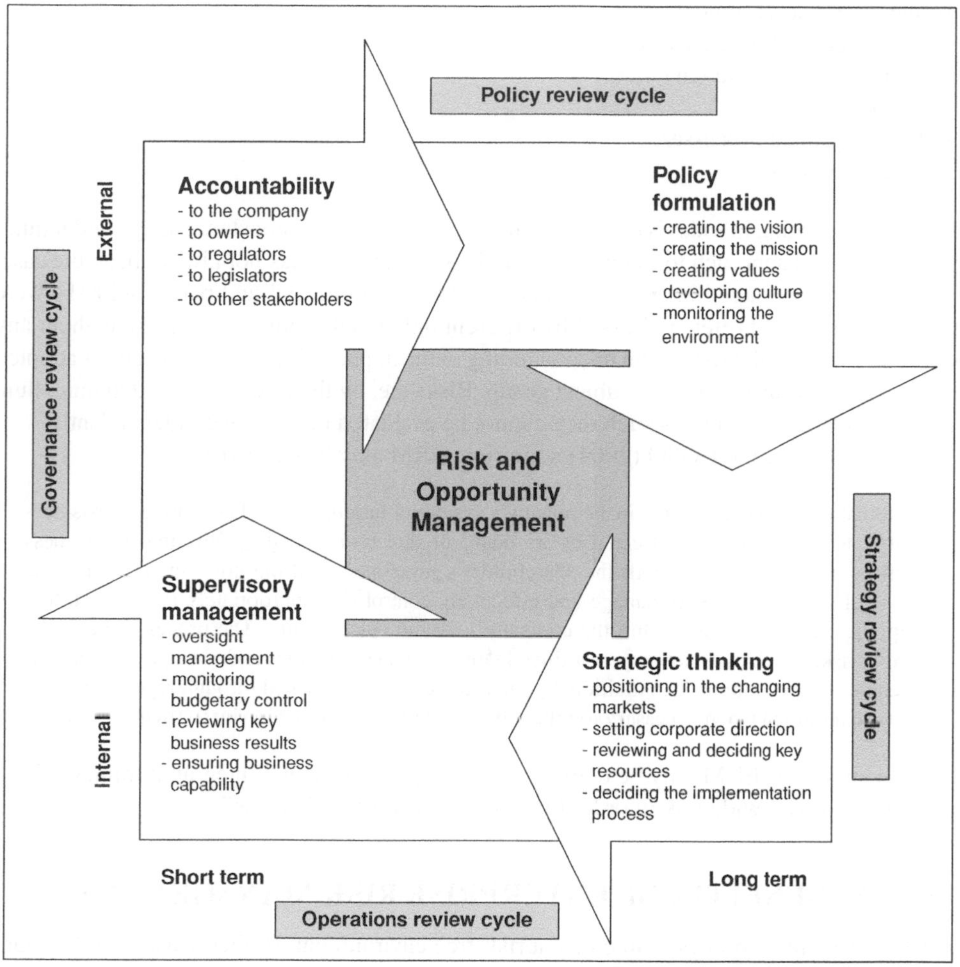

Figure 1.1 The role of the board and the integration of risk management (Garratt 2003). Reproduced with permission from *The Fish Rots from the Head*, B. Garratt, Profile Books Ltd.

Wealth maximisation takes account of risks to long-term growth, whereas profit maximisation does not.

1.7 WHAT IS ENTERPRISE RISK MANAGEMENT?

ERM has to satisfy a series of parameters. It must be embedded in a business's system of internal control, while at the same time it must respect, reflect and respond to the other internal controls. ERM is about protecting and enhancing share value to satisfy the primary business objective of shareholder wealth maximisation. It must be multifaceted, addressing all aspects of the business plan from the strategic plan through to the business controls:

- strategic plan
- marketing plan
- operations plan

- research and development
- management and organisation
- forecasts and financial data
- financing
- risk management processes
- business controls

Enterprises operating in today's environment are characterised by constant change and require a more integrated approach to manage their risk exposure. This has not always been the case, with risks being managed in "silos". Economic, legal, commercial and personnel risks were treated separately and often addressed by different individuals within a company without any cross-referencing of the risks or an understanding of the impact of management actions adopted for one subject group on another subject group. Risks are, by their very nature, dynamic, fluid and highly interdependent. As such they cannot be evaluated or managed independently.

Largely reflecting the COSO (2004) definition, ERM may be defined as:

> A systematic process embedded in a company's system of internal control (spanning all business activity), to satisfy policies effected by its board of directors, aimed at fulfilling its business objectives and safeguarding both the shareholder's investment and the company's assets. The purpose of this process is to manage and effectively control risk appropriately (without stifling entrepreneurial endeavour) within the company's overall risk appetite. The process reflects the nature of risk, which does not respect artificial departmental boundaries and manages the interdependencies between the risks. Additionally the process is accomplished through regular reviews, which are modified when necessary to reflect the continually evolving business environment.

Hence, in summary, ERM may be defined as "a comprehensive and integrated framework for managing company-wide risk in order to maximise a company's value".

1.8 BENEFITS OF ENTERPRISE RISK MANAGEMENT

No risk management process can create a risk-free environment. Rather, ERM enables management to operate more effectively in a business environment where an organisation's risk exposure profile is never static. Enterprise risk management provides enhanced capability to:

- *Increase the likelihood of a business realising its objectives.* ERM will equip organisations with techniques to identify, record and assess the opportunities they seek to proactively pursue and exploit. At the same time it will support the identification and conscious management of the risks associated with selected opportunities to ensure that bottom-line performance is enhanced rather than eroded. In this way it will enable organisations to mature and realise their stated objectives.
- *Build confidence in stakeholders and the investment community.* As a result of the global financial crisis institutional investors, rating agencies and regulators are more focused on and more eager to learn about an organisation's capabilities for understanding and managing risk. Investors in particular will wish to understand the degree of risk their investments will be exposed to and whether the returns will be adequate. Board members and managers may be called upon to explain the framework, policy and process they have in place for managing risk. ERM provides the rigour to establish, describe and demonstrate proactive risk management.

- *Comply with relevant legal and regulatory requirements.* ERM, through establishing (and subsequently monitoring) a risk management framework, requires an organisation to understand, record (and keep up to date) the business context including, but not limited to, the legal and regulatory requirements it has to comply with and, where appropriate, the implications of not doing so.
- *Align risk appetite and strategy.* Risk appetite is the degree of risk, on a broad-based level, that a business is willing to accept in pursuit of its objectives. ERM supports management's consideration of a business's risk appetite first in evaluating strategic alternatives, then in setting boundaries for downside risk.
- *Improve organisational resilience.* As the business environment continues to change and the pace of change accelerates, resilience is critical to business longevity. Organisational resilience is sometimes considered as the degree of flexibility (or capacity) of an organisation's culture to recover from and respond to change. ERM will support an organisation in understanding potential change and preparing for it through risk response planning or in deciding to be the change catalyst through opportunity exploitation.
- *Enhance corporate governance.* ERM and corporate governance augment each other. ERM strengthens governance through challenging potential excessive risk taking as occurred in the global financial crisis, encouraging board-level engagement in the high-level risk process and improving decision making on risk appetite and tolerance.
- *Embed the risk process throughout the organisation.* ERM, through the creation of a framework, policy, process, plans and training can embed risk management throughout the organisation from the board down to all elements of the organisational structure as risk exposure can emanate from any corner of the organisation (e.g. from a breach of ethics at board level to a breach of environmental legislation by production).
- *Minimise operational surprises and losses.* ERM supports businesses to enhance their capability to identify potential risk events, assess risks and establish responses, and thereby to reduce the occurrence of unpleasant surprises and associated costs or losses.
- *Enhance risk response decisions.* ERM provides the rigour to identify and select among alternative risk responses – risk removal, reduction, transfer or retention.
- *Optimise allocation of resources.* A clear understanding of the risks facing a business can enhance the effective direction and use of management time and the business's resources to manage risk.
- *Identify and manage cross-enterprise risks.* Every business faces a myriad of risks affecting different parts of the organisation. The benefits of enterprise risk management are only optimised when an enterprise-wide approach is adopted, integrating the disparate approaches to risk management within a company. Integration has to be effected in three ways: centralised risk reporting, the integration of risk transfer strategies and the integration of risk management into the business processes of a business. Rather than being purely a defensive mechanism, it can be used as a tool to maximise opportunities.
- *Link growth, risk and return.* Businesses accept risk as part of wealth creation and preservation and they expect returns commensurate with risk. ERM provides an enhanced ability to identify and assess risks and establish acceptable levels of risk relative to potential growth and achievement of objectives.
- *Rationalise capital.* More robust information on risk exposure allows management to more effectively assess overall capital needs and improve capital allocation.
- *Seize opportunities.* The very process of identifying risks can stimulate thinking and generate opportunities as well as threats. Reponses need to be developed to seize these

opportunities in the same way that responses are required to address identified threats to a business.

- *Improve organisational learning.* ERM can enhance organisational learning through the use of lessons learnt prior to embarking on new change projects and the maintenance of records of successful risk treatment plans that effectively removed risks prior to realisation.

There are three major benefits of ERM: improved business performance, increased organisational effectiveness and better risk reporting.

1.9 STRUCTURE

A structure for understanding ERM is included in Figure 1.2 and is composed of seven elements:

1. Corporate governance is required to ensure that the board of directors and management have established the appropriate organisational processes and corporate controls to measure and manage risk across the business.
2. The creation and maintenance of a sound system of internal control is required to safeguard shareholders' investment and the business's assets.
3. A specific resource must be identified to implement the internal controls with sufficient knowledge and experience to derive the maximum benefit from the process.
4. A risk management framework is required that will provide the foundations and arrangements for embedding risk management throughout the organisation at all levels.
5. A policy should be prepared describing the importance of risk management to the achievement of the organisation's corporate goals.
6. A clear risk management process is required which sets out the individual processes, their inputs, outputs, constraints and enablers.
7. The value of a risk management process is reduced without a clear understanding of the sources of risk and how they should be responded to. The framework breaks the source of risk down into two key elements labelled internal processes and the business operating environment.

1.9.1 Corporate Governance

Examination of recent developments in corporate governance reveals that they form catalysts for and contribute to the current pressures on ERM. It explains the expectations that shareholders have of boards of directors. It explains the approaches companies have adopted to risk management and the extent of disclosure of risk management practice. Corporate governance now forms an essential component of ERM because it provides the top-down monitoring and management of risk management. It places responsibility on the board for ensuring that appropriate systems and policies for risk management are in place. Good board practices and corporate governance are crucial for effective ERM. The section that follows addresses internal control, which is a subset of corporate governance (and risk management is a subset of internal control).

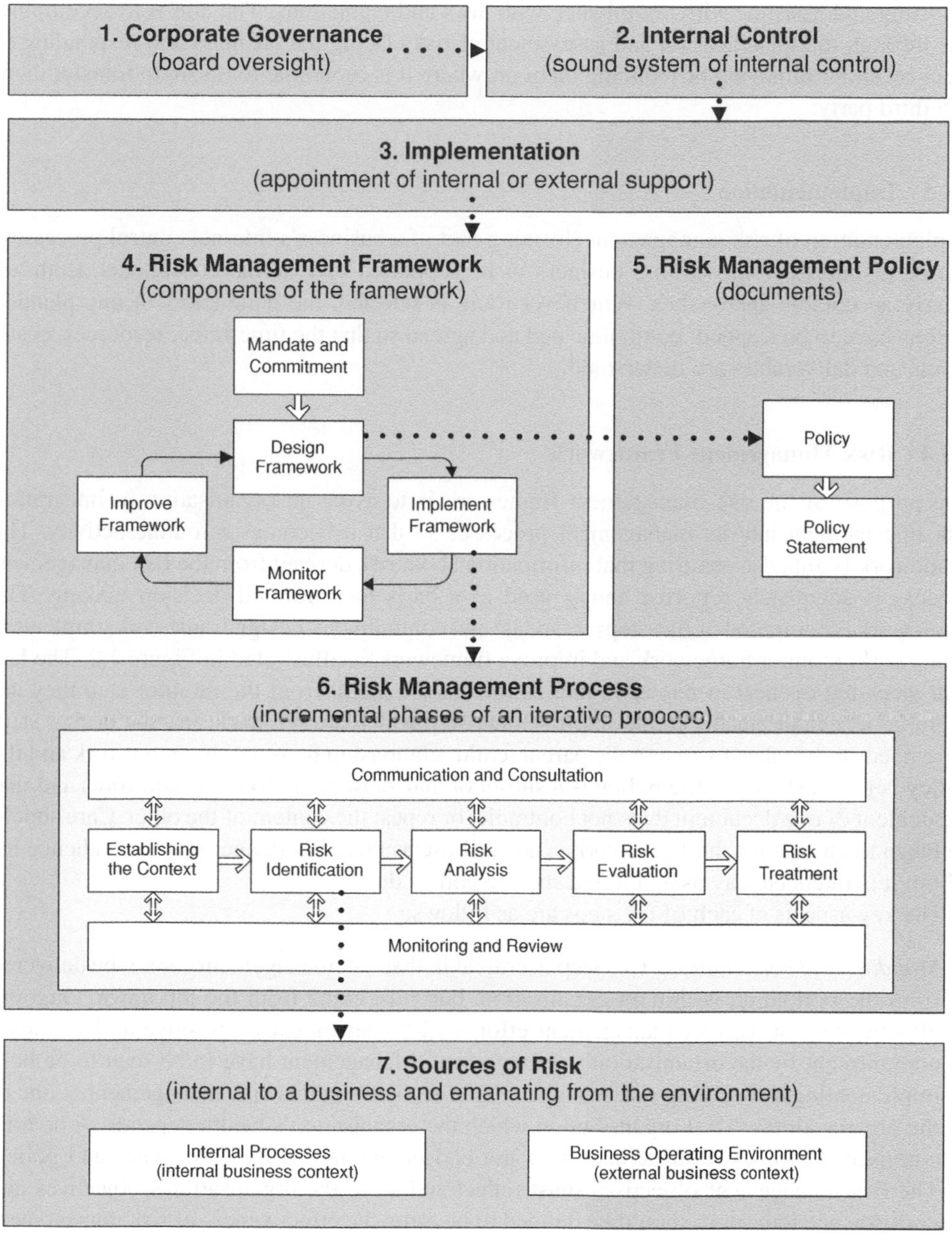

Figure 1.2 ERM structure

1.9.2 Internal Control

Examination of internal controls provides an understanding of what should be controlled and how. There is more of a focus on formal approaches. Internal controls are a subset of corporate governance. Risk management is a subset of internal controls. Risk management is aimed at facilitating the effective and efficient operation of a business, improving internal and external reporting and assisting with compliance with laws and regulations. The aim is to accomplish this through the identification and assessment of risks facing the business and responding to them by either removing or reducing them or, where it is economic to do so, to transfer them to a third party.

1.9.3 Implementation

Implementation of risk management (forming part of a business's internal control processes) can be resourced from within a business or be supported by external consultants. Both are clearly acceptable approaches. Whichever route is selected, the parameters of any planned actions have to be mapped, communicated and agreed so that the timeframe, resources, costs, inputs and deliverables are understood.

1.9.4 Risk Management Framework

The purpose of the risk management framework is to assist an organisation in integrating risk management into its management processes so that it becomes a routine activity. The framework is aimed at ensuring that information about risk derived from the risk management process is adequately reported and is used as a basis for informed decision making. The framework is composed of five steps: mandate and commitment, design framework, implement framework, monitor framework and improve framework, as illustrated in Figure 1.2. The last four steps are cyclical in nature in that as lessons are learnt from the monitor step they are captured as enhancements in the improvement step which are fed back into the design step. Care needs to be taken to ensure the parent–child relationship between the framework and the policy is established (i.e. the policy is a subset of and subservient to the framework) and that the content of one document does not contradict or repeat the content of the other. Care should be taken to ensure that the framework is not verbose as there is a danger that the audience for whom it is intended may assign it to a shelf to collect dust.

The key aspects of each of the steps are as follows:

- *Mandate and commitment*. This step is critical in that risk management cannot be delivered from the bottom up within an organisation, but must come from the top down. Ongoing effectiveness of the risk management effort will be dependent on positive and sustained commitment by the organisation's management. Management have to be seen to be both implementing and driving risk management in recognition that risk management is one of the organisation's "vital organs" upon which the organisation's health depends. A board's commitment to risk management does not end in signing off the framework and policy. The risk management objectives must reflect and serve the organisation's objectives and performance indicators should be defined to measure the effectiveness of risk management over time. The relationship with internal audit should be established so that the organisation is ensuring legal and regulatory compliance. Management should agree and endorse the risk management policy.
- *Design framework*. The design of the framework entails understanding the organisation and its context, establishing the risk management policy, determining accountability for risk management, embedding risk management in all of the organisation's practices and processes, allocating appropriate experienced and competent risk resources, and establishing tailored internal and external communication and allied reporting.

- *Implement framework.* The timing of the implementation of the framework should be planned. Introduction into the organisation should be managed with training sessions held as required. Ensure as far as possible that decision making is based on the output of the risk management processes. Develop a risk management plan (or plans) for the delivery of the risk management process, which may vary depending on where in the organisation risk management is being implemented.
- *Monitor framework.* Periodically review with internal and external stakeholders whether the risk management framework, policy, plan and process require amendment as a result of changes in the organisation's internal or external context. Assess risk management performance against pre-agreed indicators.
- *Improve framework.* Based on the results of the monitor framework, decisions should be made on whether the risk management framework step, and the policy and process which support it, should be amended with the aim of improving the effectiveness of the organisation's risk management practices.

1.9.5 Risk Management Policy

In simple terms a policy should address *why* risk management will be undertaken, *who* within and outside the organisation will undertake it, *how* it will be undertaken by reference to the framework and process and internal functions, and *what* those who are responsible will be required to undertake. Specifically, the policy should state its purpose, objectives, scope (where it applies within the organisation), related and supporting policies, its degree of confidentiality (any limitations on disclosure), the frequency of its review and the date it was last updated. The organisation should declare within its policy the importance it attaches to active risk management to support the realisation of its purpose, vision, strategic and business objectives, and implementation strategy. The policy should address the interests of all stakeholders, including shareholders, customers, suppliers, regulators and employees. It should set out and describe the accountability for risk management within the organisation. This will include describing the specific responsibilities of the board – and (depending on the size of the organisation) internal audit, external audit, the risk committee, the corporate governance committee, the central risk function, employees and third-party contractors – in implementing risk management. Where appropriate, it should describe the relationship between risk and corporate governance and internal audit. The policy is not the place to describe the risk management process; however, it should describe the policy's relationship to the process and the framework. In addition, ideally any standalone policy statement prepared for display (alongside, say, the health and safety policy and the business continuity policy), should be short, concise and lucid (and is commonly more effective when confined to a single page).

1.9.6 Risk Management Process

A way of exploring the mechanisms for implementing a risk management process is to break it down into its component parts and examine what each part should contribute to the whole. It is proposed here that the risk management process is broken down into seven stages: context, identification, analysis, evaluation, treatment, monitoring/review and communication/consultation. While activities follow a largely sequential pattern, it may be a highly iterative process over time. For instance, as new risks are identified, the earlier processes of identification and analysis are revisited and the subsequent processes are repeated through to the implementation of risk response actions.

1.9.7 Sources of Risk

A way of examining the sources of business risk is to consider that risk emanates from two primary areas: from within the business itself (relating to the actions it takes) and from the environment or context within which the business operates (and over which it has no control). In Figure 1.2 these sources are labelled "internal processes (internal business context)" and "business operating environment (external business context)", respectively, to show the relationship with the international risk standard, ISO 31000 (2009). They are a development of the traditional PEST analysis (an abbreviation for the external influences called political, economic, social and technological).

1.10 SUMMARY

All businesses in a free market are exposed to risk. This risk exposure exists from their inception. However, there would appear to be a swell of opinion that says risk is now more complex, diverse and dynamic. In particular, the source of risk is broader and the rate of change of the sources of risk has dramatically increased. The emergence of ERM has come about from the desire and need to move away from managing risk in silos and identifying and managing risk interdependencies. This is not some startling new intellectual breakthrough but rather a practical solution to a practical problem. It is clear from surveys and the press that board members believe that ERM is important to business growth. Whatever strategy boards adopt, they must decide what opportunities, present and future, they want to pursue and what risks they are willing to take in developing the opportunities selected. Hence, whatever the approach businesses adopt for risk management, they must strike a judicious balance between risk and opportunity in the form of the contradictory pressures for greater entrepreneurialism on the one hand and the limitation of downside risks on the other. In the aftermath of a series of unexpected risk management failures leading to company collapses and other corporate scandals in the UK and overseas, boards are under greater scrutiny and expectations of corporate governance have significantly increased. Board members cannot distance themselves from risk management or believe that they will not be held to account. Risk management needs to be integrated with the primary activities of the board. There are a series of clearly recognised benefits of implementing risk management practice, when applied in a systematic and methodical way. A structure was described for examining ERM to understand the pressures for its development, its composition, implementation, the overall process and the sources of risk.

1.11 REFERENCES

Boulton, R.E.S., Libert, B.D. and Samek, S.M. (2000) *Cracking the Value Code – How Successful Businesses are Creating Wealth in the New Economy*. Harper Business, New York.

Combined Code (2003) *Combined Code on Corporate Governance*. Financial Reporting Council, July.

COSO (2004) *Enterprise Risk Management – Integrated Framework*, September. Committee of Sponsoring Organisations of the Treadway Commission.

Drucker, P.F. (1979) *Management, an Abridged and Revised Version of Management: Tasks, Responsibilities, Practices*. Pan Books, London.

Economist Intelligence Unit (2001) *Enterprise Risk Management: Implementing New Solutions.* EIU, New York.

Garratt, R. (2003) *The Fish Rots from the Head. The Crisis in our Boardrooms: Developing the Crucial Skills of the Competent Director*, revised and updated edition. Profile Books, London.

Hunt, B. (2001) Issue of the moment: The rise and rise of risk management. In J. Pickford (ed.), *Mastering Risk Volume 1: Concepts.* Financial Times, Harlow.

ISO (2009) *ISO 31000:2009 Risk Management – Principles and Guidelines.* International Organization for Standardization, Geneva.

Knight, R.F. and Petty, D.J. (2001) Philosophies of risk, shareholder value and the CEO. In J. Pickford (ed.), *Mastering Risk Volume 1: Concepts.* Financial Times, Harlow.

McCarthy, M.P. and Flynn, T.P. (2004) *Risk from the CEO and Board Perspective.* McGraw-Hill, New York.

National Audit Office (2000) *Supporting Innovation: Managing Risk in Government Departments.* Report by the Comptroller and Auditor General, 17 August. The Stationery Office, London.

2
Developments in Corporate Governance in the UK

Today the management, monitoring, and governance of a business are increasingly seen as separate functions to be done by separate bodies. When these three functions are combined in one body, the short-term tends to drive out the long, with month-to-month management and monitoring issues stealing the time and attention needed for governance. The big decisions then go wrong.
(Charles Handy)

The previous chapter examined what ERM is, its benefits and its components. This chapter looks at the drive behind improvements in ERM through the examination of the incremental developments in corporate governance and their catalysts from the Cadbury Committee onwards. The purpose of corporate governance is to ensure board oversight of business operations and facilitate effective, entrepreneurial and prudent management that can deliver the long-term success of the company. For any business, governance means maintaining a sound system of internal control within its normal management and governance processes. Internal control is required to assist in: ensuring the reliability of internal and external reporting; compliance with laws and regulations; maintaining proper accounting records; and the appropriate management and control of risks. While the need for governance has always existed, corporate governance and particularly risk management has been recognised as woefully inadequate. The events of the financial crisis exposed material shortcomings in the governance and risk management of some regulated firms. Although poor governance was only one of many factors contributing to the crisis, it has widely been acknowledged to have been an important one. This has led to substantial regulatory activity in relation to governance, both in the United Kingdom and internationally. The Organisation for Economic Co-operation and Development (OECD), the Corporate Governance Task Force of the Basel Committee and the International Association of Insurance Supervisors (IAIS) are each reviewing their corporate governance principles and the FSA is engaging actively with each of these organisations as they develop this work. The financial crisis has totally overshadowed the previous high-profile business collapses in the US which triggered a clamour from investors for greater transparency of financial reporting and internal controls and more rigorous board oversight. This chapter traces the incremental developments in corporate governance in the UK, recently invigorated by the financial crisis, and the focus on deficiencies in risk management. Risk management is now firmly under the spotlight. Chapter 3 examines the developments in corporate governance in the United States.

2.1 INVESTOR UNREST

In the aftermath of a series of unexpected risk management failures leading to company collapses and other corporate scandals in the UK, investors expressed concerns about the low level of confidence in financial reporting, board oversight of corporate operations and the safeguards provided by external auditors. These concerns led to the adoption in early 1993 of the UK's first code of corporate governance: the Cadbury Code of Best Practice (Cadbury 1992). Similar initiatives were introduced overseas such as the Canadian Dey Report, published in 1994 (Dey 1994) and subsequently the two-volume report on *Corporate Governance* issued in Australia in 1997 (CG 1997). Through a continuing process of revision and amendment, subsequent reports have broadened the focus of corporate governance. The collapse of Enron in late 2001, followed by other major corporate crises in the US and elsewhere, called into question the effectiveness of many of the established concepts of corporate governance. As a result, the adequacy of governance arrangements in the US, the UK and internationally came under closer scrutiny. In the UK, this process involved a wide-ranging review, leading to the introduction of a revised Combined Code on Corporate Governance in 2003 (Combined Code 2003). In the aftermath of the global financial crisis and in particular the failure of UK banks, UK corporate governance was re-examined, culminating in the release of the UK Corporate Governance Code of 2010 (Code 2010). For ease of assimilation, the key reports, codes and guidance from Cadbury onwards are listed chronologically in Box 2.1.

Box 2.1 Reports and codes

UK corporate governance guidance, reports and codes, listed chronologically:

Cadbury Report	December 1992
Greenbury Report	July 1995
Hampel Code	June 1998
Turnbull	September 1999
Myners Review	March 2001
Smith Report	January 2003
Higgs Report	January 2003
Tyson Report	June 2003
The Combined Code	July 2003
Corporate Governance: A Practical Guide	August 2004
Internal Control: Revised Guidance	October 2005
Good practice suggestions from the Higgs Report	June 2006
The Combined Code on Corporate Governance	June 2006
The Combined Code on Corporate Governance	June 2008
A review of corporate governance in UK banks and other financial entities (The Walker Review)	July 2009
A review of corporate governance in UK banks and other financial entities: Final Recommendations (The Walker Review)	November 2009
A Stewardship Code for Institutional Investors	January 2010
The UK Corporate Governance Code	June 2010

2.2 THE PROBLEM OF AGENCY

One of the specific areas of investor disquiet emanated from the problem of agency. The function of a board in a listed company is to take responsibility for managing the company's business on behalf of its members or shareholders. Separation between membership and management has many advantages:

- decision making can be entrusted to those with the necessary skills and capacities, leaving the members to enjoy the benefits of their association with the organisation, without needing to involve themselves in matters of detail;
- it facilitates efficient aggregation and use of capital, by enabling the possessors of capital to invest in enterprise without requiring them to become involved in its operation; and
- it allows responsibility for the strategic direction and control of business to be delegated to professional managers who (it is assumed) possess the required entrepreneurial skills and management expertise.

However, the separation between management and ownership within a UK listed company may create tensions between the interests of these parties. In listed companies, these tensions are known collectively as the "problem of agency", which is essentially the potential for conflicts of interest between the shareholders, the company's owners and its directors, as their agents. According to agency theory, the managers of the company, as rational beings, will seek to maximise their own well-being through their control of the company's resources. As a result, they are likely to pursue self-serving objectives, which will not necessarily be in the best interests of the shareholders. As Cooper (2004) describes, the problem of agency may manifest itself in board decisions that promote the interests of the directors but do not necessarily enhance the value of the company for the shareholders. Cooper cites examples of such decisions as being:

- pursuit of short-term share growth, where sustained investment in the company's asset base might produce long-term benefits for shareholders;
- inappropriate expansion or diversification of the company's activities into areas which involve unwarranted risks to shareholders' investments; or
- resistance by managers to mergers or takeovers which might threaten their own job security, but which may be in the best interests of the company's shareholders.

The problem of agency in listed companies can be exacerbated by the board's ability to control the supply of information to shareholders about the company's position and performance. In extreme cases this may result in shareholders and others being seriously misled. Two recent notorious examples concern the US energy corporation Enron and the Anglo-Dutch petrochemical company Shell. Enron's directors systematically overstated profits, failed to inform shareholders about risky financing arrangements and continued to declare the corporation's financial soundness until days before filing for bankruptcy protection. The directors of Shell, the world's third largest oil company, overstated the company's oil and gas reserves. The resultant dramatic fall in share price led to investor anger, which in turn led to the departure of three of Shell's top executives. The level of reserves was restated four times. The restatements prompted investigations by both UK and US authorities.

2.3 THE CADBURY COMMITTEE

The Cadbury Committee on the Financial Aspects of Corporate Governance, a private sector initiative, was set up in 1991 by the Financial Reporting Council (FRC), the London Stock Exchange and the accounting profession, in response to concerns about the low level of public confidence in financial reporting and in the safeguards provided by external auditors. The Committee report has come to be recognised as a landmark in thinking on corporate governance and was thought to strike a chord in many countries. As explained by the Chairman of the

Committee, Adrian Cadbury, in the preface to the report, corporate governance had been the focus of public attention as a result of ongoing concerns about financial reporting, heightened by the events surrounding BCCI[1] and Maxwell[2] and the controversy over directors' pay. There was also concern over the composition of boards in relation to the balance of directors to non-executive directors (NEDs). Some company boards had no NEDs at all, and where NEDs were appointed, they were commonly outnumbered by executive directors. In addition, there was concern over the independence of NEDs as a result of their former role as executive directors of the same company, close connections with external advisers or major shareholders, or personal relationships with the chairman.

Section 1.3 of the report explained that at the heart of the Committee's recommendations was a Code of Best Practice "designed to achieve the necessary high standards of corporate behaviour". The Code of Best Practice, resulting from the Cadbury Committee's investigations, was appended to the Listing Rules in 1993 (see Section 2.19). The Cadbury Code identified generic themes of abiding concern and has had a major impact on thinking about corporate governance across the corporate and public sectors, within and outside the UK. The key recommendations of the Cadbury Code were in four main areas:

The board of directors. To ensure that the board functions as an authoritative decision-making body, rather than a formal rubber stamp for executive decisions, the Cadbury Code recommended that the full board meet regularly. In addition, it should establish a formal schedule of matters including material acquisitions and disposals, capital projects and treasury and risk management policies, specifically for its collective decision.

Non-executive directors. The Cadbury Code provided the first formal definition of the role of NEDs. It suggested that, in addition to their share in the strategic responsibilities of the board, they have explicit control and monitoring functions, which are distinct from the day-to-day managerial responsibilities of their executive colleagues.

Executive directors. The Cadbury Code referred to the treatment of executive remuneration and drew attention to the potential for conflicts of interest between shareholders and directors on matters of pay, performance and job security. Accordingly it recommended that shareholder approval should be obtained for new service contracts in excess of three years and stated that executive pay should be subject to recommendations of a remuneration committee made up wholly or mainly of NEDs.

---

[1] The collapse of the Bank of Credit and Commerce International (BCCI) which operated in 60 countries, is ranked as one of the largest banking collapses of all time, when the bank was left owing more than £10 billion ($18 billion) to its creditors. Thousands of depositors lost heavily when BCCI was wound up in 1991 amid accusations of money laundering and fraud. Those that lost included 28 UK local authorities, which had retained deposits in BCCI. The bank had lost money from its lending operations, its foreign currency dealings and its deposit accounts. It was reported that the bank was the bank of choice for money-launderers and terrorists, in that drug money from Colombia and Panama and funding for the Mujahideen in Pakistan and Abu Nidal in the Middle East all flowed through the bank. The Bingham inquiry, which looked into the BCCI collapse in 1992, was critical of the role of the Bank of England. The BCCI affair is thought to have played a role in Gordon Brown's decision, on becoming Chancellor, to remove banking supervision from the remit of the Bank of England and place it with the Financial Services Authority (FSA) as the UK's new bank regulator (BBC News (2004) Britain's biggest banking scandal. 13 January).

[2] Robert Maxwell died in mysterious circumstances at sea in 1991. Soon after his death it emerged that his empire was in serious financial difficulties and the Mirror Group pension fund was in deficit to the sum of £400 million, leaving 32 000 pensioners fearing for their future financial security. Previously, in 1980, Maxwell obtained a controlling interest in the British Printing Corporation plc, which was renamed the Maxwell Communication Corporation plc (MCC). By 1986, due to frenetic corporate expansion, the MCC became a FTSE 100 company. Maxwell ran his companies and the pension funds as if they were one. He moved assets between them as best suited his interests. Coopers & Lybrand, the auditors of nearly all of Maxwell's companies and their pension funds, were fined by the accounting profession's Joint Disciplinary Tribunal.

Reporting and controls. The Cadbury Code emphasised the board's obligation to present to shareholders a balanced and understandable assessment of the company's position. This should include a coherent narrative explanation of its performance and prospects, with details of setbacks as well as successes.

2.4 THE GREENBURY REPORT

The Greenbury Study Group on Directors' Remuneration was established in 1995 (Greenbury 1995) in response to public concern over apparently unjustified increases in the level of directors' remuneration, particularly in the then recently privatised utilities. The Study Group's remit was to establish good practice in determining directors' remuneration, particularly in the previously neglected area of performance-related pay. The resulting Code of Best Practice for directors' remuneration was annexed to the Listing Rules in 1995. The principal objectives of the Greenbury Code were to:

- prevent executive directors from setting or influencing their own remuneration;
- introduce greater rigour into the design of executive remuneration packages, with particular regard to performance incentives and rewards; and
- improve accountability to shareholders.

2.5 THE HAMPEL COMMITTEE AND THE COMBINED CODE OF 1998

The Cadbury and Greenbury Codes operated concurrently until June 1998, when a new Combined Code of Best Practice was appended to the Listing Rules. The Combined Code (Hampel 1998) was based on the recommendations of a Committee on Corporate Governance established in 1995 under the chairmanship of ICI chairman Sir Ronald Hampel. The committee's remit, which had been agreed with the Committee's sponsors (which included the London Stock Exchange, Confederation of British Industry and Institute of Directors), focused on a review of the Cadbury Code and its implementation, the role of directors (executive and non-executive), the issues arising from the Study Group on Directors' Remuneration chaired by Sir Richard Greenbury and the role of both shareholders and auditors in corporate governance. Although intended primarily as an updating and consolidation of the two earlier codes (Cadbury and Greenbury), the Combined Code represented a considerable broadening of the scope and detail of directors' obligations, particularly in the areas of internal control and risk management, accountability to shareholders and the company's relations with institutional investors. The 1998 version of the Combined Code consisted of 17 Principles of Good Governance, 14 of which were addressed to listed companies and the remainder to institutional investors. Hampel made the point that they wanted to encourage the use of the broad principles of corporate governance and their application with flexibility and common sense, adapted to the specific circumstances of a business.

2.6 SMITH GUIDANCE ON AUDIT COMMITTEES

The Smith Report (Smith 2003) provides guidance (to all UK listed companies) to assist boards in making suitable arrangements for their audit committees and to assist directors serving on audit committees in carrying out their role. The guidance includes certain essential requirements that every audit committee should meet. These requirements are highlighted in bold in the text. Compliance with these is necessary for compliance with the Combined Code (which preceded the UK Corporate Governance Code 2010). Listed companies that do not comply with these requirements are required to provide an explanation as to why they have not complied within the statement required by the Listing Rules. Section 1.4 of the guidance considers that boards should tailor their audit committee arrangements to suit the size, complexity and risk profile of the company. The audit committee is stated as having the role of acting independently of the executive, to ensure that the interests of shareholders are properly protected in relation to financial reporting and internal control. The report provides guidance on:

- the establishment and role of the audit committee, membership procedures and resources;
- relationship with the board;
- roles and responsibilities; and
- communication with shareholders.

2.7 HIGGS

In April 2002, Her Majesty's Treasury and the Department of Trade and Industry (DTI), concerned about improving the productivity performance of British industry, initiated a review of the role of the effectiveness of NEDs in publicly listed companies in the UK. The review was led by Derek Higgs (Higgs 2003), a respected investment banker and, in the eyes of the sponsor, a senior independent figure from the business world. The review was motivated by the belief that stronger and more effective corporate boards could improve corporate performance. The Company Law Review, for instance, noted "a growing body of evidence from the US suggesting that companies with a strong contingent of non-executives produce superior performance". Higgs summarised the terms of reference of the review as to build and publish an accurate picture of the status quo, to lead a debate on the issues and to make recommendations to clarify the role and increase the effectiveness of NEDs. Examining Annex K of the review, which records the terms of reference, the sponsors of the review, the government, considered that it would be valuable to build on the work of the Company Law Review and the Myners Review and undertake a review to assess such issues as the population of NEDs in the UK in terms of who they are, how they are appointed, their "independence", effectiveness, accountability and remuneration, and how to strengthen their quality, independence and effectiveness.

The summary of recommendations consisting of six pages of the report is wide ranging, reflects the terms of reference and covers such issues as independence, recruitment, appointment, induction, tenure, remuneration, resignation, audit committees, liability and relationships with shareholders. Higgs's report states that three substantial pieces of primary research were commissioned to inform his recommendations: data on the population of NEDs supplied by Hemscott Group Limited; data on the role of NEDs surveyed by MORI; and data on the relationships and behaviours that enable effective NED performance, supplied by three academics, McNulty of the University of Leeds and Roberts and Stiles of the University of Cambridge.

2.8 TYSON

The Tyson Report (Tyson 2003) on the recruitment and development of non-executive directors was published in June 2003. The report was commissioned by the Department of Trade and Industry (DTI), which was concerned to implement the recommendations included in the preceding Higgs Review on how companies might improve the quality and performance of their boards – through changes in the way they identify, select, recruit and train individuals to serve in NED positions. Laura Tyson, then Dean of the London Business School, was invited to chair the task force selected to undertake the review. The Higgs Report (see above), in Tyson's words, "raised the agenda" on boardroom effectiveness and considers that her report provides another piece of the jigsaw by highlighting how a range of different backgrounds and experiences among board members can enhance board effectiveness by exploring how a broader range of NEDs can be identified and recruited. Tyson states that diversity in the backgrounds, skills and experience of NEDs enhances board effectiveness by bringing a wider range of perspectives and knowledge to bear on issues of company performance, strategy and risk. The review report consists of 12 chapters, which cover the themes of the attributes, "sourcing" and current composition of NEDs, the benefits of diversity among NEDs, constraints on board composition and the need for ongoing training. The recommendations include:

- the selection process for each NED appointment resting on a careful assessment of the needs and challenges of a particular company;
- the broadening of selection search to include, in Tyson's words, the "marzipan" layer of management in PLC companies, professional services firms, unlisted companies and private equity firms, the non-commercial sector, and the commercial and non-commercial sectors in foreign companies;
- increasing formal training and evaluation of board members; and
- gaining greater board diversity and the development of an initiative (formation of a new organisation) to provide regular and reliable measures of board composition.

2.9 COMBINED CODE ON CORPORATE GOVERNANCE 2003

The 2003 Code replaced the Combined Code issued by the Hampel Committee on Corporate Governance in June 1998. It was derived from a review of the role and effectiveness of NEDs by Derek Higgs and a review of audit committees by a group led by Sir Robert Smith. The Financial Services Authority took the decision to replace the 1998 Code annexed to the Listing Rules with the revised Code. This new Code applied to reports issued by listed companies on or after November 2003. The preamble to the Code explained that the Listing Rules would not be amended in terms of the Code's requirement for listed companies to issue disclosure statements in two parts. In the first part of the statement they were required to state their governance policies and in the second part the company had to confirm that it had complied with the Code's provisions or, where it did not, provide an explanation. The view taken was that the "comply or explain" approach had been in operation for over ten years and would continue. The Code is broken down into five parts, covering directors, remuneration, accountability and audit, relations with shareholders and institutional shareholders.

The European Union commission did not want to enact a European code of corporate governance at that time as it saw no need; this may change. In addition, as the commission considered that the existence of different codes could cause some frictional and fragmentary cost, it encouraged a move towards greater convergence. Hence it was thought over time it may be that the EU would put pressure on the UK to modify its Combined Code.

2.10 COMPANIES ACT 2006

The aim of the Act was to reform company law and restate the greater part of the enactments relating to companies; to make other provision relating to companies; to make provision about directors' disqualification, business names, auditors and actuaries; and to amend Part 9 of the Enterprise Act 2002. The Act places a duty on directors to promote the success of the company and in the process consider the consequences of long-term decisions, employees' interests, business relationship management (including customers, suppliers, the community and the environment) and preservation of reputation. In relation to the content of director's reports, they must provide a fair review of the company's business, and a description of the principal risks and uncertainties facing the company. Any review must provide a balanced and comprehensive analysis of the development and performance of the company's business during the financial year, the position of the company's business at the end of that year and the main trends and factors likely to affect the future development, performance and position of the company. Part of the assessment of performance must include analysis of the financial key performance indicators. "Key performance indicators" are those factors by reference to which the development, performance or position of the company's business can be measured effectively.

2.11 COMBINED CODE ON CORPORATE GOVERNANCE 2008

The Combined Code, as issued in June 2008,[3] builds on the previous code and remains a well-crafted document in which the main principles of good corporate governance are described. While the months preceding the issue of this revised code had seen business failures and the exposure of possible shortcomings in the governance arrangements within certain companies, it could be argued that failures had been around the application of the Code's principles, not with the principles themselves. Minor amendments have been made to this Code. Changes were made to two provisions in order to:

- remove the restriction on an individual chairing more than one FTSE 100 company (provision A.4.3); and
- for listed companies below the FTSE 350, allow the company chairman to be a member of, but not chair, the audit committee, provided he or she was considered independent on appointment (provision C.3.1).

In addition, the following changes were made:

- Schedule C to the Code – which summarises the disclosure requirements in the Listing Rules and the Code – was updated to include the disclosure requirements under the new FSA Corporate Governance Rules, which implement EU requirements on audit committee and corporate governance statements.
- Eight provisions of the Code overlap with the new FSA Rules. The FSA Rules state that, where a company complies with these provisions of the Code, it will also be deemed to have complied with the relevant requirements in the Rules.

[3] The revised Code was published at the end of June, at the same time as new FSA Rules. The revised Code and new Rules applied to accounting periods beginning on or after 29 June 2008.

However, based on events in the UK and overseas with particular regard to the performance of banks and insurance companies, it could be argued that the Principle on Internal Control received insufficient attention. The Code currently contains an excellent principle (C2) that "the Board should maintain a sound system of internal control to safeguard shareholders' investment and the company's assets". Recent events have highlighted that company shareholders (and, in some cases, even the board members of those companies) have not been fully aware of risks associated with the business's activities, particularly risk exposure from acquisitions and certain financial instruments or risk distribution. The Code provision, C.2.1, notes that the board should, at least annually, conduct a review of the effectiveness of internal controls and should report to shareholders. The Code is then backed up by the Turnbull guidance which suggests means of applying this part of the Code. While C.2.1 refers to risk management systems, the Turnbull guidance warrants revisiting to reflect events over the last 18 months. Deloitte (2008) expressed the view that a review of the Turnbull guidance issued in 2005 was warranted to determine if it adequately handled areas such as an assessment of a company's exposure to systemic risk, procedures for risk escalation situations, recovery strategies, the adequate testing of business models and strategies and the proper consideration of risk in major transactions such as business acquisitions and complex financial instruments.

2.12 SIR DAVID WALKER'S REVIEW OF CORPORATE GOVERNANCE, JULY 2009 (CONSULTATION PAPER)

In February 2009 Sir David Walker was asked by the Prime Minister[4] to review corporate governance in UK banks in the light of the experience of critical loss and failure throughout the banking system (Walker 2009). This was a very important review in terms of enterprise risk management due to the spotlight placed on inadequacies in risk management at board level (see Box 2.2).

In his executive summary and recommendations Walker identifies that governance failures contributed materially to excessive risk taking in the lead-up to the financial crisis. In addition, he states that weaknesses in risk management, board quality and practice need to be addressed in the UK and internationally to minimise the risk of a recurrence. The review recognises that better governance will not guarantee that there will be no repetition of the recent highly negative experience for the economy and for society as a whole, but will make a repeat of these events materially less likely. Given the scale of the crisis, the size of write-downs and the precipice over which share prices fell, the language of the review appears very moderate.

The observations in the executive summary are brought out in what are called the five themes of the review. It considers that:

- The Combined Code of the FRC remains fit for purpose and the "comply or explain" regime is still appropriate. No new legislation was proposed.

[4] Walker's terms of reference were as follows (Walker 2009; emphasis added): "To examine corporate governance in the UK banking industry and make recommendations, including in the following areas: *the effectiveness of risk management at board level*, including the incentives in remuneration policy to manage risk effectively; the balance of skills, experience and independence required on the boards of UK banking institutions; the effectiveness of board practices and the performance of audit, risk, remuneration and nomination committees; the role of institutional shareholders in engaging effectively with companies and monitoring of boards; and whether the UK approach is consistent with international practice and how national and international best practice can be promulgated."

- There were major deficiencies in decision making and effective challenge was required prior to decisions on major risk and strategic issues, requiring close attention to board composition and increased commitment and involvement of experienced NEDs.
- Given the role of boards to successfully manage risk, board-level engagement in the high-level risk process should be materially increased with particular attention to the monitoring of risk and discussion leading to decisions on risk appetite and tolerance.
- There is a need for major shareholders to engage more proactively with their investee companies with the aim of supporting long-term improvement in performance.
- Given the background of defective control and serious excess in some cases, substantial enhancement is needed in board level oversight of remuneration policies.

The review makes 39 recommendations. As anticipated, in light of the focus on effective risk management within the terms of reference, numerous recommendations make reference to or discuss risk management directly (see Box 2.2).

Box 2.2 Review recommendations focusing on risk management

The review recommends that:

- Closer attention is given to the balance of the board and the experience of the individual board members to ensure an appropriate deliberation of the risk strategy (R4).
- NEDs satisfy themselves that board discussion and decision-taking on risk matters are based on accurate and appropriately comprehensive information and draw on external analysis when appropriate (R6).
- The chairman provides leadership by facilitating the informed and critical contribution of the directors in discussion and decision-taking concerning risk and strategy (R9).
- The board discloses that there is an ongoing process for identifying the skills and experience required to address and adequately challenge the key risks and decisions that confront the board (R13).

Of particular interest in the context of enterprise risk management, are recommendations 23–27 inclusive, listed under the heading "Governance and risk" and summarised in Box 2.3 (emphasis added).

Box 2.3 Governance and risk recommendations

Governance and risk

- The board of banks and other financial institutions (BOFIs) should *establish a board risk committee* (BRC) separately from the audit committee with responsibility for oversight and advice to the board on the current risk exposures and future risk strategy. In addition, the BRC, in preparing advice on overall risk appetite and tolerance, should take account of the current and prospective macro-economic and financial environment drawing on financial stability assessments (R23).
- In support of board-level risk governance, a BOFI *board should be served by a Chief Risk Officer* who should participate in the risk management and oversight process at the highest level on an enterprise-wide basis (R24).

- The BRC should draw on and take full account of *external experience* in challenging its analysis and assessment (R25).
- The BRC should oversee a *due diligence* appraisal of any proposed strategic transactions involving acquisition or disposal (R26).
- The BRC risk report should be included as a *separate report* within the annual report and accounts describing the strategy of the entity in a risk management context, identifying key exposures inherent in the strategy and the associated risk tolerance of the entity (R27).

The FRC agreed to implement those recommendations that it considered should apply to all listed companies. Some of the recommendations have been implemented through revisions to the Code (now renamed the UK Corporate Governance Code). It is understood that some recommendations will be implemented through revisions to other FRC guidance, such as the Turnbull Guidance on internal control and "Good Practice Suggestions from the Higgs Report" (the Higgs Guidance).

2.13 SIR DAVID WALKER'S REVIEW OF CORPORATE GOVERNANCE, NOVEMBER 2009 (FINAL RECOMMENDATION)

The five key themes of the Review as set out in the July consultation paper (and summarised in Section 2.12) are reiterated in the final recommendations which Walker declares attracted widespread support. Of particular significance is the second key theme described relating to the lack of the "challenge" step in board decision making. Its absence was prevalent on both sides of the Atlantic during the financial crisis, particularly within banks and other financial institutions. This key theme is repeated in full in Box 2.4. This theme should be read in conjunction with the FSA's publications and their thoughts on "group think" and the changes introduced in the UK Corporate Governance Code (see Box 2.5).

Box 2.4 Second theme of the review

Second, principal deficiencies in BOFI (banks and other financial institutions) boards related much more to patterns of behaviour than to organisation. The sequence in board discussion on major issues should be: presentation by the executive, a disciplined process of challenge, decision on the policy or strategy to be adopted and then full empowerment of the executive to implement. The essential "challenge" step in the sequence appears to have been missed in many board situations and needs to be unequivocally clearly recognised and embedded for the future. The most critical need is for an environment in which effective challenge of the executive is expected and achieved in the boardroom before decisions are taken on major risk and strategic issues. For this to be achieved will require close attention to board composition to ensure the right mix of both financial industry capability and critical perspective from high-level experience in other major business. It will also require a materially increased time commitment from the NED group on the board overall for which a combination of financial industry experience and independence of mind will be much more relevant than a combination of lesser experience and formal independence.

In all of this, the role of the chairman is paramount, calling for both exceptional board leadership skills and ability to get confidently and competently to grips with major strategic issues. With so substantial an expectation and obligation, the chairman's role in a major bank board will involve a priority of commitment leaving little time for other business activity.

The final recommendations repeat recommendations 23–27 of the July consultation document relating to governance and risk and are summarised in Box 2.3.

2.14 HOUSE OF COMMONS TREASURY COMMITTEE 2009

The House of Commons Treasury Committee undertook an inquiry to identify the lessons to be learned from the failure of the UK banks (which spanned 17 evidence sessions and 800 pages of written evidence). In its report summary, it states that the origins of the banking crisis included low interest rates, a search for yield, apparent excess liquidity and a misplaced faith in financial innovation. From its findings, the inquiry considered that "some of the banks have been the principal authors of their own demise" and that "bankers have made an astonishing mess of the financial system" (House of Commons 2009). In its evaluation of the status of the collapse of the banking sector it records that on 2 April 2007, nine banks occupied places in the FTSE 100 all-share index, whereas five of those banks – Bradford & Bingley, HBOS, Lloyds TSB, Northern Rock and RBS – are now partly or wholly in public ownership.

The report considers that the failure in the banking system relates to a series of behaviours. The first is risk appetite. Mervyn King, Governor of the Bank of England, told the inquiry that he considered that "failures in the international monetary system led to imbalances in capital flows between countries that created conditions of remarkably low interest rates and encouraged risk taking". This increased risk taking (to achieve better returns) became known as "search for yield". King had explained earlier in the year[5] that as a result of the combination of low returns and large amounts of capital (from China, Japan and oil-exporting countries) looking for a home in Western markets, a demand had been created for assets offering higher returns. The inquiry heard from a number of witnesses who had identified a change in behaviour of banks and had drawn a connection between the search for yield and an increase in the risk appetite of bank boards (see Box 2.5). John Varley, Group Chief Executive of Barclays Bank, advised the inquiry: "People with falling yields needed to increase their risk appetite to maintain yield. There was around the world a significant and very noticeable increase in risk appetite. That risk appetite was fuelled by the banks. I accept that." This increase in risk appetite was acknowledged at the inquiry by Sir Fred Goodwin, Chief Executive of the Royal Bank of Scotland until the point of the firm's failure, and Eric Daniels, Group Chief Executive of Lloyds. Lord Turner explained during his speech at the Economist's Inaugural City Lecture, that the explosion in the value and complexity of securities sold to satisfy the demand for yield uplift drove big increases in the leverage of major financial institutions – in particular, investment banks.

[5] Speech by Mervyn King, Governor of the Bank of England, to the CBI Dinner, Nottingham at the East Midlands Conference Centre, 20 January 2009.

Box 2.5 Management of increased risk exposure

Group CEO Daniels argued that the increases in risk appetite were accompanied by bespoke governance systems put in place to monitor and manage risk. Clearly some banks were better at managing risk than others. Giving evidence, Dr Jon Danielsson of the London School of Economics stated: "I think those institutions who will survive this crisis the best are the institutions with the best risk management. To my mind the crisis has shown that it is management of risk that is important." In his speech Andy Haldane[6] discussed the techniques to control risks and identified 1987 as the birth of "value at risk" reporting. Mr. Haldane argued that banks thought "risk was being held in check by a shift in the technological frontier of risk management" and this was supported by the findings of the inquiry. However, Dr Danielsson considered that the error made by banks was that they treated the economy as an "engineering system" which could be "run through a model" to give a risk. The stress testing and value at risk reporting did not take sufficient account of the fact "that people are intelligent and people react to risk".

The second behaviour related to the increase in complexity in operations on the premise that risk was being disbursed. In its findings the inquiry stated:

> We note that risk and complexity within the banking sector has increased dramatically over the last twenty years. The widespread – but at sometimes misguided – belief that risk was being dispersed and "managed" led many banks to increase the complexity of their operations and their overall risk exposure. This was manifestly a false premise. Indeed one of the factors that is key to understanding the banking crisis is that some forms of securitization, far from mitigating risk, actually obscured it.

What made risk management extremely difficult was that banks were dealing in such complex instruments. The problem articulated by Lord Turner was that some of these instruments had become "so complex that they have become very difficult to understand". As described in the FT (2009), the credit rating agency Standard & Poor's admitted in a report that it could take a whole weekend for computers to perform calculations needed to assess the risks of complex collateralised debt obligations. In fact the perception was that many products were simply incapable of being analysed.

A third behaviour was the banks' approach to leverage. The House of Commons found that among the banks that ran into trouble, there was evidence of a direct correlation between risk exposure and leverage (as banks increased their borrowings, the risk of their inability to repay their borrowings due to a fall in income increased). The Governor of the Bank of England informed the Treasury Committee that the most important common feature that distinguished the failed banks was leverage. In its findings the inquiry acknowledged that the financial sector had significantly increased its leverage over the last two decades, and those firms that showed the greatest appetite for rapid growth through leverage, particularly in the last five years, were amongst the heaviest casualties. Increased debt simply led to increased risk. An example was Northern Rock. As stated in the inquiry report, "the directors pursued a reckless business

[6]Speech by Andrew G. Haldane, Executive Director for Financial Stability, Bank of England, 13 February 2009.

model which was excessively reliant on wholesale funding".[7] Hence when wholesale funding markets tightened during the summer of 2007, Northern Rock was left unable to fund its operations.

A fourth behaviour was that of the government in its relationship with the regulator, the Financial Services Authority. As reported in the national press, the regulator had been blamed for the banking crisis which led to the near-collapse of the UK's largest banks, due to its perceived failure to intervene. The FSA has acknowledged its share of responsibility for the failures in the financial crisis, in particular in relation to firm-specific regulation, but considers that it alone would not have been able to prevent the financial crisis that unfolded, even with a different approach to supervision (FSA 2009). However, as reported in the *Daily Telegraph*, when the Chairman of the FSA, Lord Turner, appeared before the Treasury Select Committee in February 2009 he told MPs that the FSA had been put under pressure to be less heavy handed and intrusive, and have a lighter touch. Turner went on to express the opinion that there was a philosophy rooted in political behaviour which suggested the key priority was to reign back on questions (and hence not probe too deeply). According to the same report, John McFall, the Chairman of the Committee, said the remarks raised serious questions about the FSA's independence (*Daily Telegraph* 2009).

The inquiry report added richly to the body of knowledge as to why the banks now partly or wholly in public ownership had failed, but also highlighted that there was also a failure in the supervisory system designed to protect the public from systemic risk.

2.15 UK CORPORATE GOVERNANCE CODE, JUNE 2010

The UK Corporate Governance Code, formerly known as the Combined Code, sets out standards of governance for listed companies. As with the previous Codes, companies are required either to follow the Code or explain how else they are acting to promote good governance. This new edition of the Code affects businesses with financial years beginning on or after 29 June 2010. Changes (as described in Box 2.6) include a clearer statement of the board's responsibilities relating to risk, a greater emphasis on the importance of getting the right mix of skills and experience on the board, and a recommendation that all directors of FTSE 350 companies be put up for re-election every year. Introducing the new Code, Baroness Hogg, the FRC Chairman, said: "The changes we have made are designed to reinforce board quality, focus on risk and accountability to shareholders. In return, we look to see a step up

[7] Two senior directors of Northern Rock Plc were fined by the FSA on 13 April 2010 (FSA 2010a). David Baker, former deputy chief executive, was fined £504 000 and Richard Barclay, former managing credit director, £140 000. Margaret Cole, FSA director of enforcement and financial crime, said: "Baker and Barclay both failed to meet the standards we require of senior individuals within FSA-regulated firms. They both held senior positions of trust within the firm but they provided inaccurate information to the Northern Rock board and to the market." Baker was prohibited from performing any function in relation to any regulated activity. Barclay was prohibited from performing any significant influence function at an FSA-regulated firm.

As deputy chief executive, between January 2004 and March 2008, one of Baker's responsibilities was accurate internal and external reporting at Northern Rock. Despite becoming aware in January 2007 that there were 1917 loans omitted from the mortgage arrears figures, Baker failed to escalate the information internally and agreed a course of action which resulted in the loans not being reported. He also made misleading statements regarding these impaired loans to external stakeholders, including market analysts, quoting inaccurate figures.

Barclay was directly responsible for the provision of accurate management information concerning loan arrears and property possessions. He knew that the firm's arrears position enabled senior management within Northern Rock, analysts and the FSA to form a view of Northern Rock's asset quality, but failed to ensure that the management information was accurate despite warning signs at an early stage.

in responsible engagement by shareholders under the Stewardship Code, on which we have consulted and aim to publish by the end of June."

Box 2.6 Changes introduced in the 2010 Code

Changes to the Code (as described on the FRC's website[8]) include the following:

- To improve *risk management*, the company's business model should be explained and the board should be responsible for determining the nature and extent of the significant risks it is willing to take.
- *Performance-related pay* should be aligned to the long-term interests of the company and its risk policy and systems.
- To promote *proper debate* in the boardroom, there are new principles on the leadership of the chairman, the responsibility of the non-executive directors to provide constructive challenge, and the time commitment expected of all directors.
- To encourage boards to be *well balanced* and avoid "group think" there are new principles on the composition and selection of the board, including the need to appoint members on merit, against objective criteria, and with due regard for the benefits of diversity, including gender diversity.
- To help enhance the *board's performance* and awareness of its strengths and weaknesses, the chairman should hold regular development reviews with each director and FTSE 350 companies should have externally facilitated board effectiveness reviews at least every three years.
- To increase *accountability*, all directors of FTSE 350 companies should be put forward for re-election every year.

The annual re-election of all directors (see Box 2.6) was perhaps the most controversial change to the Code, attracting the greatest media attention, causing a considerable stir in boardrooms and dividing opinion between supporters and detractors. It was introduced into the Code as Provision B.7.1.

As reported in the *Daily Telegraph*, Richard Lambert, then director-general of the CBI, while praising the decision to stick with a principles-based code said he was concerned that the move to annual elections of directors could have an undesirable impact (*Daily Telegraph* 2010). Lambert considered that the code provision could promote a focus on short-term results, make boards less stable and discourage robust challenges in the boardroom. Anita Skipper, then corporate governance director at Aviva Investors, expressed an alternative opinion. Skipper recognised there were some concerns that annual elections would lead to short-term thinking on boards, but expressed hope that being more accountable to shareholders would encourage boards to foster relationships with the more important shareholders for their support of long-term strategies. In addition, evidence suggested that UK companies that already put their directors up for election each year had not behaved in a short-term manner and shareholders had not abused their right to vote directors off boards. If companies elect not to comply with Section B.7 "Re-election" (and in particular Section B.7.1), they must explain why they have not done so.

[8] http://www.frc.org.uk/press/pub2282.html

2.16 THE "COMPLY OR EXPLAIN" REGIME

A key feature of the UK's approach to corporate governance, from the Cadbury Code to the UK Corporate Governance Code of 2010, has been the avoidance of prescriptive rules. Only time will tell if statutory compliance will be introduced. This current avoidance of prescriptive rules reflects the view that different governance approaches are required for different companies, depending on their size, business activity, operating environment and ownership structure. In other words, one solution does not suit all circumstances. This stance is supported by Higgs (2003) who states: "I do not presume a 'one size fits all' approach to governance is appropriate". In consequence, successive Codes have no statutory force, but have been appended to the Listing Rules, with a requirement on listed companies to disclose in their annual reports whether or not they have complied with Code recommendations and, where they have not, to provide reasons for the areas of non-compliance. Under the resulting "comply or explain" regime, a company is under no formal obligation to comply with the best practice recommendations included within the Code. The Code states that "this 'comply or explain' approach is the trademark of corporate governance in the UK. It has been in operation since the Code's beginnings and is the foundation of the Code's flexibility" (Code 2010). However, the disclosure obligation ensures that the company's shareholders are able to monitor the extent of its compliance, consider the explanations provided by the directors for any areas of non-compliance and, if dissatisfied, express their concerns through their voting behaviour at the AGM.

2.17 DEFINITION OF CORPORATE GOVERNANCE

So what do we mean by corporate governance? A definition of corporate governance is important here to aid understanding, in terms of both its purpose and its relationship with internal control. The Institute of Directors, in its 2010 "factsheet" on corporate governance, declares that there is no single accepted definition of what the expression "corporate governance" means (IoD 2010). The definitions that do exist tend to be broad high-level statements such as that included in the Cadbury Committee report (discussed in Section 2.3) which states that "corporate governance is the system by which businesses are directed and controlled". While appealing in its simplicity, this definition is not particularly informative. The OECD (2004) expands the definition to cover issues of stakeholder management, objective setting and monitoring performance: "corporate governance involves a set of relationships between a company's management, its board, its shareholders and other stakeholders. Corporate governance also provides the structure through which the objectives of the company are set, and the means of attaining those objectives and monitoring performance are determined." The Combined Code 2010 describes one of the supporting principles of corporate governance under Section A.1, headed "The Role of the Board", as follows:

> The board's role is to provide entrepreneurial leadership of the company within a framework of prudent and effective controls which enables risk to be assessed and managed. The board should set the company's strategic aims, ensure that the necessary financial and human resources are in place for the company to meet its objectives and review management performance. The board should set the company's values and standards and ensure that its obligations to its shareholders and others are understood and met.

This introduces the themes of leadership, risk management, aims, resources, performance measurement and culture. A detailed definition is offered here, adopting the themes of earlier publications and including the elements of direction, resources and management:

> Corporate governance is the system by which companies are *directed*, in terms of (1) the company's strategic aims and (2) entrepreneurial leadership, *resourced* in terms of providing (1) the necessary financial and human resources and (2) the necessary ICT resources, and *managed* using robust, defensible and prudent controls to (1) interface with internal and external stakeholders, (2) establish risk management processes, (3) produce accurate, timely and relevant information for decision making, risk management and reporting, (4) comply with laws and regulations, (5) establish the company culture by setting the company's values and standards, and (6) reflect the perspective of the parent company as appropriate.

2.18 FORMATION OF COMPANIES

Of interest here are organisations in the private sector known as public limited companies (PLCs) which can sell their shares to the public and may be quoted on the stock exchange. While corporate governance and risk management are important to all businesses, whether they be sole traders, partnerships, PLCs, cooperatives or franchises, corporate governance and enterprise risk management have greater significance for listed companies. The main thrust of the Cadbury Committee's report, for instance, was to review the financial reporting and accountability of listed companies with a view to protecting shareholders' interests (see Appendix 1 to the Cadbury Committee report, entitled "Terms of Reference").

When a company is formed, a legal distinction is created between the existence and identity (or "personality") of the company itself and those of its members or shareholders. This distinction gives incorporated form significant advantages as a means of carrying on a business:

- As a legal person in its own right, a company can possess rights and privileges not available to its shareholders and can take action to enforce these rights.
- Only the company, not its shareholders, can be sued for breach of its legal duties.
- Property owned by the company is distinct from the property of its shareholders, with the result that the shareholders' property is unaffected by the claims of creditors in the event that the company becomes insolvent.

PLCs are limited by share and must include "PLC" in their name. This acts as a warning to those trading with such a company, because any debts it incurs from trading may not be recoverable due to the limited liability of its owners (shareholders). Where a limited company cannot pay its debts from its own financial resources, it cannot make the owners use their personal finances to meet these debts. Limited liability encourages greater investment than would otherwise take place, and ensures a demand for stocks and shares. The benefit for the economy is that it encourages people to risk owning or investing in companies, because they know their liability (losses) will be limited to the amount they have agreed to invest. The main legal provisions relevant to the formation and operation of listed companies are contained in a small number of Acts of Parliament. Requirements for the formation and operation of companies are specified in the Companies Acts of 1985, 1989 and 2006. The implementation of the Companies Act 2006 was fully completed on 1 October 2009. Arrangements for the disqualification of directors are set out in the Company Directors Disqualification Act 1986. Corporate insolvency is covered by a distinct legislative regime under the Insolvency Act 1986, and regulation of the securities markets is now contained in specific financial services legislation, the Financial Services and Markets Act 2000.

2.19 THE FINANCIAL SERVICES AUTHORITY AND MARKETS ACT 2000

The Financial Services Authority[9] is an independent non-governmental body, given statutory powers by the Financial Services and Markets Act 2000 (FSMA). It is the sole regulator for financial services in the UK. It is a company limited by guarantee and financed by the financial services industry. Her Majesty's Treasury appoints the FSA's board, consisting of a chairman, chief executive officer, two managing directors and 11 non-executive directors. This board sets the overall policy, whereas day-to-day decisions are the responsibility of the executive. The FSA is the UK Listing Authority and hence the authority in the UK for the listing of company shares and other securities for trading on public stock exchanges. The FSA (as a competent authority under Part VI of the FSMA) governs listing through its Listing Rules (which can be found in the FSA Handbook), whereby companies wishing to trade their securities must first apply for admission to the FSA, demonstrating compliance with the Rules (FSA 2010b). Companies may be required to prepare listing particulars[10] (or prospectuses) setting out the nature of their business, their management and financing arrangements and potential material risks to potential investors. In accordance with the Listing Rules private companies will not be granted admission. Once a company's securities have been listed for trading, it is required by the Listing Rules to fulfil a number of ongoing reporting requirements regarding finance, management and constitution. Additionally, all companies with a listing of equity shares in the UK are required under the Listing Rules to report on whether or not they are complying with Section DTR7 of the Corporate Governance Code, and, where not, to give reasons for non-compliance. The UK Listing Authority currently has the power to either suspend or cancel a listing. However, the regulation of financial services in the UK is evolving. In June 2010 the Chancellor of the Exchequer, George Osborne, announced that he would be dismantling the FSA and replacing it with three bodies over the next two years. He considered that the existing tripartite system (of the Bank of England, the FSA and the Treasury) had "failed spectacularly" in its mission of ensuring financial stability (BBC 2010). The three new bodies would be a Financial Policy Committee (FPC) and a Prudential Regulatory Authority (PRA), both operating under the Bank of England, addressing macro and micro prudential regulation respectively, and a Consumer Protection and Markets Agency (CPMA).

2.20 THE LONDON STOCK EXCHANGE

The London Stock Exchange (LSE) provides the bridge between issuers and the capital markets. The LSE remains by far the largest equity market in Europe. It enables companies from around the world to raise capital required for growth, by listing securities on what it claims are highly efficient, transparent and well-regulated markets. Through its two primary markets – the Main Market and Alternative Investment Market (AIM) – the Exchange provides companies with access to one of the world's largest pools of investment capital. The Main

[9] See www.fsa.gov.uk. The operating framework of the FSA is called ARROW (Advanced Risk-Responsive Operating frameWork). The current framework, ARROW II, was rolled out in 2006.

[10] See the London Stock Exchange Admission and Disclosure Standards, April 2010.

Market is the Exchange's principal market for listed companies from the UK and overseas. The AIM is the LSE's international market for young and growing companies. The AIM enables these businesses to access the capital and liquidity of the London markets. Once companies have been admitted to trading, the LSE provides expertise of the global financial markets to assist them maximise the value of their listing in London. It provides the trading platforms used by broking firms around the world to buy and sell securities. Its systems provide fast and efficient access to trading, allowing investors and institutions to tap quickly into equity, bond and derivative markets. More than 300 firms worldwide trade as members of the LSE. The LSE Issuer service works with customers before, during and after listing. As of 1 November 2010 there were over 1500 companies listed on the Main Market and over 1450 companies listed on the AIM, operating in over 100 countries. The LSE, it can be argued, is "regulated" by the Office of Fair Trading (OFT). In 2004 the OFT conducted an inquiry into increases in the annual and admission fees for the UK Main Market, resulting in the LSE settling for reduced fees. On 10 June 2010 the OFT announced plans to undertake a market study into equity underwriting and associated services. Responses received from market participants confirmed that they had concerns about the market and informed the study's final scope. The purpose of the market study was to examine the underwriting services for the different types of share issue used by listed companies to raise capital in the UK. These include rights issues, placings and other types of "follow-on" offer (but not initial public offerings). The study was limited to equity issues carried out by FTSE 350 listed companies. As a Recognised Investment Exchange (RIE), all the LSE's markets must meet standards detailed in the FSA's RIE and RCH Sourcebook. In addition to this UK standard, the Exchange has also sought to apply the EU market standards set out in the Investment Services Directive to certain of its markets.

2.21 SUMMARY

This chapter has traced the developments in corporate governance from the Cadbury Report through to the UK Corporate Governance Code 2010, examined the formation of companies and looked at the workings of both the FSA and the LSE. The Cadbury Committee and its code of best practice was first examined, as it is recognised to be the start of a formalised approach to corporate governance. One of the four main themes of the Code, executive directors' remuneration, was further developed by the Greenbury Committee, culminating in its report. Subsequently it was decided that the previous governance recommendations should be reviewed and brought together in a single code. The review was carried out under the chairmanship of Sir Ronald Hampel and the ensuing final report, known as the Hampel Report, issued in 1998 with its Combined Code on Corporate Governance, included a number of provisions relating to internal control. However, it gave little guidance on the actual implementation of internal controls. As a result the Institute of Chartered Accountants in England and Wales, in conjunction with the LSE, formed a working party to study the matter of internal control, which resulted in the Turnbull Report of 1999. For the first time there was emphasis on the creation of a system of risk management.

In 2002 the DTI asked Derek Higgs to look at how the role and effectiveness of NEDs may improve corporate performance. The Higgs Report, issued in 2003, also suggested amendments to the Combined Code. The Tyson Report, building on the Higgs Report, examined how boards may identify, select, recruit and train individuals to serve in NED positions to improve board performance. At the same time as Higgs was reporting, the FRC had asked a group chaired by Sir Robert Smith to issue guidance for audit committees. In July 2003, the revised Combined

Code, taking account of both the Higgs and Smith reviews, was published and took effect for reporting periods beginning on or after 1 November 2003.

The Walker Review of 2009 identified that governance failures had led to excessive risk taking within the banking industry. The review focused on the role of risk management in decision making. It recommended that firms in this industry should establish a board risk committee with responsibility for oversight and advice to the board on current risk exposure and the future risk strategy. In addition it recommended that the board should be served by a chief risk officer and that the annual report should contain a separate report on the risk exposure of its business strategy and assessment of its risk tolerance.

In the same year the House of Commons Treasury Committee carried out an inquiry into the failure of UK banks. It found that the combination of the increase in risk appetite of UK banks and the increase in leverage that they adopted (increased debt) significantly increased their overall risk exposure. In addition, there was a lack of appreciation of systemic risk arising from the interconnectedness of banks by the banks themselves and the regulator (FSA).

Subsequent to the publication of the Walker Review, papers prepared by the FSA, investigations by government and other reports, the UK Corporate Governance Code 2010 was published. Despite the financial crisis, the "comply or explain" approach, in existence since Cadbury, is still maintained on the premise that one corporate governance approach does not fit all circumstances and prescriptive rules should be avoided. The key governance issues addressed by these reports and codes include board structure and membership, board management, director's remuneration, financial controls, accountability, audit and relations with shareholders. In addition, the formation of PLCs, the operation of the FSA and the Listing Rules and the operation of the LSE were all examined. The link between these last three sections is that public companies, which wish to raise capital for growth on a recognised investment exchange such as the LSE, have to apply to the FSA for admission. A condition of entry is compliance with the Listing Rules, which refer to adherence to the Governance Code. Having reflected on the Codes and Reports, a definition of UK corporate governance is offered as a backdrop to internal controls and risk management.

2.22 REFERENCES

BBC (2010) Q&A: Osborne's financial regulation reforms. BBC Business News, 17 June. http://www.bbc.co.uk/news/10343900

Cadbury Committee on the Financial Aspects of Corporate Governance (1992) *Report of the Committee on the Financial Aspects of Corporate Governance: The Code of Best Practice*. Gee Publishing, London.

Code (2010) *The UK Corporate Governance Code*, Financial Reporting Council, June.

Combined Code (2003) *Combined Code on Corporate Governance,* Financial Reporting Council, July. http://www.fsa.gov.uk/pubs/ukla/lr_comcode2003.pdf

Cooper, B. (2004) *The ICSA Handbook of Good Boardroom Practice*. ICSA Publishing, London.

CG (1997) Corporate Governance – Volume One: in Principle, Corporate Governance – Volume Two: in Practice, Performance Audit Report issued by The Audit Office of New South Wales, Australia, in compliance with Section 38E of the Public Finance and Audit Act 1983.

Deloitte (2008) Review of the Combined Code. http://www.deloitte.com/view/en_GB/uk/services/audit/corporate-governance/article/0c0fbc9f55142210VgnVCM200000bb42f00aRCRD.htm

Dey, F.J. (1994) Toronto Stock Exchange Committee on Corporate Governance in Canada, *Where Were the Directors?* TSE, Canada, December.

FSA (2009) The Crisis: the role of investors. Speech by Hector Sants, NAPF Investment Conference 2009.

FSA (2010a) FSA fines and bans former Northern Rock deputy chief executive and credit director for misreporting mortgage arrears figures. 13 April. http://www.fsa.gov.uk/pages/Library/Communication/PR/2010/066.shtml

FSA (2010b) *Listing Rules*. FSA, London. http://www.fsa.gov.uk/pubs/hb-releases/rel100/rel100lr.pdf

FT (2009) Lost through destructive creation. *Financial Times*, 10 March 2009.

Greenbury Study Group (1995) *Report on Directors' Remuneration*. Gee Publishing, London.

Hampel Committee on Corporate Governance (1998) *Committee on Corporate Governance: Final Report*. Gee Publishing, London. http://www.ecgi.org/codes/documents/hampel.pdf

Higgs, D. (2003) *Review of the Role and Effectiveness of Non-Executive Directors*. Department of Trade and Industry, The Stationery Office.

House of Commons (2009) *Banking Crisis: dealing with the failure of the UK banks*. House of Commons Treasury Committee, Seventh Report of Session 2008–2009, 1 May 2009.

IoD (2010) *Corporate governance in the UK factsheet*. Institute of Directors, www.iod.com

OECD (2004) *OECD Principles of Corporate Governance*. OECD, Paris.

Smith, R. (2003) *Audit Committees Combined Code Guidance*. Financial Reporting Council, January.

Telegraph (2009) FSA head: Gordon Brown helped fuel the banking crisis. Telegraph.co.uk, 26 February.

Telegraph (2010) Code overhaul set to shake-up UK boards. Telegraph.co.uk, 28 May.

Turnbull, N. (1999) *Internal Control: Guidance for Directors on the Combined Code*. Internal Control Working Party of the Institute of Chartered Accountants in England and Wales, September.

Tyson (2003) *The Tyson Report on the Recruitment and Development of Non-Executive Directors*, London Business School (http://www.london.edu).

Walker, D. (2009) A review of corporate governance in UK banks and other financial industry entities. Walker Review Secretariat, London. http://webarchive.nationalarchives.gov.uk/+/http://www.hm-treasury.gov.uk/d/walker_review_261109.pdf

3
Developments in Corporate Governance in the US

The goal is to make certain that taxpayers are never again on the hook because a firm is deemed "too big to fail."
(Barack Obama)

The previous chapter examined the incremental developments in corporate governance in the UK from the Cadbury Committee to the UK Corporate Governance Code of 2010. This chapter examines the parallel initiatives in the US and its own unique catalysts for corporate governance and internal controls. While stock exchanges and governments around the world have clearly followed the developments in corporate governance in the UK, events in the US have directly influenced the specific course of action that it has followed. The highly publicised events surrounding the collapse of Enron and WorldCom, rather than just colouring the US's approach to corporate governance, prompted what might be termed a more radical approach than that seen in the UK, resulting in new legislation – the Public Company Accounting Reform and Investor Protection Act 2002, also known as the Sarbanes-Oxley Act. Chapter 4 describes the tumultuous global financial crisis and the changes in legislation it prompted.

3.1 CORPORATE GOVERNANCE

The US approach to corporate governance can be characterised as a regulator-led system predominantly enforced through the Securities and Exchange Commission (SEC), stock exchange listing rules and state law. The US does not have a corporate governance code similar to that of the UK, given the structure of its constitution. While the internal aspects of corporate governance such as leadership, director's duties, remuneration and shareholder rights are a matter of state law, the SEC and the stock exchanges also play a significant role in the governance of listed companies. The regulatory authorities in the US and their respective roles are described by Mark Roe in Box 3.1.

Box 3.1 Roles of regulatory authorities

The roles of the various regulatory authorities are described by Roe (2005):

> We do not have a purely state-based system of law governing the American corporation. The federal players, just to be clear, are principally four. First is the United States Congress, which passed the securities laws and approximately every 10 years or so undertakes a major update of those laws. Second is the Securities and Exchange Commission, which promulgates regulations under the securities laws and often proposes changes to Congress. The courts then are the third player interpreting those laws. And the fourth is the stock

exchange, which may look from Europe to be purely a private actor, but which makes major corporate governance law, usually when the SEC – a federal administrative agency created by the United States Congress and given by Congress, substantial power over the exchange – asks (or perhaps the better word is directs) it to make those rules.

3.2 THE SECURITIES AND EXCHANGE COMMISSION

As declared by the SEC, its mission is to protect investors, maintain fair, orderly and efficient markets, and facilitate capital formation. First and foremost, the SEC is a law enforcement agency. The SEC oversees the key participants in the securities arena, including securities exchanges, securities brokers and dealers, investment advisors, and mutual funds. The aim of the SEC is to promote the disclosure of important market-related information, maintain fair dealing and provide protection against fraud. The SEC cannot work in a vacuum and is dependent on and seeks to work with other institutions, as described in Box 3.2.

Box 3.2 Cooperation between federal departments

Although it is the primary overseer and regulator of the US securities markets, the SEC works closely with many other institutions, including Congress, other federal departments and agencies, the self-regulatory organisations (e.g. the stock exchanges), state securities regulators and various private sector organisations. In particular, the Chairman of the SEC, together with the Chairman of the Federal Reserve, the Secretary of the Treasury, and the Chairman of the Commodity Futures Trading Commission, serves as a member of the President's Working Group on Financial Markets.

To ensure all investors, whether large institutions or private individuals, have access to basic facts about an investment prior to buying it, the SEC requires public companies to disclose accurate financial and other information to the public. This provides information for all investors to use to judge whether to buy, sell or hold a particular security. Crucial to the SEC's effectiveness in each of these areas is its enforcement authority. Vigorous enforcement is required. Crime is unrelenting. Every year the SEC brings hundreds of civil enforcement actions against individuals and companies for violation of the securities laws. Typical transgressions include insider trading, accounting fraud and providing false or misleading information about securities and the companies that issue them.

3.2.1 Creation of the SEC

The SEC was created at a time when reform was vital for restoring investor confidence. Before the Great Crash of 1929, there was little support for federal regulation of the securities markets. This was particularly true during the post World War I surge in securities activity. It was considered that creating a federal government requirement for financial disclosure, to prevent the fraudulent sale of stock, was not warranted. However, this view very quickly changed when investors lost millions of dollars in the Great Crash, as described in Box 3.3.

Box 3.3 Legislation to govern business ethics

When the stock market crashed in October 1929, half of the estimated $50 billion invested in new securities (during the period of prosperity after World War I) became worthless. Public confidence in the markets plummeted. In the Great Depression which followed, investors (and the banks that had loaned to them) lost vast sums of money. There was a consensus that if the economy was to recover, the public's faith in the capital markets had to be restored. At the height of the Depression, Congress passed the Securities Act 1933. This law, together with the Securities Exchange Act 1934, which created the SEC, was designed to restore investor confidence in US capital markets by providing investors and the markets with more reliable information and clear rules for honest dealing. The primary purpose of these laws (which was a sad indictment on US business ethics) was to ensure that companies publicly offering securities tell the truth about their businesses, the securities they are selling and the risks involved in investing, and that brokers, dealers and exchanges (sellers of the trade securities) treat investors fairly and honestly.

The SEC was formed during the Great Depression in 1934 to enforce the newly-passed securities laws, to promote stability in the markets and, most importantly, to protect investors. President Franklin Delano Roosevelt appointed Joseph P. Kennedy, President John F. Kennedy's father, to serve as the first Chairman of the SEC.

3.2.2 Organisation of the SEC

The SEC is organised as a number of divisions with their own specific responsibilities, some of which are recorded in Box 3.4. The SEC's "teeth" are "bared" through the Division of Enforcement which brings both civil and criminal actions, commonly in conjunction with other agencies. The Division of Enforcement focuses on and looks for particular violations of the Acts, as described in Box 3.5.

Box 3.4 Divisions of the SEC

- The Division of Corporation Finance oversees corporate disclosure by corporations when stock is initially sold and then on a continuing periodic basis.
- The Division of Investment Management supports investor protection through oversight and regulation of America's $26 trillion investment management industry (including mutual funds professional fund managers who advise them; analysts who research individual assets and asset classes; and investment advisers to individual customers).
- The Division of Risk, Strategy and Financial Innovation supports the identification of developing risks and trends in the financial markets and provides analysis that integrates economic, financial and legal disciplines (reflecting the changes in capital markets and corporate governance and the emergence of derivatives, hedge funds and new technology).

- The Division of Enforcement executes the Commission's law enforcement function by recommending civil actions be brought before federal courts (or before an administrative law judge as appropriate), prosecuting these cases on behalf of the Commission, and working closely with law enforcement agencies both in the US and internationally to bring criminal cases when required.

Box 3.5 SEC violations

Common violations that may lead to SEC investigations include:

- misrepresentation or omission of important information about securities;
- manipulating the market prices of securities;
- stealing customers' funds or securities;
- violating broker-dealers' responsibility to treat customers fairly;
- insider trading (violating a trust relationship by trading on material, non-public information about a security); and
- selling unregistered securities.

3.3 THE LAWS THAT GOVERN THE SECURITIES INDUSTRY

3.3.1 Securities Act 1933

Often referred to as the "truth in securities" law, the Securities Act 1933 has two basic objectives:

- to require that investors receive financial and other significant information concerning securities being offered for public sale; and
- to prohibit deceit, misrepresentations and other fraud in the sale of securities.

The primary means of accomplishing these goals is the disclosure of important financial information through the registration of securities. This information enables investors to make informed judgements about whether to purchase a company's securities. While the SEC requires that the information provided be accurate, it does not guarantee it. Investors who purchase securities and suffer losses have important recovery rights if they can prove that there was incomplete or inaccurate disclosure of important information.

3.3.2 Securities Exchange Act 1934

With this Act, Congress created the Securities and Exchange Commission. The Act empowers the SEC with broad authority over all aspects of the securities industry. This includes the power to register, regulate and oversee brokerage firms, transfer agents and clearing agencies as well as securities self-regulatory organisations (SROs). The various stock exchanges, such as the New York Stock Exchange and American Stock Exchange, are SROs, as is the Financial Industry Regulatory Authority, which operates the NASDAQ system.

The Act also empowers the SEC to require annual and other periodic reports from companies with more than $10 million in assets and whose securities are held by more than 500 owners.

These reports are made available to the public through the SEC's EDGAR database. The securities laws broadly prohibit fraudulent activities of any kind in connection with the offer, purchase or sale of securities. These provisions are the basis for many types of disciplinary actions, including actions against fraudulent insider trading.

3.3.3 Trust Indenture Act 1939

This Act applies to debt securities such as bonds, debentures and notes that are offered for public sale. Even though such securities may be registered under the Securities Act, they may not be offered for sale to the public unless a formal agreement between the issuer of bonds and the bondholder, known as the trust indenture, conforms to the standards of this Act.

3.3.4 Investment Company Act 1940

This Act regulates the organisation of companies, including mutual funds, that engage primarily in investing, reinvesting and trading in securities, and whose own securities are offered to the investing public. The regulation is designed to minimise conflicts of interest that arise in these complex operations. The Act requires these companies to disclose their financial status and investment policies to investors when stock is initially sold and, subsequently, on a regular basis. The focus of this Act is on disclosure to the investing public of information about the fund and its investment objectives, as well as on investment company structure and operations.

3.3.5 Investment Advisers Act 1940

This law regulates investment advisers. With certain exceptions, this Act requires that firms or sole practitioners compensated for advising others about securities investments must register with the SEC and conform to regulations designed to protect investors. Since the Act was amended in 1996, generally only advisers who have at least $25 million of assets under management or advise a registered investment company must register with the Commission.

3.4 CATALYSTS FOR THE SARBANES-OXLEY ACT 2002

The dramatic and highly publicised failing of large corporations arising from corporate mismanagement and fraud, in particular Enron, WorldCom and Tyco International, exposed significant corporate governance shortcomings. The analysis of their complex root causes (see Box 3.6) contributed to the introduction of Sarbanes-Oxley Act in 2002. The Act takes its name from its main architects, Senator Paul Sarbanes and Representative Michael Oxley. It is generally recognised as one of the most significant market reforms since the passage of the securities legislation of the 1930s. It is intended to help protect investors and restore investor confidence by improving the accuracy, reliability and transparency of corporate financial reporting and disclosures, and reinforce the importance of corporate ethical standards. Public and investor confidence in the fairness of financial reporting and corporate ethics was seen as being critical to the effective functioning of the capital markets. The Act's requirements apply to all public companies regardless of size and the public accounting firms that audit them.

Box 3.6 Senate hearings

Prior to drafting and introducing new legislation, the Senate undertook a series of hearings to gain a broad and common understanding of the problems in the markets, as described by Senator Paul Sarbanes, who in an interview in 2004 stated:

> The Senate Banking Committee undertook a series of hearings on the problems in the markets that had led to a loss of hundreds and hundreds of billions, indeed trillions of dollars in market value. The hearings set out to lay the foundation for legislation. We scheduled 10 hearings over a six-week period, during which we brought in some of the best people in the country to testify... The hearings produced remarkable consensus on the nature of the problems: inadequate oversight of accountants, lack of auditor independence, weak corporate governance procedures, stock analysts' conflict of interests, inadequate disclosure provisions, and grossly inadequate funding of the Securities and Exchange Commission.

3.4.1 Enron

The primary catalyst of the Sarbanes-Oxley legislation was the collapse of Enron. For the unaffected observer, the story is fascinating. Until 2001, US energy trader Enron was one of the largest companies in the world. A BBC bulletin dated 8 July 2004 and entitled "Q&A, the Enron Collapse" succinctly summarises its evolution and the nature of its business:

> Enron began life as an energy producer, moved to become an energy trader, and ended up an energy "bank" providing guaranteed quantities at set prices over the long term. Enron owned power plants, water companies, gas distributors and other units involved in the delivery of services to consumers and businesses. But it was the first to realise energy and water could be bought, sold, and hedged just like shares and bonds. Enron became a huge "market-maker" in the US, acting as the main broker in energy products, also taking financial gambles far bigger than its actual core business.

The beginning of Enron's demise can be traced to an announcement in October 2001 which stated that it was taking a $544 million charge to its reported earnings, after tax, in respect of transactions with an off-balance sheet entity owned by Enron, but created and controlled by its chief financial officer (CFO). Hence, to disguise its true balance sheets, the firm had used complex financial partnerships to conceal mounting debts. In addition, shareholders' equity was being reduced by $1.2 billion. Less than a month later, the company announced that it was restating its accounts for the years 1997–2001 because of accounting errors in relation to off-balance sheet entities controlled by the CFO and other senior managers. The restatements involved reductions of between 10% and 28% in reported net income in each of the years affected, with substantial reductions in shareholders' equity and increases in reported levels of indebtedness. The company also revealed for the first time that the CFO had received personal payments from off-balance entities that he controlled. Following these announcements, investor confidence in Enron's ability to recover its position collapsed and the company filed for bankruptcy protection in December 2001.

The fallout from Enron's failure was considerable. Enron left behind $15 billion of debts, its shares become worthless and 20 000 workers around the world lost their jobs. Many banks were exposed to the firm as a result of lending money to and trading with it. JP Morgan admitted to $900 million of exposure, and Citigroup to nearly $800 million. Former high-profile bankers Merrill Lynch were charged with fraud in connection with Enron transactions.

Consultants Andersen, which failed to audit Enron accounts correctly, collapsed with the loss of 7500 jobs in the US and 1500 in the UK.

3.4.2 WorldCom

While Enron had seriously shaken investor confidence in the global financial markets, the collapse of US telecom business WorldCom gave impetus to more deep-rooted concerns over the substance of very large firms. WorldCom's bankruptcy was twice the size of Enron's, which up to that time had been America's largest. WorldCom, prior to disclosure of its problems, owned a third of the US's high-speed data cables and was the second largest long-distance phone operator with 85 000 staff in 65 countries. As a low-margin business, WorldCom needed to grow to survive and at its peak in 1997 it paid $37 billion (£24.3 billion) to take over telecommunications giant MCI, snatching the business under the jealous gaze of Britain's BT. In June 2002 the firm reported accounting irregularities which had overvalued its income by a "mere" $43.3 billion, making the firm look profitable when it was not. However, worse was to come and two months later the business revealed a further $3.3 billion improperly reported earnings.

In a court filing in New York, the SEC said that WorldCom had admitted that it concealed $9 billion in expenses, all of which had been converted into false profits. Mr Ebbers, the former chief executive, had built up the company from a small operator into the world's second biggest long-distance group. The firm filed for bankruptcy in a bid to protect it from creditors and to buy time to restructure. Interestingly Mr Ebbers said in his testimony at his trial that he never got good marks at school and dropped out of various colleges before obtaining a qualification as a physical education teacher. He bought a number of motels and only accidentally ended up running a telecoms company in the mid-1980s. William Johnson, assistant US attorney, told jurors at the trial that they should disregard the "Aw shucks, I'm not sophisticated" defence, and that Mr Ebbers was a hands-on chairman very focused on costs, citing examples. During the trial, Charna Sherman, Partner at Squire, Sanders & Dempsey, said that "Mr Sullivan [WorldsCom's former chief financial officer and main witness against Mr Ebbers] was sufficiently persuasive to substantiate the natural presumption that nobody that high up could miss such a massive fraud, so Mr Ebbers had to take the huge gamble of testifying. That decision has transformed the case into a litmus test about a CEO's responsibilities after Sarbanes-Oxley" (Van Duyn 2005).

3.4.3 Tyco International

While the actions of Tyco International board executives are less well know than Enron, they were also very significant in defining the need for tighter regulation in the US. For in 2005 former CEO and chairman, Dennis Kozlowski, and CFO, Mark Swartz, were found guilty on 22 out of 23 counts of grand larceny and conspiracy, falsifying business records and violating business law. After the verdict the government requested a mandatory jail term of 15–30 years in a state prison. In 2005, a New York State court sentenced Kozlowski and Swartz to prison terms of $8\frac{1}{3}$ to 25 years for their roles in the Tyco fraud. As a consequence of their criminal convictions, Kozlowski and Swartz paid approximately $134 million in restitution to Tyco and criminal fines of $70 million and $35 million, respectively. On 16 October 2008, the New York Court of Appeals affirmed Kozlowski's and Swartz's convictions. On 8 June 2009, the United States Supreme Court denied a petition by Kozlowski and Swartz for a writ of certiorari (i.e. asking the Supreme Court to grant the writ to review the decision of a lower court, based on declared arguments). In July 2004, former Chief Corporate Counsel and Executive Vice President, Mark Belnick, was acquitted of securities fraud and larceny in connection with his role as chief corporate counsel of Tyco. However, in May 2008 Belnick agreed to pay a $100 000 fine to settle charges with the SEC, stemming from money he borrowed from the company. In addition, Belnick was prohibited from serving as an officer or director of a public company for five years. A description of the activities of these directors reads more like pages taken from a film script rather than the actions of senior executives ranked among America's top corporate leaders responsible for managing a multi-billion-dollar international enterprise.

The Securities and Exchange Commission

This verdict was arrived at after a number of years of persistent activity by the SEC. In September 2002 the SEC filed a civil enforcement action against three of the former senior executives of Tyco International. Action was brought on the basis of violation of federal securities laws by failing to disclose to shareholders the multi-million-dollar low-interest and interest-free loans they took from the company. In a statement the SEC's Director of Enforcement, Stephen Cutler, said:

> Messers Kozlowski, Swartz and Belnick treated Tyco as their personal bank, taking out hundreds of millions of dollars of loans and compensation without ever telling investors and in some cases never paid.... Defendants put their own interests above those of Tyco's shareholders. Those shareholders deserved better than to be betrayed by the management of the company they owned.

However, in April 2004, as reported in the press at the time, the first trial against Kozlowski and Swart ended in a mistrial as a result of the sole juror holding out for an acquittal reporting receiving threats.

Rise and Fall

During his ten-year tenure as CEO, Kozlowski played an instrumental role in acquiring hundreds of companies valued at tens of billions of dollars. As a result, Tyco grew from a small manufacturing company into an industrial giant. At its peak Tyco was a manufacturer and service provider operating in over 100 countries. However, Kozlowski resigned as CEO and Chairman in June 2002 for what he declared were personal reasons, while the press speculated it was because he was being investigated for tax evasion. At the same time there were persistent allegations of accounting fraud within Tyco International which the SEC was taking seriously. In 2005, when the SEC took the executives to trial for a second time, the SEC's complaint alleged, among other things, the executives had been involved in secret loans, undisclosed compensation and transactions and fraudulent stock trading.

The Secret Loans

As reported by the SEC (and as described in Box 3.7), Dennis Kozlowski, Mark Swartz and Mark Belnick granted themselves hundreds of millions of dollars in secret low-interest and interest-free loans from the company that they used for personal expenses. They then covertly caused the company to forgive (cancel) tens of millions of dollars of those outstanding loans, again without disclosure to investors as required by the federal securities laws. Belnick was accused by the SEC of taking a $17 million bonus as a reward for helping conceal corporate wrongdoing, but was acquitted of all charges by a Manhattan jury in July 2004.

Box 3.7 Secret loans

The following are examples of loans that were taken and not disclosed to shareholders, contrary to the requirements of the federal securities laws.

- Dennis Kozlowski: from 1997 to 2002, took a net $242 million from Tyco's Key Employee Corporate Loan Program (KELP) for personal expenses, including yachts, fine art, estate jewellery, luxury apartments and vacation estates, personal business ventures and investments, all unrelated to Tyco.
- Mark Swartz: during the same period, took a net $72 million from KELP for personal investments, business ventures, real estate holdings and trusts.
- Dennis Kozlowski: from 1996 to 2002, took more than $46 million in interest-free relocation loans intended to assist Tyco employees who were required to relocate to either New York City or Boca Raton, Florida. Kozlowski used at least $28 million of those relocation loans to purchase, for example, luxury properties in New Hampshire, Massachusetts and Connecticut as well as a $7 million Park Avenue apartment for his wife.
- Mark Swartz: took more than $32 million in interest-free relocation loans. Almost $9 million of the relocation loans were used for unauthorised purposes, including purchasing a yacht and investing in real estate.
- Mark Belnick: took approximately $14 million in undisclosed interest-free relocation loans. In 1998, when Belnick joined Tyco, he took a $4 million relocation loan to purchase an apartment in New York City on Central Park West, even though he had never worked at Tyco headquarters in New Hampshire. In September 2001 he took a $10 million relocation loan to purchase a house in the Park City, Utah, ski resort, even though Tyco did not have a corporate presence in Utah and despite the fact that Belnick already owned a $2 million dollar home in Park City.

The Undisclosed Compensation

Included in Box 3.8 are examples of the very significant amounts of undisclosed executive compensation Kozlowski and Swartz awarded themselves, contrary to the requirements of the federal securities laws. In most instances Kozlowski and Swartz directed others to falsify Tyco's records to conceal this secret compensation. These actions were at the expense of the shareholders.

Box 3.8 Undisclosed compensation

The following are examples of undisclosed compensation awarded to Kozlowski and Swartz.

- In August 1999, Kozlowski authorised, and Swartz recorded in Tyco's records, a $25 million loan forgiveness against Kozlowski's outstanding KELP balance and a $12.5 million credit against Swartz's outstanding KELP balance.
- In September 2000, Kozlowski engineered the covert forgiveness of more than $33 million of his relocation loans and more than $16 million of Swartz's relocation loans. They agreed to keep these payments secret.

- In November 2000, Kozlowski and Swartz engineered a cash bonus, Tyco stock and forgiveness of relocation loans.
- In June 2001, Kozlowski and Swartz directed the acceleration of the vesting of Tyco stock for the benefit of themselves and certain other favoured employees. As a result, Kozlowski and Swartz realised profits of approximately $8 million and $4 million, respectively.
- Kozlowski and Swartz enjoyed numerous and extensive undisclosed benefits that they bestowed upon themselves. For example, Kozlowski lived rent-free in a $31 million Fifth Avenue apartment that Tyco purchased in his name, while Swartz lived rent-free in a multi-million-dollar apartment Tyco purchased in his name on New York City's Upper East Side. Moreover, Kozlowski directed millions of dollars of charitable contributions in his own name using Tyco funds.

Undisclosed Related-Party Transactions

Kozlowski and Swartz also engaged in undisclosed real estate transactions with Tyco and its subsidiaries. These included Kozlowski's purchase by Tyco (with funds borrowed under the KELP) of the $7 million Park Avenue apartment for his wife and Swartz's purchase of a New Hampshire property (by a subsidiary of Tyco's) for far more than its fair market value. These transactions were not disclosed to shareholders, contrary to the requirements of the federal securities laws.

Fraudulent Stock Trading

Kozlowski and Swartz, while preventing disclosure to shareholders and potential investors of the material facts concerning their self-dealing and fraudulent loans as set forth above, sold millions of dollars of Tyco stock back to Tyco itself through Tyco subsidiaries located in offshore bank-secrecy jurisdictions. Swartz also sold stock on the open market through family partnerships.

3.4.4 Provisions of the Act

The advent of Enron and similar cases led to President Bush signing the Sarbanes-Oxley Act, which he characterised as "the most far reaching reforms of American business practices since the time of Franklin Delano Roosevelt", into law. The aim of the Act was to protect investors and influence the behaviour and conduct of public accounting firms[1] and public companies to ensure they issue accurate, reliable and informative financial statements. Significant provisions of the Act include the following:

Audit regulation. The Act establishes a Public Company Accounting Oversight Board (PCAOB) as a private-sector non-profit organisation to oversee the audits of public companies that are subject to securities laws. The PCAOB has responsibility for the registration (Section 102), inspection (Section 104) and discipline (Section 105) of public accounting firms, together with overseeing the audit of public companies that are subject

[1] The term "public accounting firm" as defined by the Act means a proprietorship, partnership, incorporated association, corporation, limited liability company, limited liability partnership or other legal entity which is engaged in the practice of public accounting or preparing or issuing audit reports.

to the securities laws[2] and related matters in order to protect the interests of investors and further the public interest in the preparation of informative, accurate and independent audit reports.

Corporate responsibility. Under Section 302, CEOs and CFOs of public companies have to take personal responsibility for financial reports by certifying that they do not contain any untrue statement of a material fact, omit any material fact or mislead. Additionally these "signing officers" are held accountable for establishing and maintaining internal controls and evaluating their effectiveness (as of a date within 90 days prior to the report). They are also responsible for advising their auditors of all significant deficiencies in the design or operation of the internal controls and any identified weaknesses in the internal controls.

It is interesting to note that the first chief executive to be charged under the Act, Richard Scrushy, former CEO of HealthSouth, was acquitted.[3] Mr Scushy had faced 58 criminal charges including conspiracy, money laundering, perjury, mail fraud and making false statements. At the time of the trial Tim Burns, a corporate governance expert at Neal, Gerber & Eisenberg (a Chicago law firm), was of the opinion that if the prosecution failed, there would be calls from politicians and the public to make tougher regulations. The trial related to a $2.7 billion fraud where the company had overstated its earnings between 1996 and 2002. In the years 2000–2002 the company had actually suffered multi-million-dollar losses, despite reporting a profit. While the CEO was cleared, 15 former HealthSouth staff pleaded guilty, including the five former CFOs. Time will tell whether the legislation will be progressively tightened.

Management assessment. Under Section 404, public companies are required to include in each annual report an internal control report which states (1) the responsibility of management to establish and maintain an adequate internal control structure and procedures for financial reporting, and (2) an assessment, as of the end of the most recent fiscal year, of the effectiveness of the internal control structure and procedures for financial reporting. Additionally, the auditor (registered public accounting firm) preparing the audit for the public company "shall attest to, and report on, the assessment made by the management" of the public company.

Non interference in the audit process. Under Section 303, the Act states that it is unlawful for directors, officers (or persons acting under their direction) to take any action to fraudulently influence, coerce, manipulate or mislead any independent public or certified accountant auditing the firm's financial statements.

Company records. Under Section 801, the Act makes it a criminal offence for anyone to knowingly alter, destroy, mutilate, conceal, cover up or falsify any record/document, with the intent to impede, obstruct or influence the investigation of any matter. Individuals found guilty of these charges, the Act states, "shall be fined under this title, imprisoned not more than 20 years, or both".

Disciplinary sanctions. Under Section 105(c)(4) of the Act, the PCAOB has the power to impose disciplinary or remedial sanctions on a registered public accounting firm (or associated person) that it considers to have been in violation of the Act. The sanctions include suspension or permanent loss of registration status, temporary or permanent

[2] The term "securities laws" means the provisions of law referred to in Section 3(a)(47) of the Securities Exchange Act 1934 as amended by the Sarbanes-Oxley Act.

[3] BBC News (2005) HealthSouth's ex-boss not guilty. 28 June. http://news.bbc.co.uk

prohibition of an individual being able to associate with a registered public accounting firm, limitation of activities, civil money penalty, censure, requirement for additional professional education or training, or other sanction.

3.4.5 Implementation

Early experience of *Fortune 500* companies has involved dealing with:

- undeveloped reporting routines and systems;
- inadequate or inconsistent levels of documentation;
- multiple financial systems as a result of geographical spread and/or acquisitions; and
- the implications of outsourcing of certain financial functions.

3.4.6 Sarbanes-Oxley Section 404

The difference between the UK Corporate Governance Code of 2010 and Sarbanes-Oxley is that, under the Combined Code, management are not required to report on the effectiveness of their internal controls and auditors are not required to report on the assessment made by management, whereas Sarbanes-Oxley is about 100% compliance! Section 404 of Sarbanes-Oxley requires:

- management to assess the effectiveness of their internal controls and procedures for financial reporting; and
- the auditor to attest to and report on the assessment made by management.

Hence management must ask:

- Are the controls explicit? Have they been identified and documented?
- Are they consistent across the business?
- Do they address the business critical success factors?
- Do the controls include risk management?
- What procedures need to be carried out to enable them to sign off that controls are effectively working?

Sheridan (2003) considers that the requirements of Section 404 need to be implemented as a project and that the issues in Table 3.1 need to be taken into account in order to ensure the successful completion of a Section 404 project.

3.4.7 The Positive Effects of Post-Enron Reforms

The effects of the changes that have been introduced to improve corporate governance by the legislative and regulatory reforms in the post-Enron era, prior to the financial crisis, are as follows:

- independence of the audit committee;
- increase in criminal penalties for white-collar crime; and
- creation of PCAOB for enhanced regulation by auditors;
- prohibition of non-audit services;
- prohibition of loans to directors and executives;

Table 3.1 Activities for the successful completion of a Section 404 project

Issue	Response
Executive sponsorship	Executive sponsorship needs to be secured, leading to the formation of a steering committee that will provide clear and ongoing direction and support.
Project management	A Section 404 project is typically a complex and challenging exercise. A strong project management team and a clearly defined plan are essential. Typically financial controllers or internal auditors are being expected to lead the project, but it is expected that IT management be part of the management team. Care should be taken to maintain internal audit's independence.
Role of external audit	As external auditors are required to attest management's assessment, it is critical to involve them at an early stage. The extent of their involvement in detailed documentation depends on the level of perceived independence desired by the board. However, they will be unable to perform any evaluation or control testing on behalf of management.
Section 404 adviser	Experience has shown that Section 404 projects benefit from the input and support provided by independent, third-party advisers. These advisers can accelerate the start-up phase by bringing to the project tried and tested tools and documentation templates along with experience of how others are tackling their implementation.
Skills of finance, IT and internal audit staff	The project execution teams need to be the right blend of experience and capabilities and must have sound training to ensure effective application of a suitable methodology and approach.
Impact on internal audit resources	Where internal audit is contributing to project execution, management must consider the impact on its ability to deliver the audit plan. In order to meet the assurance expectations of the audit committee, any gaps left by the redeployment of internal auditors must be considered and addressed.
Identification of significant controls	Controls must be selected for the significant accounts on the basis of their appropriateness, likely effectiveness and ability to remove the risk of misstatements that individually (or when aggregated with other misstatements) could have a material effect on the financial statements (as a result of either overstatement or understatement).This includes understanding the flow of transactions commonly supported by IT functions to a sufficient degree to know where the points of misstatement may arise. (Guidance and practice have continued to evolve and both the PCAOB and SEC have provided guidance on the selection of controls).
What should documentation look like?	There isn't a required style or format. It is important to cover key information such as the risk control description, who performs the controls, types of control, frequency, evidence and results of testing from an efficiency point of view. Leveraging existing processes for risk management and internal audit documentation is sensible.

*Source:*Reproduced with permission from *Managing Business Risk*, © 2003, Kogan Page Limited. (The section on "identification of significant controls" has been modified by R.J. Chapman.)

- various kinds of disclosures to provide transparency as regards the functioning of the company;
- making attorneys liable for not reporting wrongdoing;
- attempt to eliminate any conflict of interest affecting analysts;
- prohibition on improperly influencing an audit report;
- making executives liable for false certification of a corporation statement; and
- the protection of whistleblowers.

3.4.8 Criticism of Section 404 Before the Global Financial Crisis

Section 404 is one of the more contentious elements of Sarbanes-Oxley due to the significant cost of compliance. According to a survey by Financial Executives International (FEI) that included issuers with an average revenue of \$4.7 billion, compliance costs were \$1.7 million during 2007. The total cost includes internal and external labour and auditor attestation fees. Companies continue to focus on and reduce costs in their Sarbanes-Oxley 404 efforts through top-down risk assessment and compliance software.

3.4.9 Criticism of Section 404 After the Global Financial Crisis

A question raised innumerable times is whether Sarbanes-Oxley should have prevented the global financial crisis. The Act has not been without its critics since its introduction. The American Enterprise Institute for Public Policy Research, a right-wing think-tank, has dismissed Sarbanes-Oxley as a "colossal failure, poorly conceived and hastily enacted during a regulatory panic". *The Wall Street Journal* considers that it has "imposed hundreds of billions of dollars in costs on business with no noticeable decline in financial scandals".

Sarbanes-Oxley has certainly imposed heavy burdens on companies, but it has also increased accountability and transparency, and initially helped restore investors' confidence in the market and reinforced the principle that companies are owned by their shareholders, not their managers. In addition, during the global financial crisis when it appeared companies were struggling, the Sarbanes-Oxley Act warned when companies were overleveraged. During *Fortune*'s March 2010 interview with Michael Oxley, he (not surprisingly) argued the case for the Act. He said that it is very difficult to regulate a relatively free market with a lot of innovations in a climate where some companies just did not want to abide by the regulations. However, he went on to say, that did not mean we should stop trying to regulate and prosecute wrongdoers. Expanding on his theme, he said that people kill one another every day despite laws against homicide but it did not mean you give up and stop fighting.

Perhaps the best insight into Sarbanes-Oxley's effectiveness is obtained from the Anton Valukas report of March 2010 and the subsequent letter issued by a small group of senators calling upon the SEC to require more complete and accurate corporate accounting disclosures, especially of off-balance sheet transactions. Valukas was appointed by a judge to investigate the collapse of Lehman Brothers, which occurred in September 2008. Mr Valukas found that Lehman had been insolvent for weeks before it filed for bankruptcy. In addition, executives had used a complicated transaction that enabled them to remove liabilities from Lehman's balance sheet for a short time, when results were due, and effectively hide the true level of its debts. Its auditors, Ernst & Young, allowed what the press termed "this questionable practice" to continue for a prolonged period. Interestingly, a key driver for the introduction of Sarbanes-Oxley was the collapse of Enron, which also used off-balance sheet transactions to conceal its true financial position.

As reported by a number of media agencies at the time, in August 2010 a group of senators wrote to the Chairman of the SEC, urging it to require more complete and accurate corporate accounting disclosures, especially of off-balance sheet transactions. The group of six senators noted a series of incidents in which corporations concealed their debts and financial weaknesses. They used Lehman as an example. They requested that the SEC use its existing rule-making authority from the Sarbanes-Oxley Act to require full disclosure of off-balance sheet activity. The senators wrote:

> Rather than relying on carefully staged quarterly and annual snapshots, investors and creditors should have access to a complete real-life picture of a company's financial situation. The SEC was founded on the premise that when investors and creditors have full and accurate information about companies' finances, they can allocate capital effectively. But when companies use accounting gimmicks to mislead investors and creditors, capital markets malfunction. As we attempt to recover from the latest meltdown, we hope that, in addition to aggressively investigating and prosecuting past misconduct, you will put in place these new rules that will make it harder for companies to mislead investors and creditors in the future.

They observed that Sarbanes-Oxley gave the SEC the power to require reporting of off-balance sheet activities, but it had not stopped companies from hiding their debts. "While the SEC did issue rules on off-balance sheet activity pursuant to Sarbanes-Oxley, we are troubled that despite these rules, widespread off-balance sheet accounting arrangements allowed large financial firms to hide trillions of dollars in obligations from investors, creditors, and regulators", they said. In addition to Lehman's use of Repo 105, the senators cited Citigroup, which reportedly kept $1.1 trillion worth of assets off its books in various financing vehicles and trusts that were used to handle mortgage-backed securities and issue short-term debt. "Neither of these companies adequately disclosed the risks posed by their off-balance sheet activities to investors", the senators wrote. "Had they done so, investors and creditors might have made better decisions." The senators want the SEC to use its existing authority under Sarbanes-Oxley to require that companies write detailed descriptions of all their off-balance sheet activities in their annual reports, not just those that are "reasonably likely" to affect the firm's financial condition, as the regulations currently state. The senators said: "Companies should also explicitly justify why they have not brought those liabilities onto the balance sheet." Clearly history tells us that complete disclosure of all off-balance sheet activities is particularly crucial for the largest and most interconnected companies, including both banks and non-banks. The senators' letter implies the SEC had the tools but had failed to use them adequately.

3.5 NATIONAL ASSOCIATION OF CORPORATE DIRECTORS 2008

The National Association of Corporate Directors (NACD) has proposed ten key principles to strengthen corporate governance for US publicly traded companies. The principles are described in Box 3.9 and the supporting explanation of the principles is included in a document, "Key Agreed Principles to Strengthen Corporate Governance for US Publicly Traded Companies", published in October 2008. The document states:

> The Key Agreed Principles that follow are grounded in the common interests of shareholders, boards, and management teams in the corporate objective of long-term value creation (through ethical and legal means), the accountability of management to the board, and ultimately the accountability of the board to shareholders for such long-term value creation. The Principles provide a framework for board leadership and oversight in the especially critical areas of strategic planning, risk oversight, executive compensation and transparency.

Box 3.9 NACD key agreed principles to strengthen corporate governance

1. *Board responsibility for governance.*
   Governance structures and practices should be designed by the board to position the board to fulfill its duties effectively and efficiently.
2. *Corporate governance transparency.*
   Governance structures and practices should be transparent – and transparency is more important than strictly following any particular set of practice recommendations.
3. *Director competency and commitment.*
   Governance structures and practices should be designed to ensure the competency and commitment of directors.
4. *Board accountability and objectivity.*
   Governance structures and practices should be designed to ensure the accountability of the board to shareholders and the objectivity of board decisions.
5. *Independent board leadership.*
   Governance structures and practices should be designed to provide some form of leadership for the board distinct from management.
6. *Integrity, ethics and responsibility.*
   Governance structures and practices should be designed to promote an appropriate corporate culture of integrity, ethics and corporate social responsibility.
7. *Attention to information, agenda and strategy.*
   Governance structures and practices should be designed to support the board in determining its own priorities, resultant agenda and information needs and to assist the board in focusing on strategy (and associated risks).
8. *Protection against board entrenchment.*
   Governance structures and practices should encourage the board to refresh itself.
9. *Shareholder input into director selection.*
   Governance structures and practices should be designed to encourage meaningful shareholder involvement in the selection of directors.
10. *Shareholder communications.*
    Governance structures and practices should be designed to encourage communication with shareholders.

3.6 SUMMARY

This chapter has examined the formation of the Securities and Exchange Commission in the aftermath of the Great Crash in 1929 to enforce the newly established securities laws, promote stability in the markets and, most importantly, to protect investors. In addition, this chapter has described the events surrounding the collapse of Enron, WorldCom and Tyco, the known catalysts behind the US government's decision to pass the Sarbanes-Oxley Act 2002, with the aim of protecting investors by improving the accuracy and reliability of corporate disclosures. The main thrust of the Act is to influence the behaviour and conduct of public companies and public accounting firms, to ensure they issue informative accurate financial statements. The Act establishes a Public Company Accounting Oversight Board with responsibility for the registration, inspection and discipline of public accounting firms that are subject to the securities laws. The CEOs and CFOs of public companies have to take personal responsibility for financial reports by certifying that they do not contain any untrue statement of a material fact, omit any material fact or mislead. To address a specific event in the Enron collapse, the Act makes it a criminal offence for anyone to knowingly alter, destroy, mutilate, conceal, cover up or falsify any document, with the intention of thwarting any investigation of a business, and those found guilty may be fined, imprisoned or both. In the aftermath of the global financial crisis the Sarbanes-Oxley Act was heavily criticised for failing investors. However, perhaps the focus should be on how the SEC applied the legislation and how companies strived to circumnavigate full disclosure and deliberately mislead investors through complex off-balance sheet transactions.

3.7 REFERENCES

Roe, M.J. (2005) Regulatory competition in making corporate law in the United States – and its limits. *Oxford Review of Economic Policy*, 21(2), 232–242.

Sheridan, F. (2003) Implementing Sarbanes-Oxley Section 404. In A. Jolly, *Managing Business Risk, a Practical Guide to Protecting Your Business*, pp. 81–89. Kogan Page, London.

Van Duyn, A. (2005) Jurors to take their pick over WorldCom. *Financial Times*, 4 March.

4

The Global Financial Crisis of 2007–2009: A US Perspective

The fact that people are full of greed, fear, or folly is predictable. The sequence is not predictable.
(Warren Buffett)

The previous chapter examined how the US reacted to a series of prominent and highly publicised company failures such as Enron, WorldCom and Tyco by reforming the regulatory framework and introducing both the Sarbanes-Oxley Act of 2002 and reforms to the listing rules of the New York Stock Exchange and NASDAQ. This chapter examines the global financial crisis triggered by the subprime mortgage collapse and the federal reforms that followed. In addition, it examines the criticism that the discipline of risk management received in the aftermath of the crisis and new approaches to systemic risk. As will be seen, the financial crisis is important from an enterprise risk management perspective as it has catapulted risk management into the spotlight for all the wrong reasons. Subsequent chapters examine the approach to corporate governance adopted in Australia and Canada, economies that enjoy higher governance standards and which have not suffered destructive scandals to the same degree.

4.1 THE FINANCIAL CRISIS IN SUMMARY

The world financial crisis led to the collapse of several high-profile banks, the emergency bailout of others, hundreds of billions of dollars of write-downs, departures of bank CEOs and CFOs, fraud investigations by the FBI, hearings in the US Congress and introspection by national governments. It brought about a downturn in global stock markets (as a result of damaged investor confidence), which lost approximately $32 trillion in value since their peak, equivalent to the combined gross domestic product of the G7 countries in 2008. It necessitated intervention by governments and led to the introduction of new legislation, extensive job losses, a dramatic fall in house prices and a significant decline in economic activity. It is considered by the International Monetary Fund (IMF)[1] and many economists to be the worst financial crisis since the Great Depression of the 1930s. Governments and central banks responded with unprecedented fiscal stimulus, monetary policy expansion and institutional bailouts. As expressed by Ben Bernanke (Chairman of the US Federal Reserve), the dramatic rise in mortgage delinquencies and foreclosures within the US subprime housing market triggered rather than caused the financial turmoil (Bernanke 2007). However, the failure to properly price the risky pooled subprime mortgages sold as securities (supported by unrealistically positive ratings by credit agencies) truly precipitated the crisis. When in August 2007 the markets disregarded the credit agencies' rosy ratings a blanket of uncertainty descended on the investment community. It had adverse consequences for banks and financial markets around the globe. Over a short space of time the financial crisis in America became a global economic crisis.

[1] The IMF is an organisation of 187 countries working to foster global monetary cooperation, secure financial stability and facilitate international trade.

4.2 HOW THE FINANCIAL CRISIS UNFOLDED

From the middle of 2006 there was a dramatic rise in mortgage delinquencies across the breadth of the United States. A foreclosure epidemic had been triggered. House prices had initially peaked and then begun a steep decline. At the same time interest rates had continued to rise. When the introductory period of adjustable rate mortgages (issued mostly to the subprime[2] mortgage market) ended and the low interest rate ceased, mortgage payments ballooned. Mortgage refinancing for homeowners became either more expensive or impossible. In the period from July 2006 to July 2007 the rate of foreclosures (home repossessions) had risen by 93% (RealityTrac), as homeowners continued to default on their mortgages.[3] In the same year this number grew to 1.5 million before it began to decline (Bernanke 2008b). As a result of the foreclosures, securities backed with subprime mortgages widely held by investors lost most of their value. In addition, as mortgage losses mounted, investors questioned the reliability of the credit ratings, especially those of structured credit products.[4] Since many investors had not performed independent valuations of these often complex instruments, the loss in confidence in the credit ratings led to a sharp decline in the willingness of investors to purchase these products. Liquidity dried up. Prices fell. Many of these investors had assumed significant debt burdens to invest in these securities. Rather than using the boom period to build their reserves, investors (banks) increased their risk exposure as evidenced by higher debt to equity ratios, also known as financial leverage. As a result, investors did not have a sufficiently large financial cushion to absorb extensive loan defaults when they arose. The consequence was a large decline in the capital of many US banks. However, the problems were not restricted to US banks, as outlined in Box 4.1.

Box 4.1 International effects of the financial crisis

The credit crisis quickly became international as banks around the world began reporting losses arising from their investment in US mortgage-backed securities. Germany's Deutsche Bank, France's retail bank Crédit Agricole, Belgian bank Fortis, Swiss bank Credit Suisse, English bank Lloyds TSB and Japan's Mizuho were all among those banks reporting reduced profits and bad debts arising from the subprime market. In September 2007, Northern Rock, Britain's fifth largest mortgage lender, was granted emergency funding by the Bank of England after finding itself unable to secure loans from elsewhere. The news led to the first run on a British bank for more than a century as thousands of depositors queued to withdraw their money.

As reported in the *New York Times* (Sorkin and Thomas 2008) and other media sources during March 2008, at the request of the US Federal Reserve and the Treasury Department, global investment bank Bear Stearns (previously ranked 138 in the *Fortune 500* and valued at

[2] Borrowers with certain credit characteristics that disqualified them from the prime rate were known by the term "subprime". Since subprime borrowers often had poor or limited credit histories they were typically perceived as riskier than prime borrowers. To compensate for this increased risk, lenders charged subprime borrowers a premium for mortgages by applying a higher interest rate. However, the higher interest rate was often deferred and did not commence immediately. The interest rate was "reset" typically after two or three years.

[3] Default arose in the main when homeowners lost their jobs, when homes went into negative equity or homeowners were unable to pay the higher interest-rate payments when mortgages were reset.

[4] In simple terms, structured credit products are usually issued by investment banks and commonly have two components, a note and a derivative. The note provides for periodic interest payments to the investor at a predetermined rate and the derivative component provides for the payment at maturity. However, there are many permutations.

$2 billion in 2007) was bought by JP Morgan Chase for $236 million. There was considerable shareholder disquiet as the company had been sold for $2 per share, whereas the shares had been trading at $84 per share at the end of the fiscal year. According to the International Monetary Fund, by May 2007 the aggregated write-downs of JP Morgan Chase, Royal Bank of Scotland, Morgan Stanley, Bank of America, HSBC, Merrill Lynch, UBS and the United States' largest bank, Citigroup, had surpassed $100 billion. The subprime losses of Citigroup alone had exceeded $40 billion, forcing the bank to cut 9000 jobs. In September of the same year Lehman Brothers filed for bankruptcy, the first major bank to collapse since the start of the credit crisis. Merrill Lynch was sold to Bank of America amid fears of a liquidity crisis.

Critical to the financial turmoil was the shadow banking system. The renowned economist and Nobel Prize winner Paul Krugman and US Treasury Secretary Henry Paulson Jr. explained the credit crisis by way of the implosion of the parallel or shadow banking system which had grown to nearly equal to the importance of the commercial banking sector. Without the ability to obtain investor funds in exchange for most types of mortgage-backed securities or asset backed commercial paper,[5] investment banks and other entities in the shadow banking system could not provide funds to mortgage firms and other corporations (Geithner 2008).

4.3 THE UNITED STATES MORTGAGE FINANCE INDUSTRY

The traditional mortgage model involved a bank financing a loan to the borrower/homeowner through the deposits they received from customers and retaining the credit (default) risk. With the advent of securitisation, the traditional model gave way to a new model of "originate to transfer" in which the investment banks essentially sold the mortgages and transferred the credit risk to investors through mortgage-backed securities. Since 2002 the private sector had dramatically expanded its role in the mortgage market which had previously been dominated by the government sponsored agencies[6] such as Freddie Mac.

4.4 SUBPRIME MODEL OF MORTGAGE LENDING

Figure 4.1 describes the primary parties in the subprime housing market in the situation where mortgage-backed securities are created by investment banks and sold to investors.

4.4.1 Contributing Events to the Credit Crisis

To gain a clear understanding of the interrelationship of the events embroiled in the crisis it is important to find a "window" through which to gain a clear perspective. The perspective selected is based on a number of interlinked "vicious circles" some of which have been

[5] In the global money market, commercial paper is an unsecured promissory note with a fixed maturity of 1–270 days. It is a money-market security issued (sold) by large banks and corporations to get money to meet short-term debt obligations and is only backed by an issuing bank or corporation's promise to pay the face amount on the maturity date specified on the note. The money market is a component of the financial markets for assets involved in short-term borrowing and lending with original maturities of one-year or shorter timeframes. Trading in the money markets involves Treasury bills, commercial paper, bankers' acceptances, certificates of deposit, federal funds and short-lived mortgage-and asset-backed securities (Fabozzi *et al.* 2002). It provides liquidity funding for the global financial system.

[6] In the United States the federal government created several programmes or government sponsored entities to foster mortgage lending, construction and encourage home ownership. These included the Government National Mortgage Association (known as Ginnie May), the Federal National Mortgage Association (known as Fannie Mae) and the Federal Home Loan Mortgage Corporation (known as Freddie Mac). These programmes worked by buying a large number of mortgages from banks and issuing mortgage-backed bonds to investors which are known as mortgage-backed securities.

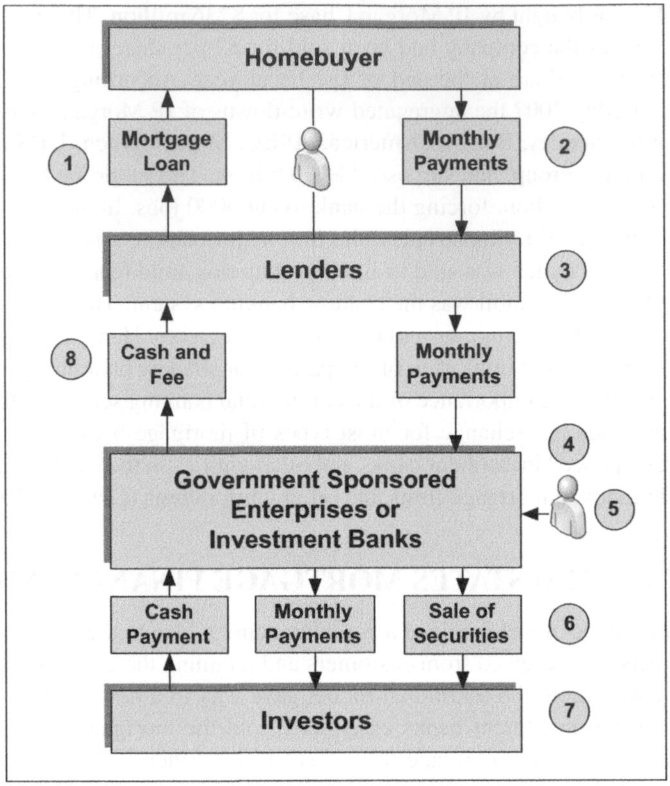

Figure 4.1 Relationship between the parties engaged in the subprime housing market

Legend

1. A *homebuyer* obtains a mortgage from a *lender* (typically a bank). The lender transfers the money into the homebuyer's account. The mortgage may be obtained through an intermediary – a broker.
2. The *homebuyer* repays the loan amount in accordance with an agreed schedule.
3. The *lender* considers holding the mortgage in its portfolio (i.e. simply collect the interest and principal payments over the next several years) or selling it. The lender elects to sell the mortgage (to raise finance to make other loans) to a government-sponsored enterprise such as Fannie Mae or Freddie Mac, or a private entity such as financial institutions or Wall Street investment firms (*investment banks*).
4. The *investment bank* groups the mortgage with similar mortgages it has already purchased (referred to as *pooling the mortgages*). The mortgages in the pool have common characteristics (similar interest rates, maturities, etc.). The *investment bank* then prepares for sale securities that represent an interest in the pool of mortgages, of which the homebuyer's mortgage is a small part (called *securitising* the pool). The products for sale are called mortgage-backed securities (MBSs).
5. Prior to the sale of the MBSs, credit rating agencies (such as Standard & Poor's, Moody and Fitch) gave ratings to every type of bond according to its risk. Letter grades mark the safety of the investments. AAA (the best rating) is given to the safest ones, for example government bonds. Of note is that the credit rating agencies awarded most MBSs an AAA rating.
6. The *investment bank* then sells the MBSs to investors in the open market. With the funds from the sale of the MBS, the *investment bank* can purchase more mortgages and create more MBSs for a fee.
7. Investors demanded MBSs as they were considered safe investments that paid a higher rate of return than Treasury bills (during the boom period) when defaults were minimal.
8. When the *homebuyer* makes monthly mortgage payments to the *lender*, the *lender* keeps a fee and sends the rest of the payment to the investment bank. The *investment bank* in turn takes a fee and passes what is left of the principal and interest payment along to the investors who hold the MBS.

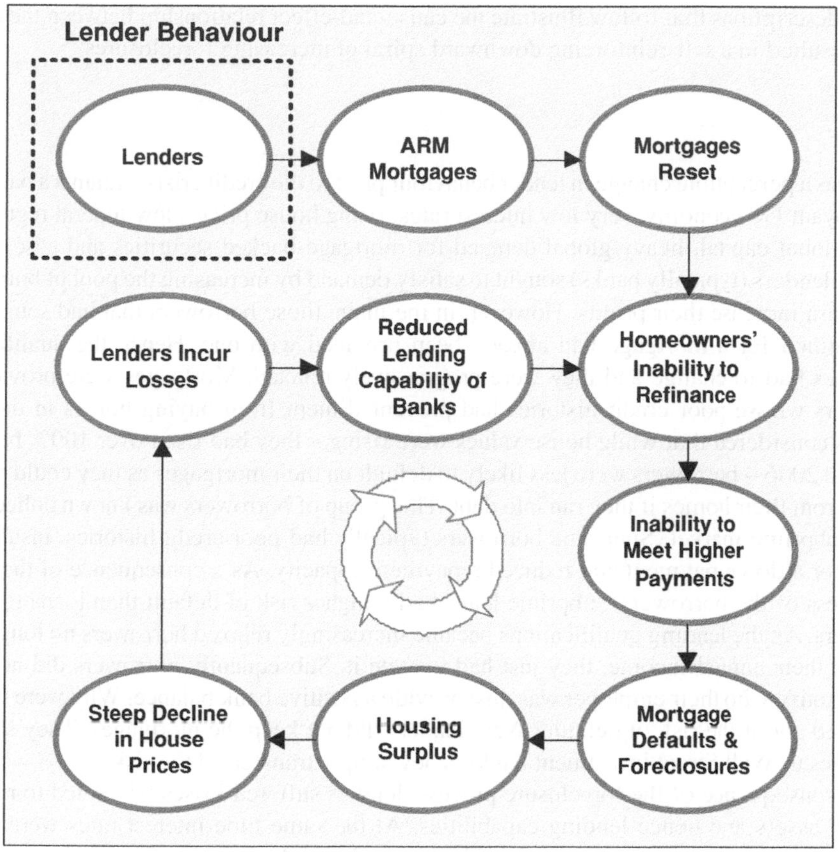

Figure 4.2 Increased foreclosures from mortgages resetting

described by writers such as Nouriel Roubini.[7] Here three integrated vicious circles are proposed, labelled "foreclosures", "negative equity" and "housing surplus". These vicious circles or self-reinforcing downward spirals were created by the behaviour and practices of participant groups, including government departments, investment banks, rating agencies, lenders, brokers, borrowers and house builders. The circles are presented as diagrams and provide a very simplistic representation of reality. Their content is not intended to be exhaustive and they do not illustrate the strength of influence of one event on another.

4.4.2 Foreclosures

The first vicious circle relates to the event of adjustable rate mortgages resetting and the inability of predominantly subprime homeowners to refinance their properties with a new mortgage with more favourable terms. As a consequence homeowners were trapped in a mortgage that they could longer afford. They defaulted and the lenders foreclosed. Figure 4.2

[7] Nouriel Roubini is the cofounder and Chairman of Roubini Global Economics, an independent, global macroeconomic and market strategy research firm. He is also a professor of economics at New York University's Stern School of Business. From 1998 to 2000, he served as the senior economist for international affairs on the White House Council of Economic Advisors and then the senior advisor to the undersecretary for international affairs at the US Treasury Department.

and the descriptions that follow illustrate the cause-and-effect relationship between the events which resulted in a self-reinforcing downward spiral of increasing foreclosures.

Lenders

There was a perceptible change in lender behaviour prior to the credit crisis. Against a backdrop of a buoyant US economy, very low interest rates, rising house prices, low federal regulation, excess global capital, heavy global demand for mortgage-backed securities and ease of risk transfer, lenders (typically banks) sought to satisfy demand by increasing the pool of borrowers and in turn increase their profits. However, in the main, those borrowers that had sought and had qualified for a mortgage had already been provided with one. Hence the qualification guidelines had to change and they were progressively relaxed. Mortgages were provided to customers whose poor credit histories had prevented them from buying homes in the past. Lenders considered that while house values were rising – they had risen over 100% between 1997 and 2006 – borrowers were less likely to default on their mortgages as they could release money from their homes if they ran into debt. This group of borrowers was known collectively as the subprime market. Subprime borrowers typically had poor credit histories, insufficient money for a down payment and reduced repayment capacity. As a consequence of the credit worthiness of the borrowers, subprime loans had a higher risk of default than loans to prime borrowers. As the lending qualifications became increasingly relaxed borrowers no longer had to prove their annual income, they just had to state it. Subsequently borrowers did not even have to notify who their employer was, just provide a positive bank balance. Why were lenders so relaxed about the risk of default? Well, lenders did not keep the mortgages. They sold the mortgages to Wall Street investment banks. They simply transferred the risk.

As a consequence of the foreclosure process, lenders suffered losses which led to reduced financial assets and hence lending capabilities. At the same time interest rates were rising, making mortgages even more expensive. This exacerbated the situation for existing mortgage owners seeking a new mortgage whose low interest introductory period was about to finish. As the rate of foreclosures increased, the ability and willingness of lenders to offer mortgages in this market decreased, creating a downward spiral of foreclosures.

Adjustable-Rate Mortgages

Approximately 80% of US mortgages issued to subprime borrowers were adjustable-rate mortgages[8] (ARMs). ARMs typically had a fixed interest rate for a period of two or three years, after which time they reset to a much higher rate. An estimated one-third of ARMs which originated between 2004 and 2006 had attractive or "teaser" interest rates of below 4% which then increased significantly after two or three years, frequently to double the initial rate. This meant that while the loan was affordable to the borrower at the initial rate, it may have become unaffordable when the rate was reset. The benefit of ARMs for homeowners was that the starter rates were lower than those of traditional, fixed-rate mortgages. That meant lower (initial) monthly payments, making home ownership more affordable and allowing borrowers

[8] Some of the creative ARM products that flourished included interest-only and payment option loans. With the former, during the introductory period a borrower only paid the interest on the loan and nothing towards the principal balance. With payment option ARMs borrowers get to choose how much they pay each month – including just the interest or less than the interest. In the case of the last scenario, the unpaid interest was added onto the principal, leaving borrowers owing more than the amount of the original loan.

to qualify for a larger loan. These products were sold to borrowers who thought their homes would appreciate (based on rising house prices over a number of years), enabling them to sell their home for a profit after a few years or refinance. As a home appreciates, even borrowers who are not paying the principal loan amount build up more equity. This would have made it easier for subprime borrowers to refinance into another loan with a low interest rate. ARMs put more homebuyers in the market, helping to raise home ownership rates to record levels in 2004, pushing up demand and house prices.

Mortgages Reset

At the time borrowers came to refinance, interest rates were rising, house prices were falling and lenders were becoming more cautious. In many cases homeowners were unable to refinance and either had to struggle with their existing mortgage or default.

Mortgage Defaults and Foreclosures

If a borrower was delinquent in making timely mortgage payments, the lender would in most cases take possession of the property in a process called foreclosure. Mortgage delinquencies can be traced back to 2004 when interest rates started to rise. In 2004 the US Federal Reserve started a cycle of interest rate rises that would lift borrowing costs from 1%, their lowest since the 1950s. In fact it went on to increase interest rates 17 times in a row as it tried to slow inflation. It paused in June 2006, setting the cost of borrowing costs at 5.25%. These increases pushed many ARMs beyond the means of borrowers. It tipped borrowers over the edge, making repayments unaffordable and contributed to the mortgage crisis. By the fourth quarter of 2007, the rate of serious delinquency as measured by credit records stood at 2% of all mortgage borrowers, up nearly 50% from the end of 2004.[9] To understand the financial scale of the problem, by November 2007 the value of US subprime mortgage payments outstanding was estimated at $1.4 trillion.[10] By October 2007, approximately 16% of subprime ARMs were delinquent (Bernanke 2007). By January 2008, the delinquency rate had risen to 21% and by May 2008 it was 25% (Bernanke 2008a, 2008b).

4.4.3 Negative Equity

The second vicious circle refers to the rising tide of homeowners with negative equity (Figure 4.3) which came about after the collapse of the US housing bubble. This bubble, as other bubbles before it, was characterised by a rapid increase in the valuations of domestic property until unsustainable levels were reached in relation to incomes and other measures of affordability. A massive rise in household mortgage debt had arisen. A housing surplus was generated by a combination of an oversupply of new homes and mortgage foreclosures. This oversupply placed downward pressure on housing prices which lowered current homeowners' equity. Homeowners with negative equity owed more on their mortgages than their properties were worth. They were incentivised to default on their mortgage.

[9] These figures are reported by Ben Bernanke (2008a) and are based on information from TrenData drawn from the credit records of a geographically stratified random sample of more than 20 million individuals for each calendar quarter beginning in 1992. "Serious delinquency" includes accounts that are 90 days or more past due or are in foreclosure.

[10] Remarks by Martin Gruenberg of the US Federal Deposit Insurance Corporation, 27 November 2007.

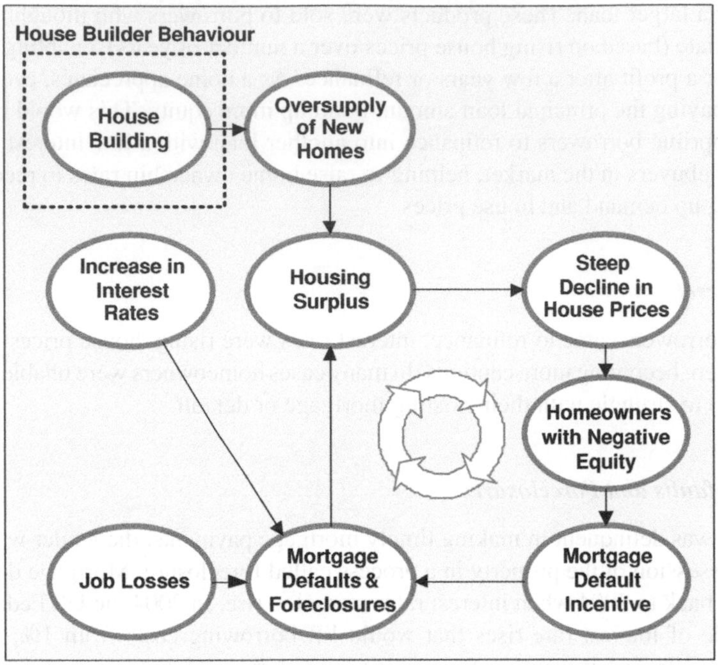

Figure 4.3 Negative equity triggers mortgage defaults

As of March 2008 an estimated 8.8 million borrowers, 10.8% of all homeowners, had negative equity in their homes, a number which was believed to have risen to 12 million by November 2008. By June 2009 the US Government Accountability Office had discovered through research conducted in 16 metropolitan areas that the percentage of nonprime borrowers with negative equity ranged from about 9% (Denver, Colorado) to more than 90% (Las Vegas, Nevada).[11] As of the third quarter of 2010, it was estimated that almost 11 million mortgages or 22.5% of all mortgage holders owed more on their mortgage than the properties were worth, with another 2.4 million mortgages approaching negative equity.[12]

Borrowers in this situation had an incentive to "walk away" from their mortgages and abandon their homes, even though doing so would damage their credit rating for a number of years. As the number of foreclosures grew, the supply of homes increased creating even greater downward pressure on house prices and the number of homeowners with negativity equity grew, creating a negative or downward spiral.

Job Losses

Job losses exacerbated mortgage defaults and foreclosures. The news media recorded the trend in unemployment as it evolved month by month. According to the US Department of Labor, the US jobless rate rose to 7.3% in December 2008, the highest monthly figure for 16 years, as a result of employers laying off 524 000 staff. This figure was surpassed many times in 2009,

[11] US Government Accountability Office. "State-Level Information on Negative Home Equity and Loan Performance in the Nonprime Mortgage Market", 14 May 2010.

[12] New York State economic report, February 2011.

with the unemployment rate reaching 10.1%. The US Department of Labor's Bureau of Labor Statistics unemployment figures record the increase in unemployment between December 2006 and December 2009 as 7.2 million. The US national press described the drop in employment as the biggest decline since the Great Depression. The total US unemployment figure for 2006 was 7 million and by 2009 it had reached 14.27 million (9.3%). Job losses had not only occurred in the housing and service industries (banks and insurance companies) which were directly affected by the subprime turmoil but had spread to retail, manufacturing, leisure and professional services. Falling house prices, combined with news of rising unemployment in other parts of the country, made homeowners both less wealthy and more cautious, contributing to a gradual decline in spending. Less consumer spending eventually weakened the economy, prompting companies to start laying off workers in a vicious cycle that caused households to become even more frugal.

4.4.4 Housing Surplus

Another vicious circle was created by the housing stock surplus (Figure 4.4) generated by a combination of an excessive production of new homes which outstripped demand, a foreclosure tsunami arising from adjustable-rate mortgages resetting, overvaluation of properties, speculators withdrawing from the market and borrowers overstretching themselves. During October 2007 the US Department of Housing recorded that the number of new building permits granted in the US (a signal of future construction activity) fell 6.6%, the biggest monthly

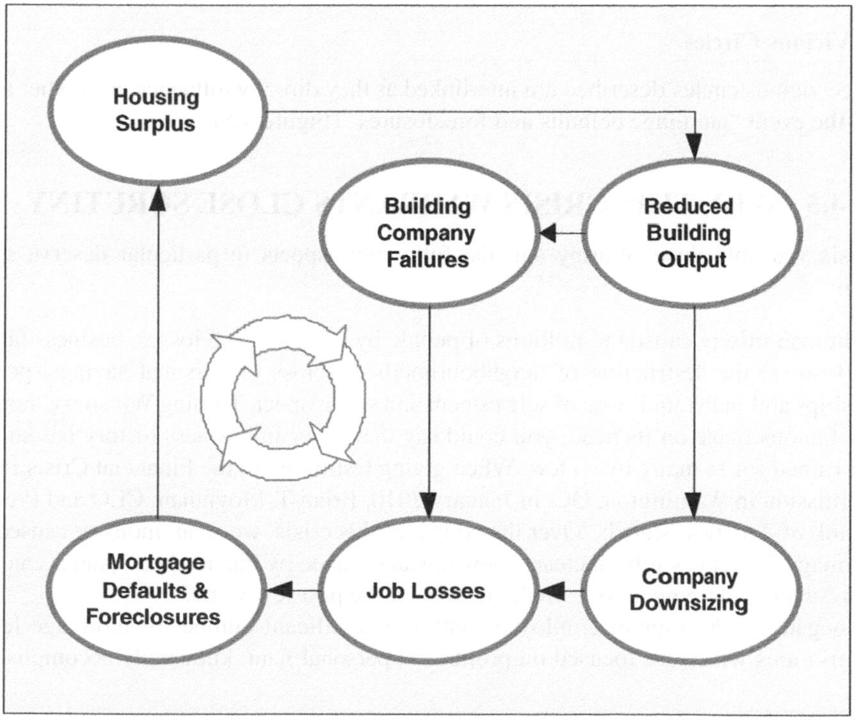

Figure 4.4 Housing surplus leads to fall in construction and job losses

decline in 14 years. This surplus created a huge and sustained downward pressure on house prices. It was exacerbated by mortgage delinquencies and foreclosures. As the volume of foreclosures rose, more houses were released onto the market, suppressing house prices further. When the property was placed back on the market by the lender it was common for the value of the property to have fallen. By early 2005 house prices had peaked and by 2006 they had started to decline. However, they continued to fall on a year-on-year basis, the first time this had occurred since the Great Depression. Average home sale prices nationwide fell 4.2% in the 12 months to September 2007.[13] In October 2007, the US Secretary of the Treasury, Henry Paulson, called the bursting of the bubble "the most significant risk to our economy". In response to this surplus building output was reduced and house builders were forced to reduce the sale price of their new properties. The reduced demand for new homes placed pressure on companies initially in the house building sector and subsequently those allied to it. A significant number of house builders either downsized or went into liquidation, contributing to the pool of unemployed. As this pool grew it expanded the number of mortgage defaults, driving up the number of vacant properties.

The housing problems continued into 2010. When the tax credits designed to boost sales ended in May 2010 house sales began to slump. The National Association of Realtors subsequently recorded a plunge in house sales in July 2010 of 27% compared with the previous month. July had been the third month in a row that sales of previously built single homes had fallen. More significantly, the monthly sales figure was the worst in ten years. On the announcement the Dow Jones Index closed down 134 points at 10 040.45, close to the perceived critical 10 000 threshold.

4.4.5 Vicious Circles

The three vicious circles described are interlinked as they directly influence each other and all contain the event "mortgage defaults and foreclosures" (Figure 4.5).

4.5 WHY THIS CRISIS WARRANTS CLOSE SCRUTINY

The crisis warrants close scrutiny and the following aspects in particular deserve special attention:

- The human misery caused to millions of people by extensive job losses, business failures, foreclosures, the destruction of neighbourhoods, the loss of personal savings, personal hardships and individuals loss of self-esteem and self-respect. Turning Winston Churchill's most famous quote on its head, you could say that never in business history has so much been ruined for so many by so few. When giving testimony to the Financial Crisis Inquiry Commission in Washington, DC, in January 2010, Brian T. Moynihan, CEO and President of Bank of America, stated: "Over the course of this crisis, we as an industry caused a lot of damage. Never has it been clearer how mistakes made by financial companies can affect Main Street, and we need to learn the lessons of the past few years."
- The negative behaviour of employees within a significant number of mortgage lending organisations who were focused on profits and personal gain, knowingly accomplished at

[13] National Association of Realtors, "Mortgage Availability Improving but Hampered September Existing-Home Sales", News Release, 24 October 2007.

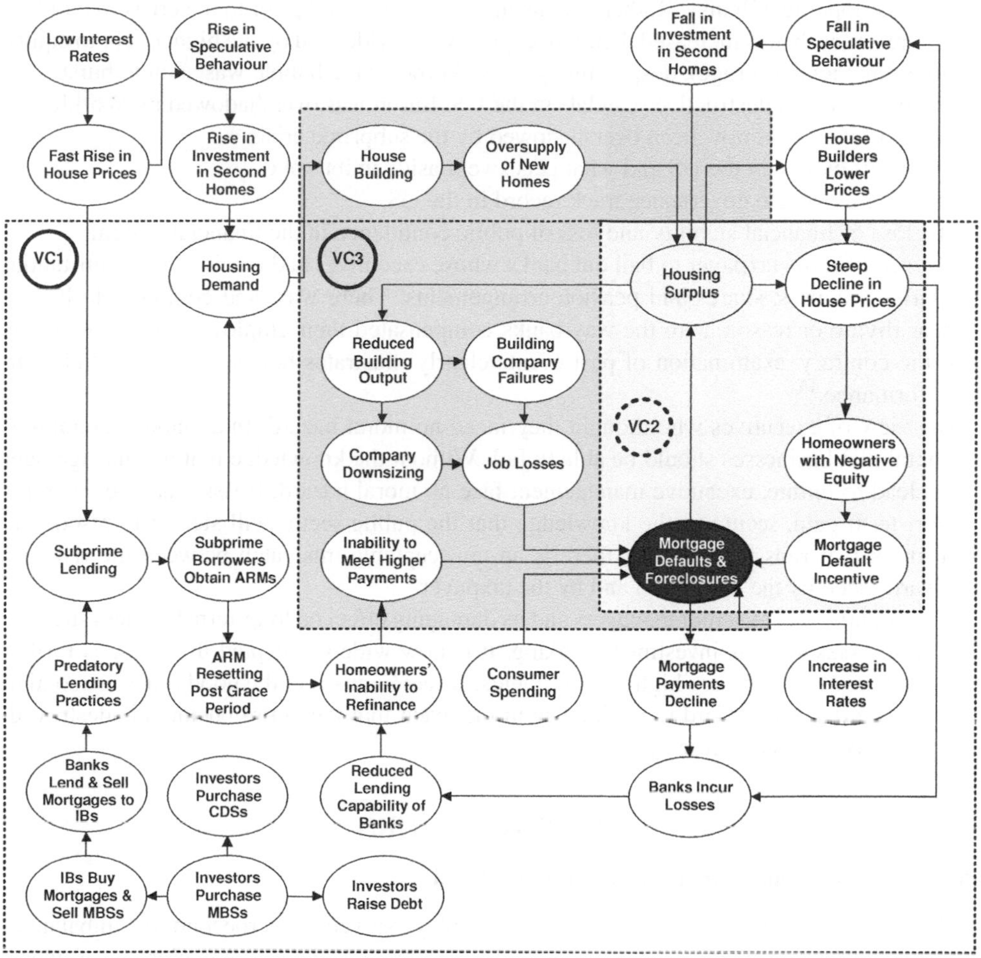

Figure 4.5 Overlapping vicious circles

Legend
IB = Investment Bank.

the expense of ill-informed, inexperienced and in some cases poorly educated individuals many of whom were in minority groups struggling to "find their feet" and establish their families in the US. As stated by Barak Obama: "And while it's true that many Americans took on financial obligations that they knew or should have known they could not have afforded, millions of others were, frankly, duped. They were misled by deceptive terms and conditions, buried deep in the fine print."[14]

- Inappropriate relaxation of lending regulations and the award of mortgages to recipients that were "bound to fail".
- The occurrence of the crisis under the full force of the internal control requirements of the US Sarbanes-Oxley Act within one of the most highly regulated industries, financial services.

[14] Speech by Barak Obama, "Remarks by the President on Wall Street Reform", 22 April 2010.

- The risk appetite of banks. Rather than using the boom period prior to the crisis to build up their reserves, banks increased their risk exposure, as evidenced by the higher debt to equity ratios (financial leverage), despite many banks knowing the bubble was soon to burst.
- The frequency of destructive scandals in the US. Enron was overshadowed by WorldCom, and WorldCom has now been overshadowed by the subprime crisis.
- The profit motive in the US and what gets swept aside in its wake.
- The poor corporate governance track record in the US.
- The loss of financial stability and loss of public confidence in the financial system.
- The need for the taxpayer to bail out banks whose executives had received very substantial salaries, bonuses, shares and pension arrangements. There was (and continues to be) no clear rhyme or reason as to the way banks compensated their employees. Despite claims to the contrary, examination of past events clearly illustrates bonuses were not linked to performance.[15]
- The view of executives who thought they faced no moral hazard. In a market economy it is vital that businesses should be able to fail. Without the knowledge that mismanagement can lead to failure, executive management face no moral hazard. If risks may be taken for short-term gain, secure in the knowledge that the public sector will step in to rescue the business if it runs into trouble, there is an unacceptable mismatch between the risk and reward faced by the institution and by the taxpayer.
- The incentivising of senior managers and its damaging effect on long-term business success. In businesses such as investment banking, it is now widely accepted that there is further moral hazard[16] from very high-risk strategies, where bonuses in the good times are so high that executives can afford to walk away in the event that the entire business is destroyed. Lehman Brothers is a case in point.

4.6 BEHAVIOURS

4.6.1 Investor Behaviour in the Search for Yield

At the core of the financial crisis was an interplay between macroeconomic imbalances and financial market developments. Since 2000 there has been an explosion of world macro-imbalances, with very large current account surpluses occurring predominantly in China, Japan and the oil-exporting countries and large, increasing current account deficits in the US, the UK and other European countries. A key driver of those imbalances had been the

[15] Andrew M. Cuomo, Attorney General of the state of New York, observed in his paper "No Rhyme or Reason: The 'Heads I Win, Tails You Lose' Bank Bonus Culture" (30 July 2009) that he could see no link between performance and the size of bonus payments. He made the observation: "Thus, when the banks did well, their employees were paid well. When the banks did poorly, their employees were paid well. And when the banks did very poorly, they were bailed out by taxpayers and their employees were still paid well. Bonuses and overall compensation did not vary significantly as profits diminished." What concerned him more was the payment of bonuses by banks that were TARP recipients. He states: "An analysis of the 2008 bonuses and earnings of the original nine TARP recipients illustrates the point. Two firms, Citigroup and Merrill Lynch suffered massive losses of more than $27 billion at each firm. Nevertheless, Citigroup paid out $5.33 billion in bonuses and Merrill paid $3.6 billion in bonuses. Together, they lost $54 billion, paid out nearly $9 billion in bonuses and then received TARP bailouts totalling $55 billion."

"For three other firms – Goldman Sachs, Morgan Stanley, and JP. Morgan Chase – 2008 bonus payments were substantially greater than the banks' net income. Goldman earned $2.3 billion, paid out $4.8 billion in bonuses, and received $10 billion in TARP funding. Morgan Stanley earned $1.7 billion, paid $4.475 billion in bonuses, and received $10 billion in TARP funding. JP. Morgan Chase earned $5.6 billion, paid $8.69 billion in bonuses, and received $25 billion in TARP funding. Combined, these three firms earned $9.6 billion, paid bonuses of nearly $18 billion, and received TARP taxpayer funds worth $45 billion."

[16] "Moral hazard" is used to define the tendency for banks to become too relaxed about financial failure because they believe they will always be rescued by the central bank to avoid contagious bank failures.

very high savings rates in countries such as China, leading to central bank reserves commonly invested almost exclusively in apparently risk-free or close to risk-free government bonds or government guaranteed bonds. This in turn drove a reduction in real risk-free interest rates[17] to historically low levels. In 1990 it was possible to invest in the US or UK in risk-free index-linked government bonds at a yield to maturity of over 3% real; between 2004 and 2009 yield had been less than 2% and at times as low as 1%. These very low medium- and long-term real interest rates in turn drove a ferocious search for yield to improve upon declining risk-free rates.

4.6.2 Mortgage Lending Behaviour

As the credit crisis unfolded and the behaviour of lenders became common knowledge, they were heavily criticised. At a hearing before the US Senate Committee on Banking, Housing and Urban Affairs on 22 March 2007, the Senate Chairman, Christopher Dodd, called the predatory lending of banks "unconscionable and deceptive". He described predatory lending as the sale to homeowners of loans they could not afford – specifically, the sale of ARMs to disadvantaged groups such as the elderly, low-income households and unsophisticated borrowers. When the low introductory interest rate of ARMs expired after two years the rate rose so steeply that borrowers were faced with what Dodd called a "kind of devil's dilemma". Borrowers were forced to make one of three stark choices: refinance at a far greater cost, sell or default on the loan. Examples of the different forms of predatory lending are presented in Table 4.1.

4.6.3 Bank Behaviour and Risk Transfer through Securitised Credit

In a speech in January 2009, Adair Turner provided an insightful assessment of the effectiveness of risk transfer through securitised credit (Turner 2009). His perspective is as follows. Securitised credit began to take off in the 1980s. A significant argument put forward in its favour was that securitisation would reduce risks for individual banks by passing credit risk to end investors, reducing the need for unnecessary and expensive bank capital. Rather than a regional bank in the US, for instance, holding a dangerously undiversified holding of credit exposures in that particular region, which created the danger of a self-reinforcing cycle between the decline in a regional economy and the decline in the capital capacity of regional banks, it was much better to package up the loans and sell them through to a diversified group of end investors. Securitised credit intermediation could reduce risks for the whole banking system, since while some of the credit risk would be held by the originating bank and some by other banks acting as investors, much would be passed through to end non-bank investors. Credit losses would therefore be less likely to produce banking system failure. However, that is not what happened. When the crisis struck, and as figures from the IMF Global Financial Stability Report of April 2008 made clear, the majority of the holdings of securitised credit, and the vast majority of the losses which arose, did not lie in the books of end investors intending to hold the assets to maturity, but on the books of highly leveraged banks and bank-like institutions.

[17] The risk-free interest rate is the theoretical rate of return of an investment with zero risk, including default risk. The risk-free rate represents the interest an investor would expect from an absolutely risk-free investment over a given period of time.

Table 4.1 Predatory lending

Forms	Explanation
Excessive fees	Frequently charging fees totalling more than 5% of the loan (on competitive loans fees of 1% were typical).
Abusive prepayment penalty	A prepayment penalty, a fee for paying the loan off early which was typically effective for more than 3 years and/or cost 6 months' interest (80% of subprime mortgages carried a prepayment penalty, in the prime market only about 2% carried a prepayment penalty).
Higher interest rates	Selling mortgages with an inflated interest rate (i.e., higher than the rate acceptable to the lender). Some lenders incentivised brokers to sell this type of loan by paying them a "yield spread premium" as a reward for making the loan more costly to the borrower.
Loan flipping	Refinancing a loan to generate a fee income without providing any net tangible benefit to the borrower. This practice is called "flipping" a borrower.
Steering and targeting	Steering borrowers into subprime mortgages even when the borrowers could qualify for a mainstream loan (i.e. one with more favourable terms).
Mandatory arbitration	Not permitting borrowers to seek legal remedies in a court if they find their home is threatened by loans with illegal or abusive terms (i.e., increasing the likelihood that the borrowers would not receive fair and appropriate remedies in case of wrongdoing).
Targeting vulnerable borrowers	Deliberately targeting senior citizens and low-income borrowers with low levels of education and little experience of mortgages to deliberately place them in unnecessarily expensive loans. In 2005, 40% of mortgage loans taken out in Latino communities and 50% in African American communities were subprime.
Costs	Making mortgage payments seem artificially low by omitting to declare that quoted mortgage payments exclude property taxes and insurance.
Adjustable-rate loans	Not explaining to ARM subprime borrowers that loan payments would rise significantly when they reset.

Sources: US National Association of Consumer Advocates and US Centre for Responsible Lending.

4.6.4 "Group Think" and Herd Behaviour

When the credit crisis unfolded there was a desperate search for the causes. Naturally board performance came in for close scrutiny. Shareholders of failed or failing companies wanted to hold responsible director's "feet to the fire" and understand why corporate leadership had failed so spectacularly. From the findings of the numerous reviews of the crisis, it is abundantly apparent that board decision making was poor – and in the case of Northern Rock "reckless" (House of Commons 2009). While numerous recommendations have been made about the appointment, experience, induction, availability and evaluation of board members, the process that boards should adopt for decision making has received scant attention. However, what brings boards and companies down is dysfunction within their social system (Cairnes 2003).

Boards are fraught with extensive interpersonal dynamics, like any other group of people (Carver 2006). Optimal board performance occurs when chairmen understand group dynamics and foster openness and debate to reach informed decisions. People differ in their comfort in confrontation, their ability to express their point of view and the personal agendas they bring to a meeting. The human social dynamic within a group of individuals sees individuals agreeing to, or failing to oppose, a group decision even though they are not satisfied with it (Cairnes, 2003). Such dysfunctional boards can fall into "group think" where members reach a consensus without critically testing, analysing and evaluating ideas. This is a form of herd behaviour that tends to be more problematical within institutions. The term "group think" was coined by psychologist Janis Irving in the 1970s. It was adopted to describe the process in which a group can make irrational or bad decisions as a result of each member of the group attempting to conform their opinion to what they believed to be the consensus of the group. It can be argued that "group think" can occur on two levels: members of a group (a board), and a group within multiple groups (boards within a specific industry, sector or market).

At *board level*, when this group dynamic is present, independent thinking is sacrificed in pursuit of group cohesiveness. Members of the group impose self-censorship, succumbing to direct pressure to conform to avoid being considered disloyal. Irving discovered this behaviour during a study of military disasters and found that there were people in decision-making groups who believed that the group was making the wrong decision and had the information to support that belief (Irving 1971). The information was not tabled or, when it was tabled, was put to one side. Individuals did not fight for their point of view of even press their evidence for that view. He discovered that when this group behaviour was present there was an "us-and-them" mentality, a view that meant that each group member was either "with us or against us" (see Box 4.2). During the financial crisis, full-time board members or NEDs may have been reluctant to act according to their own information and beliefs, fearing that any contrarian view may damage their standing and relationship with fellow board directors. Hence, directors will "follow the herd" if they are concerned about how others will assess their ability to make sound judgements. Human social relationships are a key determinant in decision-making processes. The prevalence of emotional factors in corporate success and failure means that they should be recognised as being at the heart of boardroom leadership and effectiveness (Lees 2010). Lessons should be learnt from the corporate crisis in which such behaviour played such a decisive role.

Box 4.2 Group dynamic behaviour

Lehman Brothers

An example of group dynamic and emotional boardroom behaviour was exhibited by one of the most prominent banking personalities, Dick Fuld, CEO of Lehman Brothers. In forming his executive committee in 1996, "Fuld quickly instituted an 'on-the-team-or-off' mentality at Lehman" (Fishman 2008). This behaviour had striking similarities with Irving's findings (see above). A colourful description of Fuld's behaviour is provided by Andrew Gowers, a former editor of the London *Financial Times*, who in 2006 joined Lehman Brothers in London as head of corporate communications. Gower described him as a "textbook example of the command-and-control CEO" who "inspired great loyalty and on occasion great fear" (Gowers 2008). Describing Lehman being "at war" in the market, Gowers said "his ferocity could be intimidating, his eyebrows beetling over his hard eyes, his brutally angular brow appearing to contort in rage. Even when in a relatively upbeat mood he seemed to take pleasure in violent imagery." The most revealing aspect of Gowers's article is his assessment of the operation of the board. "Here was a corporate governance structure almost pre-programmed to fail: an overwhelming CEO, a top lieutenant (Joe Gregory) eager to please and hungry for risk, an executive team not noted for healthy debate and a power struggle between two key players. Furthermore, the board of directors was packed with non-executives of a certain age and woefully lacking in banking expertise. It is small wonder that Lehman was so ill-equipped to recognise and adjust to the changes in the environment that were dramatically signalled by the collapse of Bear Stearns in March this year."

At *corporate level*, when this group dynamic is present, participants in the same market are seen to be exhibiting identical behaviour. As the scale and widespread nature of the financial crisis became apparent, particularly the common acute problems of illiquidity and capital inadequacy of banks investing in the same markets, it has been suspected that forces were present which had led to both excessive and identical "group thinking". While it was initially assumed that investment decisions had reflected rational behaviour based on an assessment of all available information in an intelligent efficient manner, a contrasting view evolved. It is believed that at times investment was driven by group psychology which weakened the link between the rigorous assessments of data and informed rational decision making. Board decisions were not made in a vacuum. They will have been keenly aware of the investments of their competitors, and the profits being realised as a result of those investments. It will have been impossible for them not to be influenced by banks in the same market. If the first investors in a new class of assets profit from rising asset values, as other investors learn about the innovation, more may follow their example, driving the price even higher as they rush to buy, in hopes of similar profits. If such "herd behaviour" causes prices to spiral up far above the true value of the assets, a crash may become inevitable. If for any reason the price briefly falls, so that investors realise that further gains are not assured, then the spiral may go into reverse, with price decreases causing a rush of sales, reinforcing the decrease in prices.

4.6.5 Banks' Behaviour and Risk Appetite

In 1961 James Stoner observed that group decisions are riskier than the previous private decisions of the group's members (Stoner 1961). Since that time several hundred studies have shown that the "risky shift" is a particularly pervasive phenomenon, and both risky and cautious shifts are special cases of a more general phenomenon – group-induced attitude polarisation (Myers & Lamm 1976).[18] Group polarisation is said to occur when an initial tendency of individual group members toward a given direction is enhanced following group discussion. So, on decisions in which group members have, on average, a moderate proclivity in a given direction, group discussion results in a more extreme average proclivity in the same

[18] "Experiments exploring the effects of group discussion on risk taking are generally consistent with a 'group polarization' hypothesis, derived from the risky-shift literature. Recent attempts to explain the phenomenon fall mostly into 1 of 3 theoretical approaches: (a) group decision rules, especially majority rule (which is contradicted by available data); (b) interpersonal comparisons (for which there is mixed support); and (c) informational influence (for which there is strong support). A conceptual scheme is presented which integrates the latter 2 viewpoints and suggests how attitudes develop in a social context" (Myers and Lamm 1976).

direction. Such behaviour traits need to be considered in conjunction with group dynamics and in particular "group think".

4.6.6 Behaviour of Regulators and the Division of "Narrow Banking" from Investment Banking

In the light of the financial crisis, regulators internationally are re-examining the need for the division of the different functions of banking – deposit taking, loan extension and payment services provision – from the more complex and risky investment banking activities or whether they can be undertaken by the same firms. The actual trend has clearly been for these functions to be combined to a greater extent than ever before – as Bear Stearns has folded into JP Morgan, Merrill Lynch into Bank of America, and part of Lehman's into Barclays. In addition, Morgan Stanley and Goldman Sachs have become bank holding companies with access to the federal discount window (both are covered by the implicit assumption that the US government would consider them too important to fail). Regardless of this trend, several commentators have argued that regulation should be designed to produce a separation of "narrow banking" from risky investment bank trading activities, a reimposition of the Glass-Steagall separation of commercial and investment banking. Following the Great Crash of 1929, the US Congress passed the 1933 Glass-Steagall Act which, among other measures, prohibited a bank holding company (a retail bank) from owning other financial institutions – such as investment banks (House of Commons 2009). This provision was repealed in 1999. However, while there is support for divorcing the "utility" functions from the "riskier" investment banking practices, the world of banking has changed and it may no longer be possible to define those activities which can be simply classified as investment banking. Nevertheless there is a rationale for carefully considering how to insulate the vital functions of retail banking from adverse impacts arising from the potential irrationality of liquid traded markets.

4.6.7 Banks' Behaviour and Misplaced Reliance of Sophisticated Mathematics and Statistics

There are many lessons to be learned from the financial crisis. There is a considerable body of opinion that considers that poor risk management lay at the heart of the credit crisis. This lack of awareness of risk exposure is reinforced by Sir John Gieve, who stated when Deputy Governor of the Bank of England, that a weakness in the British banking system "was the failure of the banks and many other investors to appreciate, price and manage risk". The Bank of England published its analysis of the vulnerabilities of the system in its Financial Stability Reports of 2006 and 2007 (Gieve 2009). It delivered its findings to the CEOs of banks in both London and New York. In particular, it described the banks' exposure to global imbalances, dependence on wholesale funding and the risk of structured credit markets seizing up in a downturn. However, these executives paid scant regard to the reviews as they took comfort from the sophistication of their risk management systems and hedging strategies and were confident they could ride out the storm. The issue though, as Gieve explained in his speech, was that the banks' systems were preparing them for a shower, not a hurricane.

The predominant assumption of the banks was that the scale and complexity of the securitised credit market had been matched by the evolution of statistically sophisticated and effective techniques for measuring and managing the resulting risks. Central to many of the techniques applied was the concept of value at risk (VaR), enabling mathematical inferences about forward-looking risk (and future price movements) to be drawn from the observation of past patterns of price movement. The financial crisis has revealed, however, severe problems with these techniques. They suggest at the very least the need for significant changes in the way that VaR-based methodologies have been applied; some, however, pose more fundamental questions about our ability in principle to infer future risk from patterns observed in the past. Four categories of problem have been distinguished in *The Turner Review* (FSA 2009a), and are described in Chapter 25.

A primary message of the financial crisis was that the very complexity of the statistical methods used to measure and manage risk made it increasingly difficult for an analyst to convey the approach adopted and the content of the analysis, and for top management and boards to assess and exercise judgement over the risks being taken. Statistical and mathematical sophistication ended up not containing risk, but providing false assurance that the emerging risks could be safely ignored.

4.7 WORLDWIDE DEFICIENCIES IN RISK MANAGEMENT

Poor risk management is succinctly summarised within the declaration issued by the G20[19] at their summit held in Washington on 15 November 2008, aimed at restoring global growth. Item 3 of the declaration (listed under the heading "root causes of the current crisis") is as follows (emphasis added):

> During a period of strong global growth, growing capital flows and prolonged stability earlier this decade, market participants sought higher yields *without an adequate appreciation of the risks* and failed to exercise proper due diligence. At the same time, weak underwriting standards, *unsound risk management practices*, increasingly complex and opaque financial products and consequent excessive leverage combined to create vulnerabilities in the system. Policy makers, regulators and supervisors, in some advanced countries, *did not adequately appreciate and address the risks* building up in financial markets, keep pace with financial innovation, or take into account the systematic ramifications of domestic regulatory actions.[20]

Apart from the perception of its negative impact on investment banks, poor risk management is considered to have led to the collapse of asset prices. As Andrew Haldane explained,[21] by early 2009 world equity prices lost more than three-quarters of their gains during the Golden Decade,[22] with bank share prices losing almost 60% of their value. In the face of these falls risk management systems across all institutions were considered to have been woefully inadequate.

4.8 FEDERAL REFORM

In October 2008 President George Bush signed the Emergency Economic Stabilization Act 2008, creating a $700 billion Troubled Assets Relief Program (TARP) to purchase failing bank assets. The declared purpose of the Act was to "immediately provide authority and facilities that the Secretary of the Treasury can use to restore liquidity and stability to the financial system

[19] The Group of 20 is formed from finance ministers and central bank governors of 20 economies (19 plus the European Union) which represent two-thirds of the world's population and 80% of the world's trade. It was established in 1999 in the wake of the 1997 Asian Financial Crisis to bring together major advanced and emerging economies to both stabilise the global financial market and to achieve sustainable economic growth and development. It is committed to coordinating expansionary macroeconomic policies.

[20] White House, "Declaration of G20", November 2008. http://georgewbush-whitehouse.archives.gov/news/releases/2008/11/20081115-1.html Retrieved 2009-02-27.

[21] Andrew G. Haldane, "Why banks failed the stress test", 13 February 2009. Basis of a speech given at the Marcus-Evans Conference on Stress-Testing, 9–10 February 2009. Haldane is Executive Director for financial stability at the Bank of England.

[22] The Golden Decade is described as the period of extraordinary growth and success for the financial system and financial markets between October 1998 and June 2007 when banks' share prices increased almost 60% and their balance sheets rose more than threefold.

of the United States".[23] Over 50 banks received funds as part of the TARP bailout.[24] This was followed in July 2010 by the Dodd-Frank Wall Street Reform and Consumer Protection Act signed into law by President Obama. The stated aim of the legislation is: "To promote the financial stability of the United States by improving accountability and transparency in the financial system, to end 'too big to fail', to protect the American taxpayer by ending bailouts, to protect consumers from abusive financial services and for other purposes".[25] Its proponents describe the objectives of the Act as restoring public confidence in the financial system, preventing another financial crisis and allowing any future asset bubble to be detected and deflated before another financial crisis ensues.

Financial Stability Oversight Council

Section 111 of the Act creates a new Financial Stability Oversight Council composed of voting and non-voting members whose task it will be to mitigate systemic risk and the maintenance of system-wide financial stability. The stated purpose of the Council is: "to identify risks to the financial stability of the United States that could arise from the material financial distress or failure, or ongoing activities of large, interconnected bank holding companies or that could arise outside the financial services marketplace".[26] The council will focus on the interconnection of highly leveraged firms and will have the authority to force companies to divest holdings if their structure poses a grave threat to US financial stability.

Paradigm Shift

The Act represents a paradigm shift in the American financial regulatory environment impacting all federal financial regulatory bodies and affecting almost every aspect of the nation's services industry. During his public address in July 2010, President Obama remarked the reform introduced by the Act "will prevent the kind of shadowy deals that led to this crisis, reform that would never again put taxpayers on the hook for Wall Street's mistakes". The change in thinking came in the wake of Congress's criticism of Alan Greenspan during his testimony to the Congressional Committee for Oversight and Government Reform hearing held in October 2008, chaired by Henry Waxman. Chairman Waxman's approach to the hearing was to carry out a direct inquisition into Greenspan's decisions during his time as Federal Reserve chairman and the degree to which the Reserve set the stage for the American (and global) financial disaster. Included in Box 4.3 is an extract of the dialogue between Henry Waxman and Alan Greenspan during the hearing over Greenspan's long-term advocacy of financial deregulation.

Box 4.3 Congressional hearing into the financial crisis

The following is an extract from the Congressional Committee for Oversight and Government Reform hearing held in October 2008, chaired by Henry Waxman.

Waxman: Dr. Greenspan, you were the longest-serving chairman of the Federal Reserve in history and during this period of time you were perhaps the leading

[23] House Committee on Financial Services, http://www.house.gov/apps/list/press/financialsvcs_dem/press092808.shtml

[24] Free Government Information (FGI), http://freegovinfo.info/node/2225

[25] Dodd-Frank Wall Street Reform and Consumer Protection Act, http://frwebgate.access.gpo.gov

[26] Ibid.

	proponent of deregulation of our financial markets. Certainly you were the most influential voice for deregulation. You have been a staunch advocate for letting markets regulate themselves. Let me give you a few of your past statements. In 1994, you testified in a congressional hearing on regulation of financial derivatives. You said "there is nothing involved in Federal regulation which makes it superior to market regulation." In 1997, you said, "There appears to be no need for government regulation of off-exchange derivative transactions." In 2002 when the collapse of Enron led to renewed congressional efforts to regulate derivatives, you wrote the Senate, "We do not believe a public policy case exists to justify this governments' intervention." Earlier this year, you wrote in the Financial Times, "Bank loan officers, in my experience, know far more about the risks and workings of their counter parties than do bank regulators." My question for you is simple: were you wrong?
Greenspan:	Well, I think that's true of some products, but not all. I think that's the reason why it's important to distinguish the size of this problem and its nature. What I wanted to point out was that – excluding credit default swaps – derivatives markets are working well.
Waxman:	Well, where do you think you made a mistake then?
Greenspan:	I made a mistake in presuming that the self-interest of organizations, specifically banks and others, were such that they were best capable of protecting their own shareholders and their equity in the firms.

Provisions of the Act

The Act, as anticipated, focuses on risk management, systemic risk, capital and liquidity adequacy, as well as the interconnectedness of banks. Regulators in the past (on both sides of the Atlantic) had been too focused on the institution-by-institution supervision of idiosyncratic risk.

New Approaches to Capital Adequacy

New approaches were required to the regulation of the capital adequacy of banks, accepting that these have been extensively revised by the introduction of Basel II, which has aimed to achieve greater sensitivity of capital levels to the different risks that banks are running. It is important to remember that the crisis developed under the Basel I regime, not Basel II, and that Basel II would have addressed some of the problems which led to it – for instance, the failure to distinguish between the capital required to support mortgages of different credit quality. However, it would appear that Basel II still needs to be adjusted for instance by introducing higher levels of bank capital to reflect the nature of the recent financial crisis.

New Approaches to Liquidity

The Act recognises that liquidity is at least as important as capital adequacy, which was overlooked in the wake of intense regulatory focus on capital adequacy. There is no Basel I or Basel II for liquidity to match the equivalents for capital. Assessment will need to be on stress-test scenarios, rather than models which seek to infer the probability distribution of risks from the observation of the past. New approaches will need to reflect the lessons learnt from the financial crisis – that market-wide collapses in the liquidity of specific asset or funding markets can have huge impacts which analysis of individual specific risks will not capture.

4.9 SYSTEMIC RISK

Looking beyond the financial crisis to improved regulation, US federal agencies need to think about what is needed to avoid a similar scenario occurring in the future. As identified by Lord Turner, the major failure, shared by bankers, regulators, central banks, finance ministers and academics across the world, was the failure to identify that the whole system was fraught with market-wide, systemic risk (Turner 2009). The key problem was not that the supervision of individual banks was insufficient, but that the regulator failed to see the wood for the trees. They failed to piece together the jigsaw puzzle of a large US current account deficit, rapid credit extension and house price rises and the purchase of mortgage-backed securities by US institutions performing a new form of maturity transformation.[27] Regulators, not only in the US, failed to realise that there was an increase in total system risk to which financial regulators, overall authorities, central banks and fiscal authorities needed to respond. To their detriment regulators had been too preoccupied with institution-by-institution supervision of idiosyncratic risk[28] rather looking at the broad horizon.

SIFMA Study

The Securities Industry and Financial Markets Association (SIFMA), in response to the credit crisis, undertook a study (in conjunction with Deloitte and Touche) to examine systemic risk in the financial sector. The outcome of their study is reported in their aptly named publication, "Systemic Risk Information Study". SIFMA hope the study will provide useful guidance on how new policies on monitoring systemic risk can be effectively implemented. SIFMA recommends the creation of a systemic risk regulator and highlights that better qualitative and quantitative information regarding the identification and mitigation of systemic risk will be critical components of any comprehensive financial regulatory reform. The study does not offer a definition of systemic risk, it states that at the time of publication there was no single agreed-upon definition but declares that the industry and regulators must have a common understanding of what the term means. The study records two contemporary definitions of systemic risk.[29,30] The study identified nine drivers of systemic risk: size; interconnectedness; liquidity; concentration; correlation; tight coupling; herding behaviour; crowded trades; and leverage. These are described in summary form in Box 4.4.

[27] Economists use the term "maturity transformation" to describe the activity of a financial intermediary that accepts deposits or investments of one term (usually short term) and places those funds with a debtor in another term (usually intermediate or long term). A common example of a financial intermediary is a bank that transforms bank deposits into bank loans.

[28] Idiosyncratic risk (also known as "unsystematic risk," "non-systematic risk" and "diversifiable risk") is generally recognised to be a company-specific risk that can be reduced (and some say removed) through appropriate diversification.

[29] B.S. Bernanke, Chairman of the Board of Governors of the Federal Reserve System: "Systemic risks are developments that threaten the stability of the financial system as a whole and consequently the broader economy, not just one or two institutions."

[30] Jean-Claude Trichet, President of the European Central Bank: "In the context of our economic environment, systemic risk is the threat that developments in the financial system can cause a seizing up or breakdown of this system and trigger massive damages to the real economy. Such developments can stem from the failure of large and interconnected institutions, from endogenous imbalances that add up over time, or from a sizable unexpected event."

Box 4.4 SIFMA drivers of systemic risk

Driver of systemic risk	SIFMA measure of risk
Size	Measured by the size of the balance sheet of the financial firm.
Interconnectedness	Generally measured by consideration of counterparty risks related to a financial institution's activities. Knowledge of the interconnectedness of a financial institution could assist in determining how many additional failures could be caused by the failure of an individual firm.
Liquidity	Measured by a financial institution's available funds, at a reasonable cost, to meet potential demands from both fund providers and borrowers.
Concentration	Measured, for instance, by the number of financial institutions that have exposures to a single counterparty, industry or product. Risk concentrations can take multiple forms including exposures to individual counterparties, groups of individual counterparties and specific products or sectors.
Correlation	Measured by the correlation of risk exposures through an understanding of the types of risk, their correlations and concentrations.
Tight coupling	Measured, for instance, by mechanistic processes. The SIFMA study makes reference to the paper by Richard Bookstaber entitled "The Myth of Non-correlation" published in the *Institutional Investor*, September 2007. He talks about the inability to intervene between the stages of tightly coupled processes, and if things were going wrong it was not possible to quickly stop the process for a committee to debate the issues.
Herding behaviour	Measured by individuals within a group performing activities without consideration of the longer-term consequences, such as investor representatives "following trends set by other organisations until contrary evidence cannot be ignored any longer". An example is given of the investment of "significant funds in securities backed by subprime mortgages".
Crowded trades	Measured by the occurrence of "a large number of financial institutions following similar trading strategies and execut[ing] comparable trades". The study also states: "If a number of financial institutions adopt like investment strategies the level of risk is increased as additional assets may be invested in the same underlying risk."
Leverage	Measured by assessing firms that are vulnerable from being highly leveraged (extent of the use of borrowed funds) and committed to extensive lending without a sufficient cushion of capital. Excessive debt may force an organisation to cut investment and renegotiate with creditors and in severe cases commit credit default. The study states: "A financially distressed firm's default can lead to the distress of its lenders, and subsequently impact on other firms, thus potentially increasing the build-up of systemic risk across the financial system."

Government Intervention

In a paper in which he examined "what is the role of government in reducing systemic risk in the financial markets", John B. Taylor (2009)[31] identified that government officials were now proposing new legislation to expand significantly the role of government in the financial sector. He identified that the heads of the US Treasury Department, the Federal Reserve Board, the Federal Deposit Insurance Corporation (FDIC) and the SEC have all proposed the creation of a "systemic risk regulator" which would be a new standalone agency, or part of the Fed or a new council of existing regulators. Taylor articulates that the future role of government in the financial markets depends on the lessons learnt about the role of government in the crisis. He believes that there are two standpoints. Firstly, "the markets did it" – the crisis was due to forces emanating from the market economy which the government did not control either because it did not have the power to do so, or because it chose not to. Secondly, "the government did it" – the crisis was due more to forces emanating from the federal government, "where actions and interventions caused, prolonged and worsened the financial crisis". Taylor's standpoint is that the government was at fault through the level of federal interest rates, the government-sponsored enterprises (Fannie Mae and Freddie Mac), the support for mortgage-backed securities, misdiagnosing the financial crisis as a liquidity problem rather than a counterparty risk problem and not articulating "a clear predictable strategy for lending and intervening into a financial sector". Taylor considers the best evidence of a lack of strategy was the confusing roll-out of thc TARP plan which, according to studies, was more likely a reason for the panic in the markets than the failure to intervene in Lehman Brothers.

4.10 THE FUTURE OF RISK MANAGEMENT

The correlation between poor business performance and correspondingly poor governance and risk management has been identified by many commentators. As observed by FSA Chief Executive Hector Sants (FSA 2009b), business owners must be active in the risk management process. He went on to say:

> The impact of companies lacking robust risk management and good governance will, as we have seen, impact negatively on company's long-term investment performance. A lesson for companies from this crisis must be that greater interrogation of how well a company is managed and the adequacy of its risk controls are all material factors fundamental to investment management. A focus of a firm's risk control framework must be an effective risk and audit committee and knowledgeable non-executives with a willingness to challenge senior management.

In light of these comments, it is encouraging that a survey conducted in October 2008 by the Economist Intelligence Unit[32] found that 92% of those surveyed had or were about to review the way they manage risk. However, concerns were expressed that despite the recent credit crisis, risk was still stigmatised as a support function, boards were still short of risk management knowledge and experience and risk was still seen as a peripheral "compliance" issue rather than an essential part of strategy. On a more positive note, the majority of respondents

---

[31] Taylor, who is Professor of Economics at Sanford University, defines a systemic risk in the financial sector as "a risk that impacts the entire financial system and real economy, through cascading, contagion and chain-reaction effects."

[32] KPMG International (2009) "Never Again? Risk Management in Banking beyond the Credit Crisis". The paper recorded the results of a survey carried out in October 2008 by the Economist Intelligence Unit and involving over 500 senior managers involved in risk management from leading banks from around the world. Respondents were asked to identify the weaknesses in the risk management that contributed to the crisis and the actions being taken to prevent a catastrophe reoccurring.

reported that they were seeking to improve the way that risk is measured and reported, a clear acknowledgement that previous models did not sufficiently measure potential risk exposure.

4.11 SUMMARY

This chapter has described the global financial crisis. A way of looking at the events contributing to the crisis was presented in the form of interlinked vicious circles labelled "foreclosures", "negative equity", and "housing surplus". In addition, the behaviour of the key participants in the crisis, namely the investors, mortgage lenders, banks and regulators, was examined. Investors, particularly those from China, were seeking higher yields than those provided by the apparently risk-free or close to risk-free US government bonds or government-guaranteed bonds, which at the time were providing very low returns. Hence there was considerable finance available for investment in what appeared to be an attractive subprime market. At the time several boards of investment banks were characterised as being dysfunctional in that they fell into or adopted "group think" where members reached consensus without critically testing and evaluating ideas. This behaviour was exacerbated and perhaps driven by dominant personalities, commonly CEOs, looking for and expecting acceptance of their ideas without question. Hence, investment took place without rigorous challenge. While it was thought that the securitised credit intermediation would reduce risks for the banks by passing risks to investors it was discovered that when the crisis struck the majority of the holdings were on the books of highly leveraged banks. As a consequence several commentators have argued that regulation should be designed to produce a separation of "narrow banking" from risky investment bank trading activities, a reimposition of the Glass-Steagall separation of commercial and investment banking. After the financial crisis a number of organisations looked to both the causes and ways of preventing a similar occurrence. Lord Turner identified the major failure, shared by bankers, regulators, central banks, finance ministers and academics across the world, as omitting to identify that the whole banking system was fraught with market-wide, systemic risk. A number of initiatives are now under way at a macro level to both comprehend and address systemic risk in the banking sector, but clearly this a monumental challenge due to its scale in terms of the number of stakeholders and geographical reach. At the micro or business level there is recognition that businesses need to improve their corporate and risk management practices. The link between poor business performance and poor governance and risk management has been made by many. For instance, Hector Sants observed: "The impact of companies lacking robust risk management and good governance will, as we have seen, impact negatively on company's long term investment performance". With regard to board conduct, Warren Buffett wrote in his 2009 Letter to Shareholders:

> In my view a board of directors of a huge financial institution is *derelict* if it does not insist that its CEO bear full responsibility for risk control. If he's incapable of handling that job, he should look for other employment. And if he fails at it – with the government thereupon required to step in with funds or guarantees – the financial consequences for him and his board should be severe.

4.12 REFERENCES

Bernanke, B.S. (2007) The recent financial turmoil and its economic and policy consequences. Speech at the Economic Club, New York, 15 October.

Bernanke, Ben S. (2008a) Financial markets, the economic outlook, and monetary policy. Speech, Washington, DC, 10 January.

Bernanke, Ben S. (2008b) Mortgage delinquencies and foreclosures. Speech at Columbia Business School's 32nd Annual Dinner, New York, 5 May.

Cairnes, M. (2003) *Boardrooms that Work: A Guide to Board Dynamics*. Australian Institute of Company Directors, Sydney, and Group of 100, Melbourne.

Carver, J. (2006) *Boards that Make a Difference: A New Design for Leadership in Non-profit and Public Organisations*, 3rd edition. Josey-Bass, San Francisco.

Fabozzi, F.J., Mann S.V. and Choudhry, M. (2002) *The Global Money Markets*. John Wiley & Sons, Hoboken, NJ.

Fishman, S. (2008) Burning down his house: Is Lehman CEO Dick Fuld the true villain in the collapse of Wall Street, or is he being sacrificed for the sins of his peers? New York Magazine.com, 30 November. http://nymag.com/news/business/52603/

FSA (2009a) *The Turner Review: A Regulatory Response to the Global Banking Crisis*. FSA, London. http://www.fsa.gov.uk/pubs/other/turner_review.pdf

FSA (2009b) The Crisis: the role of investors. Speech by Hector Sants, NAPF Investment Conference 2009.

Geithner, T.F. (2008) Reducing systemic risk in a dynamic financial system. Speech to the Economic Club of New York, 9 June.

Gieve, Sir John (2009) Seven lessons from the last three years. Speech delivered on 19 February at the London School of Economics, London.

Gowers, A. (2008) Exposed: Dick Fuld, the man who brought the world to its knees. *Sunday Times*, 14 December. http://business.timesonline.co.uk/tol/business/industry_sectors/banking_and_finance/article5336179.ece

House of Commons (2009) *Banking Crisis: Dealing with the Failure of the UK Banks*. House of Commons Treasury Committee, Seventh Report of Session 2008–2009.

Irving, J. (1971) Group think. *Psychology Today*, November, pp. 44–46.

Lees, G. (2010) Enterprise governance, restoring boardroom leadership. Discussion paper, Chartered Institute of Management Accountants, January.

Myers, D.G. and Lamm, H. (1976) The group polarization phenomenon. *Psychological Bulletin*, 83, 602–627.

Sorkin, R.S. and Thomas Jr., L. (2008) J.P. Morgan buys ailing Stearns at huge discount. *New York Times*, 16 March.

Stoner, J.A.F. (1961). A comparison of individual and group decisions involving risk. Unpublished master's thesis, Massachusetts Institute of Technology, Cambridge, MA.

Taylor, J.B. (2009) Systemic risk and the role of government. Dinner keynote speech, Conference on Financial Innovation and Crises, Federal Reserve Bank of Atlanta, Jekyll Island, Georgia, 12 May.

Turner, A. (2009) The financial crisis and the future of financial regulation. Speech to the Economist's Inaugural City Lecture, January.

5 Developments in Corporate Governance in Australia and Canada

We find that companies with better corporate governance outperform poorly governed companies, particularly in relation to earnings per share and return on assets.
(Australian Treasury Working Paper, March 2009)

The previous chapter examined the global financial crisis triggered by the subprime mortgage collapse, the criticism levelled at risk management, the federal reforms that followed the crisis and the desire to manage systemic risk. This chapter examines the evolving developments in corporate governance first in Australia and then in Canada, two countries that have not seen the same level of corporate scandals as the US. Both countries have examined overseas regulatory initiatives to improve their own corporate governance. This includes the formal US regulation introduced through the Sarbanes-Oxley Act 2002. In addition, it includes the more flexible approach (by way of guidelines and recommendations) based on the "comply or explain" principle, introduced, for instance, by the UK and German Corporate Governance Codes and the Swiss Code of Best Practice. Like elsewhere, Australian corporate governance and legislation has been coloured by its own corporate scandals. In Canada the landmark publication, the Dey Report, formed the baseline for Canadian corporate governance from which the country's modern corporate governance stems. Chapter 6 examines internal control and enterprise risk management.

5.1 AUSTRALIAN CORPORATE GOVERNANCE

5.1.1 Regulation Arising from Corporate Failures

Over time, the regulation of companies in Australia has been directly influenced by waves of corporate failures and interrelated accounting scandals. The most notable were during the 1960s, 1970s, late 1980s/early 1990s and early 2000s. The most recent round of corporate failures in Australia included Ansett, Centaur, Harris Scarfe, HIH Insurance, One.Tel and Pasminco (Bosch 2001; Clarke and Dean 2001). These more recent failures have not necessarily arisen from difficult or severe economic conditions, but more because of poor or ineffective management. Specifically, poor corporate governance, inadequate accounting and financial reporting all combined with concerns about the lack of perceived independence of auditors. History has repeatedly shown that accounting failure is a common determinant of unexpected corporate collapses. Vivid examples are the failures of Enron and WorldCom in the United States in the early 2000s and HIH Insurance in Australia (see Box 5.1). However, such phenomena are not without precedent. The more recent international corporate scandals have not been remarkable in the sense that such failures reflect reoccurring excesses of human behaviour. In the corporate world in particular, greed and arrogance can assume prominence, especially during periods of either rapid or extended economic growth. Such behaviour is pre-eminent during periods where there is little or no effective company regulation and especially where the accounting and audit provisions of relevant company legislation, if any, are discovered to be inadequate. Clearly Australia is not unique in exhibiting the dark side of corporate behaviour.

5.1.2 Corporate Governance Reforms Following the Accounting Scandals of the Early 2000s

A series of governance reforms typically follow each round of major corporate failures. Such reforms incorporate financial reporting and auditing reforms, which are intended to ensure, as far as possible, that the errors, misjudgements and negligence of the past are not repeated in the future. Following the corporate failures in Australia and elsewhere in the early 2000s, Australia, like the US and the UK, enacted major corporate laws in order to address the key deficiencies identified. These laws have been influenced by corrective legislation introduced in the UK aimed at improving corporate governance after major corporate failures. The Corporate Law Economic Reform Program (Audit Reform and Corporate Disclosure) Act 2004 was the Australian government's primary response to the accounting and corporate governance scandals of the recent past. This new legislation followed a major review of auditor independence, the Ramsay Report (Ramsay 2001) and the HIH Royal Commission Report 2003 (see Box 5.1).

Some of the key changes included in this complex piece of legislation were to:

- require rotation of audit partners of publicly listed company clients every five years;
- heighten legal protection of whistleblowers;
- increase disclosure requirements pertaining to executive remuneration;
- mandate CEO and CFO certification of financial statements;
- introduce legal underpinning of auditing standards; and
- to expand the role for the Financial Reporting Council to take responsibility for the auditing and accounting standard-setting regime as well as oversight of auditor independence.

This legislation, whilst significant in terms of scope and volume, is less onerous than the US Sarbanes-Oxley Act 2002. For example, certain consulting services that were readily provided by auditors to their audit clients were banned in the US, but this has not occurred in Australia. In addition, the International Accounting and Financial Reporting Standards of the International Accounting Standards Board were adopted in Australia from 2005 as the Australian International Financial Reporting Standards (AIFRS).

Box 5.1 Collapse of HIH Insurance Limited

HIH Insurance

The collapse of HIH Insurance Limited in 2001 was the largest ever collapse in Australia's corporate history (Westfield 2003). Its failure had a dramatic impact on many parts of industry and commerce.

Background

During 2000, HIH exhibited a declining profitability and capital base which management primarily attributed to low insurance premiums in highly competitive markets (Westfield

2003). The financial media and stock analysts were also questioning the company's performance and future prospects. Following an adverse review of HIH's financial position by KPMG in early March 2001, the HIH board appointed a provisional liquidator. The issue of a court order on 27 August 2001 placed HIH into official liquidation. As a result of HIH's collapse, investors lost millions and many consumers and companies suddenly found they were uninsured. The Australian government announced a Royal Commission to look into the HIH collapse in response to the public outrage and entered into an arrangement to indemnify HIH policyholders who were able to demonstrate genuine hardship. State governments around the country also took action to mitigate the effects of the collapse and undertook to fulfil many outstanding HIH builders' warranties and third party obligations (HIH Royal Commission Report 2003).

The trigger for the collapse

The Commission attributed the failure of the company to two key factors. First, claims arising from insured events in previous years were much greater than the company had provided for in its accounts, thus leading to an overstatement of reported profits, known as "under-reserving" or "under-provisioning". In addition, large claims against previously written insurance policies exceeded present income, making it impossible for the company to trade indefinitely in such adverse circumstances. The second factor was mismanagement of HIH through poorly conceived and badly executed acquisitions, expansion of UK operations in about 1997 into previously uncharted territory and an ill-fated commitment to re-enter the US market. Justice Owen, the head of the Commission, further noted that the suboptimal nature of the corporate culture of HIH was characterised by the board's unbending faith in the company's leadership, especially its dominant CEO. The Commission report pointed to management's incapacity to see what had to be done, to address the group's inherent problems, and highlighted the fact that unpleasant information was hidden from the board to reduce discomfort or undue questioning from the board. In addition, there was a lack of sceptical questioning and analysis by the board and, arguably, by the auditors.

Accounting matters

Three accounting matters are of significance: provisions for expected future claims, earnings management using reinsurance contracts and accounting for goodwill.

Provisions for expected future claims

It is critical that the amount allocated to the balance sheet as a provision for outstanding claims is sufficient to reflect current circumstances and conditions. If the provision is understated, profit will be overstated; if the provision is overstated, profit will be understated. HIH could have chosen to calculate its provision using a central estimate (mean value of claims liability) or by applying a prudential margin (a more conservative approach). Yet HIH almost always employed the central estimate and did not apply a prudential margin. The consequence was not only to take an overly optimistic view of claims provisions but also to continually overstate reported earnings.

Earnings management using reinsurance contracts

HIH appears to have obscured its optimistic provisioning by entering into so-called financial reinsurance arrangements with other parties. Reinsurance is a process "whereby a second insurer, in return for a premium, agrees to indemnify a first insurer against a risk insured by the first insurer in favour of an insured".

Accounting for goodwill
In acquiring the shares of FAI, HIH gave consideration, which in total amounted to A$300.5 million. This acquisition was initially recorded in 1999 in the consolidated financial statements of HIH as comprising A$25 million of net tangible assets and A$275 million of purchased goodwill. Subsequently, another A$163 million of FAI-related "goodwill" was added to this intangible asset account so that by the year 2000 this goodwill account had a balance of A$438 million (HIH Royal Commission Report 2003). Justice Owen contended that the goodwill adjustments (and reinsurance transactions referred to earlier) became techniques for concealing under-reserving problems inherent in FAI's insurance portfolio, and stated that "in the accounting practices of HIH goodwill became something of a repository for the unpleasant and unwanted consequences of poor business judgment".

Legal outcomes arising from the HIH collapse
As reported by the Australian Securities and Investment Commission (ASIC) in March 2002, two former directors of HIH, Rodney Adler and Ray Williams, were found guilty on civil charges of breaching their directors' duties over the payment by HIH of A$10 million into a trust controlled by Adler (subsequently upheld by the New South Wales Court of Appeal in July 2003). Adler was banned from acting as a company director for 20 years, while Williams was banned for ten years. Alder was fined A$900,000 and Williams was ordered to pay a pecuniary penalty of A$250,000. In addition, in 2005, as a result of their conduct as directors of HIH, Adler was given a jail sentence of $4\frac{1}{2}$ years with a non-parole period of $2\frac{1}{2}$ years and Williams $4\frac{1}{2}$ years with a non-parole period of 2 years and 9 months. The investigation by the ASIC also led to criminal prosecutions of three other former senior executives of HIH for breaches of the Corporations Law and the Crimes Act.

Role of the HIH auditor
Andersen's audit work in relation to the 1999 and 2000 audits was characterised (it was reported) by a lack of sufficient audit evidence to support its conclusions. Justice Owen was critical of the perceived lack of independence of Andersen from HIH. He highlighted the issues as being: the presence of three former Andersen partners on the board of HIH; HIH's dealings with the audit committee and non-executive directors, and intense pressure placed on Andersen partners to maximise fees from non-audit work available at HIH. In November 2002, the HIH liquidator took legal action against the auditor for negligence in the conduct of the HIH audit.

Source: HIH Royal Commission Report (2003) and reports by the Australian Securities and Investment Commission.

5.1.3 Horwath 2002 Corporate Governance Report

The Horwath Report was based on research undertaken by Associate Professor Jim Psaros and Michael Seamer from the University of Newcastle business school on the governance structures in Australia's top 250 listed businesses (Horwath and University of Newcastle 2002). The report measured the independence of each company's board and other key oversight committees, namely the audit, remuneration and nomination committees. Concern had been raised about the quality of corporate governance in Australia following the collapse of HIH, Harris Scarfe, Ansett and One.Tel. In the uncertain climate that followed, organisations needed to reassure the community, investors and government that business was being conducted fairly and in the best interests of all shareholders and stakeholders. Practically this meant companies delivering full transparency and accountability in their corporate governance structures. The report aimed to add substance to the prevailing debate on corporate governance by providing empirical evidence on the governance practices of Australia's top companies.

The corporate governance assessment model developed in the research was based upon factors identified in national and international best practice guidelines and research studies. These included the US Blue Ribbon Committee Report (Blue Ribbon Committee 1999), the UK Hampel Report (Hampel 1998), the OECD Report (OECD 2001) and the Ramsay Report (Ramsey 2001). Central to the model was the perception that company boards and their associated committees should have appropriate levels of independence. The model considered objective factors based on publicly disclosed information relating to the existence and structure of a company's board of directors and audit, remuneration and nomination committees.

5.1.4 The ASX Corporate Governance Council

Recognising the importance of corporate governance in the context of global markets, the Australian Securities Exchange (ASX) accepted a leadership role in enhancing Australian corporate governance practices by convening the ASX Corporate Governance Council in August 2002.

The Council undertook a wide-ranging work programme and identified a number of activities as part of its role, including:

- producing corporate governance guidelines for listed entities;
- assisting to build understanding about corporate governance practices on the part of listed companies;
- reviewing the legislative and rules framework for Australian corporate governance practices;
- providing information about corporate governance to investors and the wider community; and
- regularly reviewing the disclosure of corporate governance practices of listed entities to promote a high standard of transparency.

The overriding objective of the Council was to develop and deliver a flexible, industry-wide framework for corporate governance that could provide a practical guide for listed companies, to enable them to improve their existing corporate governance practices. The Council is formed of representatives of 21 different business, shareholder and industry groups, each offering valuable guidance and information specific to their constituencies and industry. In order to promote investor confidence and to assist companies to meet stakeholder expectations, the Council has developed and released corporate governance guidelines for Australian listed entities.

The Council released the first edition of its *Principles of Good Corporate Governance Practice and Best Practice Recommendations* on 31 March 2003 (ASX Corporate Governance Council 2003). Within this document there are ten key corporate governance principles for publicly listed companies to follow, relating to areas such as ensuring a balance of executive and non-executive directors on the board of directors, maintaining well-documented risk management processes and ensuring proper oversight by management, and ensuring the adoption of a properly constituted and well-qualified selection process in appointing individuals to the audit committee. These guidelines, while not mandatory for publicly listed Australian companies, are highly influential as any companies that decide not to follow aspects of these principles are required to outline their reasons in their annual report, in accordance with ASX Listing Rule 4.10.3.

The second edition, *Corporate Governance Principles and Recommendations*, was released in August 2007 (ASX Corporate Governance Council 2007). These guidelines took effect for listed entities at the start of the financial year for 2008 (1 January or 1 July).

On 30 June 2010, the Council released amendments to the second edition in relation to diversity, remuneration, trading policies and briefings. These apply to listed entities from 1 January 2011.

5.1.5 Financial Statements

It is difficult to generalise as to why some companies' management teams choose to employ misstatement strategies in their financial reports. It is apparent from looking at failed companies, such as the Bond Corporation and HIH, that a desire to portray a false or incomplete view of current financial performance (in the face of the reality of declining financial fortunes) was the driver behind such misleading accounting practices and financial reporting. Clearly as time passes both the increase in complexity of accounting practices and the length of financial reports (as a result of the ever-expanding number of accounting standards) will lead to the potential for misstatements. This is relatively simple to comprehend. However, it does not explain or excuse the practices of overstating income, understating expenses or recording disclosed total assets and net assets at inflated levels in audited financial reports, frequently exhibited by failed companies. The misleading disclosures that have been fuelled by exuberant optimism surrounding future recovery, in retrospect, often appear to have been misguided at best or fraudulent at worst. It should be noted that, yet again, corporate governance reforms have been enacted to rein in the excesses of corporate behaviour and address perceived deficiencies in governance mechanisms and processes in the hope that they may not arise, at least in the same form, in the foreseeable future. It is suggested here that corporate failures and accounting scandals in Australia will undoubtedly recur and will continue to reflect the excesses of human behaviour, specifically in the corporate world, where greed and dishonesty have been prevalent, especially during periods of extended or rapid economic development. History has a habit of repeating itself, as lessons are forgotten or not passed on and the darker side of human nature resurfaces. Shareholders' investments will remain vulnerable. The degree of vulnerability will depend on rigorous governance and strict regulation and human ingenuity applied to legally (or illegally) circumnavigating both.

5.2 CANADA

5.2.1 Dey Report

In 1994 the Toronto Stock Exchange (TSE) published a study of Canadian corporate governance, prepared by its own Corporate Governance Committee, chaired by Peter J. Dey QC. The report, called *Where Were The Directors*? and subsequently known as the "Dey Report", came to be recognised as a landmark in the development of Canadian corporate governance (Dey 1994). The title of the report reflected public sentiment of the time. In the early 1990s there had been a growing feeling of dissatisfaction among Canadian investors and interested parties with regard to the performance of boards of directors. Although most public companies were well governed, the highly visible failure of several poorly managed public companies (aggravated by a recession) demonstrated a need to make corporate governance more of a concern in Canada. The report made 14 recommendations focused on the board of directors and its relationship with shareholders and management. The following year the TSE adopted the recommendations included in the report as "best practice" guidelines and required every listed company to disclose annually their approach to corporate governance using the guidelines as a reference point, together with an explanation of any differences between the company's approach and the guidelines. Like the Cadbury Code, discussed in Chapter 2, the guidelines were not mandatory. The TSE recognised that there is no "one size fits all" solution. Included within the recommendations was the proposal that the TSE adopt, as a listing requirement, the disclosure by each listed corporation of its approach to corporate governance, on an annual basis commencing with companies with 30 June 1995 year ends.

5.2.2 Dey Revisited

The Committee that produced the Dey Report recommended that "a successor committee ... monitor developments in corporate governance, and evaluate the continued relevance of our recommendations". Following that lead, the TSE and the Institute of Corporate Directors (ICD) commissioned a review to assess how much progress had been made in the quality of governance. The result was the report entitled *Report on Corporate Governance 1999, Five Years to the Dey* (TSE/ICD 1999). The aim of the report was to assess the extent to which corporate governance of public companies reflected the earlier TSE guidelines and to identify opportunities for the TSE and the ICD to support sound practices. The principal component of the research behind the report was a survey of CEOs. A total of 1250 TSE listed companies were invited to participate in the survey, representing 95% of issuers listed on the TSE. The 636 replies constituted a response rate of 51%, which the report justifiably claims generated a highly reliable set of results. In summary, the survey found that the highest levels of compliance appeared to be in controlling board size, participation in strategic planning for the corporation and achieving a majority of unrelated directors. The earlier TSE guidelines had advocated that boards assume responsibility for "the identification of the principle risks of the organisation's business, ensuring the implementation of appropriate systems to manage these risks". From the survey it appeared that risk management was one of the less well-developed governance activities, with 39% of participating companies having no formal process. This percentage rose to 55% in the gold and precious metals sector. This is noteworthy in that in the preceding five years there had been spectacular corporate failures in the entertainment, electronics and mining sectors.

Figure 5.1 shows an extract of the survey questions specific to risk management and the corresponding number of answers received, expressed as a percentage of the total number of responses.

5.2.3 Kirby Report

Following on from the governance regulation initiated by the Dey Commission, the Senate Standing Committee on Banking, Trade and Commerce of the federal parliament held hearings and released a report in 1998 (Kirby 1998). The Kirby Report, named after the Committee Chairman, Senator Michael J.L. Kirby, made a number of recommendations, including new measures to improve the governance practices of institutional investors. These included that

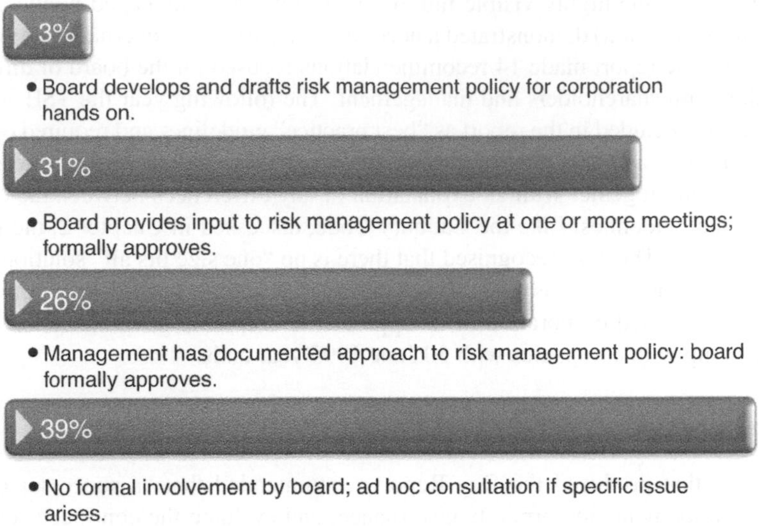
3%
• Board develops and drafts risk management policy for corporation hands on.
31%
• Board provides input to risk management policy at one or more meetings; formally approves.
26%
• Management has documented approach to risk management policy: board formally approves.
39%
• No formal involvement by board; ad hoc consultation if specific issue arises.

Figure 5.1 Risk management survey questions and their responses

boards of pension plans must be knowledgeable and communicate with pension holders through an annual report and other means of communication, making sure to explain the risk management and governance practices of their fund manager. The report also recommended that every mutual fund be required to have a majority of independent directors and to adopt a corporate rather than a trust structure.

5.2.4 Saucier Committee

More recently the Canadian Institute of Chartered Accountants (CICA), the TSE and the TSE Venture Exchange (the then Canadian Venture Exchange) established the Joint Committee on Corporate Governance in July 2000 (the Saucier Committee – named after the Committee's chair, Guylaine Saucier). The mandate of the Saucier Committee was to review the state of corporate governance in Canada and make recommendations for improvements. In preparing its guidelines, the Committee examined the corporate disclosure regimes of stock exchanges in the US, Australia and the UK. The Saucier Committee's final report, released in November 2001 (Saucier 2001), recommended that the TSE amend its corporate governance guidelines for listed issuers in a series of ways to align them with international developments. Since the publication of the Dey Report there had been an unprecedented surge in global trade and investment and a number of high-profile corporate scandals. On 26 April 2002, the TSE proposed changes to its guidelines for effective corporate governance in response to the Saucier Committee's recommendations.

5.2.5 National Policy and Instrument (April 2005)

The Policy and the Instrument were initiatives of the members of the Canadian Securities Administrators. The National Policy 58-201 *Corporate Governance Guidelines* had been made, or was expected to be made, as a policy in every jurisdiction in Canada. The National

Instrument 58-101 *Corporate Governance Practices* had been made, or was expected to be made, a rule, a commission regulation, a regulation, a policy or a code, in the different jurisdictions across Canada.[1]

The Policy

The Policy provides guidance on corporate governance practices. Although the Policy applies to all reporting issuers (other than investment funds), the guidelines in the Policy are not intended to be prescriptive. Rather, issuers are encouraged to consider the guidelines in developing their own corporate governance practices.

The following corporate governance guidelines are contained in the Policy:

- maintaining a majority of independent directors on the board of directors;
- appointing a chair of the board or a lead director who is an independent director;
- holding regularly scheduled meetings of independent directors at which non-independent directors and members of management are not in attendance;
- adopting a written board mandate;
- developing position descriptions for the chair of the board, the chair of each board committee and the CEO;
- providing each new director with a comprehensive orientation, and providing all directors with continuing education opportunities;
- adopting a written code of business conduct and ethics;
- appointing a nominating committee composed entirely of independent directors;
- adopting a process for determining the competencies and skills the board as a whole should have, and applying this result to the recruitment process for new directors;
- appointing a compensation committee composed entirely of independent directors;
- conducting regular assessments of the board effectiveness, as well as the effectiveness and contribution of each board committee and each individual director.

The Instrument

The Instrument applies to reporting issuers other than investment funds, issuers of asset-backed securities, designated foreign issuers, SEC foreign issuers, certain exchangeable security issuers, certain credit support issuers and certain subsidiary issuers. The Instrument establishes both disclosure requirements and a requirement to file any written code that the issuer has adopted. The Instrument requires an issuer to disclose those corporate governance practices it has adopted. The specific disclosure items are set out in Form 58-101F1.

5.2.6 TSE Corporate Governance: Guide to Good Disclosure 2006

Each listed issuer subject to National Instrument 58-101 *Disclosure of Corporate Governance Practices*, is required to disclose its corporate governance practices in accordance with that instrument, or any replacement of that instrument. In addition the TSE monitors corporate

[1] http://www.osc.gov.on.ca/documents/en/Securities-Category5/rule_20050415_58-201_gov-practices.pdf

governance disclosure of listed issuers and outlines the measures it will take where there is noncompliance. The Guide addresses the following subjects:

1. Board of directors
2. Board mandate
3. Position descriptions
4. Orientation and continuing education
5. Ethical business conduct
6. Nomination of directors
7. Compensation
8. Other board committees
9. Assessments

5.3 SUMMARY

Following a series of corporate failures within and beyond their shores (the collapse of HIH Insurance Ltd in 2001 being a particular catalyst), Australia enacted major corporate laws, including the Corporate Law Economic Reform Act 2004. In addition, it adopted the International Accounting and Financial Reporting Standards in 2005. In 2002 the Horwath Report (based on research by the University of Newcastle) was published, focusing on the need for company boards and their associated committees to have appropriate levels of independence. Subsequently the Australian Securities Exchange Council adopted a leadership role in enhancing Australian corporate governance practices through a number of initiatives, including delivering an industry-wide corporate governance framework. In 2003 they issued guidance in the form of the *Principles of Good Corporate Governance Practice and Best Practice Recommendations*, subsequently updated in 2007.

In Canada it might be argued that corporate governance and guidance developed earlier. In 1994 the Toronto Stock Exchange published a study of Canadian corporate governance, prepared by its own Corporate Governance Committee (the Dey Report). This report laid out best practice guidelines for corporate governance for listed companies. Five years later, the Institute of Corporate Directors issued a follow-up report describing the degree of adoption of the Dey Report's recommendations. In 1998 the Kirby Report described new measures to improve risk management and corporate governance practices of institutional investors. In 2001 the Saucier Committee's report made recommendations regarding the TSE's corporate governance guidelines for listed issuers to reflect international developments in governance. These reports were followed by the Canadian Securities Administrators policy on corporate governance in 2005 and the TSE guide to good disclosures in 2006.

5.4 REFERENCES

Bosch, H. (2001) Introduction. In *Collapse Incorporated Tales, Safeguards & Responsibilities of Corporate Australia*, pp. 71–98. CCH Australia, North Ryde, NSW.

Clarke, F. and Dean, G. (2001) Corporate collapses analysed. In *Collapse Incorporated: Tales, Safeguards & Responsibilities of Corporate Australia*, pp. 71–98. CCH Australia, North Ryde, NSW.

Dey, P.J. (1994) *Where Were the Directors?* Toronto Stock Exchange Committee on Corporate Governance in Canada, December.

Hampel Committee on Corporate Governance (1998) *Committee on Corporate Governance: Final Report*. Gee Publishing, London. http://www.ecgi.org/codes/documents/hampel.pdf

HIH Royal Commission Final Report (2003) *The Failure of HIH Insurance*. The HIH Royal Commission (The Hon Justice Owen Commissioner), April. Commonwealth of Australia, Canberra.

Horwath and University of Newcastle (2002) *Horwath 2002 Corporate Governance Report*. Horwath (NSW) Pty Ltd, Sydney. http://www.ecgi.org/codes/documents/horwath_cg_02.pdf

Kirby, M. (1998) *The Governance Practices of Institutional Investors*. Standing Committee on Banking, Trade and Commerce, Government of Canada, Ottawa, November. http://www.parl.gc.ca/36/1/parlbus/commbus/senate/com-e/bank-e/rep-e/rep16nov98-e.htm

OECD (2001) *OECD Guidelines for Multinational Enterprises: Global Instruments for Corporate Responsibility*. OECD Publishing, Paris. http://www.oecdbookshop.org/oecd/display.asp?lang=EN&sf1=identifiers&st1=212001171e1

Ramsay, I. (2001) *Independence of Australian Company Auditors: Review of Current Australian Requirements and Proposals for Reform*. Report to the Minister for Financial Services and Regulation, October. Department of Treasury, Parkes, ACT. http://treasury.gov.au/documents/296/PDF/ramsay2.pdf

Saucier, G. (2001) *Beyond Compliance: Building a Governance Culture*. Final Report, Joint Committee on Corporate Governance, November. Sponsored by the Canadian Institute of Chartered Accountants (CICA), the Canadian Venture Exchange (CDNX) and the Toronto Stock Exchange (TSE). http://www.ecgi.org/codes/documents/beyond_compliance.pdf

TSE/ISD (1999) *Report on Corporate Governance 1999, Five Years to the Dey*. Principal author Ruth Corbin. Toronto Stock Exchange and the Institute of Directors, Toronto, Ontario.

Westfield, M. (2003) *HIH: The Inside Story of Australia's Biggest Corporate Collapse*, John Wiley & Sons Australia, Milton, Qld.

6
Internal Control and Risk Management

The man who is denied the opportunity of taking decisions of importance begins to regard as important the decisions he is allowed to take.
(C. Northcote Parkinson)

Chapters 3 and 5 examined developments in corporate governance in the US, Australia and Canada, as a logical progression of the examination of corporate governance in the UK. This chapter continues to establish the context of enterprise risk management by examining first what is meant by the term "internal control", a subject that permeates the vast majority of the guides and publications on governance, and second how internal control relates to risk management. The governance study reports and guides briefly examined in Chapter 2 are re-examined here to establish the intention and composition of internal controls and their relationship with risk management. Chapter 7 examines developments in risk management in the public sector.

6.1 THE COMPOSITION OF INTERNAL CONTROL

What is "internal control"? While published descriptions of corporate governance make it clear that it is multifaceted and that internal controls are a subset of corporate governance, specific internal controls cannot be readily distilled. The Cadbury Committee (Cadbury 1992) stated under Section 4.31 that "Directors are responsible under s.221 of the Companies Act 1985 for maintaining adequate accounting records. To meet these responsibilities directors need in practice to maintain a system of internal control over the financial management of the company, including procedures designed to minimise the risk of fraud." This requirement has been maintained in the Combined Code 2003, which states that a function of the board is to maintain a sound system of internal control. Controls are described in Section C.2.1 of the Code as including financial, operational and compliance controls as well as risk management systems. Hence the Code assumes a broader definition of internal controls than the Cadbury Committee. The Code cross-refers to the Turnbull Report[1] (Turnbull 1999) which (in Section 20) provides guidance on the composition of an internal control system as follows:

> An internal control system encompasses the policies, processes, tasks, behaviours and other aspects of a company that, taken together:
>
> - facilitate its effective and efficient operation by enabling it to respond appropriately to significant business, operational, financial, compliance and other risks to achieving the company's objectives. This includes the safeguarding of assets from inappropriate use or from loss and fraud, and ensuring that liabilities are identified and managed;

[1] When the Combined Code of the Committee on Corporate Governance was published in 1998, the Institute of Chartered Accountants in England and Wales agreed with the London Stock Exchange that it would provide guidance to assist listed companies to implement the requirements of the Code relating to internal control. The result was the creation of the Turnbull Working Party which published its report in September 1999.

- help ensure the quality of internal and external reporting. This requires the maintenance of proper records and processes that generate a flow of timely, relevant and reliable information from within and outside the organisation;
- help ensure compliance with applicable laws and regulations, and also with internal policies with respect to the conduct of business.

Section 17 of the report provides further guidance on the composition or content of a sound system of internal control in the following statement:

> In determining its policies with regard to internal control, and thereby assessing what constitutes a sound system of internal control in the particular circumstances of the company, the board's deliberations should include consideration of the following factors:
>
> - the nature and extent of the risks facing the company;
> - the extent and categories of risk which it regards as acceptable for the company to bear;
> - the likelihood of the risks concerned materialising;
> - the company's ability to reduce the incidence and impact on the business of risks that do materialise; and
> - the costs of operating particular controls relative to the benefit thereby obtained in managing the related risks.

It addresses: the nature and extent of risks facing a company; acceptable risks; the likelihood of occurrence; risk reduction; and the cost of risk management actions. Hence, it is not all-embracing and does not address such issues as methods of identification, methods of assessment, risk evaluation, risk appetite, risk transfer and secondary risks.

6.2 RISK AS A SUBSET OF INTERNAL CONTROL

Turnbull describes a company's system of internal control as having a key role in the management of risks that are significant to the fulfilment of its business objectives and states that financial records help ensure that the company is not unnecessarily exposed to avoidable financial risks. Section 10 of the guidance describes one of the main functions of internal control as follows: "A sound system of internal control contributes to safeguarding the shareholders' investment and the company assets."

Turnbull states that a company's objectives, its internal organisation and the environment within which it operates are continually changing. A sound system of internal control therefore depends on a thorough and regular evaluation of the nature and extent of the risks to which the company is exposed. He argues that as profits are in part the reward for successful risk taking in business, the purpose of internal control must be to help manage and control risk appropriately.

Figure 6.1 illustrates the relationship of corporate governance (in the form of the 2003 Combined Code) to internal control, its subsets and specifically risk management.

6.2.1 The Application of Risk Management

On completion of the guidance on internal control produced by the Working Party led by Nigel Turnbull, the Institute of Chartered Accountants (ICA) published a briefing to aid its implementation. It provides clear, unambiguous guidance on how to implement risk management within a business. In the foreword to *Implementing Turnbull* (Jones and Sutherland 1999), Sir Brian Jenkins (the then chairman of the Corporate Governance Group of the ICA) stated

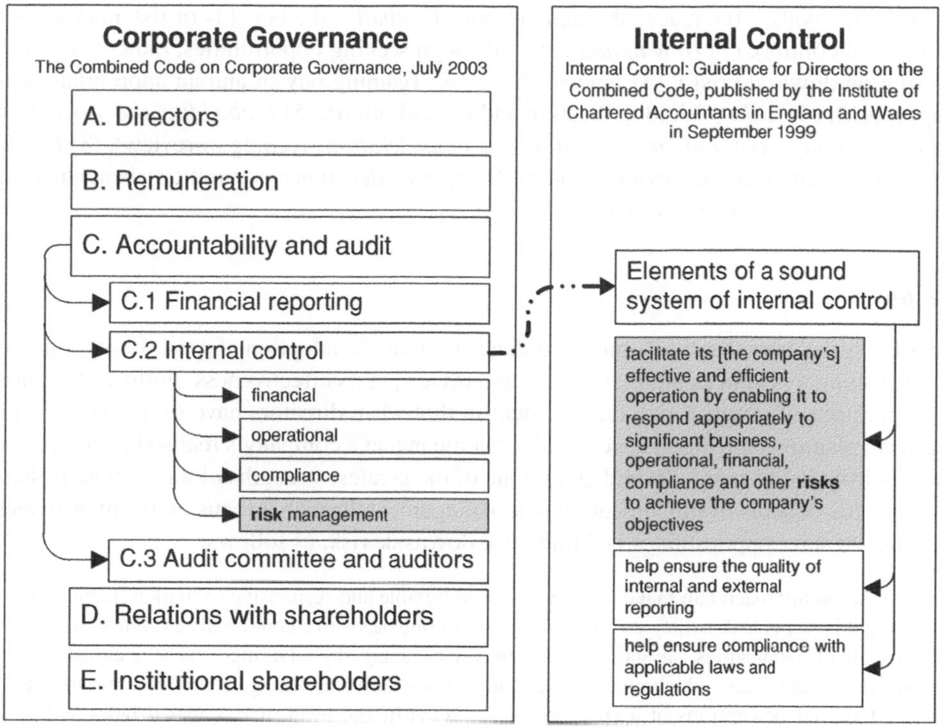

Figure 6.1 Composition of the Combined Code 2003 and its relationship to the Turnbull guidance

that the aim of this briefing was to be a source of timely, practical help to those directors who wished to take steps to implement the new guidance in a straightforward way, which would bring business benefits. The executive summary (echoed in the foreword) stated that the briefing had been prepared for directors who wished to take straightforward steps towards achieving Turnbull or who are interested in the practicalities of good risk management and internal control and in getting added value for their companies from the guidance. The key messages of the briefing are:

- Do not delay in implementing Turnbull
- Obtain management buy-in at all levels of the organisation
- Prepare a plan
- Identify clear company objectives
- Prioritise the risks to the achievement of the objectives
- Establish a clear risk management policy and control strategies
- Consult throughout the business
- Improve the business culture where appropriate
- Keep it simple and straightforward
- Monitor continuously
- Avoid audit committee overload
- Incorporate Turnbull in your management and governance processes
- Aim to obtain business improvement

The briefing "walks" the reader through (1) *Why Turnbull?* (the benefits of risk management and internal control), (2) *How to add value* (through seeking opportunities, rather than solely focusing on downside risk), (3) *Immediate actions* (gaining buy-in and an appropriate scale of approach), (4) *Risks* (risk identification and prioritisation), (5) *Embedding the process,* (6) *Monitoring and internal audit,* (7) *Board level considerations* (timing of review), (8) *Disclosures* (the content of annual reviews) and (9) *Other considerations* (committees, benchmarking performance and pitfalls to avoid).

Benefits

The briefing explains that the Turnbull guidance is about the adoption of a risk-based approach to establishing a system of internal control and reviewing its effectiveness. Further, it explains the importance of effective risk management in that when directors have set goals as part of long-term planning, the emergence of risks can mean that a company's realised goals are very different from its intended, desired goals. One of the greatest strengths of the briefing is that it spells out the benefits of implementing risk management through a focus on the management of change to seize opportunities and minimise downside risk, as follows:

> A risk based approach can make a company more flexible and responsive to market fluctuations making it better able to satisfy customers' ever-changing needs in a continually evolving business environment. Companies can gain an early-mover advantage by adapting to new circumstances faster than their rivals, which again could lead to competitive advantage in the medium to long term. External perceptions of a company are affected by the level of risk that it faces and by the way its risks are managed. A major risk exposure and source of business failure and/or lack of opportunity success has been the failure to manage change. Companies need to be aware of changing markets, service delivery (e.g. e-commerce) and morale. Effective risk management and internal control can be used to manage change, to all levels of people in the company in meeting its business objectives, and to improve a company's credit rating and ability to raise funds in the future, not to mention its share price over the longer term.

The briefing states the following potential benefits of effective risk management:

- Early mover into new business areas
- Greater likelihood of achieving business objectives
- Higher share prices over the longer term
- Reduction in management time spent "fire fighting"
- Increased likelihood of change initiatives being achieved
- More focus internally on doing the right things properly
- Lower cost of capital
- Better basis of strategy setting
- Achievement of competitive advantage
- Fewer sudden shocks and unwelcome surprises

Risks

The briefing provides guidance on the process of the identification of risks, understanding risk appetite, whether detailed quantification should be carried out and how risks should be prioritised.

Table 6.1 Risk matrix (Jones and Sutherland 1999, Figure 7)

Business
Wrong business strategy
Competitive pressure on price/market share
General economic problems
Regional economic problems
Political risks
Obsolescence of technology
Substitute products
Adverse government policy
Industry sector in decline
Takeover target
Inability to obtain further capital
Bad acquisition
Too slow to innovate

Financial
Liquidity risk
Market risk
Going concern problems
Overtrading
Credit risk
Interest risk
Currency risk
High cost of capital
Treasury risk
Misuse of financial resources
Occurrence of types of fraud to which the business is susceptible
Misstatement risk related to published financial information
Breakdown of the accounting system
Unrecorded liabilities
Unreliable accounting records
Penetration and attack of IT systems by hackers
Decisions based on incomplete or faulty information
Too much data and not enough analysis
Unfulfilled promises to investors

Compliance
Breach of Listing Rules
Breach of financial regulations
Breach of Companies Act requirements
Litigation risk
Breach of competition laws
VAT problems
Breach of other regulations and laws
Tax penalties
Health and safety risks
Environmental problems

Operational and other
Business processes not aligned to strategic goals
Failure of major change initiative
Loss of entrepreneurial spirit
Stock-out of raw materials
Skills shortage
Physical disasters (including fire and explosion)
Failure to create and exploit intangible assets
Loss of intangible assets
Breach of confidentiality
Loss of physical assets
Lack of business continuity
Succession problems
Year 2000 problems
Loss of key people
Inability to reduce cost base
Major customers impose tough contract obligations
Overreliance on key suppliers or customers
Failure of new products or services
Poor service levels
Failure to satisfy customers
Quality problems
Lack of orders
Failure of major project
Loss of key contracts
Inability to make use of the Internet
Failure of outsource provider to deliver
Industrial action
Failure of big technology related project
Lack of employee motivation or efficiency
Inability to implement change
Inefficient/ineffective processing of documents
Poor brand management
Product liability
Inefficient/ineffective management process
Problems arising from exploiting employees in developing countries
Other business priority issues
Other issues giving rise to reputational problems
Missed business opportunities

The briefing also provides a risk matrix (see Table 6.1), which it describes as setting out the various risks to consider, while at the same time providing a cautionary note that the matrix should not be regarded as comprehensive. The matrix provides a useful guide to the types of issues to be thought about and, where relevant, addressed. A way of framing the risk exposure of a business and developing a risk taxonomy is discussed in Chapter 9.

6.3 ALLOCATION OF RESPONSIBILITY

It is now commonly recognised that as part of its accountability to shareholders for the strategic direction of a company and the safeguarding of its assets, the board of directors has ultimate responsibility for ensuring the existence of an appropriate system of internal control and risk management. The responsibilities for a company's system of internal control and risk management have become more formalised over time within the progressive development of corporate governance.

6.3.1 Cadbury Committee

The Cadbury Committee is remembered for making three principal recommendations on the subject of internal control. The first (as described in Section 4.13) was that it considered that directors should be responsible for maintaining a system of internal control, including procedures designed to minimise fraud. It considered this requirement to be implicit under the Companies Act 1985. The second (as described in Section 4.32) was that directors should make a statement in the report and accounts on the effectiveness of their system of internal control. The third (as described in Section 5.16) was that the accounting profession should take the lead in developing:

- criteria for assessing effective systems of internal control;
- guidance for companies on the form in which directors should report; and
- guidance on relevant audit procedures and the form in which auditors should report.

Section 4.39, headed "Internal Audit", makes the distinction between external and internal auditing and considers the establishment of internal audit functions to undertake regular monitoring of key controls and procedures as good practice. It regards such regular monitoring as an integral part of a company's system of internal control and considers monitoring as helpful to ensuring its effectiveness. Additionally, it considers internal audit is well placed to undertake investigations on behalf of the audit committee and to follow up on any suspicion of fraud.

6.3.2 Hampel Committee

The Hampel Committee report challenged the practicalities of the recommendations of the Cadbury Committee in terms of its recommendations regarding internal control (Hampel 1998). In Section IV, entitled "Internal Control", the Hampel Committee refers to the Cadbury recommendation that the accountancy profession should take the lead in developing criteria for assessing the effectiveness of a company's system of internal control and in developing guidance for both directors and auditors to assist in reporting on internal control. As a result of Cadbury, the accounting profession established a working group to develop criteria for assessing effectiveness and guidance for directors on reporting. The group reported back in December 1994. Hampel explains that the word "effectiveness" had proved difficult for both directors and auditors alike in the context of public reporting. Concern had been expressed that the concept of the existence of a process that could determine effectiveness and offer absolute assurance against misstatement or loss was inappropriate as no system of control was foolproof against human error or deliberate override.

What was at the root of the problem was that directors and auditors were concerned that those who confirmed the effectiveness of a company's control system may be exposed to legal liability if unintentional misstatement or loss of any kind was found to have occurred. The report of the working group therefore recommended, possibly through self-interest, that the director's statement should acknowledge the board's responsibility for the internal financial control system, but explain that such a system could provide only reasonable assurance against material misstatement or loss; should describe the key procedures established in order to provide effective financial controls; and should confirm that the directors had reviewed the system's effectiveness. Hampel concurred that auditors should not be required to report publicly on directors' statements. The working group recommended that directors review and report on all aspects of internal control, including controls to ensure effective and efficient operations. Hampel concurred it was difficult in practice to distinguish financial from other controls. Hence, Hampel considered that directors should maintain and review controls addressing all relevant control objectives. He also believed that these controls should include business risk assessment and response planning, financial management, compliance with laws and regulations and the safeguarding of assets, including minimising risk and fraud.

6.3.3 Turnbull

The guidance was prepared to assist boards of UK incorporated listed companies complying with specific aspects of the 1998 Code and in particular:

- assessing how the company has applied Code principle D.2;
- implementing the requirements of the Code provisions D.2.1 and D.2.2; and
- reporting on the matters to shareholders in the annual report and accounts.

Principle D.2 of the Code states that "The board should maintain a sound system of internal control to safeguard shareholders' investment and the company's assets". Principle D.2.1 states that "The directors should, at least annually conduct a review of the effectiveness of the group's system of internal control and should report to shareholders that they have done so. The review should cover all controls, including financial, operational and compliance controls and risk management." Principle D.2.2 states that "companies which do not have an internal audit function should from time-to-time review the need for one".

The objective of the guidance, as it states in Section 8, is to:

- reflect sound business practice whereby internal control is embedded in the business processes by which a company pursues its objectives;
- remain relevant over time in the continually evolving business environment; and
- enable each company to apply it in a manner which takes account of its particular circumstances.

The guidance describes the responsibility of management and employees in the implementation of risk management as follows:

> It is the role of *management* to implement board policies on risk and control. In fulfilling its responsibilities, management should identify and evaluate the risks faced by the company for consideration by the board and design, operate and monitor a suitable system of internal control which implements the policies adopted by the board.

Additionally, clause 19 (repeated below) makes reference to the responsibility of employees. Importantly it states that to establish, operate and monitor the system of internal control individuals will have to know:

- the company;
- the company's objectives;
- the industries the company operates in; and
- the markets the company operates in.

This knowledge is vital if employees are to be effective in identifying the risks facing the company as comprehensively as possible.

> All *employees* have some responsibility for internal control as part of their accountability for achieving objectives. They, collectively, should have the necessary knowledge, skills, information and authority to establish, operate and monitor the system of internal control. This will require an understanding of the company, its objectives, the industries and markets in which it operates, and the risks it faces.

The Code describes (in Section C.3.2) responsibility for reviewing the company's financial controls and (unless addressed by others) the company's internal controls and risk management systems, as lying with the audit committee.

6.3.4 Higgs Review

Further to the introduction to the Higgs Review (Higgs 2003) included in Chapter 2, this section describes the Higgs perspective on internal control. In his introduction, Derek Higgs (Item 1.5, p. 11) describes corporate governance as providing an architecture of accountability – the structures and the process to ensure companies are managed in the interests of their owners. However, he states that architecture in itself does not deliver good outcomes and that, as non-executive directors are the custodians of the governance process, they have a crucial part to play in the success of companies. Item 6.6, referring to the role of non-executive directors, suggests they must ensure they are confident of the adequacy of financial controls and risk management. Hence, Higgs proposed the inclusion of the following in the Code: "Risk: Non-executive directors should satisfy themselves that financial information is accurate and that financial controls and systems of risk management are robust and defensible." (This wording is now included in Section A.1 of the Combined Code, 2003.) It was interesting to learn that Higgs, during his research, observed that some of the non-executive directors were concerned about the increasing amount of technical knowledge necessary in order to fulfil their roles on board committees and noted the value of training "on issues such as risk management". In this context "training" was understood to be continued professional development rather than formal training.

6.3.5 Smith Review

The Smith Review (Smith 2003), discussed in Chapter 2, provides guidance designed to assist company boards in making suitable arrangements for their audit committees. To set this particular review in context, its guidance was incorporated in the Combined Code on Corporate

Governance published in 2003. Smith considered that the role of the audit committee should include reviewing the company's internal financial control system and, where appropriate, risk management systems. Of interest here, Section 2.1 of the report states one of the main roles and responsibilities of the audit committee is "to review the company's internal financial control system and unless addressed by a separate risk committee or by the board itself, risk management systems". In Section 5, "Roles and Responsibilities", the guidance reiterates the role of the audit committee, stating that it should monitor the integrity of the company's internal financial controls and, in the absence of the existence of a risk committee, for instance, assess the scope and effectiveness of the systems established by management to identify, assess, manage and monitor financial and non-financial risks. Further, in Section 5.7, it states that management is responsible for the identification, assessment, management and monitoring of risk, for developing, operating and monitoring the system of internal control and for providing assurance to the board that it has done so. Additionally, it considers that, where the board or a risk committee is expressly responsible for reviewing the effectiveness of the internal control and risk management systems, the audit committee should receive reports from management on the effectiveness of the systems they have established and the results of any testing carried out by internal or external auditors.

6.3.6 OECD

The Organisation for Economic Co-operation and Development (OECD) was formed to promote policies designed to achieve economic growth and employment and raise the standard of living while maintaining financial stability. It published revised principles of corporate governance in 2004 (OECD 2004). The principles are aimed at providing non-binding standards, good practices and guidelines of corporate governance for tailoring by member countries to their specific circumstances. As a measure of the widespread adoption of the principles, they now form the basis of the corporate governance component of the World Bank/IMF Reports on the Observance of Standards and Codes. The principles are another source document providing a view on the relationship between internal controls and risk management. In Part 1, Section VI, headed "The Responsibilities of the Board", the principles refer to board responsibilities including "Ensuring . . . that appropriate systems of control are in place, in particular systems for risk management, financial and operational control and compliance with law and relevant standards". It is therefore reasonable to conclude that this statement implies that risk management is a subset of systems of internal control. The principles make minimal reference to risk management. The OECD does recommend, however, in Part 1, Section V, entitled "Disclosure and Transparency", the disclosure of both "foreseeable risks" and "governance structures and policies". Additionally, in Part 2 of the principles, called "Annotations to the OECD Principles of Corporate Governance", under "Disclosure and Transparency" (item A6), the principles describe what is meant by foreseeable risks:

> Users of financial information and market participants need information on reasonably foreseeable risks that may include: risks that are specific to the industry or the geographical areas in which the company operates; dependence on commodities; financial market risks including interest rate or currency risk; risk related to derivatives and off-balance sheet transactions; and risks related to environmental liabilities.

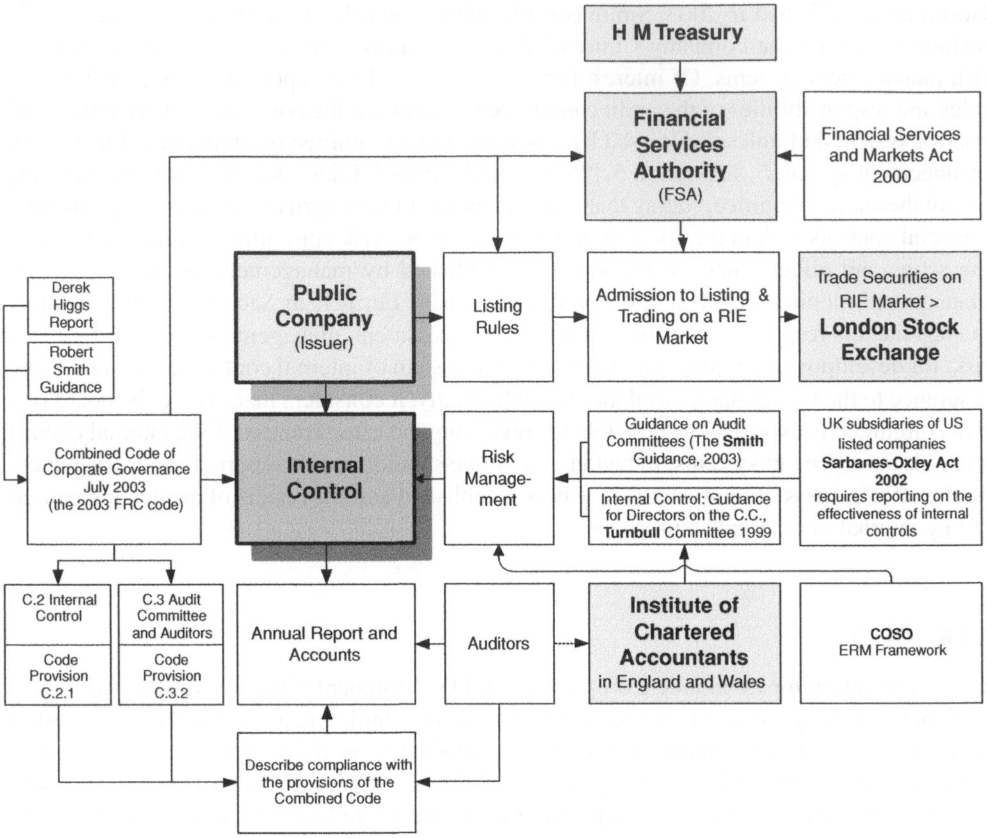

Figure 6.2 Internal control and risk management in context (based on the Combined Code of 2003)

6.4 THE CONTEXT OF INTERNAL CONTROL AND RISK MANAGEMENT

Figure 6.2 describes the context of internal control and risk management as exercised by a public company (the issuer). Financial regulation is a web involving Her Majesty's Treasury (HMT), the Financial Services Authority (FSA), the Financial Reporting Council (FRC),[2] the London Stock Exchange (LSE) and the Institute of Financial Accountants (IFA). The relationship between these organisations is as follows:

- The Financial Services and Markets Act 2000 provides the FSA, an independent non-governmental body, with statutory powers.
- HMT appoints the FSA's board, consisting of a chairman, CEO, three managing directors and 11 non-executive directors. This board sets the overall policy.
- The FSA is the UK Listing Authority and hence the authority in the UK for the listing of company shares and other securities for trading on public stock exchanges.

[2] The FRC is the UK's independent regulator for corporate reporting and governance. The FRC is funded by the UK Government, the accounting profession and by the business community (currently through listed companies).

- The FSA governs listing by the application of the Listing Rules, which control the trade by issuers of their securities on the LSE.
- The Listing Rules are published by the FSA.
- The Combined Code of Corporate Governance, dated July 2003, published by the FRC, is annexed to the Listing Rules. The Code reflects guidance produced by Higgs, Smith and Turnbull.

6.5 INTERNAL CONTROL AND RISK MANAGEMENT

It has long been recognised that directors have a primary responsibility for the stewardship of investors' assets and the protection of their investment. A company's *system of internal control* has as its main purpose the identification and management of risks that might impede the achievement of the company's business objectives and thereby reduce the value of shareholders' investments. Internal control and risk management are thus inextricably linked and are integral to the discharge of the directors' responsibilities.

6.6 EMBEDDING INTERNAL CONTROL AND RISK MANAGEMENT

The systems of internal control and risk management adopted by different companies will vary according to company-specific factors, including the nature of the business, the markets and regulatory environments in which the company operates and the attitudes of shareholders and directors towards risk taking. In all cases, however, the board is responsible for ensuring the development and maintenance of an embedded system of internal control and risk management, based on a continuous cycle of activities encompassing the company's policies and procedures, the allocation of tasks and responsibilities, its communication processes and its cultural and behavioural norms.

6.7 SUMMARY

It is the duty of directors to establish and maintain a system of internal control (Cadbury, OECD and the Combined Code 2003). It is generally accepted that an internal audit system encompasses the politics, processes, tasks and behaviours, which combined facilitate its effective and efficient operation. The Hampel Committee challenged the practicalities of the Cadbury recommendation that the accounting profession initiates the development of criteria for assessing the effectiveness of a company's system of internal control and concluded auditors should not be required to report publicly on directors' statements. The Higgs Report considered that it was the role of the non-executive directors to satisfy themselves that financial controls and systems of risk management are robust and defensible. Smith explained that it is the role of boards to establish an audit committee whose responsibilities include the review of a company's internal control system and, where appropriate, risk management systems. A company's system of internal control has a key role in the management of risks that are significant in accomplishing the business objectives (Turnbull).

Jones and Sutherland (1999) explain the benefits of risk management and its ability to make companies more flexible and responsive to market fluctuations. It is now commonly recognised that as part of its accountability to shareholders for the strategic direction of the company and the safeguarding of its assets, the board of directors has ultimate responsibility for ensuring the existence of an appropriate system of internal control and risk management. The responsibilities for a company's system of internal control and risk management have become more formalised over time with the progressive development of corporate governance. Additionally, Turnbull states that a company's objectives, its internal organisation and the environment in which it operates are continually changing. Hence, a sound system of internal control depends on a thorough and regular evaluation of the nature and extent of the risks to which the company is exposed. He argues that as profits are in part the reward for successful risk taking in business, the purpose of internal control is to help manage and control risk appropriately.

6.8 REFERENCES

Cadbury Committee on the Financial Aspects of Corporate Governance (1992) *Report of the Committee on the Financial Aspects of Corporate Governance: The Code of Best Practice*, Gee Publishing, London.

Combined Code (2003) *Combined Code on Corporate Governance,* Financial Reporting Council, July. http://www.fsa.gov.uk/pubs/ukla/lr_comcode2003.pdf

Hampel Committee on Corporate Governance (1998) *Committee on Corporate Governance: Final Report*. Gee Publishing, London. http://www.ecgi.org/codes/documents/hampel.pdf

Higgs, D. (2003) *Review of the Role and Effectiveness of Non-Executive Directors*, published by the Department of Trade and Industry, The Stationery Office.

Jones, E.M. and Sutherland, G. (1999) *Implementing Turnbull, A Boardroom Briefing*. Centre for Business Performance, Institute of Chartered Accountants in England and Wales, London, September.

OECD (2004) *OECD Principles of Corporate Governance*. OECD, Paris.

Smith, R. (2003) *Audit Committees Combined Code Guidance*. Financial Reporting Council, January.

Turnbull, N. (1999) *Internal Control: Guidance for Directors on the Combined Code*. Internal Control Working Party of the Institute of Chartered Accountants in England and Wales, September.

7

Developments in Risk Management in the UK Public Sector

It is always wise to look ahead, but difficult to look further than you can see.
(Winston Churchill)

The previous chapter examined developments in internal control and the relationship between risk and internal control. This chapter examines the developments in risk management in the public sector. Government departments are responsible for services such as the provision of health care and education, protecting the environment, regulating industry and the payment of social services. All involve some degree of risk. Governments have always been concerned with the protection of their citizens from risk. However, it may be argued that they now have to deal with risks from a more diverse range of sources, emanating from the broad spectrum of public services currently provided. But this is no real surprise, as there has been a similar broadening of business risk exposure.

Governments, like multinational mining and petrochemical companies, for instance, have to deal with terrorism, the vulnerability of IT systems and exposure from outsourcing, along with environmental, safety, project and reputational risk. Also, like the increase in expectations of listed company stakeholders, the government would appear to be acutely aware of the increase in citizens' expectations of government performance. In particular, the public has come to expect fewer external risks to health, together with financial and physical security. Yet this is not the full story. Government is aware of the change in the nature of risk. There are two main drivers, developments in science and technology, and global interconnectedness. The government has to make judgements about issues as diverse as cloning, genetically modified food and drugs, mobile phones, nuclear energy and the stability of banks. Additionally, it has to protect its citizens from events emanating from the other side of the world, such as virus attacks on IT networks, diseases carried by travellers, economic downturn and civil unrest.

7.1 RESPONSIBILITY FOR RISK MANAGEMENT IN GOVERNMENT

The clearest picture that would appear to be available of who is responsible for what within government, in terms of risk management, is provided by Figure 7.1. This figure illustrates the roles that the Cabinet Office, HM Treasury and the Office of Government Commerce undertake in the provision of leadership, guidance and advice to government departments, in the implementation of risk management. However, the figure does not mention the cross-government Risk Management Steering Group[1] or the Interdepartmental Liaison Group on

[1] The Risk Management Steering Group was established by the Treasury and the Cabinet Office and chaired by the Treasury. The group included representation from the Treasury, the Cabinet Office, NAO, ILGRA and departments to advise on and facilitate the consistent and coordinated development of policies and guidance relating to risk across central government. It first met in November 2000.

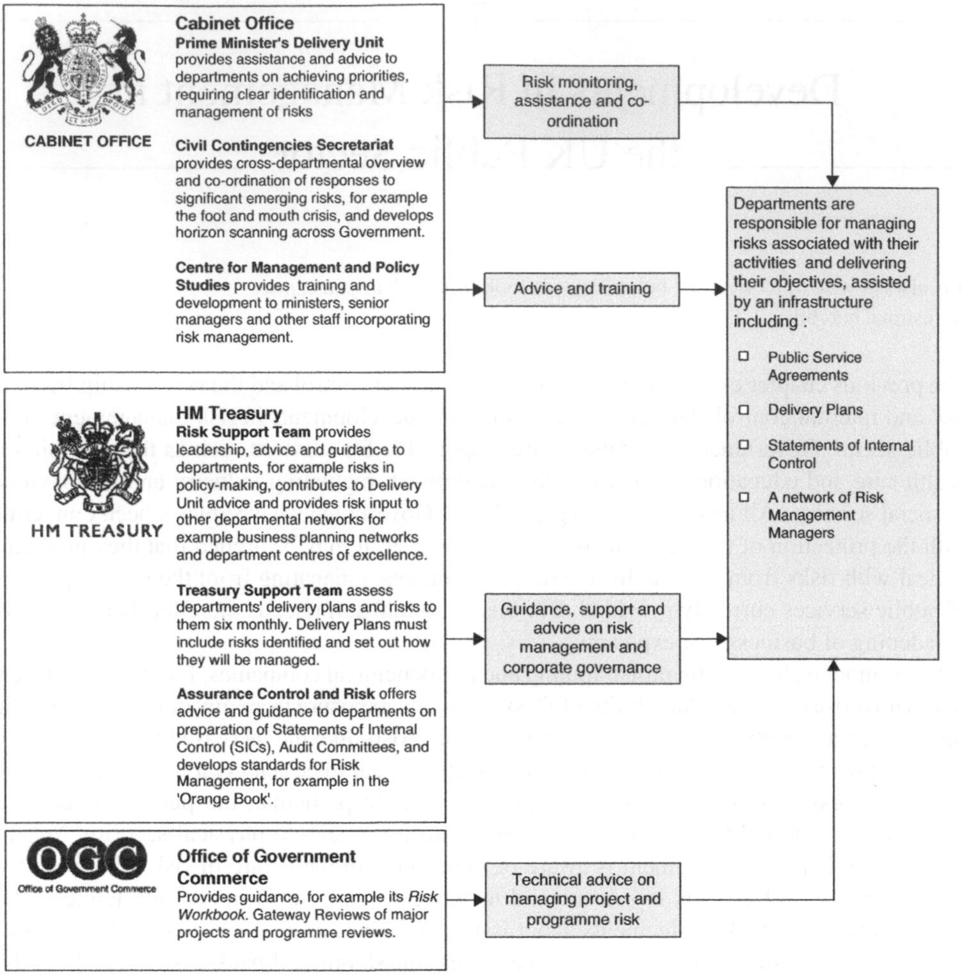

Figure 7.1 Parties responsible for risk management in government (National Audit Office 2004). An earlier version of this diagram appeared in National Audit Office (2000), which also made reference to ILGRA[2]

Risk Assessment (ILGRA), which were active in 2001. To place Figure 7.1 in context, this section gives a description of the primary role of the Cabinet Office, Treasury, Office of Government Commerce and the National Audit Office.

7.1.1 Cabinet Office

The Cabinet Office monitors departments responding to the Modernising Government Action Plan, reports to ministers on progress and is responsible for guidance on the content of

[2] The *Risk Workbook* referred to under the Office of Government Commerce is part of the latter's Successful Delivery Toolkit™ and can be found at http://www.org.gov.uk/sdtoolkit/workbooks/risk/index.html

risk frameworks. In terms of strategy and planning, the Cabinet Office oversees and coordinates policy-making across government, enabling departments to work together to achieve joint objectives. The role of the Cabinet Office's Strategy Unit is to improve policy-making at a strategic level within and between government departments. The Strategy Unit provides the Prime Minister and government departments with the capacity for longer-term thinking, cross-cutting studies and strategic policy work. The Unit was set up in 2002, bringing together the Performance and Innovation Unit, the Prime Minister's Forward Strategy Unit and parts of the Centre for Management and Policy Studies. The unit has three key roles:

- to support the development of strategies and policies in key areas of government in line with the Prime Minister's priorities;
- to carry out occasional strategic audits to identify opportunities and challenges facing the UK and UK government;
- to develop as a "centre of excellence" to enhance strategy across government.

7.1.2 Treasury

The Treasury is responsible for providing guidance to departments on risk management needed to support the production of the annual statement of internal control, by accounting officers. The Treasury, having previously decided to incorporate the principles included within the Turnbull Report (see Chapter 6) into central government, requires departments and other bodies to produce a statement of internal control as part of their annual accounts, the first time being for the financial year 2001/2002. The statement of internal control has to be signed off by the accounting officers. As part of this statement, departments have to report on their risk management processes. To assist departments to develop their risk management processes, the Treasury issued the *Orange Book* (its formal title being *Management of Risk: A Strategic Overview*) in 2001 (HM Treasury 2001), subsequently revised (HM Treasury 2004). The Treasury is also working with departments on the improvement of risk management and internal control as part of the corporate governance agenda.

7.1.3 Office of Government Commerce

The Office of Government Commerce (OGC) is an independent office of the Treasury with its own chief executive appointed at permanent secretary level. The OGC has responsibility for delivering corporate governance across government and providing guidance and advice on risk management, appraisal and evaluation as well as policy for internal audit activity. It plays an important role in the government's efficiency and modernisation agendas. It comprises four service areas and the OGC buying solutions trading arm. The service area labelled "Better Projects" includes gateways, embedding centres of excellence, mission-critical reporting, intervention and support. For all high-risk projects, "Better Projects" mobilises highly experienced review teams, independent of the client department, to provide advice on how to achieve project success. For medium-risk reviews, team leadership and logistics are provided by the OGC and departments adopt the process for their own reviews for low risk projects. "Better Projects" is helping centres of excellence to be able to undertake medium and low-risk reviews for themselves and for their agencies and non-departmental public bodies.

7.1.4 National Audit Office

The National Audit Office (NAO),[3] while not being part of the government, is appropriate to mention here, due to its close involvement with the workings of government and the large number of risk management reports it produces relating to the delivery of public services. As Gus O'Donnell (Permanent Secretary, HM Treasury) stated in the NAO's 2003 annual report:

> Risk management is a critical business process, not a tick in the box exercise. The NAO has made an important contribution to our thinking on how this works in large and complex sector organisations.

Each audit is planned to obtain sufficient, appropriate evidence on which to base the audit opinion. As part of their financial audits they review statements of internal control and, as declared in their 2003 report:

> These statements flow from the recognition that effective risk management lies at the heart of improving organisational performance.

7.2 RISK MANAGEMENT PUBLICATIONS

The UK government has a very chequered history over recent years, in the management of risk, with several highly publicised failures, from the handling of the BSE epidemic to expensive IT failures. As a result, the government has issued a wealth of risk management publications in the last ten years to aid the delivery of services. The central government drive for improvement in the delivery of local government services, the requirement for financial statements to be accompanied by statements on internal control and the achievement of value for money have also led to a series of publications for local authorities. For ease of assimilation the key reports and guides are listed chronologically in Box 7.1.

[3] The NAO is independent of government (including its finances) but it has no corporate status. The NAO's independence is derived from the unique position of the Comptroller and Auditor General (C&AG), the head of the NAO. The role of the C&AG is to report to Parliament on the spending of central government money. The C&AG has ultimate discretion as to his work programme and how it is executed. The budget of the NAO is determined by the legislature on a recommendation from the C&AG. The Public Accounts Commission, a committee of Members of Parliament established in 1983, considers the NAO's plans and budget. The Commission then makes a recommendation to the House of Commons to accept the budget. As a consequence of his office, the C&AG is an officer of the House of Commons, is independent of the executive and the judiciary, and has no relationship with investigating agencies. The staff of the NAO are not civil servants. The role of the NAO is to audit the financial statements of all government departments and agencies and many other public bodies, to provide an independent opinion. As well as providing accountability to Parliament, the NAO aims to bring about real improvements in the delivery of public services. Each audit is planned to obtain sufficient, appropriate evidence on which to base the audit opinion. The C&AG has no powers to disallow expenditure, to impose surcharges or to take punitive action, or to follow public money wherever it goes.

Box 7.1 Risk management guides and reports

• Cabinet Office	*Successful IT: Modernising Government in Action*	2000
• NAO	*Supporting Innovation: Managing Risk in Government Departments*	2000
• HM Treasury	*The Orange Book*	2001
• Audit Commission	*Worth the Risk*	2001
• CIPFA/SOLACE	*Corporate Governance in Local Government – A Keystone for Community Governance: The Framework*	2001
• OGC	*Management of Risk: Guidance for Practitioners*	2002
• DEFRA	*Risk Management Strategy*	2002
• Cabinet Office	*Risk: Improving Government's Capability to Handle Risk and Uncertainty*	2002
• OGC	*Procurement Guide 04 Risk and Value Management*	2003
• HM Treasury	*The Green Book*	2003
• CIPFA	*Guidance on Internal Control and Risk Management in Principal Local Authorities and other Relevant Bodies to Support Compliance with the Accounts and Audit Regulations 2003*	2003
• NAO	*Managing Risks to Improve Public Services*	2004
• HM Treasury	*Management of Risk – Principles and Concepts (The Orange Book – updated)*	2004
• OGC	*Management of Risk: Guidance for Practitioners*	2007
• NAO	*Managing Risks in Government*	2011

These publications provide businesses with another source of information and reference on: sources of risk; procedures; processes; methods of assessment; problems of embedding risk management; allocation of responsibility; and communication. In addition, businesses do not exist in a vacuum. They have to engage with the environment within which they operate. A key component of the structure of the environment is government. The way governments deal with risk is reflected in legislation, regulation and their handling of domestic and global crises. This behaviour directly impacts on businesses.

A number of these publications are examined below to describe the thinking behind the government's approach to risk management, as a backdrop to understanding the development of risk management in the UK and to aid management of risk in the business sector. The following review of the publications is not to provide a detailed critique but an understanding of their focus, drivers, lessons learnt, tools and techniques and to distil issues relevant to business.

7.3 SUCCESSFUL IT

The government, in its commitment to modernising and delivering improvements in public services, wished to harness information technology (IT). The e-government strategy, published in April 2000, set out its commitment to using IT to deliver services in new ways. IT was seen as a tool to deliver services faster, more effectively and in innovative ways. However, it was not finding it easy to implement, and projects were often complex and fraught with risk. The aim of the report *Successful IT: Modernising Government in Action*, published by Cabinet Office in May 2000, was to improve performance by avoiding past mistakes (Cabinet Office

2000). The report states that, in the past, government IT projects had too often missed delivery dates, run over budget or failed to fulfil requirements. The report was based on evidence from extensive research undertaken in the UK public and private sectors and abroad, which showed that there were a great many reasons why failures occurred. The report concluded that these failures could not be addressed by one or two catch-all measures and, accordingly, the report made a series of recommendations.

The review of projects found that the quality of risk management varied widely across government with its application ranging from simple lists (without ownership of risks or actions to mitigate them), to the allocation of full-time risk managers with comprehensive risk registers. Some of the reasons for poor risk management were considered to include those listed in Box 7.2.

Box 7.2 Reasons for poor risk management (Cabinet Office 2000)

- Having a narrow focus looking only at the inward-facing project risks that are tangible and within the project manager's control, without considering risks to the organisation's business as a whole
- Relying too much on tabulating numerous risks in a register without prioritising them or considering the extent to which they may be correlated with each other
- Failing to understand that the ultimate risks of not meeting the business objectives or realising the business benefits, or ending up with an unsatisfactory delivery of services to the public, cannot be transferred to a partner or supplier
- Failing to understand or define the boundary between the responsibilities of the supplier and the purchasing department or agency
- Depending on the contract or its penalty clauses to mitigate risk rather than taking action or forming effective contingency plans
- Failing to monitor the effectiveness of mitigating action and contingency plans, or to refer risks, which fall outside of tolerance, to the appropriate level, in good time.

Additionally the report concluded that there was evidence of:

- a failure to take end users' needs into account;
- inadequate guidance to ensure good risk management;
- a lack of support for projects at the highest level, in terms of an awareness of the importance of business risk management;
- a missing link between the effective analysis and management of risk;
- inadequate reporting and upward referral of problems inundating senior management with too much detail; and
- the omission of tolerance levels for cost, time and functionality being set at the start of the project, beyond which the project manager cannot go without seeking approval.

The report recommends the use of a *project profile model* and a *summary risk profile* to aid risk management. The former is described as being used to determine the risk profile and corresponding risk strategy of a project, whereas the latter is a simple mechanism to increase the visibility of the risks and facilitate the prioritisation of risk management action. Both are described in Appendix 1 of this book.

7.4 SUPPORTING INNOVATION

The aim of the report entitled *Supporting Innovation: Managing Risk in Government Departments* (NAO 2000) is to (1) promote improvements in risk management by departments, by identifying examples of good practice (for both public and the private sector), and (2) convey the findings of a survey conducted to provide an overview of the extent and practice of risk management across organisations responsible for the delivery of public services. The report sets out why risk management is important, how well risk management is understood and implemented by departments, agencies and non-departmental bodies, which for the purposes of the report are collectively described as "departments", and what more needs to be done to improve risk management.

7.4.1 Part 1: Why Risk Management is Important

The report suggests that risk management can lead to better service delivery, more efficient use of resources, better project management, can help minimise waste, fraud, poor value for money and promote innovation. The report also advises that the reputation of departments can suffer when services fail to meet the public's expectations.

7.4.2 Part 2: Comprehension of Risk Management

Understanding. While respondents agreed that risk management was important to the achievement of their objectives, they expressed their lack of awareness of how it could address the risks that threaten the delivery of services.

How it is implemented. Thirty-eight per cent of departments did not routinely assess risks. The most commonly identified risks were financial, project, compliance and reputational.

Actions implemented by departments. While departments stated they were managing risks, the absence of early warning indicators for alerting senior managers to changing risks and regular reports to senior management, it was thought, might have meant that key risks were not being identified or identified too late for effective action.

7.4.3 Part 3: What More Needs to be Done to Improve Risk Management

The survey found acceptance and recognition of the importance of risk management but less certainty over implementation. This was being addressed in a number of ways. Examination through case studies of good practice suggested that six essential requirements needed to be in place if risk management was to be effective (Box 7.3).

Box 7.3 Six essential requirements for effective risk management

1. Risk management policies and the benefits of effective risk management should be clearly communicated to all staff.
2. Senior management need to support and promote risk management.
3. The department's culture should support well-thought-through risk taking and innovation.

4. Risk management should be embedded in management processes.
5. The management of risk should be closely linked to the achievement of objectives.
6. Risks associated with other organisations should be assessed and managed.

7.5 THE ORANGE BOOK

The first edition of the *Orange Book* (HM Treasury 2001) declares that its aim is to provide some pointers to the development of a strategic framework for the organisational consideration of risk. Additionally, it describes tools and techniques which may be adopted by organisations to guide them in the development of their risk management processes across the spectrum of risk encountered in day-to-day business. This publication, and its successor, are widely referred to in other government publications and provide an introduction for those new to the subject. It describes a process that had been in use in the private sector for some considerable time. It describes the risk management process as a cycle composed of the following steps:

- identify the risks and define a framework;
- evaluate the risks;
- assess the appetite;
- identify suitable responses to risk;
- gain assurance about the effectiveness;
- embed and review.

7.5.1 Identify the Risks and Define a Framework

The *Orange Book* describes three important principles for analysing risk, which are universally relevant: (1) adopting a consistent approach throughout the organisation; (2) ensuring that there is a clear structure to the process; and (3) establishing a framework approved at senior level. It goes on to say that "a strategic approach to risk management depends on identifying risks against key organisational objectives". Interestingly, it says that framework implementers have found it beneficial to confine the objectives to a small number. Certainly with too many objectives the process becomes unwieldy and difficult to manage.

It says that in order to identify risk it is appropriate to adopt an appropriate tool and that the two most commonly used tools are (1) commissioning a risk review and (2) risk self-assessment. The first tool is described as the formation of a team, which supports conducting a series of interviews with key staff to identify the risks to the operations and activities undertaken to achieve its objectives. The second tool is described as a bottom-up approach where each level and part of the organisation is invited to identify the risks to its activities through a questionnaire or a facilitated workshop approach. These "tools" are common ways of identifying risk and could more readily be described as processes. For a summary of the most common categories of risk and the risk self-assessment tool, see Table A2.6 in Appendix 2.

7.5.2 Assign Ownership

The book describes risk management being most effective when ownership of risk is allocated to an appropriate senior official. It describes alternative means of risk ownership to suit different organisation structures such as allocation to (1) the risk management committee members reporting to an accounting officer, (2) board members or (3) executive members of the audit committee. It states that regardless of the means of allocation, a mechanism must be established to report to the accounting officer who has ultimate reporting responsibility and that committee membership does not diminish a member's individual risk management responsibilities.

7.5.3 Evaluate

The experience of the authors permeates the text when it says that evaluation is important to prioritise risk and that while some risks such as financial risks lend themselves to numerical assessment, reputational risk, say, can only be assessed subjectively. In addition, this experience is evident when it declares that when creating models to evaluate risks in combination, it is often necessary to undertake an iterative process in their development, particularly when the initial results do not withstand scrutiny.

7.5.4 Assess Risk Appetite

As part of an overall risk strategy, it is important to understand an organisation's risk appetite when considering response to risk. The *Orange Book* defines risk appetite as "the amount of risk to which the organisation is prepared to be exposed before it judges that action is necessary". It goes on to say that a balance has to be struck to accomplish value for money when considering the degree of risk exposure and the cost of the risk response.

7.5.5 Response to Risk

The four common responses are described as transfer, tolerate, treat and terminate in lieu of, say, transfer, retain, reduce and remove. The term "terminate" is used here to describe the situation where a risk is too high to be acceptable to the organisation and so the activity that is generating the risk has to be stopped or terminated. In the private sector the term "remove" relates to undertaking an activity in another way or not undertaking it at all. The term "treat" is described as implementing an "internal control", an action undertaken from within the organisation, designed to contain the risk to an acceptable level. "Control" is described as "any action, procedure or operation undertaken by management to increase the likelihood that activities and procedures achieve their objectives". Four types of control are described as follows:

- detective controls – to identify if undesirable outcomes have occurred;
- directive controls – to ensure that a particular outcome is achieved;
- preventive controls – to limit the possibility of an undesirable outcome being realised;
- corrective controls – to correct undesirable outcomes which have been realised.

A proportionality of control is recommended where the control put in place is proportional to the risk. Other than matters which relate to human life, it is suggested that it is normally sufficient to design a control to give a reasonable assurance of confining likely loss within the risk appetite of the organisation.

7.5.6 Gain Assurance

The book suggests that assurance is accomplished by reporting and internal audit. In addition, reporting enables senior management to understand the effectiveness of risk management, whereas internal audit is described as being required to provide assurance on the adequacy of the embedded risk and control mechanism.

7.5.7 Embed and Review

The case is made that the objectives within an organisation are hierarchical, cascading down from directorates to divisions to teams and hence that responsibility for risk management should also be hierarchical, reflecting the same structure. Following the same theme, a parallel delegation of risk responsibility should exist at each level of objectives. The risk management process should be intrinsic to the way an organisation operates and incorporated in day-to-day activities. Risk management should be dynamic, and regularly updated to reflect changes within the environment within which the organisation operates, to reflect corporate governance requirements.

7.6 AUDIT COMMISSION

The Audit Commission[4] is an independent body responsible for ensuring that public money is spent economically, efficiently and effectively, to achieve high-quality local and national services for the public. Its remit covers more than 15 000 bodies, which between them spend nearly £125 billion of public money every year. Its work covers local government, housing, health, criminal justice and fire and rescue services. It describes itself as both an independent watchdog, providing important information on the quality of public services, and a driving force for improvement in those services, providing practical recommendations and sharing best practice.

The Audit Commission's publication *Worth the Risk* was written to address either the absence or minimal formal activity within councils to evaluate and manage risk (Audit Commission 2001). It aims to raise the awareness about the need to address key strategic risks and to provide good practice guidance for councils to manage such risks in a more effective and formalised way. The paper considers that while formal systems of risk management were being established across all parts of the private and public sectors, risk management developments in local government were totally dependent on initiatives taken by individual authorities rather than as a sector-wide response. As a consequence the Commission considered local government to be in danger of falling behind best practice. The paper describes its aim as being to help local government bodies in England and Wales improve the way in which they identify, evaluate and manage significant risks. Additionally, the paper is aimed at helping local government members and officers assess whether their current risk management activities are satisfactory and are developing in line with the best value initiative. The paper states that it is

[4] As an independent auditor, the Commission monitors spending to ensure public services are good value for money. Its mission is to be a driving force in the improvement of public services. It promotes good practice and assists those responsible for public services to achieve better outcomes for members of the public, with a focus on those people who need public services most.

primarily written for elected members and officers of local government bodies in England and Wales. The paper is composed of five chapters: the first provides background to the practice of risk management, the second looks at the application of risk management to local government, the third and fourth examine the relevance to members and officers respectively, and the last proposes pitfalls to be aware of.

Interestingly the Commission finds it necessary to place risk management in context with governance and internal control. It defines governance as "the system by which local authorities direct and control their functions and relate to their communities. In other words, the way in which organisations manage their business, determine strategy and objectives and go about achieving those objectives. The fundamental principles are openness, integrity and accountability." It defines internal control as "those elements of an organisation (including resources, systems, processes, culture, structure and tasks) that, taken together, support people in the achievement of business objectives. Internal financial systems form part of the wider system of internal controls." In addition, the Commission states that an authority's system of internal control is part of its risk management process and has a key role to play in the management of significant risks to the fulfilment of its business objectives.

Risk management is seen by the Commission as an integral part of good governance and the constituents of good risk management are seen to be in evidence when:

- there is a shared awareness and understanding within the authority;
- there is regular and ongoing monitoring and reporting of risk, including early warning mechanisms;
- an appropriate assessment is made of the cost of operating particular controls relative to the benefit obtained in managing the related risk;
- the authority conducts, at least annually, a review of the effectiveness of the system of internal control in place; and
- the authority reports publicly on the results of the review, and explains the action it is taking to address any significant concerns it has identified.

The report includes a number of case studies. Case study 2 relates to the application of risk management by Liverpool City Council to its main areas of concern, which include:

- an excessive number of committees and member groups;
- a failure to produce effective corporate plans;
- poor communications, especially with the workforce;
- poor-quality high-cost services;
- disengagement from local people;
- weak or non-existent corporate and strategic management as a result of chronic departmentalism; and
- hostility and mutual distrust between councillors and senior officers.

In response the council produced a strategic framework of 42 risks, which were prioritised with control strategies being developed to address the higher-order risks.

7.7 CIPFA/SOLACE CORPORATE GOVERNANCE

In 2001 the Chartered Institute of Public Finance and Accountancy (CIPFA),[5] in conjunction with the Society of Local Authority Chief Executives and Senior Managers (SOLACE),[6] produced a framework for use by local authorities to review their existing corporate governance arrangements and to prepare and adopt an up-to-date local code of corporate governance. This guidance, called *Corporate Governance in Local Government – A Keystone for Community Governance: The Framework*, is intended to be followed as best practice for establishing a locally adopted code of corporate governance and for making adopted practice open and explicit (CIPFA/SOLACE 2001). The framework uses the terms "principles", "dimensions", "local codes" and "elements" within the text, and it would have been helpful if these had been explained at the outset, together with guidance on how they relate to each other. The framework states that authorities must be able to demonstrate that they are complying with the underlying principles of good governance (openness and inclusivity, integrity and accountability) by translating them into a framework which seeks to ensure that they are fully integrated into the conduct of the authority's business. The framework is subdivided into four sections:

1. Introduction to corporate governance in local government
2. Framework for a code of corporate governance for local authorities
3. The elements of corporate governance
4. Annual review and reporting.

The guidance argues in section 2 that the fundamental principles of corporate governance need to be reflected in the five different dimensions of a local authority's business (which I interpret as the aims or goals). These dimensions are described as community focus, service delivery arrangements, structures and processes, risk management and internal control and standards of conduct.

Of interest here is dimension 4, "risk management and internal control", which states that an authority needs to establish and maintain a systematic strategy, framework and process for managing risk (again without spelling out what these terms are intended to mean). Together, the framework says, these arrangements should:

- include making public statements to stakeholders on the authority's risk management strategy, framework and process to demonstrate accountability;
- include mechanisms for monitoring and reviewing effectiveness against agreed standards and targets and the operation of controls in practice;
- demonstrate integrity by being based on robust systems for identifying, profiling, controlling and monitoring all significant strategic and operational risks;
- display openness and inclusivity by all those associated with planning and delivering services, including partners; and

[5] CIPFA is one of the leading professional accountancy bodies in the UK and the only one that specialises in the public services. It is responsible for the education and training of professional accountants and for their regulation through the setting and monitoring of professional standards. Additionally CIPFA provides courses, conferences and publications and a range of advice, information, training and consultancy services. It is a membership organisation with more than 15 000 members and is part of the accountancy profession within the UK and internationally. It is a key stakeholder in the public services where governments across the world are seeking to engineer major reforms.

[6] SOLACE is the representative body for senior strategic managers working in local government. Like other vocational organisations, its members are drawn from a variety of professional backgrounds.

- include mechanisms to ensure that the risk management and the control process is monitored by continuing compliance to ensure that changes in circumstance are accommodated and that it remains up to date.

Section 3 of the framework calls for local authorities to develop local codes of corporate governance, which comprise the following elements (which I interpret to be the activities). This section uses the same headings as section 2, commencing with community focus again. The element entitled "risk management and internal control" lists what I call the activities to be undertaken to satisfy this framework:

- develop and maintain robust systems for identifying and evaluating all significant risks which involve the proactive participation of all of those associated with planning and delivering services;
- put in place effective risk management systems, including systems of internal control and an internal audit function – these arrangements need to ensure compliance with all applicable statutes, regulations and relevant statements of best practice and need to ensure that public funds are properly safeguarded and used economically, efficiently and effectively and in accordance with the statutory and other authorities that govern their use;
- ensure that services are delivered by trained and experienced people;
- put in place effective arrangements for an objective review of the effectiveness of risk management and internal control, including internal audit;
- maintain an objective and professional relationship with external auditors and statutory inspectors; and
- publish on a timely basis, within the annual report, an objective, balanced and understandable statement and assessment of the authority's risk management and internal control mechanisms and their effectiveness in practice.

In the final section of the framework, section 4, it states that every local authority should publish a statement annually in its financial statements on how it is complying with the principles set out in the framework and how it is complying with its own local code of corporate governance. In section 1 of the framework it repeats this same statement, but also adds that arrangements should be made by authorities for their local code of governance to be in place by 31 March 2002.

7.8 M_o_R 2002

The full title of this publication is *Management of Risk: Guidance for Practitioners*, though it was branded by the authors, the OGC IT Directorate, as M_o_R (Office of Government Commerce 2002). The guide declares that its purpose is to help organisations to put in place effective frameworks for taking informed decisions. It is subdivided into eight chapters. Following an introduction, chapter 2 describes the key principles underpinning risk management and chapter 3 the management of risk. Chapters 4–7 describe managing risk at the strategic, programme, project and operational levels respectively. Each of these four chapters includes the common headings of: types of risk, where to apply risk management, when to do it, who is involved and policy for risk management. The final chapter, chapter 8, discusses the range of techniques available to support the risk management process. A series of annexes provide supporting information.

Chapter 2 examines where risk occurs in an organisation in terms of decision making and splits decision making into four types or levels: strategic or corporate, programme, project

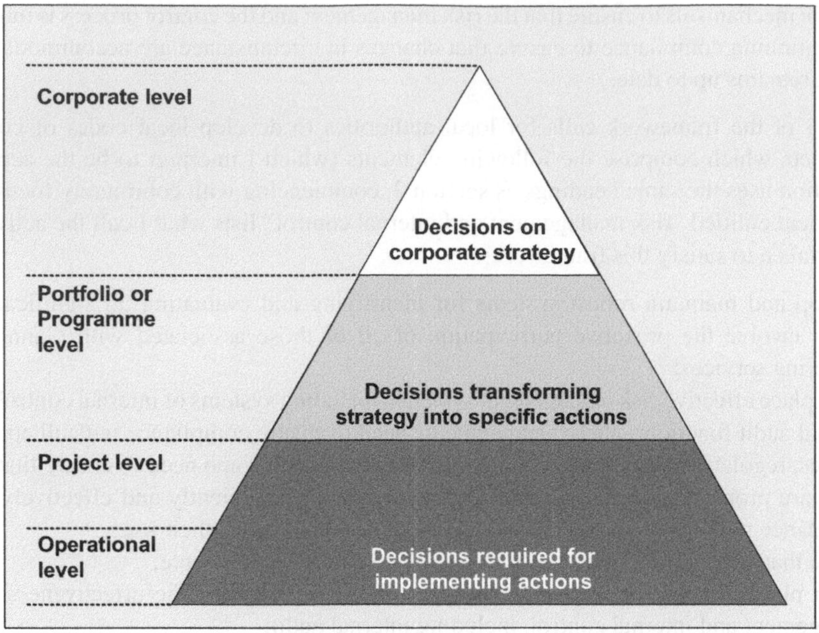

Figure 7.2 Decision making within the management hierarchy of an organisation

and operational (Figure 7.2). The guide correctly makes the point that a risk may materialise initially in one level but subsequently have a major impact at a different level.

Chapter 8 describes a series of techniques that can be used to support the management of risk, which are reproduced in Box 7.4. The guide makes the observation that experience in managing risk is a more critical factor for success than the choice of tools and techniques.

Box 7.4 Consequence categories

Strategic/corporate level	Programme level	Project level	Operational level
NPV (net present level)	Decision trees	Simulations	Simulations
IRR (internal rate of return)	CPA (critical path analysis)	LCC	LCC
ROI (return on investment)	Cost/benefit analysis	Decision trees	Performance analysis
Cash flow analysis	Sensitivity analysis	Risk tables	Reliability analysis
Currency analysis	Stakeholder risk analysis	PERT (Programme Evaluation and Review Technique)	Queuing analysis
SWOT analysis	Simulations	Performance analysis	Algorithm analysis
Scenarios	Scenarios	Reliability analysis	Capability analysis
Cost/benefit analysis	LCC (life cycle analysis)	Capability analysis	Top-down analysis

Strategic/corporate level	Programme level	Project level	Operational level
Decision trees		Monte Carlo simulation	HAZOP (HAZard OPerability, analysis, risk registers and databases)
CRAMM[7] for business impact security requirements		Influence diagrams CRAMM	CRAMM

7.9 DEFRA

The Department of the Environment, Food and Rural Affairs (DEFRA) is a large department within the UK government and, as it declares in its five-year strategy (HM Government 2004), "brings together the interests of farmers, the countryside, the environment and the rural economy, the food we eat, the air we breathe and the water we drink". Hence, the breadth of subjects the department has to address is vast, from reducing carbon emissions, via emergency planning to address flooding and animal disease (such as BSE),[8] to agricultural policy. It has a large agenda of regulation that seeks to improve the environment, protect public health and deliver high standards of animal welfare.

7.9.1 Risk Management Strategy

DEFRA in 2002 published its *Risk Management Strategy* (DEFRA 2002). While the approach is now very familiar in terms of the stages of risk management and the risk response categories adopted, careful thought has been given to what risk management means for the department. Consideration has been given to the sources of risk, the consequences should it materialise, how risk management will be embedded in the organisation and the potential benefits. In accordance with the foreword to the strategy written by Brian Bender (Permanent Secretary), the department required a clear understanding of how the risks facing it should be managed. Doing this properly, Brian Bender declared, "is central to planning to succeed and avoiding failure; to meeting our key objectives and targets; to create confidence in a watchful public; and meeting the demands of good corporate governance". Significantly Bender declared: "doing this needs to be a living process, not a 'tick-box' exercise". The strategy is divided into six

---

[7] In 1985 the UK government's Cabinet Office tasked the Central Computer and Telecommunications Agency (CCTA) with investigating the risk analysis and management methods currently in existence within central government for information security. Following this investigation, a new method was developed by the CCTA, which drew upon all of the existing best practices under the title of CCTA Risk Analysis and Management Method (CRAMM). In 1987 a software tool was launched that automated the CRAMM process. Due to its success within government, a commercial version of CRAMM soon followed in 1988 that was tailored specifically to the needs of commerce. At the same time, an independent users group was formed which continues to provide a valuable forum for users to meet and help shape the development of the method. Insight Consulting currently manages CRAMM on behalf of the Crown.

[8] Bovine spongiform encephalopathy (BSE), known as "mad cow disease", is a relatively new disease of cattle, first recognised and defined in the UK in 1986. The disease reached epidemic proportions, peaking in 1992, after which time was a steady decline. It is a neurological disease which lasts for several weeks and is invariably progressive and fatal. BSE is one of the family of transmissible spongiform encephalopathies. Creutzfeldt–Jakob disease is an example of the spongiform encephalopathies found in humans. Like the animal disorders, they are progressive and universally fatal. Variant CJD (vCJD) is distinguished from other forms of CJD by its symptoms, duration of illness, age of onset and other characteristics.

sections: (1) introduction and purpose; (2) aim, principles and implementation; (3) identifying risks; (4) assessing risks; (5) addressing risks; and (6) reviewing and reporting risks. Each of these areas is discussed in Appendix 3, as they describe implementation issues pertinent to establishing and embedding risk management systems within businesses. To ignore processes and experience of government is to deprive businesses of useful and pertinent background knowledge.

7.10 STRATEGY UNIT REPORT

In July 2001, the then Prime Minister, Tony Blair, announced the Strategy Unit[9] study on risk and uncertainty. A clear aim of the study and the accompanying report was to improve government's handling of risk. The study was concerned that there was a danger that risk management would be seen as a mechanical process with the potential for real issues being missed. The report arising from the study was called *Risk: Improving Government's Capability to Handle Risk and Uncertainty* (Cabinet Office 2002). Of all the government reports on the subject of risk management produced, this report is one of the most lucid, erudite and well researched. In his foreword, Tony Blair makes four points which will strike a chord with most organisations that have suffered adverse risk events and have grappled with embedding risk management. The report states that the desire of government is to:

- do more to anticipate risks so that there are fewer unnecessary and costly crises (citing BSE and failed IT contracts as examples);
- ensure that risk management is part of delivery plans;
- get the right balance between innovation and change on the one hand and avoidance of shocks and crises on the other;
- improve the management of risk and its communication.

While one of the catalysts for the study was the Phillips Report (Phillips 2000) on BSE, the study was established with a remit to look broadly across the whole of government's involvement in managing risk. The report produced by the study made recommendations for government action to improve its handling of risk. It built on a number of reports such as those issued by the National Audit Office (2000) and the Public Accounts Committee (2001), which highlighted the need for improvement. The study did not seek to provide detailed technical advice on how to undertake risk management but rather to develop a broad framework for understanding risk, managing risk and risk allocation, together with proposals for organisational coherence and cultural change. The study, while recognising that at the heart of the handling of risk is judgement, strove to explore where analysis should end and judgement commence. With some joined-up thinking, the report (in the main) adopts the risk management language contained within the government OGC guidance (Office of Government Commerce 2002). It does, however, make some departures, an example being the term "handling risk" which embraces not only risk management but also wider issues such as the government's approach, its roles and responsibilities and its organisational culture. The report is structured into five areas: (1) the government's role and responsibilities, (2) improving government's handling of risk, (3) improving capacity, (4) handling the communication of risk and (5) the role of leadership and culture change. Each of these areas is discussed in Appendix 4, as

[9] The Strategy Unit was formerly known as the Performance and Innovation Unit.

there are strong parallels between the problems encountered by governments and businesses in embedding risk management.

7.11 RISK AND VALUE MANAGEMENT

Procurement guide 04 of the Achieving Excellence in Construction series, entitled *Risk and Value Management*, explains how risk and value are managed (Office of Government Commerce 2003). These processes, according to the guide, are fundamental to the successful delivery of projects and are recommended for use throughout the life of a project. This short guide summarises the key principles of risk and value management in the context of construction projects and describes the practical steps to be undertaken. The guide provides key messages on the management of risk which are included in Box 7.5.

Box 7.5 Key messages about managing risk (Office of Government Commerce 2003)

- Process: a common risk management process should be understood and adopted at all levels within the integrated project team;
- Register: the risk register is regularly reviewed and updated throughout the project life cycle;
- Maintenance and demolition: risks inherent in the maintenance and demolition of the facility should be considered during design development and the decision about risks kept on the register for future reference;
- Facilities management: the FM risks should be considered and owned early on – usually by the client;
- PLC: the project life cycle runs until the facility is demolished or disposed of (whoever acquires it in a disposal will need to know the risks on transfer);
- Time: there must be adequate time and effort early on to identify and analyse the risks and to develop a risk management plan governing how they will be managed and funded (the calculation of the risk allowance);
- Allocation: risks should be allocated to individual risk owners within the integrated project team who should fully understand the risks for which they are responsible;
- Commitment: clients should not make any financial commitment to a project or a major change, unless the integrated project team has identified and assessed the risks, allocated them and ensured that management action is in place;
- Ongoing management: the risks should be managed actively throughout the life of the project in accordance with the risk management plan;
- Retention: the plan should deal with all risks, whether retained by the client or transferred to others in the integrated project team;
- Business case: the business case should include a time element and the risks of that changing should be kept under review; and
- Procurement: risk management and the procurement route are interrelated. Risk allocation should be considered a part of the procurement route, as different routes will entail differing degrees of risk transfer.

In addition, the guide provides guidance on the timing of the application of risk management (in terms of the OGC gateway process), describes the stages of the risk management process (common to other publications) and suggests a traffic light probability impact risk matrix for scoring risks. Risk responses are described using the headings avoidance, reduction, transfer and retention/acceptance.

Risk feedback. The guide recommends that feedback should be encouraged from all those involved in the delivery of the project on how well risks were managed, and how this could be improved. This information can be used to improve risk management performance in future projects and should be part of the post-project review.

Project execution plan. The guide recommends that the project execution plan should include the risk register and the risk management plan.

Risk allowance. The guide explains that the budget (for a project) should be composed of two elements of cost – the estimate and risk allowance. This risk allowance it considers should be included in the budget to cover the potential financial impact of the client's retained risks as estimated in the risk analysis. Additionally, it should not be based on an arbitrary percentage of the total project budget, but be costed out as accurately as possible. Also this allowance should be used exclusively to mitigate the impact of identified risks.

7.12 THE GREEN BOOK

The *Green Book* is published by HM Treasury, and its formal title is *Appraisal and Evaluation in Central Government* (HM Treasury 2003). It declares that its purpose is to ensure that no policy, programme or project is adopted without first having the answer to the questions "are there better ways to achieve this objective?" and "are there better uses for these resources?". The *Green Book* aims to be a best practice guide and make the appraisal process throughout government more consistent and transparent. The book is directed at, as it declares: "anyone required to conduct a basic appraisal or evaluation of a policy, project or programme; and people seeking to expand their knowledge in this area". It is subdivided into seven chapters and after an introduction it provides an overview of appraisal and evaluation, the need to establish the requirement for government intervention in a market, the setting of objectives, and appraisal of options, the development and implementation of a solution, and evaluation.

7.12.1 Optimism Bias

What might be considered to be unique to the *Green Book* is the subject of "optimism bias". In chapter 5 the authors address the subject of adjusting for bias and risks. The authors state that there is a demonstrated systematic tendency for project appraisers to be overly optimistic and that this is a worldwide phenomenon affecting both the private and public sectors. It is considered that optimism is not just confined to initial capital expenditure, but also extends to benefits, time and operating costs. To address this tendency it is considered that appraisers should make explicit adjustments for this bias. This, the book advises, is accomplished by increasing estimates of the costs together with decreasing and delaying the receipt of estimated benefits. The authors state that these adjustments should be empirically based using data from past projects or similar projects elsewhere, taking account of the unique characteristics in hand. Where past data is not available, a separate publication is referred to (Mott MacDonald 2002). Due to projects commonly being substantially different and there being a distinct shortage of information on completed projects, data to inform optimism bias is in short supply. Attempts to extract meaningful data from project records are commonly hampered by "project morphing" (the project you start with is not the project you finish with), changes in personnel and poor record keeping. Hence it is unfortunate that the proposed method of calculation is not included in the *Green Book* for ease of reference. This subject is explored in more detail in Section 18.10 and Appendix 16.

7.12.2 Annex 4

In annex 4, entitled "Risk and Uncertainty", the *Green Book* provides guidance on risk management, transferring risk, optimism bias, Monte Carlo analysis, irreversibility and the cost of variability in outcomes. This annex provides more information on the subject of optimism bias. Optimism bias, it explains, is "the demonstrated systematic tendency for appraisers to be over-optimistic about key project parameters". The book calls for optimism bias to be accounted for explicitly in all appraisals and states it can arise in relation to: capital costs; works duration; operating costs; and under-delivery of benefits. The authors consider that the two main causes of optimism bias in estimates of capital costs are:

- poor definition of the scope and objective of projects in the business case, due to poor identification of stakeholder requirements, resulting in over omission of costs during project costing; and
- poor management of projects during implementation, so that schedules are not adhered to and risks are not mitigated.

The authors suggest that appraisers should adjust for optimism bias in the estimates of capital costs in the following way:

- Estimate the capital costs of each option.
- Apply adjustments to these estimates, based on the best available empirical evidence relevant to the stage of the appraisal.
- Subsequently reduce these adjustments according to the extent of confidence in the capital costs' estimates, the extent of management of generic risks and the extent of work undertaken to identify and mitigate project specific risks.

To minimise the level of optimism bias in appraisal, best practice guidance (Mott MacDonald 2002) suggests that the following actions should be taken:

- Project managers, suitably competent and experienced for the role, should be identified.
- Project sponsor roles should be clearly defined.
- Recognised project management structures should be in place.
- Performance management systems should be set up.
- For large or complex projects: simpler alternatives should be developed wherever possible; consideration should be given to breaking down large, ambitious projects into smaller ones with more easily defined and achievable goals; and knowledge transfer processes should be set up, so that changes in individual personnel do not disrupt the smooth implementation of a project.

7.13 CIPFA GUIDANCE ON INTERNAL CONTROL

To compensate for the lack of guidance on the implementation of internal control, CIPFA published *Guidance on Internal Control and Risk Management in Principal Local Authorities and Other Relevant Bodies to Support Compliance with the Accounts and Audit Regulations 2003* (CIPFA 2003). The publication provides an explanation of what internal control is, why it is required, when it should be implemented, how it should be implemented and by whom it should be implemented.

The publication explains why internal control is required as a matter of regulatory compliance. The introduction states:

> The purpose of this guidance is to outline the elements needed for the establishment, maintenance and review of a system of internal control and risk management in local authorities in the context of the requirements of the Audit Regulations 2003, to establish proper practices for the publication of a statement of internal control . . . and to provide a proforma statement on internal control.

Specifically the guidance responds to Regulation 4 of the Accounts and Audit Regulations 2003, which required from 1 April 2003 that:

> The relevant body shall be responsible for ensuring that financial management of the body is adequate and effective and that the body has a sound system of internal control which facilitates the effective exercise of that body's functions and which includes arrangements for the management of risk.

This statement explains the relationship between internal control and risk management, in that risk management is a subset of internal control.

The guidance explains when internal control is implemented by again referring to the Accounts and Audit Regulations 2003, which state:

> The relevant body shall conduct a review at least once a year of the effectiveness of its system of internal control and shall include a statement of internal control, prepared in accordance with proper practices, with (a) any statement of accounts it is obliged to publish in accordance with Regulation 11, or (b) any income and expenditure account, statement of balances or record of receipts and payments it is obliged to publish in accordance with Regulation 12

It is understood that this stipulation on frequency emanates from the recognition that an organisation's external environment never stands still and is constantly evolving, as described by the Turnbull Report.

As to what internal control is, the guidance describes internal control as being a series of controls designed to ensure:

- the authority's policies are put into practice;
- the organisation's values are met;
- laws and regulations are complied with;
- required processes are adhered to;
- financial statements and other published information are accurate and reliable; and
- human, financial and other resources are managed effectively and efficiently.

The guidance explains how to set up a system of internal control, referring the reader to *Effective Internal Control – A Framework for Public Service Bodies*, published by CIPFA in 1994.

Turning to the question of who should implement internal control, the guidance describes each local authority being responsible for establishing, maintaining and reviewing its own system. In practice each local authority is likely to take assurance from the work of internal audit. The Code of Practice for Internal Audit in Local Government in the United Kingdom (the Code) defines internal audit as:

> an assurance function that primarily provides an independent and objective opinion to the organisation on the control environment, comprising risk management, control and governance by evaluating its effectiveness in achieving the organisation's objectives. It objectively examines, evaluates and reports on the adequacy of the control environment as a contribution to the proper, economic, efficient and effective use of resources.

While internal audit looks after the system of internal control, it is common for aspects of risk management to be devolved to different parties:

- risk identification is delegated to a corporate risk management group;
- individual managers are assigned the responsibility of managing individual risks;
- members of the appropriate member committee establish procedures whereby they can attest that the local authority has "a sound system of internal control"; and
- the head of internal audit has overall responsibility for the system of internal control (including risk management).

Additionally, in terms of the responsibility for Standard 9 ("Reporting") of the Code, the head of internal audit is required to include in the annual internal audit report to the local authority an opinion on the overall adequacy and effectiveness of the authority's control environment, providing details of any weaknesses that should be considered in the preparation of the statement of internal control.

7.14 MANAGING RISKS TO IMPROVE PUBLIC SERVICES

The declared purpose of *Managing Risks to Improve Public Services* (National Audit Office 2004) was to assess the progress which government departments had made since the previous survey results recorded in *Supporting Innovation* (National Audit Office 2000). It focuses in particular on the resilience of departments' risk management to prevent adverse impacts on service delivery or value for money. The report findings were based on a survey of the 20 main Whitehall departments, three departmental focus groups, comparisons with private sector organisations and five case studies of government organisations. The general conclusion was that, while significant progress had been made, departments had further to go in demonstrating that they had made effective risk management a central part of their day-to-day general management processes that would deliver improved performance. Of interest here is that while this publication does not focus on tools and techniques it does describe through case study evidence how departments can secure the benefits of risk management in practice (and avoid it being seen as purely an administrative process). The report considers that good risk management has four key benefits: delivering better public services, improving efficiency, making more reliable decisions, and supporting innovation. Extracts from the case studies, which illustrate benefits derived from risk management, are included in Box 7.6.

Box 7.6 Case study evidence: how risk management can deliver tangible benefits

Department	Problem	Risk management solution
BENEFIT 1: Deliver better public services		
• HM Customs and Excise	A series of high profile High Court trials in which prosecutions collapsed due to mistakes and omissions in procedure.	Customs and Excise created a new programme of professional standards training to reduce the risk of officers making costly mistakes. The aim is to maximise the likelihood of a conviction by ensuring that when intercepting smuggled goods Customs Officers follow precise legal rules and procedures.
• Prescription Pricing Authority	Pharmacists send prescriptions monthly following dispensing to the Prescription Pricing Authority (PPA), which calculates and authorises payments. A postal dispute could cause financial hardship for the small pharmacy business whose cash flow is dependent on payments from the PPA.	A partial response to the PPA secured a contract with an alternative provider of collection and delivery services to help ensure that dispensers would receive prompt payments in the event of postal disruption. The PPA is also working towards e-prescribing, now included as part of the NHS National Programme for IT.
BENEFIT 2: Improve efficiency		
• HM Customs and Excise	By 2000, one in five cigarettes smoked in the UK was smuggled, costing around £2.5 billion in lost tax revenue, creating serious law and order problems and undermining government health objectives.	The department identified the risks to achieving a reduction in illegally imported tobacco and invested £209 million over three years to tackle the problem. The department refined its risk assessments on the basis of new intelligence analysis, which enabled it to refocus resources to disrupt smuggling and reduce its profitability by directing its interventions to supply routes, activities and ports of entry where illegal importation was most likely.
BENEFIT 3: Make more reliable decisions		
• National Savings and Investments	To address known weaknesses of its elderly IT systems, NS&I agreed that its partner Siemens Business Services (SBS) should transfer the Premium Bonds database, with records representing an investment value of £24 billion, to another IT system.	To mitigate the risk to its reputation and potential loss of sales if errors were made in customer data in the transfer, NS&I devoted two-and-a-half years to implementing the migration and timed it for the Easter weekend 2004, when fewer people would be making Premium Bonds transactions.

Department	Problem	Risk management solution
• National Savings and Investments	In 2004 NS&I launched a new type of savings account, the Easy Access Savings Account, which involved major changes to NS&I's business, including creating a system for customers to access the new account through automated teller machines.	Staff with experience of launching financial products in the private sector were aware of the risks of over-stimulating demand and not being able to deliver the product to customers in a timely fashion. Good risk management enabled NS&I to achieve an effective product launch.

7.15 THE ORANGE BOOK (REVISED)

The revised *Orange Book* entitled *Management of Risk – Principles and Concepts*, has a shift in emphasis (HM Treasury 2004). As Mary Keegan (Managing Director, Government Financial Management Directorate) states in the foreword, with all government organisations now having basic risk management processes in place, the main risk management challenge no longer lies in the initial identification, analysis and management of risk, but rather in the ongoing review and improvement of risk management. This revised guidance aims to reflect this change. It now includes guidance on issues such as "horizon scanning" for the emergence of new risks or changes in existing risks affecting the organisation's risk profile. Scanning is thought to be dependent on maintaining a good network of communications with relevant contacts. A series of issues surrounding "horizon scanning" are described which are captured in Box 7.7.

Box 7.7 Issues relating to horizon scanning (HM Treasury 2004)

- Periodically/regulatory: Horizon scanning must be continuous (in an organisation like the Civil Contingencies Secretariat (CCS) which continually searches for potential future disruptive challenges) or periodic (e.g. weekly or annually);
- Timescale: Policy makers could well be interested in developments over the next 25 years whilst horizon scanning that supports operational decision making may be restricted to a six month timeframe;
- Scope: Some organisations may be fairly insular in their risk identification processes if they perceive that the major element of risk arises from within the organisation; others may need to consider a much wider scope if they consider that they may face risks from a wider environment. Depending on the nature of the organisation's business this element of risk identification may range from almost exclusively internal activity to activity that depends on international networks of technical information;
- Opportunity threat: Some horizon scanning is concerned mainly with spotting potential problems, but it can equally be used to scan for opportunities ("positive risks") and many problems may be translatable into opportunities if spotted early enough;

- Rigour/technicality: Horizon scanning varies in the extent to which it is structured and supported by technology. Some organisations use sophisticated assessment schemes and information search technologies; other organisations rely almost entirely on informal networks of contacts and good judgement.

The book also focuses on both internal processes for risk management and consideration of the organisation's risk management in relation to the wider environment in which it functions. It makes reference to the Treasury's "risk management assessment framework" which provides a means of assessing the maturity of an organisation's risk management processes. In lieu of the risk management cycle previously described, the book now uses a model developed from the Strategy Unit's report (Cabinet Office 2002) to describe the risk management process within its context. This context is described as being composed of an "extended enterprise" and a "risk environment". The extended enterprise is described as being the source of additional risks emanating from, say, other organisations with which interdependencies exist, the context of being a "parent" to or a "child" of another organisation or third-party arrangements with, say, a contractor to whom risks have been deliberately transferred. In addition, this revised addition offers a revised schedule of risk categories intended to help organisations check that they have considered the range of potential risks which may arise.

7.16 M_o_R 2007

In 2007 the second edition of *Management of Risk: Guidance for Practitioners* was published (Office of Government Commerce 2007). (The first edition was described earlier in Section 7.8). This edition is virtually a total rewrite of the first edition. The description of the individual stages within the risk process is based on the structure adopted in Part II of this book. It provides useful guidance on implementation, for example chapter 5 describes embedding and reviewing management of risk. In addition appendices A–F provide guidance on document guidelines, common techniques, a health check, risk maturity, risk specialisms and selecting risk management software tools.

7.17 MANAGING RISKS IN GOVERNMENT

Following on from its report *Managing Risks to Improve Public Services* (see Section 7.14), the National Audit Office revisited approaches to risk management in departments to understand the challenges that they face in making the most effective use of risk management with a particular focus on three themes: the culture around risk management, value for money in risk management and the benefits of better risk management (National Audit Office 2011). Boards (and, where appropriate, their subcommittees) are encouraged to use the report (and in particular a series of questions which are repeated below) to challenge whether the adopted risk management arrangements are being used effectively to improve service delivery outcomes.

- How do we ensure that our focus is on managing the things that matter? Are we content that management's assessment of risk is not overly optimistic?
- Are we clear about where we are prepared to tolerate differing levels of risk and, in turn, how this influences and drives the actions of management?

- How confident are we that risks are being managed appropriately and that we will be informed of the most significant risks to our business?
- What information do we need both to take decisions and to challenge the rigour with which risk is managed throughout the organisation?
- How do we ensure that our decisions are based on a clear and balanced evaluation of the costs and impacts associated with risks and mitigations?
- How do we learn from successes and failures both within our own and other organisations?

From the findings, the NAO identified (in both government and the private sector) six key principles which underpin and support the use of risk management to improve decision making:

- An engaged Board focuses the business on managing the things that matter.
- The response to risk is most proportionate when the tolerance of risk is clearly defined and articulated.
- Risk management is most effective when ownership of and accountability for risks is clear.
- Effective decision making is underpinned by good quality information.
- Decision making is informed by a considered and rigorous evaluation and costing of risk.
- Future outcomes are improved by implementing lessons learnt.

The report explains why it considers each of the principles to be important and describes the findings of the research conducted by the NAO. Of interest here is principle 1: "An engaged Board focuses the business on managing the things that matter" (see Box 7.8). This principle and the associated findings are disappointing not inasmuch as the report (as regards this issue) tells us nothing new but more in that the adoption of risk management at board and senior management level remains an uphill struggle. Direct experience of failure can be the greatest catalyst for risk management but by then the organisation, whether it is in the private or public sector, has suffered reputational loss at the very least, most likely financial trauma or, in the worst case scenario, failure of one or more of its organisational objectives.

Box 7.8 Championing risk management at Board level (National Audit Office 2011)

Principle 1

The tone at the top of the organisation has an impact on the priority that management and staff give to risk management. The behaviour and actions of the Board and the senior management team, particularly how they communicate with and challenge the business, reinforces the importance of risk management, and drives and encourages a consistent approach to safeguarding the business.

Why is an engaged Board so important?

- The Board sets the agenda and priorities for the organisation. If management and staff believe that the Board views risk management as a key part of successful management they are more likely to buy into and understand its importance to the organisation. The Board is able to challenge management, to ensure that their views are not overly optimistic in both the assessment of risk and the effectiveness of mitigating actions.

- Transparent communication by the Board of the key threats to the organisation's ability to deliver successful outcomes helps staff to understand and engage with managing these risks.
- The Board sets the tone and can foster a climate of trust, developing a culture where staff feel comfortable in openly highlighting risks, which can then be managed, without fear of blame.

Our findings

- Leadership and ownership of risk management in government departments is inconsistent. Some departments have risk management champions at Board-level whilst others rely on specialist risk managers to build momentum and develop understanding. One recent internal audit report observed that "risk management activities are currently undertaken for compliance purposes only". An emphasis on compliance is acceptable if it encourages the right kind of behaviour but compliance alone may be detrimental to the way that risk management is perceived within the business, as opposed to a culture of pro-active risk management.
- Whilst we found that consideration and discussion of risk is a standard agenda item at Board and sub-committee meetings in most departments, discussions do not always focus on those significant issues which could pose the biggest risk to the department. Departments have improved risk management and most acknowledge that they have some way to go to embed risk management internally and through their delivery bodies but have plans in place to improve.

7.18 SUMMARY

This chapter has examined the parties within government with the responsibility for embedding risk management and internal controls within departments. Additionally, a number of government-sponsored risk management publications were examined in chronological order. Each publication enriches our understanding of the discipline of risk management, commonly through examination of a combination of the following: the process; difficulties of implementation; embedding risk management; and tools and techniques. Aspects of each publication have an application in business or they reinforce existing experience in terms of what constitutes a workable approach. Table 7.1 records the publications together with subjects in them (listed as bullet points) which would be relevant to business.

Table 7.1 Government publications and their contribution to the risk body of knowledge

- The Cabinet Office report *Successful IT: Modernising Government in Action* sought to improve the performance or success rate in government IT projects by learning from the past.
  - Common risks to IT projects
  - Tool – summary risk profile (risk map)
  - Tool – project profile model (scoring table for project evaluation)
- *Supporting Innovation: Managing Risk in Government Departments*
  - Essential requirements for effective risk management

Table 7.1 (*Continued*)

- *The Orange Book* (HM Treasury)
  - Tools: commissioning a risk review, risk self-assessment and Control & Risk Self Assessment (CRSA)
  - Categories of risk
  - Four responses: transfer, tolerate, treat and terminate
  - Four controls: detective, directive, preventive and corrective
- *Worth the Risk* (Audit Commission)
  - The constituents of good risk management
- *Corporate Governance in Local Government – A Keystone for Community overnance: The Framework* (CIPFA/SOLACE)
  - Outputs of a systematic strategy, framework and process for managing risk
  - Elements (or activities) of risk management and internal control
- *Management of Risk: Guidance for Practitioners* (M_o_R) 2002
  - Decision making within the management hierarchy of an organisation
  - Techniques to support risk management
- *Risk Management Strategy* (DEFRA)
  - Consequence activities
  - Likelihood rating definitions
  - Impact category definitions
  - Measures for determining the currency of the RM process
- *Risk: Improving Government's Capability to Handle Risk and Uncertainty* (Cabinet Office)
  - The government's role and responsibilities in handling risk
  - Poor past project performance
  - Improving capacity to handle risk, communicating risk to win trust and specific action
- *Risk and Value Management* (OGC)
  - Key messages about managing risk
  - Gateway reviews
  - Risk feedback
  - Risk allowance
- The *Green Book* (HM Treasury)
  - Optimism bias
- *Guidance on Internal Control and Risk Management in Principal Local Authorities and other Relevant Bodies to Support Compliance with the Accounts and Audit Regulations 2003*
  - The publication provides an explanation of what internal control is, why it is required, when is it required, how it should be implemented and by whom
- *Managing Risks to Improve Public Services* (NAO)
  - Four key benefits: deliver better public services, improve efficiency, make more reliable decisions and support innovation
  - Benefits derived from risk management
- *The Orange Book* revised (*Management of Risk – Principles and Concepts*) (HM Treasury)
  - Horizon scanning
  - Risk management assessment framework
  - The extended enterprise
- *Management of Risk: Guidance for Practitioners* (M_o_R) 2007
  - Embedding risk management
  - Document outlines
  - Common techniques
  - Health checks
  - Maturity model
  - Selecting software
- *Managing Risks in Government* 2011
  - Six key principles which underpin and support the use of risk management to improve decision making
  - Six questions for boards for introspection about the effectiveness of their risk management practices

7.19 REFERENCES

Audit Commission (2001) *Worth the Risk.* Audit Commission, London.

Cabinet Office (2000) *Successful IT: Modernising Government in Action*, May. Cabinet Office, HM Government, London.

Cabinet Office (2002) *Risk Improving Government's Capability to Handle Risk and Uncertainty*. Strategy Unit, Cabinet Office, HM Government, London.

CIPFA (2003) *Guidance on Internal Control and Risk Management in Principal Local Authorities and other Relevant Bodies to Support Compliance with the Accounts and Audit Regulations 2003*. Chartered Institute of Public Finance and Accountancy, London.

CIPFA/SOLACE (2001) *Corporate Governance in Local Government – A Keystone for Community Governance: The Framework*. Chartered Institute of Public Finance and Accountancy, London.

DEFRA (2002) *Risk Management Strategy*. Department for Environment, Food and Rural Affairs, London.

HM Government (2004) *Delivering the Essentials of Life – Defra's Five Year Strategy*, Cm. 6411, December. Stationery Office Books, Norwich.

HM Treasury (2001) *Management of Risk: A Strategic Overview*. London.

HM Treasury (2003) *Appraisal and Evaluation in Central Government*. The Stationery Office, Norwich.

HM Treasury (2004) *Management of Risk – Principles and Concepts*. The Stationery Office, Norwich.

Mott MacDonald (2002) *Review of Large Public Procurement in the UK*. Mott MacDonald, Croydon.

National Audit Office (2000) *Supporting Innovation: Managing Risk in Government Departments*. Report by the Comptroller and Auditor General, 17 August. The Stationery Office, London.

National Audit Office (2004) *Managing Risks to Improve Public Services*, October. The Stationery Office, London.

National Audit Office (2011) *Managing Risks in Government*, June. National Audit Office, London.

Office of Government Commerce (2002) *Management of Risk: Guidance for Practitioners*. The Stationery Office, London.

Office of Government Commerce (2003) *Risk and Value Management*. Procurement Guide 04 of the Achieving Excellence in Construction series.

Office of Government Commerce (2007) *Management of Risk: Guidance for Practitioners*, 2nd edition. The Stationery Office, London.

Phillips, N.A. (2000) *The BSE Inquiry, Volume 1, Findings and Conclusions*. The Stationery Office, London.

Public Accounts Committee (2001) *First Report Session 2001–2002, Managing Risk in Government Departments*, November. The Stationery Office, London.

Part II
The Risk Management Process

Part II is subdivided into seven chapters, each of which describes one of the core risk management stages: context, identification, analysis, evaluation, treatment, monitoring/review and communication/consultation. Collectively these stages form a logical sequence of activities necessary for a robust approach to the implementation of enterprise risk management. All of these stages are present in most guides on the subject of risk management published over the last 15 years, albeit that some of the stages may have different labels. The COSO (2004)[1] guide uses similar but not identical stages. The table included in Appendix 2 records what might be described as the key guides and publications, as a source of reference.

To describe these stages and their interrelationships, the chapters in Part II are structured on early process mapping[2] initiatives sponsored by the United States Air Force during their ICAM Program.[3] The goal of the Program was to develop a baseline for generic subsystem process planning that could be developed through the cooperative effort of a large number of industry partners. In essence the baseline was to act as a communication tool. The catalyst for adopting process mapping to describe the enterprise risk management process is the excellent book by Hall called *Managing Risk: Methods for Software Systems Development* (Hall 1998).

The rationale for adopting process mapping here is to enable the individual stages in the overall risk management process to be both structured and, similarly, readily communicated. As risk management involves group participation, the overall process and its constituent incremental stages must be readily understood. The overall risk management process (and its relationship to the risk management framework) is described in the form of a process map and is illustrated in Figure P2.1. The stages labelled A31–A37 are described in Chapters 8–14, respectively, as individual processes with their own unique goals, subgoals, inputs, outputs, controls, mechanisms and actions.

The process map in Figure P2.1 illustrates the hierarchical order between the risk management framework and the other risk management documents. The design framework activity within the overall risk management framework includes the task "establish the risk management policy" as one of the tasks to be completed as part of this step. The risk policy has the node number A22 to illustrate that it is a child of the parent A2 (design framework). Similarly, the implement framework activity within the overall risk management framework includes the task "implement the risk management process" as one of the tasks to be completed as part

[1] The Committee of Sponsoring Organisations of the Treadway Commission (COSO) is a voluntary private sector organisation dedicated to improving the quality of financial reporting through corporate governance, effective internal controls and business ethics, effective internal controls and corporate governance. COSO was originally formed in 1985 to sponsor the National Commission on Fraudulent Financial Reporting, an independent private sector initiative which studied the causal factors leading to fraudulent financial reporting and developed recommendations for public companies and their independent auditors. The chairman of the National Commission was James Treadway Jr (a former commissioner of the SEC), hence COSO is sometimes simply referred to as the "Treadway Commission".

[2] "Process mapping" is now a widely recognised management tool, initially developed and implemented by General Electric as part of a strategy to significantly improve their bottom-line business performance.

[3] The US Air Force Integrated Computer Aided Manufacturing (ICAM) Program adopted the structured analysis and design technique (SADT) originally developed in 1972 by Douglas T. Ross of SofTech as the "architecture method" (or process design).

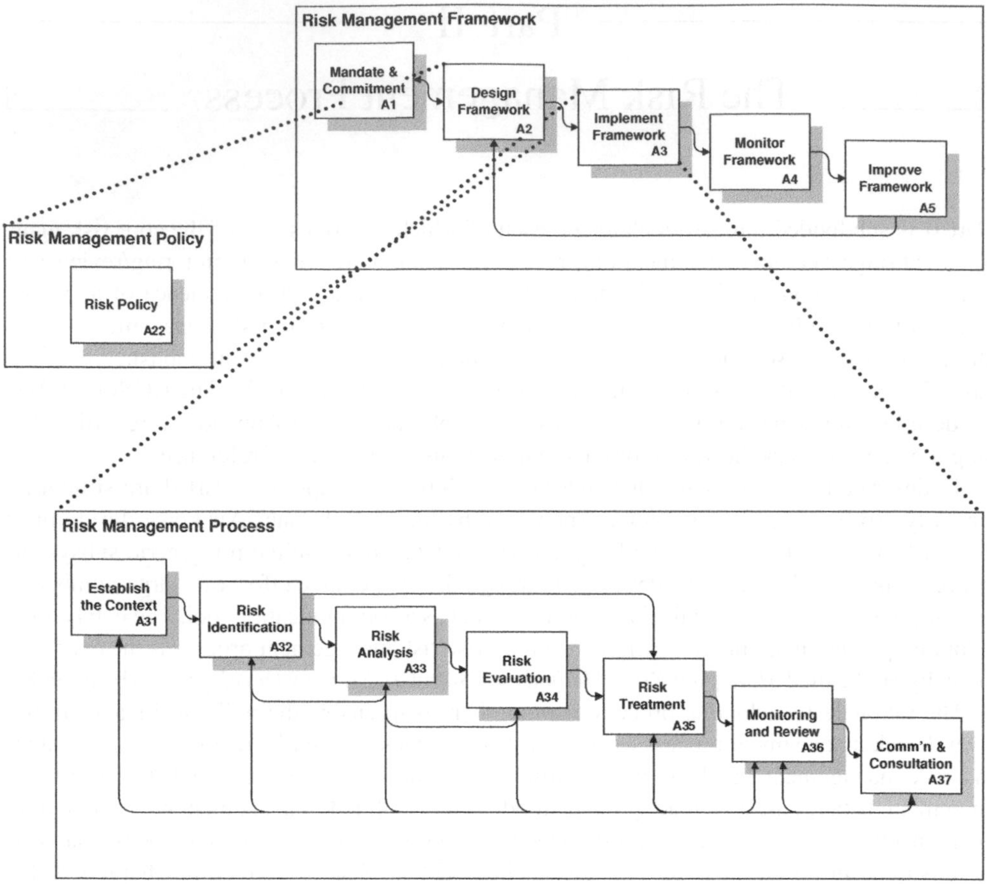

Figure P2.1 Stages in the risk management process

of this step. The risk management process has the node numbers A31–A37 to illustrate that it is a child of the parent A3 (implement framework). The content of the risk management framework was discussed in outline in Chapter 1.

Within the risk management process the monitor and review stage and the communication and consultation stage occur throughout the risk management process and hence arrows connect each of these activities to all of the other stages.

The major elements of this functional process map adopted by the ICAM Program later came to be referred to as the "integration definition for function modelling" (IDEFO). The identifying characteristics of the IDEFO technique are that it is based on the organised and systematic combination of graphics and text, to provide understanding and analysis. It provides a structure and logic for implementing potential changes. An IDEFO process map is composed of a hierarchical series of diagrams that gradually display increasing levels of detail describing functions and their interfaces across all processes. The building blocks of the IDEFO process mapping technique are the individual processes that have their own unique data flows. IDEFO adopts "box and arrow" graphics as a visual communication tool to facilitate the planning, development and implementation of process systems. The rules of the IDEFO process map include syntax rules for graphics (boxes and arrows, where boxes are processes and arrows

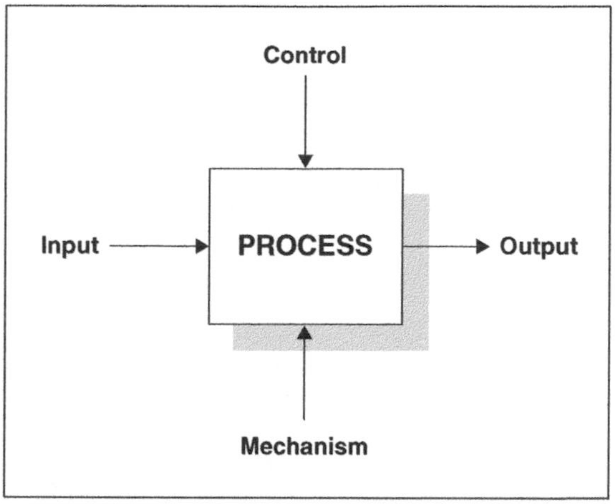

Figure P2.2 IDEFO process design notation: process elements are described by IDEFO using inputs, outputs, controls and mechanisms

are data flows) and data connectivity (input, control, output and mechanism (ICOM)) codes. Controls can be considered as constraints and mechanisms can be considered as enablers. A simple process is included in Figure P2.2, which illustrates the four modes of data connectivity. These modes of data connectivity are used to describe how the risk management stages are implemented and how the output of one process forms the input to the subsequent process.

REFERENCES

COSO (2004) *Enterprise Risk Management – Integrated Framework*, September. Committee of Sponsoring Organisations of the Treadway Commission.

Hall, E.M. (1998) *Managing Risk: Methods for Software Systems Development*. Addison-Wesley Longman Inc., Reading, MA.

8

Establishing the Context: Stage 1

Get the habit of analysis – analysis will in time enable synthesis to become your habit of mind.
(Frank Lloyd Wright)

Establishing the context is the first stage in the overall seven-stage process of enterprise risk management. Establishing the context is concerned with gaining an understanding of (1) the background to the business as a whole, in general terms, and (2) the specific business activity, process or project, forming the subject of the risk management study. It provides a basic foundation for everything that follows. How well this process is completed will determine the quality of the remainder of the risk management process. The objective of this first stage is to discover timely and accurate data. However, its degree of usefulness will depend on its relevance, breadth, depth and currency, in terms of providing sufficient insights to create a prompt tool with substance and teeth. It is not uncommon for representatives of the business under examination to either regard activities such as investigation, research or diagnosis as expensive and wasteful, or be frustrated by the time required to carry them out. Typically these representatives fail to recall that their own knowledge has been built up over months or even years. But even when time is of the essence, discipline and rigour need to be applied so that important issues are not overlooked. The three most important aspects of risk management are preparation, preparation and preparation. Before data gathering can commence, a decision must be made as to the approach to be adopted and the information to be examined. This will largely be dictated by the focus of the study, such as whether it is analysis of:

- a single business activity or project;
- a single department's planned or ongoing activities within a business;
- a proposed merger or acquisition;
- a new production facility oversees;
- a "health check" of existing risk management procedures;
- a "health check" of an individual business case model, investment proposal or quantitative project risk model.

The next chapter describes the risk identification stage (Stage 2). The ability of the risk facilitator to question and challenge the participants engaged in the risk identification process will (to a large part) be dictated by the depth and breadth of analysis carried out in the context stage (Stage 1) and the facilitator's own knowledge and experience of the business type or project under examination. The structure of Chapter 8 is illustrated in Figure 8.1.

8.1 PROCESS

Adopting the philosophy of process mapping, each stage within the overall risk study exists to make a contribution to one or more of the risk management goals. Each stage is a process in its own right. Hence each process should be measured against specific process goals that reflect the contribution that the process is expected to make to the risk management study. Processes

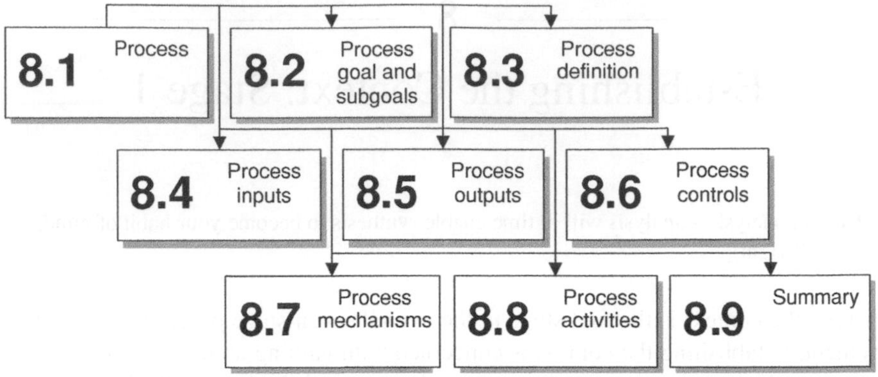

Figure 8.1 Structure of Chapter 8

are simpler to comprehend when they have primary goals and subgoals. Hence, analysis is described here as having a primary process goal which is accomplished by a series of subgoals, as described below. Any one process is accomplished within a context and might be considered to have two perspectives. The *external view* examines the process inputs, outputs, controls and mechanisms. The *internal view* examines the process activities that transform inputs to outputs, applying the mechanisms and being influenced by the controls.

8.2 PROCESS GOAL AND SUBGOALS

The primary process goal of establishing the context (the first stage in the risk management process) is to understand the business processes in order to inform the following incremental stage in the overall risk management process. While recognising that the stage of establishing the context will be tailored to suit the particular requirements of the assignment or study, when an overview of the whole business is being obtained, this stage will be sufficient when it satisfies the following subgoals:

- the business objectives were established;
- an organogram of the business structure was obtained or constructed;
- the business process map was examined, or, where one does not exist, a high-level process map is constructed;
- the existing internal controls established were examined;
- all of the primary business functions were examined;
- the existing corporate risk management plan was reviewed, along with the remit of the audit committee;
- the business risk appetite was made explicit;
- the existing risk register was reviewed (if one had been prepared);
- personnel were involved from all appropriate company departments and no departments were excluded or forgotten;
- department representatives participating in the analysis process were senior enough to be knowledgeable in their own area of specialisation and were aware of both corporate lessons learnt and company risk exposure; and
- consideration was given to consultation with non-executive directors and, where appropriate, they are included in the risk identification process.

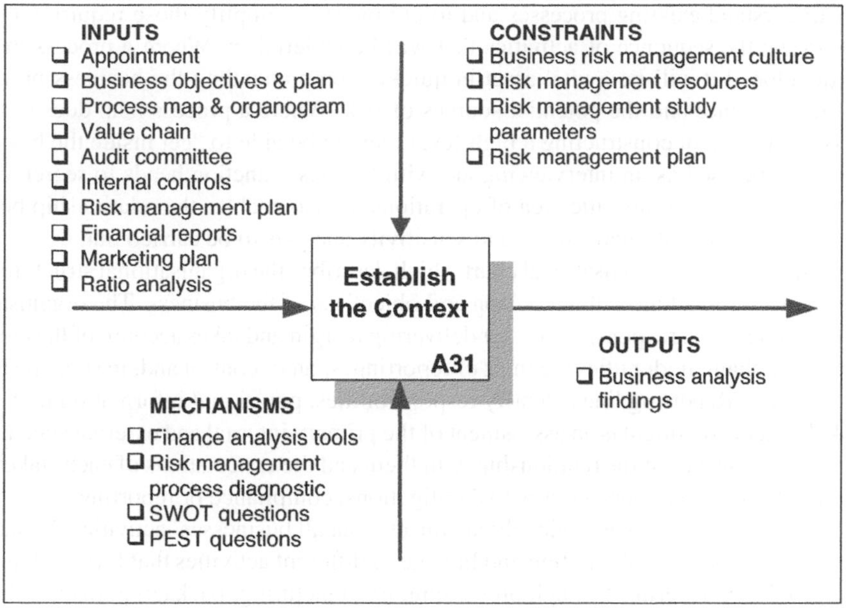

Figure 8.2 The "establish the context" process illustrating the inputs, outputs, constraints and mechanisms

8.3 PROCESS DEFINITION

The process of establishing the context is described by an IDEFO diagram (see Figure 8.2). The diagram describes a process with inputs entering on the left of the box, outputs leaving on the right of the box, constraints entering from above and mechanisms (or enablers) entering from below.

8.4 PROCESS INPUTS

The following are suggested inputs to the process of establishing the context, but should be tailored to suit the study requirements. An explanation is provided for each input for the sake of clarity.

- Appointment details are the specific issues recorded in the assignment prepared by the sponsor/client that the assignment must address/examine.
- Business objectives are statements of the business goals against which success will be measured. They must be concise, easy to understand and enduring. They should ideally consist of no more than five bullet points so that they can be easily memorised and recalled.
- The business plan is, in simple terms, a statement of how the business will accomplish its business objective(s). The style, length and content of a business plan will depend on the business decision or activities the plan is designed to support and the audience for whom the plan is prepared. It documents why the forecast effort and time will be worth the expenditure to achieve the change and the anticipated benefits.
- A business process map is a management tool used to describe, in workflow diagrams and supporting text, every vital step in a business's processes. It is a communication tool to

better understand existing processes and to eliminate or simplify those requiring change. It sets down the sequence of activities that will be undertaken. Where a process map has been developed it will be a vital asset in quickly comprehending the business processes, how they interface and the potential sources of risk. Where a process map does not exist, time is well spent in constructing a high-level map, to be able to "get inside the business". A map can be used as an interviewing aid with business function heads to search out the risks pertinent to their specific area of operations. It also enables the relationship between risks to be more readily understood and sensitivity analysis to be carried out.

- An organogram is an organisational chart which describes the organisational structure of the business. It is consistent with the vision and objectives of the business. The organisational structure reflects the responsibilities for delivering margin and takes account of the elements of the value chain. It identifies the lines of reporting, span of control and, in some instances, staff numbers. Reporting lines identify responsibilities, power and information flow.
- A stakeholder assessment is an assessment of the primary internal and external stakeholders in terms of the nature of the relationship with them and the expectations of each stakeholder in terms of, say, behaviours, contractual obligations, compliance or reporting.
- A value chain is a consistent policy thread throughout all businesses' activities. Value chain analysis explores the configuration and linkage of different activities that form a chain from the original raw materials through processing, manufacturing, packaging, distribution and retailing to the end customer. Analysis is used to identify a strategy mismatch between different elements of the value chain. If a company competes on the basis of low cost, then every part of the value chain should be geared towards low cost. If a policy is to keep stocks to a minimum (in order to respond quickly to changing customer tastes and not be left with extensive redundant stock), each element of the business activity chain should be geared around just-in-time manufacturing.
- The audit committee is responsible for monitoring the integrity and completeness of a company's financial statements and, in particular, for establishing whether management has adopted appropriate accounting policies and supported them with realistic estimates and judgements. Where no alternative arrangements have been made by the board, the audit committee may also be closely involved in reviewing the effectiveness of the company's risk management system.
- Internal controls are the controls designed to ensure the business's policies are put into practice; the organisation's values are met; laws and regulations are complied with; required processes are adhered to; financial statements and other published information are accurate and reliable; and human, financial and other resources are managed effectively and efficiently.
- A risk management plan is a "map" of the intended implementation of risk management to support a project or a business activity. The plan will typically describe the objectives of the risk study, project overview, the risk study timeframe, resources to be deployed, the risk management process, the responsibilities of the parties and the study deliverables.
- Projected financial statements portray the predicted financial outcomes of pursuing a particular course of action. By showing the financial implications of certain decisions, managers should be able to allocate resources in a more efficient and effective manner. The projected financial statements will normally comprise a cash flow statement, profit and loss account and balance sheet.
- The marketing plan contains a detailed description of the marketing mix and guidelines for the implementation of the business's marketing programmes. The marketing mix is

defined by product, price, promotion and place. All elements of the marketing mix together constitute the "offer". The offer is more than the product. It is a value proposition that satisfies customer needs. The attributes of the offer are defined by the marketing mix.
- Ratios analysis provides a picture of a firm's performance, liquidity, profitability, efficiency and vulnerability.

8.5 PROCESS OUTPUTS

The output is simply the stage findings. These findings should be recorded and included in any report prepared on the conclusion of the study. An appendix should be included in the report listing the documents referred to, their title, date and author (should further reference need to be made to them). The findings will act as prompts to inform the identification process.

8.6 PROCESS CONTROLS (CONSTRAINTS)

From Figure 8.2, the business risk management culture, resources, study parameters and plan (where one exists) are described as regulating/constraining the risk management process.

- The business risk management culture will constrain risk management activities as a result of the degree of importance, commitment and enthusiasm attached to the process and the extent of support provided when the risk management process is initiated.
- The risk management resources will constrain risk processes in terms of time. When cost is a constraint, particularly when external support is being commissioned, less expensive and most likely less experienced staff may be allocated to the assignment. When time is limited and risk management activities are accelerated, there is a strong likelihood the quality of the output will diminish. All of these constraints are likely to compromise process effectiveness, particularly the breadth of risk identification, potentially leaving "blind spots".
- The risk management study itself will constrain the risk identification process if:
  - the study lacks a clear focus;
  - the activities are too ambitious for the timescale;
  - inadequate notice is provided to attendees of interviews or workshops;
  - inadequate notification is given to attendees of the purpose of the study, timetable of events and/or their expected involvement;
  - inadequate experience on the part of the facilitator;
  - inadequate preparation on the part of the facilitator and the attendees;
  - participants are unfamiliar with the process, terminology and products of risk management;
  - key participants are not available to suit the timetable; and
  - participants bring additional personnel of their own volition, without consulting the sponsor.

  This can lead to an inadequate study from a number of perspectives, the most serious of which is that the risk study is too superficial or shallow, leading to a series of blind spots across the potential sources of risk.
- The risk management plan will also constrain the risk identification process if roles and responsibilities are not clearly defined, business objectives are not captured and disseminated to participants, studies are not timetabled and diarised in advance and the purpose of the process is unclear.

8.7 PROCESS MECHANISMS (ENABLERS)

Four of the common process mechanisms are:

- financial analysis tools (ratios);
- risk management process diagnostic;
- SWOT questions
- PEST questions.

8.7.1 Ratios

Financial ratios can be used to examine various aspects of financial position and performance and are widely used for planning, control and evaluation purposes. They can be used to evaluate the financial health of a business and can be utilised by management in a wide variety of decision making involving such areas as profit planning, pricing, working capital management, financial structure and dividend policy. Financial ratios provide a quick and relatively simple means of examining the financial condition of a business. A ratio simply expresses the relation of one figure appearing in the financial statements to some other figure appearing there (e.g. net profit in relation to capital employed) or perhaps some resource of the business. Ratios can be grouped into certain categories, each of which reflects a particular aspect of financial performance or position. The following broad categories provide a useful basis for explaining the nature of the financial ratios to be considered:

Profitability. Businesses come into being with the primary purpose of creating wealth for their owners. Profitability ratios provide an insight into the degree of success of the management in achieving this purpose. They express the profits made in relation to other key figures in the financial statements or to some business resource.

Efficiency. Ratios may be used to measure the efficiency with which certain resources have been utilised within the business. These ratios are also referred to as activity ratios.

Liquidity. Liquidity is an important measure of risk exposure. It is vital to a business that there are sufficient liquid resources available to meet maturing obligations. Certain ratios may be calculated which examine the relationship between liquid resources held and creditors due for payment in the near future.

Gearing. Gearing is an important issue which managers must consider when making financing decisions. The relationship between the amount financed by the owners of the business and the amount contributed by outsiders has an important effect on the degree of risk associated with a business.

Investment. Certain ratios are concerned with assessing the returns and performance of shares held in a particular business.

Key aspects of calculating ratios to aid risk analysis are as follows:

- Where there was considered to be a risk relating to the inability to repay amounts owing in the short term, the liquidity ratio would be of interest.
- Where there was considered to be a risk to returns on investment, the profitability, investment and gearing ratios would be of interest.
- In the event that there was concern by long-term lenders over the long-term viability of the business, the profitability and gearing ratios would be likely to be of interest.

The ratios that may be useful for business analysis are included in Appendix 5.

8.7.2 Risk Management Process Diagnostic

Difficulties of Embedding Risk Management

ERM studies that focus on the effectiveness of existing risk management processes will need to establish how well developed risk management processes are and how effectively they have been embedded into the organisation. Risk management is a fundamental building block of business management. However, its integration into an organisation is not straightforward. The Office of Government Commerce (2002) provides an insight into this complexity by describing eight key areas, which it sees as critical to the establishment of an effective risk management process, as follows:

- Clearly identified senior management support, own and lead on risk management.
- Risk management policies and the benefits of effective management clearly communicated to all staff.
- Existence and adoption of a framework for management of risk that is transparent and repeatable.
- Existence of an organisational culture which supports well-thought-through risk taking and innovation.
- Management of risk fully embedded in management processes and consistently applied.
- Management of risk closely linked to achievement of objectives.
- Risk associated with working with other organisations explicitly assessed and managed.
- Risks actively monitored and regularly reviewed on a constructive "no-blame" basis.

Hillson (1997), in a paper proposing a risk maturity model, describes the difficulties of embedding risk management into an organisation. He states that the implementation of risk management into an organisation is not a minor challenge and that it cannot be undertaken in a short period of time. In addition, it is:

> not a simple process of identifying techniques, sending staff on training courses, buying software and getting on with it. Risk capability is a broad spectrum, ranging from the occasional information application of risk techniques to specific projects, through routine formal processes applied widely, to a risk-aware culture with proactive management of uncertainty.

The difficulties of integrating risk management into an organisation are also highlighted in a paper produced by the Risk Management Research and Development Program Collaboration (2002) as follows:

> effective implementations of risk management processes into organisations and projects are not common. Those who have tried to integrate risk management into their business processes have reported differing degrees of success, and some have given up the attempt without achieving the potential benefits. In many of these uncompleted cases, it appears that expectations were unrealistic, and there was no clear vision of what implementation would involve or how it should be managed. Organisations attempting to implement a formal structured approach to risk management need to treat the implementation itself as a project, requiring clear objectives and success criteria, proper planning and resourcing, and effective monitoring and control.

The process of embedding *project* risk management within an organisation is described in Chapter 18 – not surprisingly, there are striking similarities with the challenges that face those wishing to introduce enterprise risk management into business processes.

Establishing the Effectiveness of Existing Risk Management Processes

The collaboration (referred to above) suggests an assessment can be made of the effectiveness of existing formal risk management procedures by undertaking a review. The objective of such a review would be to benchmark the organisation's present maturity and capability in managing risk, using a generally accepted framework such as a risk maturity model (sometimes referred to as a process diagnostic). Fully developed risk maturity models are useful tools in understanding the degree of sophistication of a business risk management process, its reliability and effectiveness in identifying, assessing and managing risks and opportunities. Risk maturity models provide guidance to organisations who wish to develop or improve their approach to risk management, allowing them to assess their current level of maturity, identify realistic targets for improvement and develop action plans for increasing their risk capability. Risk maturity models proposed by Hillson (1997) and Hopkinson (2000) are described in Appendix 6. In addition, a business risk maturity model is proposed by the author for assessing business risk management processes based on four levels of maturity called initial, basic, standard and advanced.

8.7.3 SWOT Analysis

SWOT is an acronym for "strengths, weaknesses, opportunities and threats". The SWOT analysis headings provide a framework for reviewing a business as a whole or a series of issues such as: a strategic option; an opportunity to make an acquisition; a potential partnership; a new product; a business proposition; or outsourcing an activity. A SWOT analysis is a subjective assessment of data, which is organised by the SWOT format into a logical order that helps understanding, presentation, discussion and decision making. The SWOT analysis template is normally presented as a grid, comprising four sections, one for each of the SWOT headings. A SWOT analysis can be used specifically for risk identification, as discussed in Section 9.7.6.

A SWOT analysis can be considered as bringing together a strategic review of a business and in particular (Friend and Zehle 2004):

- the analysis of the firm (internal elements);
- the market analysis (internal and external elements);
- the product, portfolio and matrix analysis (internal and external elements);
- the analysis of the general environment (external elements).

Guidance on implementing a SWOT analysis is included in Appendix 7.

8.7.4 PEST Analysis

A PEST analysis is a useful tool for analysing a business and, in particular, understanding market growth or decline. PEST is an acronym for "political, economic, social and technological" factors, which are used to assess the market for a business or organisational unit. It is a business measurement tool. Businesses are continually reacting to changes in the environments in which they operate. Proactive businesses try to anticipate change in their external environment by monitoring trends through, say, market research. This means that they can plan and be prepared. Reactive businesses wait until change has happened and then have to decide what to do. They are taken by surprise and tend to move from one crisis to another. As a result decision making is rushed and tends to be less effective. To make effective decisions, businesses should be constantly scanning their environment to identify changes and potential risks and prepare for them. A PEST analysis will reveal many of the external environmental influences on a business's performance. These influences are part of the macro environment over which a business has no control. An initial PEST analysis can be undertaken as part of a desktop study and then used again with a project team or business group to gain a broad consensus of senior managers of the external influences.

Guidance on implementing a PEST analysis is included in Appendix 8.

8.8 PROCESS ACTIVITIES

The Stage 1 risk study process activities will be dictated by the objectives of the study and hence will have to be tailored to suit the information that needs to be gathered. Depending on the breadth of the risk study, the activities undertaken will consist of a selection of or possibly all of the following activities:

- clarifying and recording the business objectives or the subset thereof under examination;
- examining the business plan;
- examining the industry (business position/market context/regulatory framework);
- business processes;
- projected financial statements;
- resources;
- change management;
- marketing plan;
- compliance systems.

If the desire is to construct a high-level process map of the business activities or risk breakdown structure to aid risk identification, a thorough understanding of the business operations and of the context within which they operate will be required.

Process activities may also include internal controls, the role of the audit committee and examination of the existing risk management processes. Figure 8.3 describes the likely activities that will be undertaken as part of the process.

8.8.1 Business Objectives

The first and most important activity is recording and taking time to understand the business objectives. In the next stage of the risk management process, the objective of the study will be to identify the risks and opportunities to these objectives. The business objectives will be the criteria against which the success of the business strategy will be measured – the business strategy being the overall plan aimed at achieving sustainable competitive advantage to produce healthy profits. The objectives should be SMART:

- Specific
- Measurable
- Achievable within the timeframe included in the business strategy
- Relevant in the context of the business vision (broad direction such as the leading provider of mobile phones)
- Time bound.

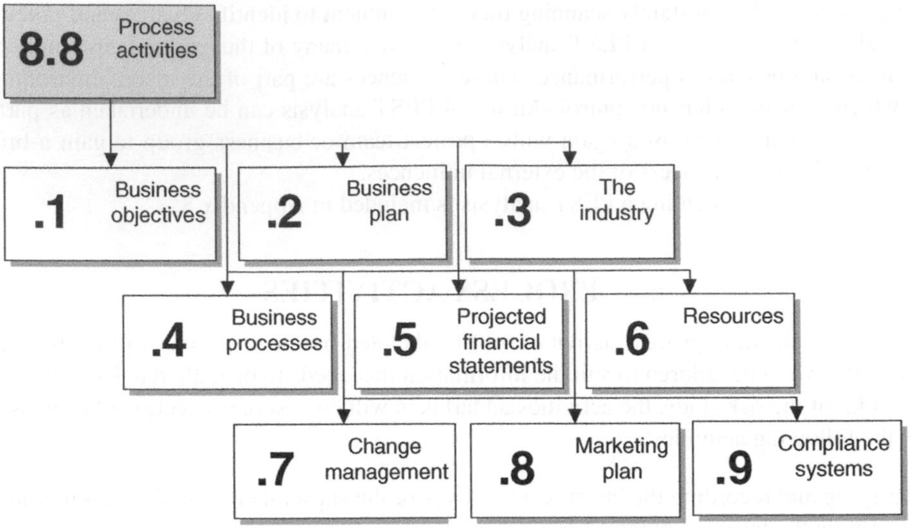

Figure 8.3 Structure of Section 8.8

8.8.2 Business Plan

Examination of the business plan is important as it should provide a "story". It should explain how the business will achieve its objectives in a coherent, consistent and cohesive manner. The "story" should be focused on the customer. The plan should identify the market, its growth prospects, the target customers and the main competitors. It should be based on a credible set of assumptions and should identify the assumptions to which the success of the business is most sensitive. It should identify the risks facing the business, their likely impact should they materialise and the actions planned to reduce or remove the risks. As the blueprint for the business, it should describe what makes the business different from its competitors, its unique selling point, and how it will maintain its competitive advantage in the long term. It should describe the experience and track record of the management team and, for larger organisations, provide similar details for those holding key support posts in implementing the business. Additionally, it should identify the source of funding and the cost of that funding. When viewed as part of a risk study, the currency of the plan should be established, particularly the market analysis. A good business plan should:

- tell a coherent and cohesive, customer-focused story;
- clearly define the market, customers, suppliers and competitors;
- contain credible planning assumptions and sales forecasts;
- describe how the business will achieve sustainable competitive advantage;
- identify the assumptions within the business case to which its validity is most sensitive;
- identify the risks that the business faces and the planned or ongoing response actions;
- identify the opportunities that it plans to exploit;
- contain a summary of the experience of the managers and key staff involved in managing the business; and
- identify the funding requirements and the source of funding.

A study by the risk analyst of the business plan against this checklist will provide an early indication as to whether there are any significant gaps in the business strategy which in themselves may pose either threats or the loss of potential opportunities.

8.8.3 Examining the Industry

To understand the risks that a business faces it is necessary to understand the industry within which the business is operating and the competitive forces within that industry. Questions to be answered include the following:

- What is the current size of the industry?
- What are the major trends and changes in the industry?
- Who are the competitors and what are their current strengths?

Techniques for industry analysis that may be used to obtain an understanding of the contexts of the business are an industry overview, the industry life cycle, structural analysis and main competitor analysis.

- *Industry overview*. The first task is to collect some basic data about the sector relevant to the business under examination. This may include such relevant metrics as:
  - annual sales in value for the last three years;
  - annual unit or volume sales for the last three years;
  - trend in prices for the last three years; and
  - a measure of capacity and possibly capacity utilisation.

  Competitors should be identified by name and their market share should be listed. This should be combined into a measure of concentration – it may be found, for example, that the top 20% of competitors serve 80% of the market.
- *The industry life cycle*. The industry life cycle is measured in total industry sales over time. The structure of the industry and the competitive forces that shape the environment in which a business operates will change throughout the life cycle.
- *Structural analysis*. Understanding the structure of an industry is the basis for the formulation of competitive strategy. An industry is an open system and is affected by potential entrants, suppliers, buyers and the threat of competition from substitutes.
- *Main competitor analysis*. A company of any substance will have undertaken market analysis. The intensity of competition or rivalry will have a significant impact on the ability of the business under examination to generate adequate margins.

8.8.4 Establishing the Processes

Process mapping is recognised as a proven analytical and communication tool to improve understanding of a business's existing processes and eliminate or simplify those requiring change (Hunt 1996). A business "process is a series of steps designed to produce a product or service". A business enterprise is only as successful as its processes. Process mapping provides a proven tool with which to understand business processes to help improve bottom-line performance and competitive position. Without knowing where you are at a given moment it is hard to determine how to get to your destination.

Table 8.1 Examples of business processes

Generic customer processes
- Marketing and sales
- Product/service development and introduction
- Manufacturing
- Distribution
- Billing
- Order processing
- Customer service

Industry-specific customer processes
- Loan processing (banking)
- Claim adjudication (insurance)
- Grant allocation (government)
- Merchandise return (retail)
- Food preparation (restaurants)
- Baggage handling (airline)
- Reservation handling (hotels)

Generic administrative processes
- Formal strategic and tactical planning
- Budgeting
- Training
- Facilities management
- Purchasing
- Information technology (IT) management

A roadmap (process map) is a communication tool which enables a business to better understand its processes. A business process may be defined as a series of steps to produce a product or service. Rummler and Brache (1994) describe three primary processes:

- Processes which result in a product or service that is received by an organisation's external customer. These are known as "customer processes".
- Processes which produce products or services which are invisible to the external customer but essential to the effective management of the business. These are known as "administrative processes".
- Processes which are actions taken by managers to support the business processes. These are known as "management processes".

Examples of customer and administrative processes are included in Table 8.1.

Hunt (1996) argues that business enterprise performance is driven by three process variables: (1) process goals, (2) process design and (3) process management. He states that each customer and administrative process exists to make a contribution to one or more business enterprise goals. Hence each process should be measured against process goals that reflect the contribution that the process is expected to make to one or more of the business enterprise goals. Processes need to be designed to achieve those goals efficiently.

To create a process map, a high-level map is developed initially to obtain a manageable overall picture of the key processes, showing the complete chain of activities within the business. The high-level process mapping stage enables businesses to (1) determine where their process starts and ends, (2) to identify what is included in the process, (3) name the process, (4) state the purpose of the process, (5) create several map flow charts of the process at a high level and (6) identify the products and services of the process. More detailed lower-level process mapping is then performed to assist business managers to examine specific process steps for each high-level process step.

A process can be seen as a "value chain". By its contribution to the creation or delivery of a product or service, each step in a process should add value to the preceding step. The subject of value chains is discussed further under Section 8.8.6.

8.8.5 Projected Financial Statements

Projected financial statements portray the predicted financial outcomes of pursuing a particular course of action. By showing the financial implications of certain decisions, managers should be able to allocate resources in a more efficient and effective manner. The projected financial statements will normally comprise a *cash flow statement*, *profit and loss account* and a *balance sheet*. Where there are competing options, projected statements can be prepared for each of the options being considered. They will set out the expected revenues and costs associated with each option and will reveal the impact of these items on the future profitability, liquidity and financial position of the business. Where managers are considering only one course of action, projected financial statements can still be extremely useful. The preparation of projected statements will still provide a useful insight into the impact of a particular course of action on the future financial position of the business. For example, where a business is considering a strategy to increase market share, managers need to satisfy themselves that adequate resources are available to sustain the planned growth in sales. Projected financial statements will also help managers to strike an appropriate balance between sales, operating capability and finance levels.

For most businesses, the starting point for preparing projected statements will be the forecast for sales. The ability to sell the goods or services produced will normally be the key factor which decides the overall level of activity for the business. A reliable sales position is, therefore, essential as many other items including certain costs, stock levels, fixed assets and financing requirements will be determined partially or completely by the level of sales for the period. As can be imagined, forecasting the future level of sales is both a difficult and critical task. Future sales will be influenced by a number of factors, including the degree of competition, the planned expenditure on advertising, the quality of the product or service, changes in consumer tastes and the condition of the economy. Some of these issues will be under the control of the business, others will not. The sales forecasts must take account of all of the relevant factors if reliable figures are to be produced. Sales projects may be based on market research, statistical techniques or economic models.

A projected cash flow statement is useful because it helps to identify changes in liquidity of a business over time. Cash can be described as the "life blood" of a business. It is vital for a business to have sufficient liquid resources to meet its ongoing obligations. Failure to meet an adequate level of liquidity can have disastrous consequences for a business. The projected cash flow statement helps to assess the impact of expected future events on the cash balance. It will identify periods where there are cash surpluses and cash deficits and will allow managers to plan for these occurrences. In terms of forecasting costs, while some costs will vary directly and proportionately with the level of sales, other costs are unaffected by the level of sales in the period. Cost of sales, materials consumed and sales force commission are examples of variable costs which vary directly with sales output. Other costs such as depreciation, rent, rates, insurance and salaries may stay fixed during the period irrespective of the level of sales generated and are referred to as fixed costs. Some costs have both a fixed and a variable element, may vary partially with sales output and are referred to as semi-variable costs.

A projected profit and loss account helps to provide an insight into the expected level of profits. When preparing the profit and loss account, all revenues that are realised (achieved) should be included within the relevant period. All expenses (including non-cash items such as depreciation) that relate to the revenues realised in the period must be shown in the profit and loss account in which the sales appear. The timing of the cash outflows for expenses is also irrelevant.

A projected balance sheet reveals the end-of-period balances for assets, liabilities and capital and should normally be the last of the three statements to be prepared. This is because the previous statements prepared will produce information to be used when preparing the projected balance sheet. The projected cash flow statement reveals the end-of-period cash balance for inclusion under the "current assets". The projected profit and loss account reveals the projected profit (loss) for the period for inclusion under the "share capital and reserves" section of the balance sheet. In terms of forecasting balance sheet items, the numbers of items appearing on the balance sheet of a business are likely to increase automatically with an increase in the level of sales. An increase in the level of sales should lead to an increase in the level of current assets where a business is likely to need:

- more cash to meet the increased costs incurred;
- higher levels of trade debtors as a result of higher sales; and
- higher levels of stock to meet the increase in demand.

Additionally, an increase in the level of sales should also lead to an increase in the level of current liabilities. A business is likely to incur more trade creditors as a result of increased purchases and more accrued expenses as a result of increased overhead costs. The projected financial statements once prepared should be critically examined by managers. There is a danger that the figures contained within the statements will be too readily accepted by those without a financial background. Questions should be asked such as:

- How reliable are the projections which have been made?
- What underlying assumptions have been made and are they valid?
- Have all relevant items been included?

The projected statements can be examined to find answers to a variety of questions concerning the future performance and position of the business. These questions may include any or all of the following:

- Are the cash flows satisfactory?
- Is there need for additional financing?
- Can any surplus funds be profitably reinvested?
- Is the level of profit satisfactory for the risks involved?
- Are the sales items at a satisfactory level?
- Is the financial position at the end of the period acceptable?
- Is the level of borrowing acceptable?
- Is the dependency on borrowing acceptable?

8.8.6 Resources

One method of analysing a business is to identify its resources and explore how they are used to competitive advantage. Businesses that allocate and deploy their resources in the most efficient manner are likely to achieve a greater return on capital employed than those that do not. There are three aspects to the analysis of a business' resources:

- *The resources themselves*. Resources can be a competitive advantage, as rivals may not have access to the same resources and may not be able to duplicate similar resources within their own business in terms of experience and number. This is the central principle of the resource-based view of competitive advantage and can be analysed using tools such as VRIO (see Appendix 9).
- *The configuration of the resources*. Resources can be a source of competitive advantage – an opportunity. If a business configures its resources optimally, it will have a competitive advantage over its rivals. This view is central to the value chain and value system concept of competitive advantage. An analytical tool for adding value throughout a business, value chain analysis, is described in Appendix 10.
- *The resource audit*. This covers operational, human and financial resources. The objective of the resource audit is to identify resources and ascertain how effectively resources are utilised and deployed. The contents of a resource audit are discussed in Appendix 11.

8.8.7 Change Management

No one solution fits all situations. Hence, change leaders cannot afford the risk of blindly applying a standard change recipe and hoping it will work. Successful change takes place on a path that is appropriate to the specific situation. The factors common to successful change projects are described in Appendix 12.

8.8.8 Marketing Plan

Rivalry among firms is the central force determining a business's competitive position. It is therefore necessary to analyse competitors. The elements of a competitor analysis are:

- current strategy or positioning;
- strengths, weaknesses, opportunities, threats;
- possible changes in strategy;
- financial strength;
- operational strength;
- resource strength;
- research and development strength.

Markets are becoming more complex and unpredictable and technology and information flows permit companies to sense and react to competitors at a faster rate. This accelerated competition means it is no longer possible to wait for a competitor to make a move before deciding how to react. The new watchwords are anticipation of and preparation for every eventuality. A competitor's every move is met with a rapid countermove to ensure any advantage is temporary. Sony, a household name in digital consumer electronics, was both surprised and severely affected by the inroads Apple made with its iPod, in a market that Sony had pioneered with its Walkman. The most intense rivalries spawned the cola wars, where every move Coca

Cola made was met by Pepsi Cola and every initiative by Pepsi was quickly countered by Coke. More recently in the world of telecommunications, every advertisement by MCI immediately stimulated a response by AT&T, and vice versa. As soon as Kodak launches a new disposable camera, it appears Fuji will have a similar model ready for the market. Currently banks and building societies are fighting credit card wars, where offers are quickly matched. BSkyB, BT, Cable & Wireless, Carphone Warehouse, O2 (Telefónica), Orange (France Telecom) and Virgin Media now compete for the lion's share of the broadband market. No company can afford to let its rivals gain an obvious advantage for long. Hence, a company that brings a good new product to the market cannot be complacent. For example, when in the 1970s Okidata released an excellent dot-matrix line printer and won a significant share of the market, Hewlett-Packard responded by offering the laserjets. These were a family of highly reliable printers based on a technological breakthrough that made them faster and quieter, with greater resolution. Okidata stubbornly continued to market its dot-matrix printer, losing significant market share to Hewlett-Packard. Research and development is critical to compete in the market place and developments by competitors can be so radical that they change the market landscape shifting consumer needs. These are what are termed industry breakpoints – the laserjet printer represented such a breakpoint. Breakpoints are discussed in more detail in Appendix 13.

8.8.9 Compliance Systems

Any analysis of a business should understand the regulatory framework within which the business operates – this will be particularly important for the pharmaceutical, utility, defence, nuclear and financial sectors. Business pressures, arising from the regulatory regime and the repercussions for failing to comply, need to be understood and captured.

8.9 SUMMARY

The implementation of the analysis process of establishing the context (the first stage in the overall risk management process) is critical to the quality of any risk management study. Its execution will have a direct bearing on the relevance, breadth, depth and currency of the information available to provide insights into the business so that identification and analysis can be conducted in a meaningful and not a superficial way. Assessment of the context will be vital in acting as a prompt to interrogate the sources of risk, determine the essential participants in any identification process and identify the subjects that will warrant closer inspection. Any context assessment will be tailored to suit the objectives of any risk management study. The process activities are likely to look at the business objectives, plan and processes, financial statements, resources, change management and marketing plan.

8.10 REFERENCES

Friend, G. and Zehle, S. (2004) *Guide to Business Planning*, p. 48. Profile Books Limited, London.

Hillson, D. (1997) Towards a risk maturity model. *International Journal of Project and Business Risk Management*, 1(1), 35–45.

Hopkinson, M. (2000) Risk maturity models in practice. *Risk Management Bulletin*, 5(4).

Hunt, V.D. (1996) *Process Mapping: How to Reengineer Your Business Processes*, John Wiley & Sons, Inc., New York.

Office of Government Commerce (2002) *Management of Risk: Guidance for Practitioners*. The Stationery Office, London.

Risk Management Research and Development Program Collaboration (2002) Risk Management Maturity Level Development, April. Formal Collaboration between INCOSE Risk Management Working Group Project Management Institute Risk Management Specific Interest Group and UK Association for Project Management Risk Specific Interest Group. http://www.pmi-switzerland.ch/fall05/riskmm.pdf

Rummler, G.A. and Brache, A.P. (1994) *Improving Performance: How to Manage the White Space on the Organization Chart*. Jossey-Bass, San Francisco.

9

Risk Identification: Stage 2

To win big sometimes you have to take big risks.
(Jim Collins)

The previous chapter examined Stage 1 – establishing the context of the business – which is a prerequisite to undertaking risk identification. This chapter examines the risk identification process through the lens of the IDEFO process mapping technique. All risk management process frameworks state a need to identify risk events (upside and downside) at the outset of activities. Identifying risks requires undertaking two key activities: thinking through and recognising the *source* of the risks and opportunities (upside risks) and *searching* out and identifying both the risks and opportunities. What makes the identification process interesting is that the market place is in a constant state of flux. The risks identified to a business last week will not be entirely the same as the risks identified to the same business next week. For global businesses there is greater immediacy, where the risks and opportunities identified yesterday may well be different than those identified tomorrow. The structure of this chapter is illustrated in Figure 9.1. The next chapter examines the process of risk analysis and the attendant benefits.

9.1 PROCESS

Risk identification is a transformation process (commonly facilitated by a risk practitioner) where experienced personnel generate a series of risks and opportunities, which are recorded in a risk register. As risk identification is one process in the overall risk management process it is useful to adopt the philosophy of process mapping, where each process exists to make a contribution to one or more of the risk management goals. Hence, each process should be measured against specific process goals that reflect the contribution that the process is expected to make to the risk management study. Processes are simpler to comprehend when they have primary goals and subgoals. Therefore, risk identification is described here as having a primary process goal which is accomplished by a series of subgoals (see below). Any one process is accomplished within a context and might be considered to have two perspectives. The *external view* examines the process inputs, outputs, controls and mechanisms. The *internal view* examines the transformation process, where inputs are transformed to outputs through the application of the process mechanisms. This process is subject to modification and influence by potential controls.

9.2 PROCESS GOAL AND SUBGOALS

The primary process goal of risk identification is to identify both the *risks* to the business, which would reduce or remove the likelihood of the business reaching its objectives, and the

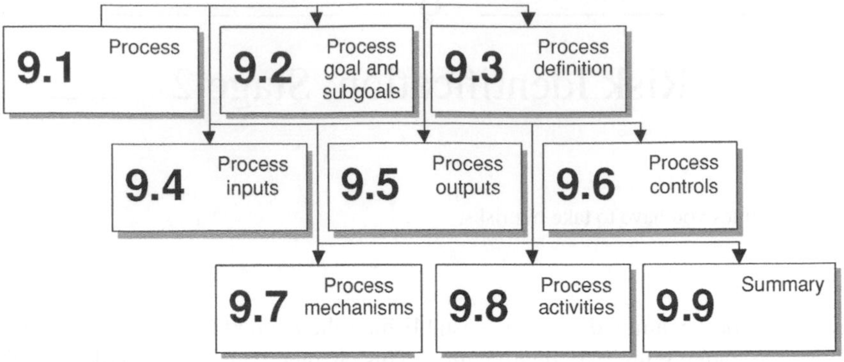

Figure 9.1 Structure of Chapter 9

opportunities, which could enhance business performance. The risk and opportunity identification process will have been sufficient when it has satisfied these subgoals:

- The overall management of the business activity was understood.
- The risk identification process was not commenced before the business objectives (or the objectives of the activity under examination) were made explicit. (Risks are only threats to objectives. Without understanding the objectives it is not possible to undertake risk identification.)
- Risk identification was not commenced until the business objectives, deliverables and success criteria were aligned.
- Risk identification was not commenced prior to a "map" or flow chart of the business process being prepared.
- The risk identification process was comprehensive, examining all primary sources of risk and opportunity. The process of identification was undertaken with the aid of a process map or risk breakdown structure (RBS), which included all core business activities.
- Personnel were involved from all appropriate company departments and no departments were excluded or forgotten.
- Department representatives participating in the identification process were senior enough to be knowledgeable in their area of specialisation and were aware of both corporate lessons learnt and company risk exposure.
- Consideration had been given to consultation with non-executive directors and where appropriate they had been included in the risk identification process.
- Risk descriptions were comprehensive and comprehendible to all the participants in the identification process.
- The risk descriptions were carefully considered to avoid them becoming a mixture of causes and effects. Programme overrun, for example, is not a risk but the effect of a risk.
- The interdependencies between the risks were identified.
- Consideration was given to the three views of risk proposed by NASA: known knowns, known unknowns, and unknown unknowns.

9.3 PROCESS DEFINITION

The risk identification process is described by an IDEFO diagram (see Figure 9.2). The diagram describes a transformation process with inputs entering on the left of the box, outputs leaving

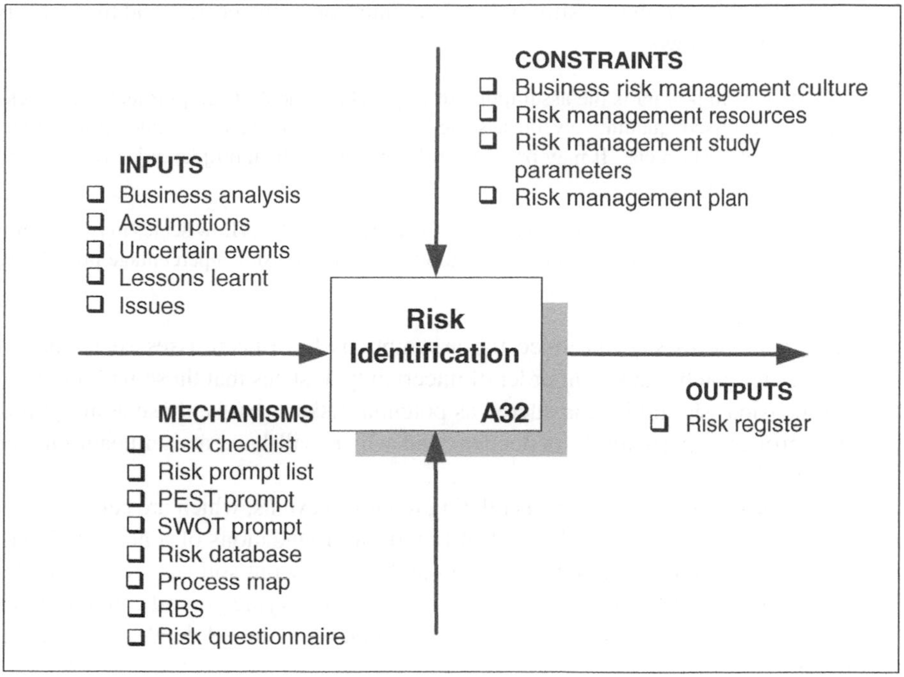

Figure 9.2 Risk identification process

on the right of the box, controls (constraints) entering from above and mechanisms (enablers) entering from below.

9.4 PROCESS INPUTS

The inputs to the identification process are those listed in Figure 9.2. There is clear disagreement in the literature regarding the definition of risk and uncertainty. The debate, discussed by Chapman (1998a), continues to unfold without resolution. The definitions of uncertainty and risk adopted below have been applied successfully to risk commissions across the rail, water, health, construction, heritage, media and petrochemical industries. From experience, the chosen definitions of the terms need to be explained to the participants at the outset of the study and their use consistently applied throughout the course of the study.

- *Assumptions* are statements of belief concerning the outcome of future events. These future events may be aspects of a project or business venture. They are assessments made at a point in time, which are assumed to be correct but are not borne out by facts and hence not proven. Assumptions can be classified as risks to an activity until such time that sufficient information has been obtained either to confirm a supposition or disprove it. Assumptions can cover such issues as competitor behaviour, size of market, potential changes in the market, the availability of resources and increases in fuel or energy costs. For projects, one particular area where assumptions are commonly made is the performance of contractors, subcontractors and suppliers. In the context of projects, the Project Risk Analysis and Management (PRAM) Guide (Simon *et al.* 1997) describes a test to establish whether an assumption should be considered as part of risk identification. The test is composed of two questions which relate to the sensitivity of the assumption to the project and the stability of the assumption as follows:

> Sensitivity: how important is the assumption to [the] achievement of [the] project objectives? This can be assessed qualitatively on a scale such as: insignificant, of minor importance, important, vital, [or] crucial. It may prove useful to identify which project objectives depend on each assumption.
>
> Stability: how likely is the assumption to prove false? High/Medium/Low scales may be useful for this, or a more direct assessment of likelihood can be made using a percentage probability estimate.

The Guide goes on to say that once the assumptions have been assessed in these two dimensions, they can be ranked in order of uncertainty. It states that those with high scores in both dimensions should be considered as potential risks – that is, those assumptions on which the project is particularly dependent and which will have a significant chance of proving to be false.

- *Uncertain events* are described here as relating to known events, which are certain to occur, but their magnitude is unclear. The cost of bored pile foundations of a new headquarters building is an uncertain event. While it is possible to carry out soil surveys, calculate the ground-bearing ratio and complete a piling design, due to varying ground conditions across a site, it is not until the auger has bored the hole that the required depth of a pile can be determined.
- *Lessons learnt* are the accumulated knowledge drawn from a completed activity or project in terms of what went well and should be repeated and what did not go well and should be improved upon, should a similar activity be embarked upon in the future. Capturing lessons learnt is essential for informed decision making and business improvement to avoid the repetition of actions that had an unfavourable outcome and capitalise on the positive outcomes.
- *Issues* are matters that are identified and recorded as unresolved, in terms of the approach to be adopted. Over time an issue may be removed or translated into a fixed cost, an uncertainty or a risk.
- *Business analysis* is the group of findings arrived at from conducting the activities described in the previous chapter.

9.5 PROCESS OUTPUTS

The output is a risk register. The content of the register needs to be tailored to the task undertaken. The typical contents of a risk register are included in Table 9.2 (see Section 9.7.10). The register should be structured using the business case section headings, an RBS, a risk taxonomy, or, where the risk study relates to an investment decision such as a new building, the work breakdown structure or project life cycle. The risk register is a key communication tool as it is referred to and incrementally developed throughout the overall risk management process.

9.6 PROCESS CONTROLS (CONSTRAINTS)

From Figure 9.1 it is suggested here that the business risk management culture, resources, risk management system and plan (where one exists) regulate/constrain the risk identification process. These controls were discussed in Section 8.6.

9.7 PROCESS MECHANISMS (ENABLERS)

9.7.1 Risk Checklist

A risk checklist, as described by the PRAM Guide (Simon *et al.* 1997), is an in-house list of risks "that were identified on previous projects". Projects in the context of enterprise risk are either capital investment projects or business activities. Risk checklists are often developed from managers' past experience. Checklists permit managers to capture lessons learnt and assess whether similar risks are relevant to the business activities of today.

9.7.2 Risk Prompt List

A risk prompt list, as described by the first edition of the PRAM Guide (Simon *et al.* 1997), is a list which "categorises risks into types or areas". The HM Treasury guide (HM Treasury 2001) known as the *Orange Book*, which aims to provide "pointers to developing a strategic framework for the organisational consideration of risk", provides a schedule of what it considers the most common categories of risk with examples of source and effect for each category (see Appendix 2). Cooper (2004), Day (2001) and Holliwell (1998) also suggest risk categories, as does BSI PD (2000) (see Appendix 2). The *Orange Book* (HM Treasury 2004) revised categorisation of risk is based on the PESTLE model as described at www.strategy.gov.uk, where the acronym stands for "political, economic, social, technological, legal and environmental" (see Appendix 8). Prompt lists have to be used with caution, as described in the second edition of the PRAM Guide, as by their nature they may be too exhaustive or too project-specific (Risk Management Specific Interest Group 2004).

9.7.3 Gap Analysis

Gap analysis can be used to draw out the main risks to an activity or project and is commonly carried out by calling upon department heads to complete a questionnaire. An extract of a sample questionnaire is included in Figure 9.3.

- The questionnaire calls for an assessment to be made as to the current status of an activity in terms of how well it has been completed to date.
- The questionnaire identifies the two extremes in terms of the worst position for the activity (column headed "1") and best (column headed "5").

Ref	Process	Issue	Worst Condition	1	2	3	4	5	Best Condition
2.1	Planning	Organisation	Decision-making process and its requirements (within the client organisation) not clearly communicated		X			Y	Decision-making process and its requirements clearly communicated

Figure 9.3 Structure of questionnaire

1	2	3	4	5
Activity not commenced/ undertaken	Activity commenced but in outline only	Activity partially completed	Activity almost completed	Activity completed satisfactory
Critical Risk	Major Risk	Significant Risk	Minor Risk	No risk to the project

Figure 9.4 Definition of categories of risk

- Recipients of the questionnaire are requested to score each row/line item by inserting an "X" in the column which denotes their perception of the current project position and a "Y" in a second column to denote the realistically achievable position.
- Recipients of the questionnaire are required to comment on all issues (as far as possible). If they are unable to comment on an item they are requested to write "IK" in column 1, indicating that they have either no or insufficient knowledge to comment on this activity/issue.

The meaning behind columns 1–5 is explained in Figure 9.4.

9.7.4 Risk Taxonomy

The business risk taxonomy (BRT) proposed here provides a structured checklist that organises known enterprise risks into general classes subdivided into elements and attributes. Attributes can be further subdivided into features if this is found to be productive. A taxonomy enables risk and opportunity to be broken down into manageable components that can then be aggregated for exposure measurement, management and reporting purposes. The BRT is based on a software risk taxonomy developed by the Software Engineering Institute of Carnegie Mellon University, Pittsburgh, Pennsylvania, USA (Carr *et al.* 1993); see Figure 9.5.

The BRT provides a framework for studying business management issues and is a structure for eliciting risks from commonly recognised risk sources in the business environment. Source information for the BRT is included in Appendix 2. The taxonomy proposed here organises business risks into four levels (as opposed to the three levels in the Carr *et al.* model) – class, element, attribute and feature – and is illustrated in part in Table 9.1. The taxonomy is organised into twelve major classes, which are further divided into elements. For example, the "operational" class has five elements. Each element in turn is broken down into its attributes. An example of features is included under "strategy" in Table 16.2. The proposed taxonomy incorporates and builds on Annex 7, entitled "Detailed Loss Event Type Classification", sometimes called the Basel matrix, included in the revised framework for measuring capital adequacy known as Basel II. Annex 7 relates solely to operational risk.

Each of the sources of risk is examined in Part III, commencing with Chapter 15. For risk identification to be effective it needs to be comprehensive. So while the assignment of elements to classes varies between regulators, authors and practitioners, the important thing to remember is that their inclusion within a taxonomy is more important than their precise location, as unidentified risks are unmanaged risks. Due to the nature of risk, the boundaries

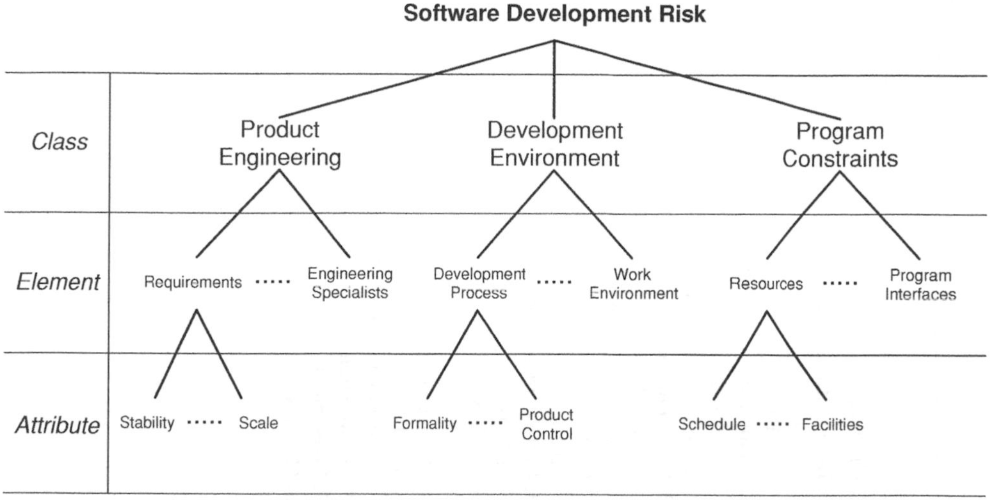

Figure 9.5 Software development risk taxonomy (Carr *et al.* 1993). Special permission to reproduce *Taxonomy-Based Risk Identification*, © 1993 by Carnegie Mellon University, is granted by the Software Engineering Institute

between classes are sometimes not clear and each business must decide for itself where sources of risk will reside in its bespoke taxonomy. Information risk is not listed as a class of its own as information risk is assumed to be inherent in all of the elements, in that the quality of the information gleaned when investigating these risks will directly relate to the quality of the decisions made when using the information. Each business should develop its own taxonomy. The taxonomy for a bank will be markedly different from that for an aircraft manufacturer, which in turn will be markedly different from that for a petrochemical company. Any taxonomy produced should be maintained as a live document, to reflect the changing business environment.

9.7.5 PEST Prompt

PEST analysis was previously discussed in Section 8.7.4. It is referred to here as it is another tool for uncovering risk exposure. Completing a PEST analysis is very simple and can be done during a workshop or as part of a brainstorming session. A PEST analysis measures a market (a SWOT analysis measures a business unit, a proposition or an idea). The PEST model can be expanded to seven factors by adding ecological, legislative and industry analysis (the model is then known as PESTELI). To be effective prior to the commencement of a PEST analysis, the subject must be made clear to the participants so that they properly understand the goals. Hence, the PEST subject should be a clear definition of the market being addressed, which might be from any of the following standpoints: a company looking at its market; a specific business unit; a product looking at its market; a brand in relation to its market; a strategic option or an investment opportunity. Common and beneficial applications of PEST are providing greater understanding and insights into competitors and market position. Guidance on implementing a PEST analysis is included in Appendix 8.

Table 9.1 Business risk taxonomy

Internal Processes		
Financial	**Operational** *(Class)*	**Technological**
1. Liquidity risk	**1. Strategy** *(Element)*	**1. Information technology**
2. Credit risk	(a) objectives (*attribute*)	(a) software
(a) default	(b) business plan	(b) MISs
(b) exposure	(c) new business development	(c) intranets
(c) recovery	(d) resources	(d) telematics
(d) counterparty	(e) stakeholder interests	(e) information assets
3. Borrowing risk	(f) corporate experience	**2. Communication**
4. Currency risk	(e) reputation	(a) broadband
5. Funding risk	**2. People**	(b) video conferencing
6. Foreign investment risk	(a) HR management practices	(c) e-commerce
(e) country risk	(b) salaries	(d) e-mail
(f) environment risk	(c) regulatory and statutory requirements	**3. Control technology**
7. Derivatives	(d) staff constraints	(a) CAD
	(e) staff dishonesty	(b) CAM
	(f) risk management system	(c) FMS
	(g) health and safety	(d) Mechatronics
	3. Processes and systems	(e) MRP
	(a) controls	(f) operational research
	(b) regulatory and statutory requirements	**4. IT governance**
	(c) continuity	**5. Investment**
	(d) indicators of loss	**6. IT projects**
	(e) transactions	
	(f) computer/IT system	
	(i) knowledge management	
	(j) project management	
	4. External events	
	(a) change management	
	(b) business continuity/interruption	
	5. Outsourcing	

Project	Ethical	Health and Safety
1. Embedding project risk management	**1. Source**	**1. Health and safety system**
2. Risk management process	**2. Recognition**	**2. Workplace precautions**
3. Project team	**3. Factors that affect ethics**	**3. Contribution of human error to disasters**
4. Optimism bias	**4. Risk events**	**4. Improving human reliability in the workplace**
5. Tools	**5. Implementation**	**5. Risk management best practice**
6. Techniques		
	Business Operating Environment	
Economic	Environmental	Legal
1. Macroeconomics	**1. Energy sources**	**1. Companies**
2. Microeconomics	**2. Use of resources**	**2. Intellectual property**
3. Government policy	**3. Pollution**	**3. Employment law**
4. Aggregate demand	**4. Global warming**	**4. Contracts**
5. Aggregate supply	**5. Levies/emission controls**	**5. Criminal liability**
6. Employment levels	**6. Environmental sustainability**	**6. Computer misuse**
7. Inflation		
8. Interest rate		
9. House prices		
10. International trade + protection		
11. Currency risk		
Political	Market	Social
1. Contracts	**1. Market structure**	**1. Education**
2. Transition economies	(a) number of firms	(a) general level
3. UK government fiscal policies	(b) barriers to entry	(b) language skills
4. Pressure groups	(c) new entrants	**2. Population movements**
5. Terrorism and blackmail	(d) homogeneous goods	(a) location
	(e) knowledge	(b) age mix
	(f) relationships	(c) pensions
	2. Product life cycle stage	(d) "grey market"
	3. Alternative strategic directions	**3. Socio-economic patterns**
	4. Acquisition	**4. Crime**
	5. Game theory	(a) business vulnerability
	6. Price elasticity	(b) staff relocation
	7. Distribution strength	**5. Lifestyles and social attitudes**

9.7.6 SWOT Prompt

A SWOT analysis can be used to draw out the risks and opportunities facing an enterprise and has the advantage of being quick to implement and readily understood. Analysis of the strengths, weaknesses, opportunities and threats brings together the results of both analysis of the company (internal) and environmental analysis (external). The results of a PEST analysis (discussed above) can be used to inform the environmental analysis. The process of creating a SWOT analysis is valuable because it involves discussion among the key managers in the business. It stimulates thinking that is not overly structured or restrictive. Implementation of a SWOT analysis is discussed in Appendix 7.

9.7.7 Database

A risk database has a number of benefits. A newly constructed database can be used to capture information in a controlled and consistent way. Depending on its construction, its availability on a computer network and the access rights established, it can permit multiple users to enter data or view the current information held. The fields that a database would commonly hold are those listed below. Databases provide an audit trail of identification, assessment and implementation of management actions. Where risk information is collected on the completion of a project as part of a lessons learnt study it can be used to inform subsequent projects. This is accomplished by recording the risks that materialised, how they were addressed, budget and outturn costs, contingency allocation and spend and achievement against objectives.

- Risk ID
- Date of entry
- Status of the risk
- Originator
- Owner
- Actionee
- Manager
- Risk category
- Risk description
- Probability
- Consequence (cost)
- Consequence (time)
- Consequence (business activity)
- Risk date
- Project
- Phase
- WBS
- Business function
- RBS element
- Risk response category
- Risk action
- Cost of response
- Indicator
- Trigger

A populated risk database established during previous analysis, provided it relates to a similar subject and is comprehensive in terms of the fields that it contains, may be used as an enabler in a variety of ways:

- It may help the construction of an RBS.
- It can provide a register of risks to use as a prompt during interviews.
- It can be used as an interrogation tool to learn about risks associated with a particular subject.
- It can assist with possible risk responses to the risks identified during the current analysis.
- It can tell you who in the organisation dealt with a particular category of risk in the past.

9.7.8 Business Risk Breakdown Structure

A business RBS is defined here as "A hierarchical decomposition of the business environment through to business processes, assembled to illustrate potential sources of risk. It organises and defines the total extent of business operations established to accomplish the business objectives. Each descending level represents an increasingly detailed definition of sources of risk to the business". Business RBSs have their roots in project management work breakdown structures (WBSs). The WBS is considered a key planning tool used to define projects in terms of their deliverables while providing a method for breaking these deliverables into meaningful subsets. By defining projects in this way, the WBS enables project managers to clearly describe the hierarchical nature of the work to be done and provides a consistent language or code to be used in other elements of formal project planning such as the resource plan, budget, organisational plan and master schedule. The Project Management Institute (PMI) of the USA defines a WBS as "A deliverable-orientated hierarchical decomposition of the work to be executed by the project team to accomplish the project objectives and create the required deliverables. It organises and defines the total scope of the project work. Each descending level represents an increasing detailed definition of the project work" (PMI 2008). An earlier definition of the WBS provided by PMI was the basis of Hillson's definition of a project RBS. Hillson (2002) describes a risk breakdown structure (RBS) in the context of projects as "A source orientated grouping of project risks that organises and defines the total risk exposure of the project. Each descending level represents an increasingly detailed definition of sources of risk to the project." Early hierarchical structures were developed for software development (Dorofee *et al.* 1996), construction (Chapman 2001) and a high-voltage transmission line (Tummala and Burchett 1999). Chapman has produced specific RBS structures for different project types within the rail, pharmaceutical, heritage and construction sectors.

9.7.9 Risk Questionnaire

A risk questionnaire aims to elicit, through a series of questions, issues that are unresolved, incomplete, giving rise for concern, delayed, uncoordinated, appear to be in a rapid state of change, uncertain and so on. The objective is to elicit as much information as possible while at the same time not deterring the recipients from responding. The questionnaire should be structured in a logical manner, such as reflecting a project life cycle, process map or a similar structure, which the recipients will immediately understand and be able to relate to.

9.7.10 Risk Register Content/Structure

The value of a risk register is its ability to capture information in a consistent manner and to simplify communication. The layout of the register (reading from left to right) should reflect the sequence in which information is captured. For document management purposes the register should carry a title (reflecting the project or business activity), date, version number, author and file reference. The typical content of a risk register is recorded in Table 9.2.

Table 9.2 Typical content of a risk register

Document control	
Title	Title of the register
Author of the register	Originator of the register and point of contact for any questions regarding the content
Date register compiled	Date of issue
Issue number	Unique issue number
File reference	The location at which document can be found on the server
Register content	
Risk identifier	Unique number to identify the risk
Risk category	Risk subject area
Risk cause	The activity or event which will trigger the risk event
Risk description	Full description of the risk which will be readily understood by all of the business leaders (or project team) on completion of the identification process and 12 months later
Risk impact	The impact of the risk should it materialise such as delay, increase in costs, reduction in quality, increase of cost-in-use, environmental incident and so on
Risk status	The commonly adopted descriptions are "active" (or "open"), "closed" and "OTBE" (overtaken by events)
Probability	Assessment of how likely the risk is to happen. The probability can be recorded as a percentage, a category or both
Impact	Impact can be measured in terms of cost, duration, quality or any other business or project objective
Proximity	Reflects the timing of the threat of the risk. Is its threat strongest at a particular point in time? Does its probability or impact change over time?
Risk response category	The terms adopted here are reduce, retain, remove or transfer
Owner	"Owner" refers to business entity that will be affected by the risk should it materialise. For instance, if the register is related to an investment decision such as a new office building, the *owner* column would most probably be populated with "client", "contractor", "insurer" or "nominated supplier"
Manager	The individual responsible for agreeing and overseeing the implementation of the risk response action
Actionee	The individual responsible for implementing the risk response action under the direction of the risk manager
Risk response action	A description of the specific action or actions decided upon to address the identified risk to remove, reduce, retain, transfer or share the risk
Planned start	Planned commencement date of the risk response action
Planned finish	Planned completion date of the risk response action
Expected value	Calculated by multiplying the average impact by the probability percentage

9.8 PROCESS ACTIVITIES

The activities of the risk identification process are the tasks necessary to capture risks and uncertainty and record them in a risk register, log or list. These consist of the following:

- Clarifying and recording the business objectives or business objectives subset under examination. You cannot identify the risks to the objectives without knowing the objectives in the first instance. All very obvious, but it is surprising how often the objectives are not clear or project team members have differing views on what the objectives are.
- Reviewing the business analysis (see previous chapter).
- Identifying the risks and opportunities to the objectives as comprehensively as possible using the information gained from the business analysis to act as prompts.
- Gaining a consensus on the risks and opportunities, their description, their interdependencies and how they would impact on the business.
- Documenting the risks and opportunities.

9.8.1 Clarifying the Business Objectives

At the outset of the risk identification process the business or activity objectives must be made clear, as the primary objective of the process is to identify the threats or opportunities to those objectives. Where objectives, deliverables and success criteria are stated they must be aligned. The deliverables and success criteria must spring directly from the objectives. Lists of objectives should not be a mixture of primary and secondary objectives.

9.8.2 Reviewing the Business Analysis

This process activity will examine the findings from the business analysis described in the previous chapter. Depending on the study objectives, one or a combination of the following areas may be examined for sources of risk and/or opportunity:

- Business plan
- Market
- Change management
- Acquisition
- Regulatory compliance
- Resources
- Risk processes
- Value chain
- Financial ratios
- Audit committee roles and responsibilities
- Process map

9.8.3 Need for Risk and Opportunity Identification

The need for systematic risk and opportunity identification, as described in this risk management process, is predicated on the following assumptions:

- All business activities including change management, capital projects, acquisitions, counterparty contracts and supply chain management are exposed to risk.
- Business risks are generally known by management but are poorly communicated.
- A structured and repeatable method of risk identification is necessary for consistent and auditable risk management.
- Ad hoc approaches lead to "blind spots" and unidentified risks.

- A formal non-judgemental, non-attributive environment is required to provide a setting whereby alternative or controversial views can be heard.
- The identification and existence of risk is not a criticism of management performance.
- Opportunity identification is as important as the identification of risks.

9.8.4 Risk and Opportunity Identification

Risk identification can be conducted in a number of ways and is a facilitated process typically adopting one or a combination of the following: questionnaires (including the Delphi technique), interviews or interactive workshops using brainstorming, scenario analysis, systems dynamics or the nominal group method. Risk and opportunity identification is commonly a group-oriented approach that draws on the combined knowledge and experience of the individuals selected to participate. Depending on the size of the business and particularly where business premises are geographically dispersed, identification may be carried out by e-mail, video conferences and questionnaires. Dangers to be avoided are where the identification process mirrors the structure of the organisation and the interdependencies between departments and hence between the risks are missed. This pattern can be reflected in risk management. While many firms have invested in enterprise risk management, they frequently view risk in silos, often leaving themselves blind to relationships between risks (Kambil *et al.* 2005). Kambil *et al.* suggest that the first essential step in gaining a comprehensive view of risk interdependencies is to build an integrated risk management function, championed and supported by senior management that sits above all divisions and departments. The purpose of this group is to identify the key risks across the corporation, understand the connections between them and develop a risk management strategy that takes into consideration the organisation's appetite for risk. Kambil *et al.* cite the example of a large multinational bank, which integrates risk at the time strategies are developed, rather than planning for risk after a strategy has been established. Central to their approach is the examination of risks holistically, rather than in isolation. For instance, when considering risks in the underwriting process, the bank assesses how its business strategy, sales practices and business development practices affect the risk profile. Figure 9.6 illustrates one organisation's view of the merits of the alternative approaches to risk identification. Circumstances, however, may dictate the approach to be adopted due to time, geographical or personnel constraints.

9.8.5 Facilitation

The facilitator's role involves planning the means of eliciting and recording the risks to the business activity or project under examination and their assessment. It involves controlling and leading a team through a process using analytical, arbitration, guiding and influencing skills (Kelly *et al.* 2004). For ERM this entails recognising the constraints of the study and selecting the best technique for identifying risks and opportunities to suit the circumstances. Consequently, for interactive workshops (and interviews), the responsibilities of the facilitator as follows:

- *Timing*. Agreeing with the study sponsor the date and time of the workshop.
- *Physical environment*. Selecting an appropriate room which will comfortably accommodate the agreed number of attendees in an appropriate seating arrangement such as a horseshoe formation. The room should have appropriate fixtures and fittings such as blackout facilities,

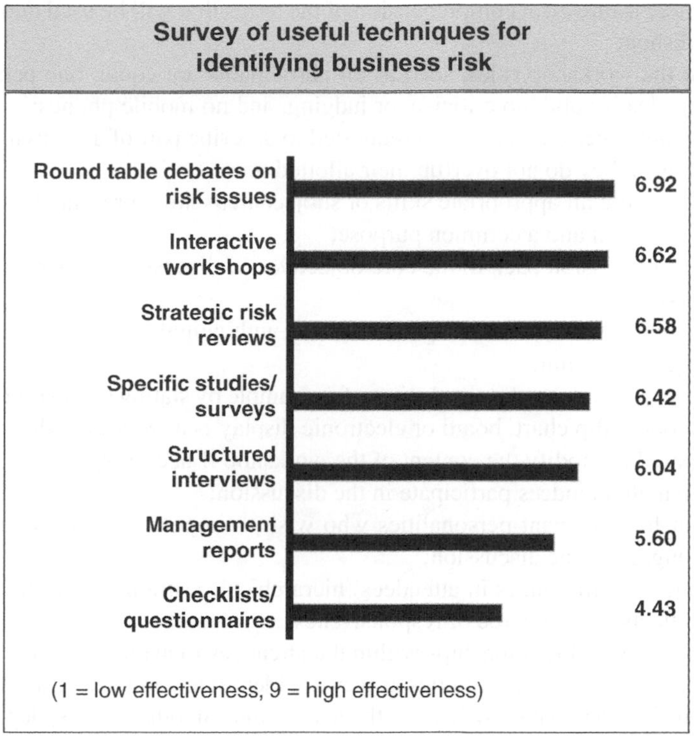

Figure 9.6 Techniques for identifying business risk (Jones and Sutherland 1999)

power outlets and pin-up space. Additionally, when required, the room should be furnished with a white wall, whiteboard or screen to accommodate the combined use of a notebook computer with a projector and/or flip charts. The room should be sufficiently remote from attendees' normal place of work that there is no possibility of interruption. For this reason hotel conference rooms are a common choice.

- *Arranging attendees to maximise an effective outcome.* Agreeing with the study sponsor the appropriate workshop attendees to ensure that the right skills and experience are present to strive to make identification as broad as possible to avoid "blind spots".
- *Producing an agenda.* A structured realistic agenda should be developed which is not overambitious, reflects the objectives of the workshop, includes appropriate breaks and examines the aspects of interest.
- *Preparing and forwarding a briefing pack.* The briefing pack should contain the time, date, location, purpose of the workshop, attendees, agenda, workshop rules, preparation required, risk management terms that will be used, status and background to the business activity or project, presentations, outputs and any other relevant information.
- *Managing the process of the workshop.* This involves:
  - stating the objective of the workshop at the outset;
  - gaining consensus to the workshop outputs;
  - "walking through" the agenda at the outset to give attendees a clear understanding of the intended course of events;

- ensuring there is universal comprehension of the terms that will be used during the course of the workshop;
- setting out the workshop rules, such as all participants are equal, one person talks at a time, every idea is valid, no criticism or judging, and no mobile phones;
- ensuring, where speakers have been requested to describe part of an activity or project to "set the scene", they do not overrun their allotted time;
- planning to ensure all appropriate skills or subject areas are represented;
- providing direction and a common purpose;
- bringing the discussion back to the core objectives, if discussion centres for too long on a detailed issue;
- preventing individuals from having a discussion with neighbours;
- maintaining momentum;
- ensuring everyone agrees the conclusions, for example by stating that everything recorded on the notebook, flip chart, board or electronic display is a record of what is agreed;
- being prepared to modify the content of the workshop if necessary;
- ensuring that all attendees participate in the discussion;
- looking out for dominant personalities who wish to impose their views and ensure all attendees engage in the discussion;
- being attuned to differences in attendees' hierarchical position within the company and how this reflects in the degree of responsiveness;
- sensing interpersonal relationships within the attendees to avoid point scoring;
- recording the risks and opportunities as they are identified and gaining a consensus to the descriptions in a way which will stand the test of time (in other words, department heads will understand what was meant by the risk description three months later).

Facilitation is distinguishable from meeting chairmanship in that the facilitator is not normally a business employee or a member of the project team, contributes nothing more than facilitating skills and has no vote and certainly no casting vote in decision making. There are distinct advantages in not selecting a facilitator from a business function (or the business as a whole) as it avoids problems of bias, lack of independence, hidden agendas and distortion of focus to permit pursuit of personal or departmental goals. To accomplish the aims of facilitation it is common for the facilitator to adopt one of the seven techniques described below, commencing with brainstorming.

Brainstorming

The brainstorming process, borrowed from business management and not specifically created for risk management, involves redefining the problem, generating ideas, finding possible solutions, developing selected feasible options and conducting evaluation (Chapman 1998b). Originated by Osborn (1963) in the early 1950s, brainstorming was proposed as a problem-solving method which would produce a much larger quantity of ideas in less time than existing group problem-solving techniques. In the third revised edition of his text entitled *Applied Imagination* originally published in 1953, Osborn argues that the effectiveness of brainstorming is derived from two essential components. These components are described succinctly by Johnson (1972): group thinking is more productive than individual thinking; and the avoidance of criticism improves the production of ideas. Concerning the first component, Osborn explains that the generation of more ideas by group activity as opposed to individuals working on their own is the result of "free associations" – that is, the generation of suggestions triggered by suggestions voiced by other group members, a process Osborn refers to as "social facilitation". The second component is what psychologists recognise as "reinforcement", where "correct" answers are reinforced by "rewards" in the form of immediate corroboration. In brainstorming, "reinforcement" is provided by "rewarding the suggestions with receptiveness, or suspending criticism". Osborn states that "deferment of judgement is the essence of group brainstorming". Osborn argues that the creative "climate" of brainstorming is only achieved by strict adherence to the four rules of brainstorming:

- criticism is ruled out – evaluation of ideas must be withheld until later;
- "free-wheeling" is encouraged – "the wilder the idea the better";
- quantity is wanted – the greater the number of ideas, the greater the chance of having useful ones;
- combination and improvement – try to "build" on other people's ideas.

Structured or Semi-Structured Interviews

Structured interviews are where individual interviewees are asked a set of prepared questions from a prompt sheet which encourage the interviewee to view a situation from a different perspective and thus identify risks from that perspective. A semi-structured interview is similar, but allows more freedom for a conversation to develop to explore issues as they arise. Structured and semi-structured interviews are useful where it is difficult to get people together for a brainstorming session – for instance, where they are geographically dispersed or where free-flowing discussion in a group is not appropriate for the situation or individuals involved. Interviews may be adopted at any stage of a project or process. They are a means of providing stakeholder input to risk assessment. They are commonly used to identify risks or to assess effectiveness of existing response plans as part of ongoing risk management.

Prior to the interviews taking place there must be a clear definition of the interview objectives, a list of interviewees selected from relevant stakeholders and a set of questions prepared to guide the interviewer. The questions should ideally be open-ended, readily understood, in an appropriate language for the interviewee (i.e. using familiar terms and not abundant acronyms) and cover one issue at a time. Possible follow-up questions to seek clarification are also prepared. Questions are then posed to the person being interviewed. When seeking elaboration, questions again should be open-ended. Care should be taken not to "lead" the interviewee. Responses should be considered with a degree of flexibility in order to provide the opportunity of exploring areas into which the interviewee may wish to go. The outputs are the stakeholder's views on the issues which have been raised by the interview questions. Interviews of this nature they have the advantage of allowing people time to consider their responses, allow more in-depth dialogue and engage with a greater number of stakeholders than, say, brainstorming. Their drawback is that they can be very time-consuming.

Nominal Group Technique

The nominal group technique (NGT) was developed by Delbecq (1968). It was derived from social-psychological studies of decision conferences, management science studies of aggregating group judgements and social work studies. Delbecq *et al.* (1975) describe the operation of the NGT method as commencing with the group members (between seven and ten) without discussion, writing ideas related to the problem down on a pad of paper. After 5–10 minutes, each individual in turn briefly presents one of his or her ideas. These are recorded on a flip chart in full view of the group members. Round-robin listing continues until all members indicate that they have no more ideas. Discussion does not take place until all the ideas are recorded. Then each one is discussed. Finally, each individual writes down their evaluation of the most serious risks, by rank-ordering or rating. Then these are mathematically aggregated to yield a group decision. Delbecq *et al.* summarise the NGT decision-making process as follows:

- silent generation of ideas in writing;
- round-robin feedback from group members to record each idea in a short succinct phrase on a flip chart;
- discussion of each recorded idea for clarification and evaluation;
- individual voting to prioritise the ideas generated with the group decision being mathematically derived through rank ordering or rating.

Scenario Analysis

Scenario analysis emerged in the 1970s. It is a strategic thinking tool that can aid decision making and help an organisation prepare for the future, even if that future is not known. The events and trends that will determine an organisation's future and its success are with us now. To thrive, even to survive, an organisation must find a way to recognise them, understand their potential impact, and direct itself appropriately. Scenario analysis takes its name from the development of descriptive models of how the future might turn out. The use of scenarios is based on the assumption that in a rapidly changing and uncertain world, by considering a range of possible futures, an organisation's strategic horizons can be broadened, managers can be opened up to new ideas and perhaps the right future for the organisation can be identified. Scenarios allow organisations to exercise strategic choice in terms of whether to try shaping the future, adapting to events as they emerge or keeping their options open by investing in a range of products, technologies and markets. Essentially building scenarios can be regarded as making different pictures of an uncertain future through the construction of different potential outcomes. Scenario building utilises an organisation's knowledge of both the past and current discernable trends. A scenario group builds a set of visions of how their world might look in the future. Their task is then to "look back" from that future to today and explain how they got there by telling a story.

Scenario analysis can be used to identify risks by considering possible future developments and exploring their ramifications for an activity or project. Sets of scenarios reflecting, for example, "best case" (optimistic), "expected case" (most likely) and "worst case" (pessimistic) may be used to analyse a risk, including both the probability of occurrence and potential consequences. It can be used to look back over a fixed period and examine, for instance, major shifts in technology, transportation and property development with a view to considering future change. Scenario analysis, however, is not a crystal ball and cannot predict the degree (extent), likelihood (probability) and timing (when) of such changes. It can consider possible trends and their consequences and in that way help organisations develop strengths and the resilience needed to adapt to predictable change as it materialises.

Scenario analysis can be used to assist planning future strategies as well as to consider existing activities. It can assist in both the identification and assessment stage of overall risk management by considering what may occur, how likely, when and to what degree. Scenario analysis may be used to anticipate how both opportunities and threats might develop and for both short- and long-term time horizons. With short time horizons and reliable data, likely scenarios may be extrapolated from the present. For longer timeframes or with poor unreliable data, scenario analysis becomes more of a reach, a "guesstimate", speculation or conjecture. Conducting scenario analysis is dependent on assembling a team of people who between them have an understanding of the nature of relevant potential changes (e.g. possible advances in technology) and imagination to think into the future without necessarily extrapolating from the past. Access to literature and data about changes already occurring is also useful.

To commence scenario analysis, the area of concern or the decision that needs to be made is defined. Next a group of individuals with a stake in the future of the organisation is established, with appropriate knowledge and experience relating to the area of interest. Once the group is assembled they brainstorm all of the factors or variables that will determine the nature of the changes that might occur. This will be accomplished with the aid of research into the major trends of the past, the probable timing of future changes and imaginative thinking about the nature of potential changes. The context of the change may need to be considered, such as the external environment, which may include consideration of movement in fuel prices (increase in oil prices and increased transportation costs offsetting the use of cheap labour overseas), political unrest (Arab Spring), technology (advances in the microchip), the World Wide Web (social networking websites), demographics (arising from climate change) or politics (change in political party). Sometimes a change may occur due to a ripple effect, being at the tail end of a series of events. Key factors or trends are mapped against each other to show areas where scenarios can be developed. A series of scenarios is proposed with each one focussing on a plausible change in some way.

The group then splits into small teams, and each team writes a "story" for one of the scenarios identified. Each story describes how circumstances may change and move from the present day towards the subject scenario. The stories may include plausible details that add value to the scenarios. The teams must develop a story line that plausibly and interestingly tells what major events took place, who the main characters were, and where the major twists to the plot occurred which led to a particular future. When each team has a story to tell, they rejoin the larger group and share their stories. Other teams are encouraged to ask for clarifications, explanations and to question the reasoning behind any part of the story.

The scenarios can then be used to test or evaluate the original question. The test takes into account any significant but predictable factors and then explores how "successful" the potential change would be in this new scenario, and "pre-tests" outcomes by using "what if" questions based on model assumptions. When the question or proposal has been evaluated with respect to each scenario, it may be obvious that it needs to be modified to make it more robust or less risky. It should also be possible to identify some leading indicators that show when change is occurring. Monitoring and responding to leading indicators in the external organisational environment can provide opportunities to modify planned strategies. Since scenarios are only one possible outcome among many, it is important to make an attempt to qualify, or express the probability of each scenario occurring.

Scenario analysis can provide a range of possible futures where there is little current knowledge on which to base predictions or where risks being considered relate to the longer term. Having established a range of possible outcomes it will be necessary for management to be fluid and monitor developments over time. The organisation will be equipped with a tool to be more resilient and respond more quickly as opportunities or threats emerge. Like all approaches to strategic planning, there are a number of implementation problems:

- There is lengthy debate on how many scenarios to construct, on selecting the scenarios to develop and how they should be used.
- People's ideas of the future are informed by their knowledge and experience of the past. Since the past is not always the best indicator of the future, scenarios can be based on false assumptions.
- Team members can be strongly influenced in their preference of a scenario by their own personal interests, which can be to the detriment of the organisation.
- The process cannot be carried out by novices, and therefore can be time-consuming and expensive in terms of senior management time and outside experts.
- Scenarios often require strong visionary leaders, who are in short supply.
- Dominant personalities do not always have the best ideas.

The health warning that comes with using scenario analysis as a decision-making tool is that unrealistic scenarios built on sand are not recognised as such. More critically, planning hours and ultimately investment decisions are based on scenarios assumed to have a high probability but that do not materialise.

Delphi Technique

A less commonly used method of assessing expert opinion is the Delphi technique, originally developed by Dalkey, Helmer and others of the Rand Corporation primarily for technological forecasting, which has seen a wide variety of applications. It is a method for the systematic collection and collation of judgements from domain experts on a particular topic. These experts, working independently of each other, are requested to respond anonymously to a set of carefully designed sequential questionnaires. The second and any subsequent questionnaires are preceded by feedback in the form of summarised information assembled from earlier responses, with the aim of arriving at a consensus. Turoff (1970) suggests that at least three separate groups of individuals are required to perform three different roles:

- decision maker(s) – the individual or individuals expecting data or results to use for their purposes (client/sponsor);
- a "staff" group – those who design the initial questionnaire, summarise the data received and prepare the feedback information and subsequent follow-up questionnaires (facilitator);
- a respondent group – those whose judgements are being sought and who are asked to respond to the questionnaire (leading business representatives).

The basic principles of the multistage method are the elimination of direct social contact providing unattributed contributions, the provision of feedback and the opportunity for the revision of opinions. The participants are asked individually, usually by mailed questionnaires and more recently by e-mail, for their estimates concerning the variables under examination. These are then aggregated and summarised in such a way as to conceal the origin of the original estimates. The results are then circulated and participants are asked if they wish to revise their earlier forecasts. These rounds can continue until the estimates stabilise, though in practice no more than three questionnaires are issued. Results produced from these interrogations may be amenable to statistical treatment with the view to deriving assessments.

Cross Impact Method

This is a variation of the Delphi technique described above. It uses essentially the same interrogation method as the Delphi, i.e. a panel of experts; the difference, however, lies in what they are asked to do. The cross impact method asks its panel of experts to assign subjective probabilities and time priorities to the list of potential events and developments supplied by the organisation. The emphasis is on identifying event relationships and indicating the importance of specific events.

Systems Dynamics

The appeal of systems dynamics (a variant of systems thinking) to the modelling of any activity, such as bringing about change within a business, arises from its focus on the interrelationships of the component parts and their influence upon the effectiveness of the total process. This approach is particularly relevant to a change project as its success depends on the way its contributors work together. What each of them achieves individually depends on the actions of the other contributors. They are totally reliant on each other for the satisfactory completion of the change project. In organisations that are strongly differentiated (with numerous discrete disciplines such as information technology, marketing, human resources and finance) but at the same time all highly interdependent, the key to a successful outcome is the extent of interactive effort invested (Chapman 1998c). An understanding is required of the interrelation between planned activities or potential events.

Systems dynamics is concerned with creating models or representations of real-world systems of all kinds and studying their dynamics (or behaviour). In particular, it is concerned with improving (controlling) problematic system behaviour. The purpose of applying system dynamics is to facilitate an understanding of the relationship between the behaviour of a system over time and its underlying structure and decision rules (Wolstenholme 1990). The use of system dynamics causal-loop diagrams to structure, analyse and communicate ill-defined situations can be considered as a free-standing methodology, having much in common with the soft system problem solving approaches of Checkland (1987) and Ackoff (1978). This type of visual representation of events is useful as the relationship between events can be quickly grasped.

Risk Metalanguage

Metalanguage is a form of language or set of terms used for the description or analysis of another language. The value of risk identification metalanguage is well known and understood and has been in existence for over fifteen years (Central Computer and Telecommunications Agency 1995). It enhances communication and improves the effectiveness of risk management. It helps us distinguish and isolate risks from their causes and impacts so that risk response planning can take place. This is accomplished by getting contributors to the risk identification process to adopt the kind of language that allows them to say that "as a result of..." some activity or event "there is a risk that..." an adverse or positive event might occur "which may result in..." a negative or positive effect on an activity's objectives. It avoids risks becoming a confused mixture of causes and impacts. This subject is explored in more detail in Section 18.6.2.

Implementation

The examples below describe different identification methods adopted during risk management assignments as a result of constraints imposed by the client organisation. The constraints may vary considerably depending on the organisational culture and the degree to which risk management processes are embedded.

Example 1

The research and development arm of a major international pharmaceutical company wished to obtain internal investment approval for a research and development facility to be constructed in India. The decision was taken to conduct a risk analysis to determine a risk contingency based on the actual risks and uncertainty. The author undertook the analysis. Project personnel were located in India and two European countries. Due to the constraints of cost and time, in terms of bringing personnel together for a risk workshop, an alternative approach was selected. Risk identification and assessment was conducted by a combination of (1) an e-mailed questionnaire (structured as a gap analysis) to draw out the risks, (2) a draft risk register, based on the findings of the gap analysis to stimulate a review of the risks, and (3) a video conference (between India and England) to gain consensus on the risk descriptions and their assessment. While the video conference saved travel time and cost, as a communication tool it had its shortcomings in terms of the sound quality, the time delay and the inability to always see who you were talking to or who was talking. The time difference (between India and England) placed pressure on the time available for the video conference. Despite these difficulties a comprehensive risk register was produced.

In this instance the key risks identified were that the procurement route was not aligned to the project objectives, a formal business continuity plan had not been prepared and reflected in the design, the insurer's requirements had not be ascertained and reflected in the design, and roles and responsibilities of the different client representatives had not been made explicit.

Example 2

A unitary authority wished to understand the risks to its organisational change project, which had the laudable aims of improving its corporate working and comprehensive performance assessment rating (CPA), as assessed by the Audit Commission.[1] The author undertook the risk study. Due to the nature of councils, the study was carried out in a political context which influenced the behaviour of officers. The study was commenced by examining the project documentation. The first task was to strive to understand the objectives of the change project and how these were aligned to both the deliverables and the success criteria. From the project documents, this was not readily discernible and information had to be prepared to compensate. A table of five columns was prepared with the column headings reading from left to right as follows: "objectives identification number", "objectives", "deliverable identification number",

[1] The Audit Commission is an independent body responsible for ensuring that public money is spent economically, efficiently and effectively, to achieve high-quality local and national services to the public. It provides practical recommendations for improvement and promulgating best practice. There are five categories or ratings of performance ranging from excellent, through good, fair and weak, down to poor. Each rating under the CPA framework is derived by combining scores for: performance of each key council service; the council's overall performance as a service provider; and a score for the council's ability to lead its community and improve services.

"deliverable" and "success criteria". This table was populated as far as possible and reviewed with and amended to reflect comments from the council. The table was then used to construct the risk register inasmuch as the project deliverables were included in the register so that risks could be identified against the deliverables and objectives.

Implementation of the study was constrained in a number of ways. Officers were in the unenviable position of operating in an environment where there was a tension between officers and members, between political parties and between members within the same party. In addition, the relationship with the local press was poor. It appeared that members were leaking information to the press to pursue their own ends and certain senior officers felt their position was vulnerable as a result of member behaviour. There was a genuine concern by certain officers that risks should not be identified against activities already completed, as this information could be used as "ammunition" by members against officers. Against this background, access to officers was restricted, as multiple workshops planned were not permitted to take place. A single workshop was arranged with the project implementation group (which included officers, members and union representatives) to identify the risks to the deliverables. Prior to the workshop, a briefing pack was prepared and sent to the attendees, which explained the aim of the workshop, what aspects of the risk management process would and would not be undertaken in the workshop and the schedule of deliverables. In addition to the briefing pack, a draft register or "straw man" was prepared with a small number of officers and issued to stimulate thinking prior to the workshop. During the workshop, the draft register was used to gain a consensus on the risks facing the project, including agreement to the wording of both previously identified risks and additional risks. The limited workshop duration and limited access to officers were clear constraints on the risk study.

In this instance the key risks identified were (in summary) that the planned improvements in project and risk management were slow to produce results, projects exceeded budgets, option appraisals were inadequate, lessons learnt on capital projects were not reflected in ongoing operations, corporate working was not enabled, and projected savings were not realised.

Example 3

A UK media company was preparing to enter into contract negotiations with a third party with regard to the provision of support services. The media company wished to make an informed decision about the terms of the contract to be entered into with regard to the balance of risk ownership between the two parties, their own degree of retained risk and the financial reward to be sought commensurate with the degree of retained risk. The author undertook the analysis. The company wished to look at alternative scenarios and their corresponding risk profiles. It was clear that the third party would wish to impose financial penalties on the media company if performance fell below pre-agreed levels of service. Over a single year the potential penalties, aggregated together, could exceed several million pounds. Losses would be calculated monthly and paid out yearly. The risk analysis was problematical from six main perspectives: (1) gaining initial agreement to the focus of the risk management study, (2) the amount of time allocated to the study, (3) the amount of time media company representatives could devote to participating in the study due to ongoing commitments, (4) the timing of the study as the media company was going through significant organisational change, (5) lack of clarity over the ownership of the assets to be managed as part of any contract, and (6) the amount of information available regarding the services to be offered by the media company and the associated penalties.

In this instance the key issues that the study highlighted were: the likely legal costs in administering a contract which involved a detailed penalty regime, the high level of excess currently paid on insurance (and hence the limited protection it offered), the possibility of the UK media company being paid on a cost and margin basis rather than a penalty regime with both parties sharing the risk, the services required could vary significantly year by year, and finally, if back-to-back arrangements were required by the third party (to pass on their own risks), significant contract drafting would be required.

9.8.6 Gaining a Consensus on the Risks, the Opportunities and their Interdependencies

To be able to assign risks to risk owners and managers downstream in the risk management process, it is important to have a consensus and a buy-in to the risks and opportunities, their descriptions and the interdependencies.

9.8.7 Risk Register

The risk register is populated with the findings of the process activity. The typical content of a register was discussed in Section 9.7.10 above. Its usefulness will depend on whether it is placed on a shelf to collect dust or is used as a proactive tool to manage the business.

9.9 SUMMARY

This chapter has examined the risk identification process using the IDEFO process mapping technique to understand the inputs, outputs, the constraints which might inhibit conducting identification and the mechanisms that will support it. The primary process goal was described in terms of identifying both risks and opportunities, and the subgoals were identified as good practice steps to enhance the overall identification process. The inputs were described as: assumptions, uncertain events, lessons learnt, issues, business analysis, a business plan and a business process map. The risk identification process is a key foundation stone in the overall process of risk management, for risks not identified will not be managed. The mechanisms are used to attempt to avoid blind spots so that the identification process is as thorough and comprehensive as possible.

9.10 REFERENCES

Ackoff, R.L. (1978) *The Art of Problem Solving*. John Wiley & Sons, Inc., New York.

BSI PD 6668 (2000) *Managing Risk for Corporate Governance* (2001 reprint), p. 19. British Standards Institution, London.

Carr, M.J., Konda, S.L., Monarch, I., Ulrich, F.C. and Walker, C.F. (1993) *Taxonomy-Based Risk Identification*, Technical Report CMU/SEI-93-TR-6 ESC-TR-93-183, June. Software Engineering Institute, Carnegie Mellon University, Pittsburgh.

Central Computer and Telecommunications Agency (1995) *An Introduction to Managing Project Risk*. HMSO, London.

Chapman, R.J. (1998a) An investigation of the risk of changes to key project personnel during the design stage. Unpublished doctoral thesis, Department of Construction Management & Engineering, Faculty of Urban and Regional Studies, University of Reading.

Chapman, R.J. (1998b) The effectiveness of working group risk identification and assessment techniques. *International Journal of Project Management*, 16(6), 333–343.

Chapman, R.J (1998c) The role of system dynamics in understanding the impact of changes in personnel on design production within the construction industry. *International Journal of Project Management*, 16(4), 235–247.

Chapman, R.J. (2001) The controlling influences on effective risk identification and assessment for construction design management. *International Journal of Project Management*, 19(3), 147–160.

Checkland, P.B. (1987) The application of systems thinking in real world problem situations: the emergence of soft systems methodology. In M.C. Jackson and P. Keys (eds), *New Directions in Management Science*, pp. 87–96. Gower, Aldershot.

Cooper, B. (2004) *The ICSA Handbook of Good Boardroom Practice*. ICSA Publishing, London.

Day, A.L. (2001) *Mastering Financial Modelling, a Practitioner's Guide to Applied Corporate Finance*, p. 219. Pearson Education, London.

Delbecq, A.L. (1968) The world within the span of control: Managerial behaviour in groups of varied size. *Business Horizons*. 11(4), 47–57.

Delbecq, A.L., Van de Ven, A.H. and Gustafson, D.H. (1975) *Group Techniques for Program Planning*. Scott Foresman, Glenview, IL.

Dorofee, A.J., Walker, J.A., Alberts, C.J., Higuera, R.P., Murphy, R.L. and Williams, R.C. (1996) *Continuous Risk Management Guidebook*. Carnegie Mellon University Software Engineering Institute.

Hillson, D. (2002) Use a risk breakdown structure (RBS) to understand your risks. In *Proceedings of the Project Management Institute Annual Seminars & Symposium*, San Antonio, TX, 3–10 October.

HM Treasury (2001) *Management of Risk: A Strategic Overview*. HM Treasury, London.

HM Treasury (2004) *Management of Risk – Principles and Concepts*. The Stationery Office, Norwich.

Holliwell, J. (1998) *The Financial Risk Manual: A Systematic Guide to Identifying and Managing Financial Risk*. Pearson Education, UK.

Johnson, D.M. (1972) *Systematic Introduction to the Psychology of Thinking*. Harper & Row, New York.

Jones, E.M. and Sutherland, G. (1999) *Implementing Turnbull, A Boardroom Briefing*. Centre for Business Performance, Institute of Chartered Accountants in England and Wales, London, September.

Kambil, A., Layton, M. and Funston, R. (2005) It is critical to model and manage interdependencies between risks. *Strategic Risk*, June.

Kelly, J., Male, S. and Drummond, G.D. (2004) *Value Management of Construction Projects*. Blackwell Science, Oxford.

Osborn, A.F. (1963) *Applied Imagination. Principles and Procedures of Creative Problem Solving*, 3rd revised edition, 14th printing. Charles Scribner's Sons, New York.

PMI (2008) *A Guide to the Project Management Body of Knowledge (PMBOK® Guide)*, 4th edition. Project Management Institute, Newtown Square, PA.

Risk Management Specific Interest Group (2004) *Project Risk Analysis and Management Guide*, 2nd edition. APM Publishing, High Wycombe.

Simon, P., Hillson, D. and Newland, K. (1997) *Project Risk Analysis and Management Guide*. Association for Project Management, Norwich.

Tummala, V.M.R. and Burchett J.F. (1999) Applying a risk management process (RMP) to manage cost risk for an EHV transmission line project. *International Journal of Project Management*, 17(4), 223–235.

Turoff, M. (1970) The Design of a Policy Delphi. *Technological Forecasting and Social Change*, 2.

Wolstenholme, E.F. (1990) *System Enquiry: A Systems Dynamics Approach*. John Wiley & Sons, Ltd, Chichester.

10
Risk Analysis: Stage 3

Chance favours the prepared mind.
(Louis Pasteur)

The previous chapter examined risk identification. This chapter examines the risk analysis stage. The purpose of the risk analysis stage is to provide a judgement of the likelihood of the risks and opportunities occurring and their impact, should they materialise. The benefit of undertaking this activity is that analysis provides an order of pain or gain for each risk and opportunity, respectively. I use the words "an order of" because quantitative analysis is not a precise science, based as it is on subjective estimates. While some question the merit of quantitative analysis as a result of this subjectivity, it makes sense to articulate these perceptions of likelihood and impact in order to aid decision making. Even when there is considerable uncertainty about the business outlook, quantitative techniques provide a framework for thinking about the problems. Decision making becomes much "tighter" as soon as the risks are quantified, and the assessment is progressively refined as more information becomes available. Risk management leads to rational, defensible decisions. Without risk assessment:

- how is a preferred option to be selected from a number of possible solutions?
- how is risk management activity to be prioritised?
- how is a manager to judge whether it is more economic to retain a risk or transfer it to a counterparty?
- how is a manager to judge whether to enter a new market?
- how is a business to decide if it wishes to increase market share through acquisition?

The structure of this chapter is reflected in Figure 10.1. The following chapter examines the assessment of risks and opportunities in combination.

10.1 PROCESS

As described in the preceding chapter, adopting the philosophy of process mapping, each process exists to make a contribution to one or more business enterprise goals. Hence, each process should be measured against specific process goals that reflect the contribution that the process is expected to make to the overall enterprise goals. Processes are simpler to comprehend when they have primary goals and subgoals. Hence, risk analysis is described here as having a primary process goal which is accomplished by a series of subgoals, as described below. Any one process is accomplished within a context and might be considered to have two perspectives. The *external view* examines the process inputs, outputs, controls and mechanisms. The *internal view* examines the process activities that transform inputs to outputs using the mechanisms.

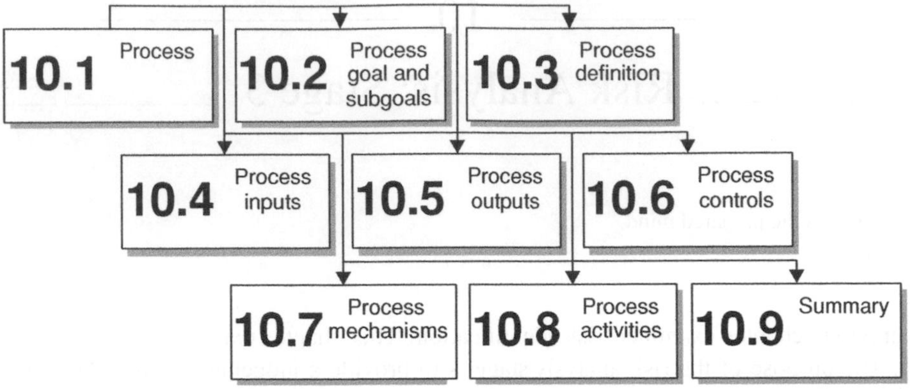

Figure 10.1 Structure of Chapter 10

10.2 PROCESS GOAL AND SUBGOALS

The primary process goal of risk analysis is to assess both the *risks* and the *opportunities* to the business, in terms of their probability and impact. The risk analysis process is sufficient when it satisfies these subgoals:

- The risk analysis process was comprehensive and included, as far as possible, an assessment of all of the risks in the risk register developed in the "identification" stage.
- Personnel were involved who could make an informed and well-reasoned analysis of the risks.
- Sufficient time was allocated to the analysis process.
- Consistent definitions of probability and impact were adopted.
- The analysis was supported by risk management expertise.
- Where a probability impact matrix was used, the financial banding adopted for each risk was appropriate and not too broad or open-ended (in terms of the upper band).

10.3 PROCESS DEFINITION

The risk analysis process is described by an IDEFO diagram (see Figure 10.2). The diagram describes a process with inputs entering on the left of the box, outputs leaving on the right of the box, controls entering from above and mechanisms or enablers entering from below.

10.4 PROCESS INPUTS

The inputs to the analysis process will be dictated by the risk study parameters. They may include the risk register, profit and loss account, balance sheet and industry betas.

- *Risk identification*. During the previous risk management stage, "identification", the risks will have been discussed and captured in the risk register.
- *Risk register*. The risk register is an output of the proceeding process, which at this stage of development should contain as a minimum a full description of the risks and the risk categories. Each risk should be assigned a unique reference number. The risks should be listed under the risk category to which they relate. Additionally, where possible the risk

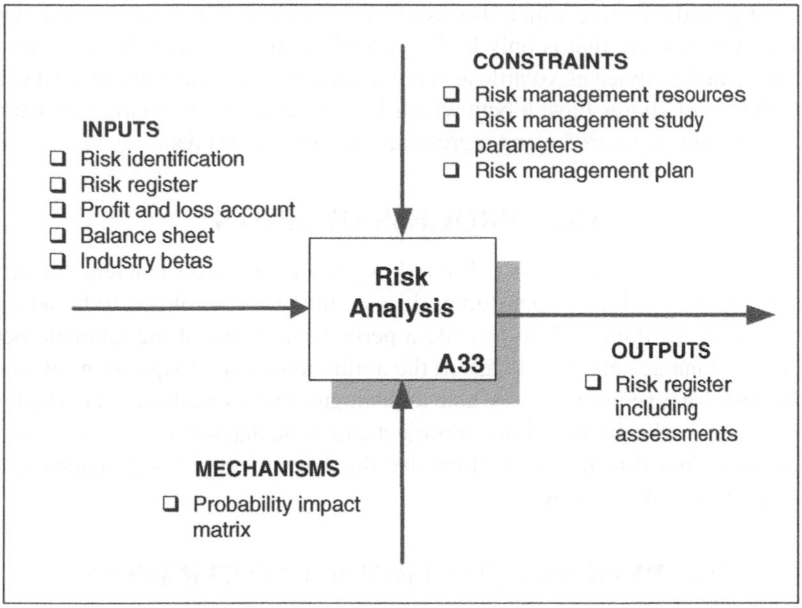

Figure 10.2 Risk analysis process

owner and risk manager should be identified. If considered helpful the register can include additional columns, such as a notes column, to provide background information on the risk, and an impact column, so that the impact on the business of each individual risk can be described.

- *Profit and loss account*. The projected profit and loss account was discussed in Chapter 8. This account provides information on the expected levels of profit for a particular period. Low levels of projected profit will expose a business to a series of related risks in terms of operating practicalities and business longevity.
- *Balance sheet*. Likewise the projected balance sheet was discussed in Chapter 8. This statement should be critically examined to establish the reliability of the projections with regard to the validity of the underlying assumptions and whether the input data was comprehensive. The balance sheet will provide an indication of the vulnerability of a business to late payments or bad debts, for instance.
- *Industry betas*. The non-diversifiable risk element for a particular share can be measured using beta. This is a measure of the non-diversifiable risk of the share in relation to the market as a whole. A risky share is one that experiences greater fluctuations with movements in the market as a whole and therefore has a high beta value. It follows that the expected returns for such a share should be greater than the average returns of the market. Atrill (2000) offers a method of calculating the required rate of return for investors for a particular share as follows:

$$K_{\mathrm{O}} = K_{\mathrm{RF}} + b(K_{\mathrm{m}} - K_{\mathrm{RF}}),$$

where K_{O} is the required return for investors for a particular share, K_{RF} is the risk-free rate on government securities, b is the beta of the particular share, K_{m} is the expected returns to the market for the next period, and $K_{\mathrm{m}} - K_{\mathrm{RF}}$ is the expected market average risk premium for the next period. A share which moves in perfect step with the market will have a beta measure of 1.0. A share that is only half as volatile as the market will have a beta of 0.5, and a share which is twice as volatile as the market will have a beta of 2.0. Past experience suggests that most shares have a beta which is fairly close to the market measure of 1.0. Betas are normally measured using regression analysis on past data.

10.5 PROCESS OUTPUTS

- *Risk register including assessments.* The risk register is updated to include the probability and impact of each risk and opportunity. It is common for a column to be added on the register headed "justification", to provide a permanent record of the rationale behind the probability and impact selected, to afford the ability to respond to questions at a later date. The very least that is required is to be able to distinguish between those events that will have a minor impact on a business activity or project and those that will have a major impact. For financial institutions this activity will involve the assessment of losses against commonly recognised sources of exposure.

10.6 PROCESS CONTROLS (CONSTRAINTS)

The business risk management culture, risk management resources, risk management study and risk management plan (where one exists) regulate the risk identification process. These controls were discussed previously in Section 8.6.

10.7 PROCESS MECHANISMS (ENABLERS)

10.7.1 Probability

Assessing the likelihood of a risk or opportunity occurring requires an understanding of probability. Probability is expressed on a sliding scale from 0 to 1. If there is no chance of an event happening, it has a probability of zero. If it must occur, it has a probability of one. The simplest way to describe this is that an unbiased coin can land on a flat surface in one of only two ways. There is a 50% chance of either. Hence, there is a 0.5 probability of a head and a 0.5 probability of a tail. If four businesses are equally likely to be selected for a contract, there is a 1 in 4, or 25%, chance that any one will be selected. Hence, they each have a 0.25 probability of success. It is important to understand that an unavoidable event has a probability of 1. By looking at the coin example, there is a 0.5 probability of a tail and 0.5 probability of a head. One of these two events must happen, so the probabilities must add up to one. If the probability of something happening is known, then by definition the probability of it not happening is also known. This is an important concept to be aware of when constructing decision trees. Logic, objective and subjective probability, together with probability relationships (including Bayes' theorem) are described in Appendix 14. This appendix forms a source of reference when a deeper understanding of probability is required.

- *Probability distributions.* The probability distributions commonly selected where there is very little data to model, are the rectangular/uniform and triangular distributions. Where historical data is available it may be possible to use the normal, binomial and/or Poisson distributions. There are many more distributions to select from, and their use will be dictated by the circumstances.

Table 10.1 Quantitative probability impact matrix

	Probability	Cost £	Time	Brief
Very High	>70%	>1m	> 8 wks	Major shortfall in the brief
High	50–69%	500–999k	6–8 wks	Significant shortfall in the brief
Medium	30–49%	250–499k	4–6 wks	Shortfall in the brief
Low	10–29%	10–249k	2–4 wks	Major specification issue
Very Low	< 10%	<10k	< 2 wks	Minor specification issue

- *Probability impact matrix.* The very least that is required of risk assessment is to be able to distinguish between those risks that will have a minor impact on business activities and those that will have a major impact. For financial institutions this activity will involve assessment of losses against commonly recognised risks.

Example

An example of a probability impact matrix is included in Table 10.1. The matrix has five levels of severity, from very low to very high. Opinion varies on the most appropriate number of levels. Too few levels and there is insufficient granularity to differentiate the probability and impact of different risks. Too many levels and participants in the assessment process lose patience and/or interest as they have insufficient information to support the selection of one scale over another. This matrix was used on a risk assignment conducted by the author. The project was the refurbishment of existing rolling stock undertaken by a rolling stock company in conjunction with a train operating company. The matrix was used as part of the assessment process to aid project team members to assign a probability of occurrence (expressed as a percentage) and a cost and time impact for each of the identified risks. The matrix was not used rigidly. In other words if the cost impact of a risk was considered to be medium, the financial impact did not have to lie between £250 000 and £500 000. The figures were a guide only. Values were selected to reflect the individual nature of a risk and so an impact may have ranged between £250 000 and £550 000. The assessments made were defined as pre-mitigation assessments. In other words, if no risk response action was taken and the risks materialised, these would be the impacts on the project.

10.8 PROCESS ACTIVITIES

The activities of the risk assessment process are the tasks necessary to capture the likelihood of the risk occurring and its impact, should it materialise, and record them in a risk register, log or list. These consist of:

- understanding and assessing the likelihood or probability of the risk or opportunity arising;
- assessing the impact of the risk or opportunity in terms of the business or project objectives;
- understanding and taking account of the interdependencies between the risks – whether they would occur sequentially (i.e. one risk potentially triggers another risk) or happen concurrently (i.e. in parallel);
- documenting the findings;
- updating the risk register, log or list.

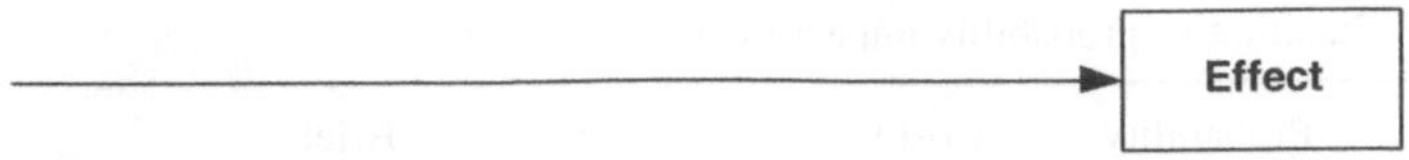

Figure 10.3 Cause and effect

10.8.1 Causal Analysis

Causal analysis shows the relation between an effect and its possible causes, to get to the root cause of a risk. Its purpose is to prevent problems by determining the problem's root cause. The premise behind causal analysis is that if an error (or risk) has occurred it may happen again unless something is done to stop it. Hence, learning from past errors prevents future errors. Among the techniques of causal analysis is the cause-and-effect diagram. It does not have a statistical basis but is excellent for uncovering the sources of risk and mapping their relationships. This diagram is also sometimes called a fishbone diagram (as its appearance resembles the skeleton of a fish) or an Ishikawa diagram (after its inventor, Professor Kaoru Ishikawa[1]). The tool can produce a quick identification of major causes, usually indicates the most fruitful areas for further investigation and always leads to a better understanding of the problem. It explores the relationship between the problem and its causes (by category) visually. It is commonly developed in a brainstorming session. Its appeal stems from its simplicity and its adaptability. It allows the group or individual to broaden their thinking about potential or real causes of the problem, and it then facilitates further analysis and examination of these causes.

The diagram is commenced by writing the potential effect on the far right-hand side, and drawing a long horizontal arrow pointing towards it (see Figure 10.3). From the horizontal arrow, the major risk events are created as branches off the spine and labelled. These will form the main "bones" attached to the "backbone arrow". They are added incrementally and it is common to start with just a few. Brainstorming is continued until about six to eight main categories are defined. It does not have to be an even number (Figure 10.4).

Having decided on the main categories, a group can continue to brainstorm in more detail, and capture the contributory causes to the main categories, described as level 1 and level 2 causes, as illustrated in Figure 10.5. It is best to expand the diagram as far as possible so that no potential root cause is overlooked.

Example

The example in Figure 10.6 was produced by the author for a petrochemical company wishing to invest in support facilities for oil and gas exploration in Russia. Level 2 causes were excluded from the diagram, for ease of assimilation by the audience.

10.8.2 Decision Analysis and Influence Diagrams

Decision analysis is used to structure decisions and to represent real-world problems by models that can be analysed to gain insight and understanding. The elements of a decision model are the decisions, uncertain events and values of outcomes. Once the elements of the decision have been identified, a model can be constructed using the influence diagram technique. An

---

[1] Ishikawa, formerly of Tokyo University, was a management leader who made significant and specific advancements in quality improvement and pioneered the quality circle movement in Japan during the 1960s. He first used the cause-and-effect tool in 1943.

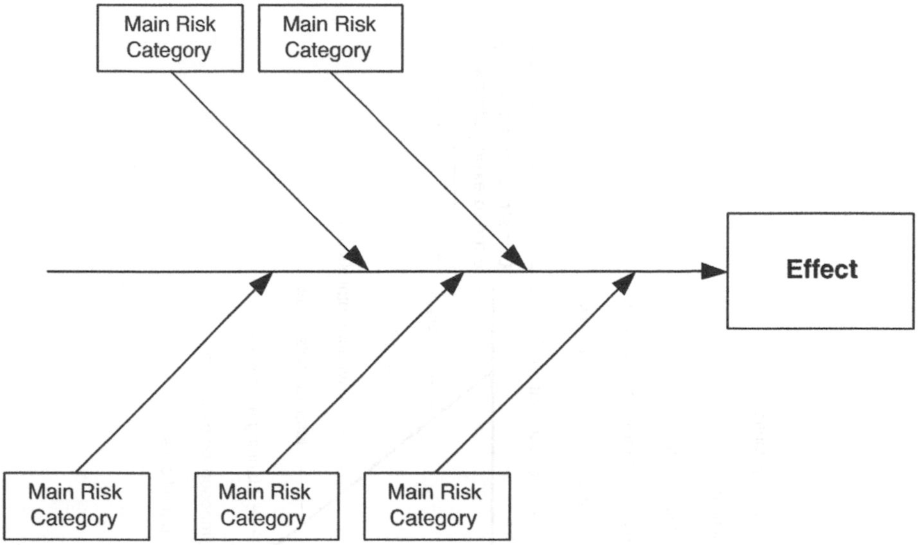

Figure 10.4 Main causes of effect

influence diagram is a graphic representation of a model and is used to assist in model design, development and understanding. An influence diagram provides visual communication to the model builder or development team. It also serves as a framework for expressing the exact nature of the relationship of the variables. The term "influence" refers to the dependency of a variable on the level of another variable. As with all modelling techniques, an influence diagram provides a snapshot of the decision environment at one point in time. Influence diagrams are

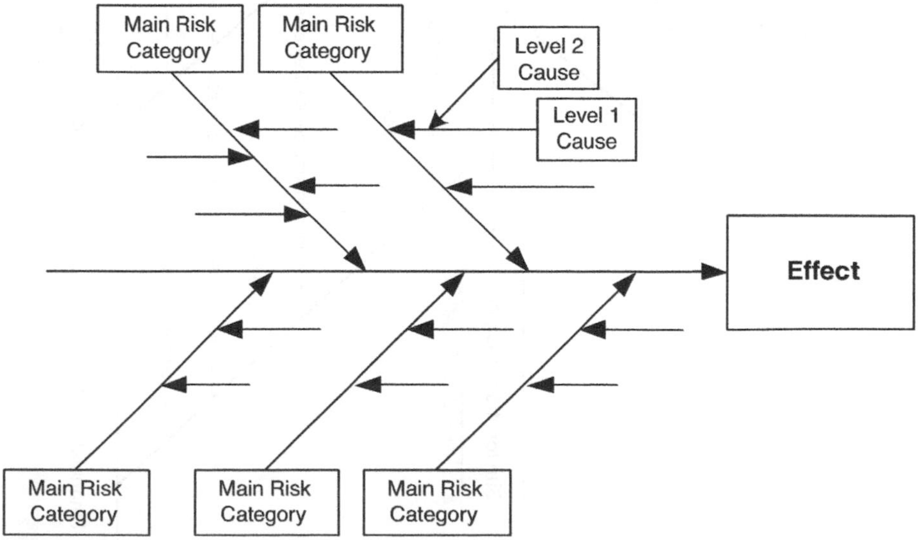

Figure 10.5 Main, level 1 and level 2 causes

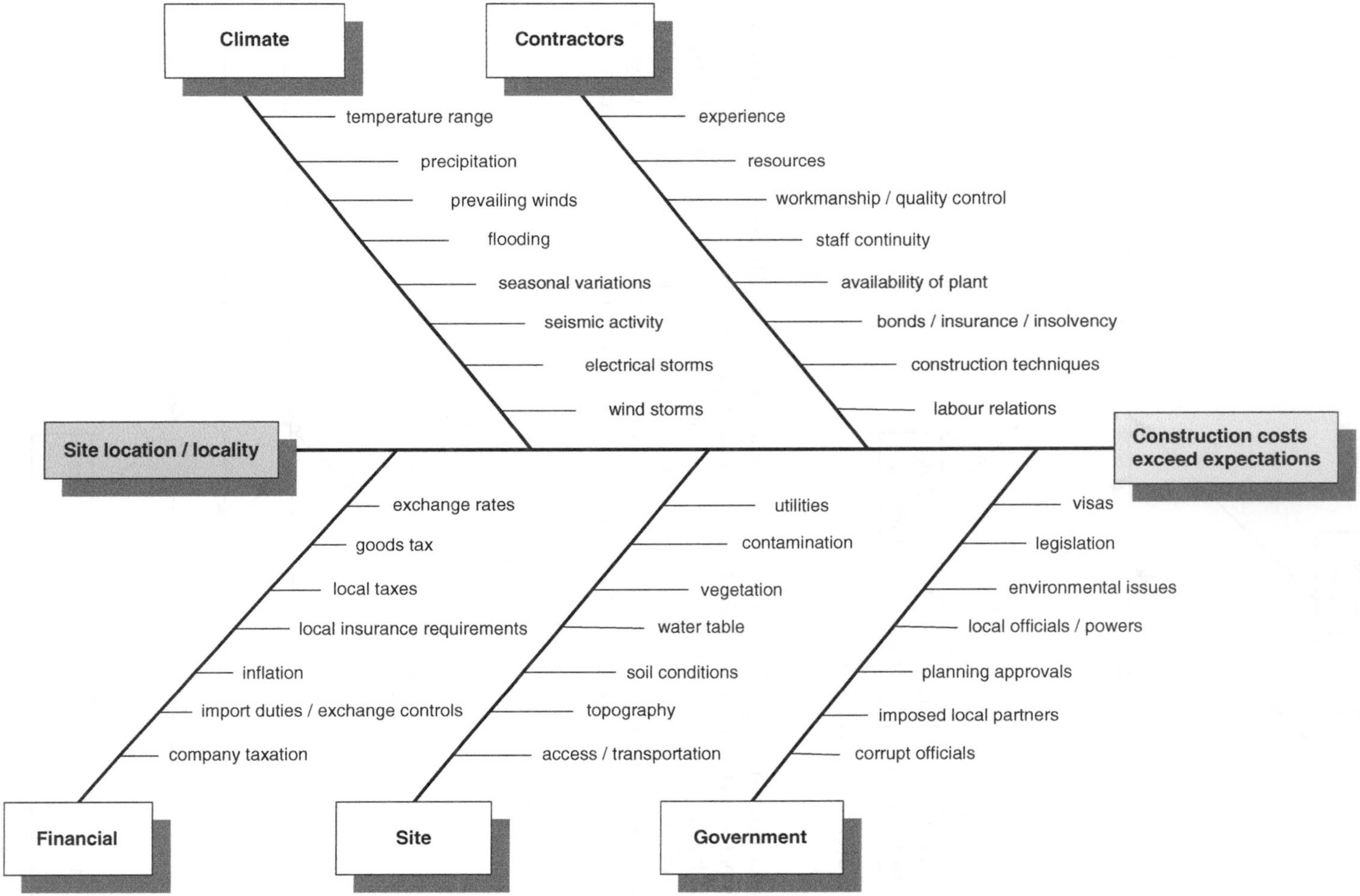

Figure 10.6 Cause and effect diagram for a petrochemical company

an attempt to incorporate decisions as well as chance events into a single diagram that is easy to understand. They can represent the value or likely utility of pursuing a given course of action. These diagrams are designed to deliberately provide as simple a model as possible of a decision process which contains uncertainties, so that management can understand the process involved and, if necessary, alter the model to more accurately reflect the management process that takes place. The influence diagram is also formal and can be transformed into an equivalent decision tree and evaluated. A common notation is used:

- Squares represent decision nodes.
- Circles represent chance or uncertain events.
- Diamonds represent values.
- Double circles represent outcomes known when the inputs are given.
- Arcs (arrows) represent influences between variables. The direction of an arc is vital, as the arc specifies that the value of the node at its head (arrow end) depends directly on the value of the node at its tail.

The advantages of influence diagrams are that (1) they provide a framework in which experts and decision makers can discuss the interdependencies of decisions and events and the management of the problem, without requiring any formal mathematical, probabilistic or statistical notation, (2) they provide a significant contribution towards reducing large volumes of data to those parts that are essential to the decision-making process, and (3) they can provide a degree of sensitivity analysis to show how much influence particular decisions or uncertain events have upon the final outcomes.

The rapid development of computer technology has led to increasingly powerful and inexpensive computer systems. As a consequence it has made the computer an easily accessible tool for decision making and the construction of influence diagrams. However, while current decision analysis software has succeeded in constructing and evaluating influence diagrams, the software requires the user to have a certain amount of knowledge in the area of influence diagrams to formulate the decision problem accurately.

10.8.3 Pareto Analysis

Pareto analysis is used to focus management effort on those risks that have the potential to have the greatest detrimental impact on a project or a business's objectives. Pareto analysis is the expression given to the simple process of ranking or ordering risks once they have been assessed, to determine the order in which they should be managed. The Queensland Government State Department (2002) states: "It is important to focus on the important risks – follow the Pareto Principle of placing emphasis on, and allocating resources to, the significant few, rather than the insignificant many." Once the risks have been ranked they can be represented pictorially by a bar chart. When the bars are arranged in descending order of height from left to right with the most frequently occurring cause appearing first, the bar graph is called a Pareto diagram (Department of Trade and Industry 2000; Office of Government Commerce 2008). Microsoft Excel includes a Pareto function for formatting histograms, which enables data to be presented in an output table in descending order of frequency. Commonly Pareto diagrams reveal that 20% of the risks within an analysis contribute some 80% of the overall risk exposure/impact, following the Pareto principle or 80/20 rule, as it is known. The originator of the 80/20 rule was Vilfredo Pareto (1848–1923), an Italian economist who observed an unequal distribution of the nation's wealth and power in a relatively small proportion of the total population. The

80/20 rule has subsequently been found to hold for a host of issues outside of economics, and its principle was suggested by quality management pioneer, Dr Joseph Juran. Working in the US in the 1930s and 1940s, he recognised a universal principle he called the "vital few and trivial many" which built on the work of Pareto. Pareto Pro, supplied by SigmaXL, is an add-in to Microsoft Excel, which is one example of available software that enables Pareto diagrams to be produced.

The Pareto principle can be used in a wide variety of problem-solving and continual improvement activities. Here are some examples:

- 80% of benefit comes from the first 20% of effort.
- 80% of complaints are about the same 20% of your services.
- 80% of the decisions made in meetings come from 20% of the meeting time.
- 80% of innovation comes from 20% of the staff.
- 80% of success comes from 20% of the business efforts.
- 20% of advertising yields 80% of the campaign results.
- 20% of customers account for 80% of sales volume.
- 80% of an equipment budget is spent on 20% of the items.

By combining EMV calculations with a Pareto diagram, risk significance can be readily communicated to the business team involved in the risk analysis.

10.8.4 CAPM Analysis

The capital asset pricing model (CAPM) relates the expected return on an asset to its risk, while giving a precise definition of what we mean by risk. The key aspect of the CAPM is that investors can expect a reward for investing in an asset with a high-risk profile. There can be no expected reward for exposure to risks that can be easily diversified away. The required rate of return should be higher for investments that have a larger element of non-diversifiable risk, otherwise there would be no incentive to invest in assets other than those providing a modest return.

An investor holding shares in a holding is exposed to equity market risk. There is a tendency for the value of the share to move with general stock market movements. In the CAPM, market risk is measured by its beta. A stock with a beta of 1.0 tends to move broadly in line with the equity market; a share with a beta of 1.5 tends to move up or down by 1.5% for each percentage point movement in the market. In the past the Lloyds TSB Group has had a beta of just under 1.5% and Cadbury Schweppes had a beta of just over 0.5%.[2] Some companies have a beta over 1.5. If the market goes up these shares can be expected to outperform others; in a bear market they can be expected to fall by more than average. Other shares have betas of 0.5 or less, and these defensive companies are likely to do relatively well in a bear market while being left behind when the share prices surge ahead.

[2] Source: London Business School, *Risk Management Service*, 2001. The *Risk Management Service* (RMS) is a quarterly publication designed for use by investment professionals and corporate executives. Beta values quoted were drawn from the RMS web page.

Required Rates of Return

To estimate the required rate of return for an investment, it is important to know the beta for the capital project. This is essentially easy to do if the project essentially replicates, probably on a smaller scale, the existing business of the company. It is also easy if the project is typical of an industry sector for which betas are published. A capital project with a beta of 0 would be riskless and its cash flows should be discounted at the risk-free rate of interest. An investment in an equity index fund would have the same risk as the market, namely a beta of 1.0. This investment would have a required rate of return equal to the riskless rate of interest plus the expected equity market risk premium.

In general, the CAPM tells us that the required rate of return on an investment is equal to the risk-free rate of interest plus a premium for risk. The premium for risk is equal to beta multiplied by the equity market risk premium. To use the CAPM to calculate the required rate of return, three items of data are required: the risk-free interest rate (which may be obtained from the Currencies and Money page of the *Financial Times*), the beta of the project (which may be estimated using the RMS), and the equity market risk premium, which historically has averaged around 8%.

10.8.5 Define Risk Evaluation Categories and Values

The process activities for the risk assessment stage involve assessing the risks identified in the previous identification stage using subjective judgements, combined with historical data when it is available. This can be qualitative or quantitative. Qualitative assessments describe the size of the impact of a risk, whether this is in terms of, say, time or cost, simply with the aid of categories such as "high", "medium" or "low". Where more granulation is required, the categories can be extended to "very high", "high", "medium", "low" or "very low". Quantitative assessments provide numerical assessments of, say, a financial or time risk. A probability impact grid, as described above, may be used to provide a scale for these numerical impact assessments. Quantitative assessments are more reliable when historical data is available. The objective of assessing the risks by whatever method is to ensure that management action is prioritised to respond to the most serious risks first. The purpose of using quantitative (numerical) assessments is so that the true likely outturn cost or duration of a business activity is used by decision makers, rather than basing decisions on information which takes no account of uncertainty and risk.

10.9 SUMMARY

This chapter has examined the implementation of the analysis process (the third stage in the overall risk management process), which builds on the information gained in the preceding stages. It has examined the inputs and outputs to the analysis process together with the constraints and enablers, which modify the success of the process. As analysis includes the assignment of probabilities to anticipated events (risks), assessing the process mechanisms included, examination of probability including logic, objective and subjective probabilities, probability relationships, conditional probability and Bayes' theorem. The process activities looked at causal analysis, Pareto and CAPM analysis. There is no exact science to the assignment of probabilities and impact assessments for embryonic business organisations where historic data does not exist, unlike the insurance industry, for instance, where historical records provide strong guidance on the assessments to adopt.

10.10 REFERENCES

Atrill, P. (2000) *Financial Management for Non-Specialists*, 2nd edition, p. 259. Pearson Education, Harlow.

Department of Trade and Industry (2000) *Tools and Techniques for Process Improvement.* Prepared by Professor John Oakland, European Centre of Business Excellence. http://www.businessballs.com/dtiresources/TQM_process_improvement_tools.pdf

Office of Government Commerce (2008) Category management toolkit. http://www.ogc.gov.uk/documents/Data Analysis(1)

Queensland Government State Department (2002) Public private partnerships guidance material: Business case development. http://www.sd.qld.gov.au/dsdweb/docs-bin/v2/major_proj/ppp_bus_case_dev.pdf

11

Risk Evaluation: Stage 4

The previous chapter examined risk analysis and the assignment of probabilities and impacts to risks and opportunities. This chapter examines the risk evaluation stage within the overall risk management process. As its name suggests, this involves evaluation of the results of the analysis stage. This stage is central to understanding the likely risk exposure or potential opportunity arising from a business activity. It involves the important step of getting to grips with the relationship between the individual risks and opportunities, so that when they are combined together their true net effect is portrayed. When the results are obtained from the first set of calculations (the first output or "first pass"), questions are likely to be raised about the inputs and so it is highly probable that it will be necessary to revisit the previous process. In this way completion of the evaluation stage is an iterative process of both challenge and refinement of the information captured during the analysis process and its integration into the evaluation process. The structure of this chapter is described in Figure 11.1. The next chapter examines the process of risk treatment to determine the approach to be adopted to respond to the risks and opportunities identified, assessed and evaluated.

11.1 PROCESS

As described in the preceding chapters, adopting the philosophy of process mapping, each process exists to make a contribution to one or more business enterprise goals. Hence, each process should be measured against specific process goals that reflect the contribution that the process is expected to make to the overall enterprise goals. Processes are simpler to comprehend when they have primary goals and subgoals. Hence, risk evaluation is described here as having a primary process goal which is accomplished by a series of subgoals, as described below. As described in previous chapters, any one process is accomplished within a context and might be considered to have two perspectives. The *external view* examines the process inputs, outputs, controls and mechanisms. The *internal view* examines the process activities that transform inputs to outputs using the mechanisms.

11.2 PROCESS GOAL AND SUBGOALS

The primary process goal of risk evaluation is to assess both the *risks* and *opportunities* to the business, in terms of their aggregated impact, on either the business as a whole, or specific projects. The risk evaluation process will be sufficient when it has satisfied these subgoals:

- The aim of the aggregation process had been made explicit.
- The limitations (if any) of the aggregation process were recorded and stated alongside the results.
- Personnel were involved who could make an informed and well-reasoned assessment of the relationship between the risks.
- Sufficient time was allocated to the evaluation process.

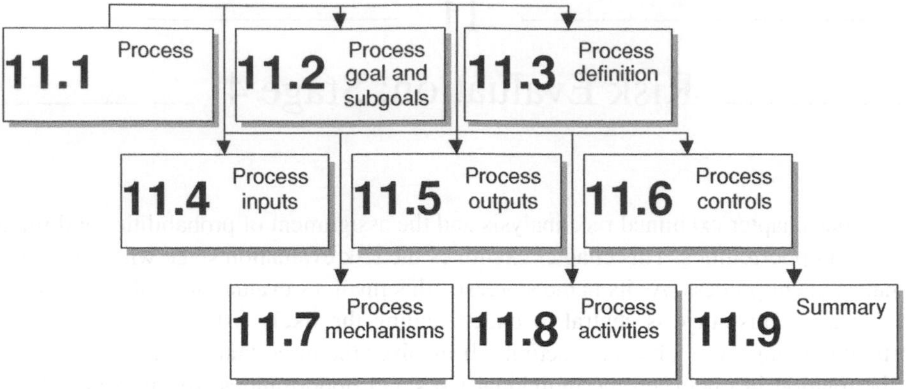

Figure 11.1 Structure of Chapter 11

- A recognised and reliable method of aggregation was adopted.
- The evaluation was supported by risk management expertise.
- Assumptions in the evaluation process were made explicit.
- Sensitivity analysis could be conducted on the results. This is the name given to the step of rerunning the model to carry out a what-if analysis to see what the outcome would be if any one figure is changed.

11.3 PROCESS DEFINITION

The risk evaluation process is described by an IDEFO diagram (see Figure 11.2). The diagram describes a process with inputs entering on the left of the box, outputs leaving on the right of the box, controls (constraints) entering from above and mechanisms or enablers entering from below.

11.4 PROCESS INPUTS

- *Risk register*. The risk register is an output of the preceding process, risk analysis. From the risk identification process the risk register will as a minimum contain a full description of the risks and the risk categories. Each risk will have been assigned a unique reference number. The risks should have been listed under the risk category to which they relate. Additionally, where possible the risk owner and risk manager will have been identified. If considered helpful, additional columns such as a notes column will have been added to provide background information on each risk. From the risk analysis process the impacts and probabilities will have been added.

11.5 PROCESS OUTPUTS

The outputs will be dictated by the objectives of the evaluation process, the mechanisms used and the data collected. Outputs commonly consist of a combination of the following:

- investment model results;

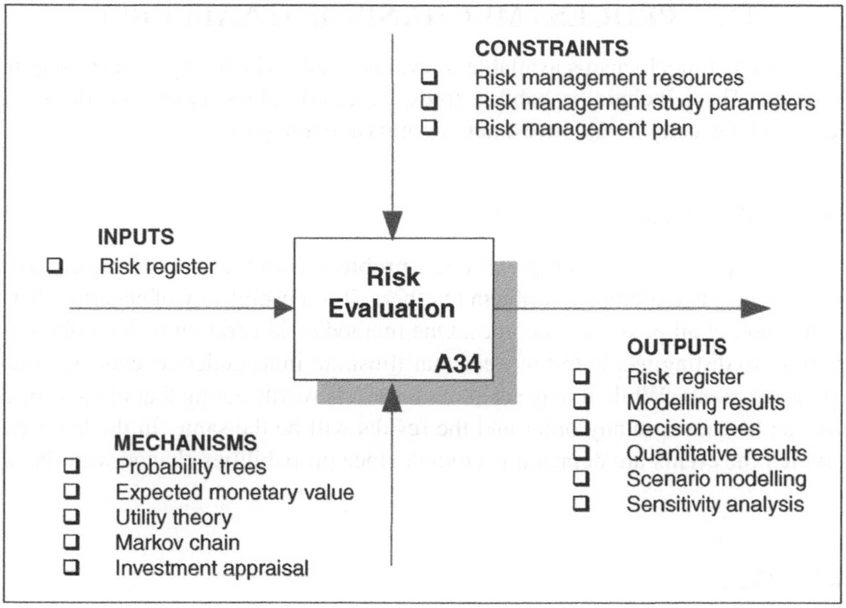

Figure 11.2 Risk evaluation process

- decision support framework such as a completed decision tree (of hedging strategies, investment options);
- quantitative schedule risk analysis results;
- quantitative cost risk analysis results;
- sensitivity analysis;
- scenario modelling;
- revised risk register.

The outputs will, in many instances, be accompanied by explanatory text to describe:

- the objectives of the study;
- the background to the study;
- the participants;
- the inputs;
- the process adopted to derive the findings;
- the findings;
- recommendations for further action;
- terminology used.

11.6 PROCESS CONTROLS (CONSTRAINTS)

The business risk management culture, risk management resources, risk management study and risk management plan (where one exists) regulate the risk identification process. These controls were discussed previously in Section 8.6.

11.7 PROCESS MECHANISMS (ENABLERS)

There are a series of mechanisms available to evaluate risk, which may be used singularly or in combination. These include probability trees, expected values, utility functions, decision trees, Markov chain analysis and investment appraisal techniques.

11.7.1 Probability Trees

Diagrams are a very useful way of representing a problem. In the case of probability, a diagram may be used to help to explain the problem to others. It is a useful way of ensuring that a team has taken account of all possible outcomes. One method of illustration is the probability tree. It is important to distinguish between trees that illustrate independent events, and those that show dependent events. While it may seem obvious, it is worth stating that in the former case, the events can be shown in any order and the results will be the same. In the latter case, the order in which the events are depicted is crucial, since probabilities change with the order of events.

Independent Event

As an example of independent events (which is included for illustration purposes only), consider people's risk appetite preferences and their eye colour. The two things are not related in any way. Suppose that we know that 30% of people have hazel eyes, 40% have green eyes and 30% have blue eyes. Suppose also that we have asked for their risk appetite preferences and found that 20% are risk seeking, 70% are risk neutral and 10% are risk averse. Now if we wish to know the proportion of people with a certain eye colour who have a particular risk appetite, we can use the rule for independent events and multiply the probabilities. For example, those who are risk seeking and have hazel eyes:

$$\text{P(risk seeking)} \times \text{P(hazel)} = 0.2 \times 0.3 = 0.06.$$

We could use a diagram such as Figure 11.3 to show all the possibilities and their probabilities.

Dependent Event

If we have three groups of people, a red team of five men and five women, a blue team of seven men and three women and a green team of four men and six women, then using a two-stage selection procedure, first selecting a team and then selecting an individual, the probability of selecting a woman will be dependent on which team is selected. In Figure 11.4 the probabilities of selecting a red, blue or green team are respectively 0.4, 0.4 and 0.2. The probabilities of being a man or a woman were derived from the numbers in each team. Since individuals can only belong to one team, then we can treat the teams as mutually exclusive. This means that we can add the probability of selecting a woman in the red, blue and green teams to get the overall probability of selecting a woman:

$$\begin{aligned}\text{P(woman)} &= \text{P(woman | red)} + \text{P(woman | blue)} + \text{P(woman | green)}\\ &= 0.2 + 0.12 + 0.12\\ &= 0.44.\end{aligned}$$

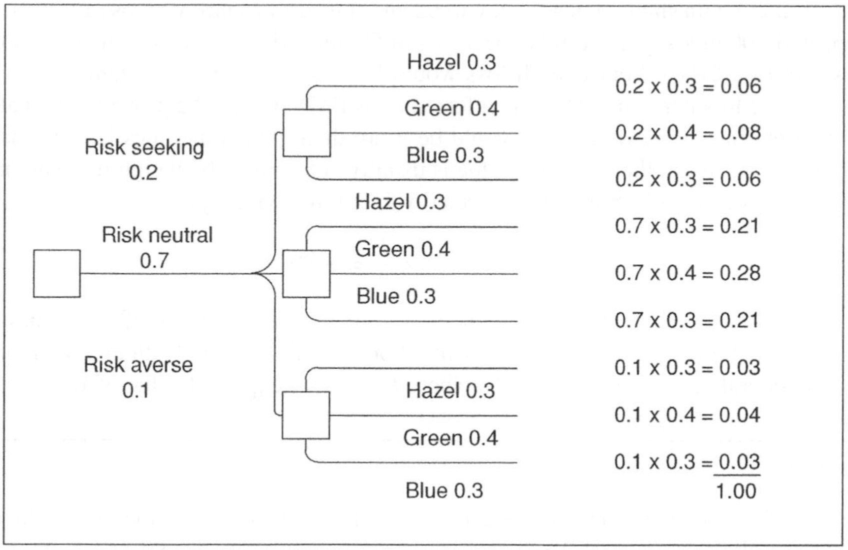

Figure 11.3 Probability tree

11.7.2 Expected Monetary Value

Most decision situations can clearly have more than one outcome. When evaluating problems or situations where there is uncertainty about the outcome, the concept of expected values is particularly important. If it is possible to assign a probability of each outcome being achieved, then the combination of the weighted outcomes can be calculated. It is this sum of the weighted outcomes that is known as the *expected monetary value* (EMV). Considering a simple game

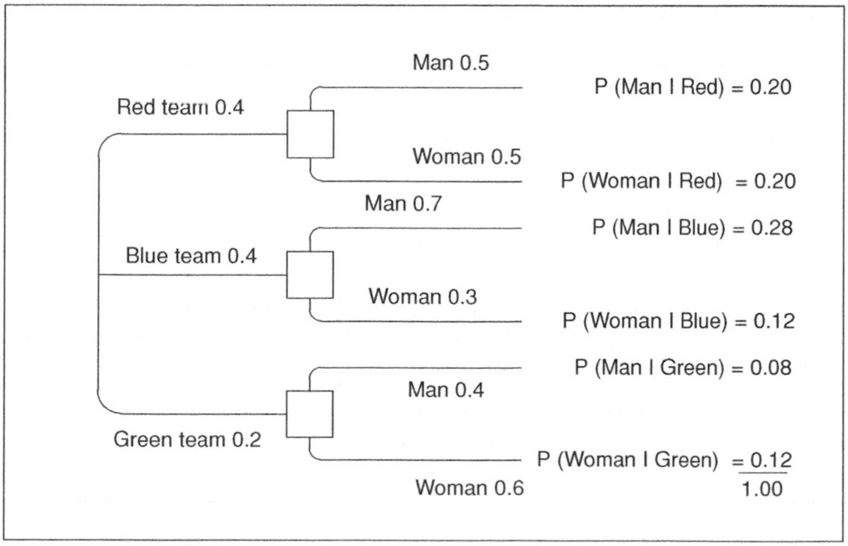

Figure 11.4 Dependent events

of chance, if a fair coin shows a head you win £2 and if it shows a tail, you lose £3. If the game were repeated 100 times, you would expect to win 50 times (i.e. win £100) and expect to lose 50 times (i.e. lose £150). Your overall loss would be £50 or 50 pence per game *on average*. It is this average loss per game which is referred to as EMV. Given the probabilistic nature of the game, sometimes the overall loss would be more than £50, sometimes less. Rather than working out frequencies, the expected value is usually determined by weighting outcomes by probabilities. In this simple game, the expected value of the winnings is:

$$£2 \times \frac{1}{2} + (-£3) \times \frac{1}{2} = -£0.50$$

or – 50 pence (where – £0.50 represents a negative win or loss). You will of course never lose 50 pence in a single game, you will either win £2 or lose £3. Expected values give a long-run average. In general, $E(x) = \sum (x \times \mathrm{P}(x))$, where $E(x)$ is the expected value of x.

Example 1

If three possible yields for an investment are 8%, 10% and 14% and the probability that these yields will be achieved is 0.3, 0.4 and 0.3 respectively, then the expected value of this yield is:

$$\begin{aligned}\text{EMV} &= (0.3 \times 8) + (0.4 \times 10) + (0.3 \times 14)\\ &= 10.6\%.\end{aligned}$$

In general terms, if the probabilities of $O_1, O_2, O_3, \ldots, O_n$ are $p_1, p_2, p_3, \ldots, p_n$ respectively, then

$$\text{EMV} = p_1O_1 + p_2O_2 + p_3O_3 + \ldots + p_nO_n.$$

However, it should be noted that not every outcome will be positive and due allowance must be made in the arithmetic for positive or negative results.

Example 2

Contractors tendering for construction contracts know that they will not be successful all of the time and from experience recognise that they will only obtain a certain proportion of the work for which they tender. Suppose that the contractor believes that this proportion is 1 in 10 for a particular section of the industry. The probability that he will be successful with any one tender is 0.1 and the probability of failure is 0.9. To simplify the example, it is assumed the size, duration and complexity of projects tendered for is similar, tenders regularly cost £8000 to prepare and a profit of £100 000 is commonly obtained; then the contractor's expected return for *each* submission is:

$$\begin{aligned}\text{EMV} &= (0.1 \times 100000) - (0.9 \times 8000)\\ &= 10\,000 - 7200\\ &= £2800.\end{aligned}$$

It is important to understand the concept behind this calculation. First of all, it is not *expected* that any one situation will arise in which the contractor will actually receive a £2800 return.

On each occasion, when a decision is made to tender for work, it is expected that the outcome will either be a loss of £8000 (arising from the abortive cost of preparing the tender) or a profit of £100 000 (as a result of being awarded the contract). Hence, the evaluation of the expected monetary value can only be used with effect where the decision situation to which it applies is one of many similar situations of similar character. The theory of probability is derived on the basis that a large number of trials will take place and the EMV is the weighted average of the outcomes. The expected monetary value decision rule selects the decision alternative with the largest monetary value (EMV).

11.7.3 Utility Theory and Functions

Although EMVs can be readily calculated with a rudimentary understanding of probability, commonly the decision alternative with the highest EMV is not the most desirable or most preferred alternative for the decision maker. For example, suppose that there is an option to buy either of the two companies listed in the payoff table (Table 11.1) for exactly the same price.

The payoff values listed in Table 11.1 represent the annual profits expected from these two businesses. So in any year, there exists a 50% chance that company A will generate a profit of £150 000 and a 50% chance that it will generate a loss of £30 000. On the other hand, in any one year, there exists a 50% chance that company B will generate a profit of £60 000 and a 50% chance that it will generate a smaller profit of £40 000. If the EMV rule were followed unquestioningly, a buyer should consider company A, as it has the highest EMV. However, company A represents a far riskier investment than company B. Although it is likely company A would generate the highest EMV over the long run, our buyer may not have the financial resources to withstand the potential losses of £30 000 per year that could occur in the short term with this alternative. With company B, our buyer can be sure of making at least £40 000 each year. Although company B's EMV over the long run might not be as great as that of company A, for many decision makers this is more than offset by piece of mind associated with company B's relatively stable profit level. However, other decision makers may be willing to accept the greater risk associated with company A in the hope of achieving the higher potential payoff this alternative provides. As this example illustrates, the EMVs of different decision alternatives do not necessarily reflect the relative attractiveness of the alternatives to a particular decision maker. Utility theory offers a model for understanding this behaviour. Personal attitudes to risk are measured by studying individual trade-offs between gamblers and certain payoffs.

Table 11.1 Probabilities and outcomes

	Financial position			
Company	1	2	EMV	
A	150 000	−30 000	60 000	← max EMV
B	60 000	40 000	50 000	
Probability	0.5	0.5		

Utility Function of Executives

While obtaining a utility function for an individual is relatively straightforward, obtaining a utility function for a group of executives is more complex. Moore and Thomas (1988) describe a number of studies carried out to measure the utility functions of executives. All used the variations of the standard gamble methods described above. Two of these studies are described below.

Study 1. Oil executives were presented with a series of hypothetical drilling opportunities and they were asked whether they would accept or reject each one. There was only moderate success at introducing the oilmen to the formal concept of utility, as they seemed to distrust the idea of using formal graphs to replace judgement.

Study 2. Sixteen executives from a chemical company were questioned both as individuals and as business executives. This study showed (along with similar studies) that different executives have utility functions of varying shapes so that obtaining a corporate utility function is particularly difficult.

Moore and Thomas also describe one of the difficulties which emerged from the use of standard gamble methods: many managers found it hard to distinguish between similar probabilities, such as between 0.1 and 0.2 at one extreme and between 0.8 and 0.9 at the other extreme. To these managers it appears these were simply classifiable as very unlikely and very likely events.

Utility Functions

Utility theory assumes that every decision maker uses a utility function that translates each of the possible payoffs in a decision problem into a non-monetary measure known as utility. The utility of a payoff represents the desirability (total worth or value) of the outcome of a decision alternative to the decision maker.

Different decision makers have different attitudes and preferences towards risk and return. Those who are "risk neutral" tend to make decisions using the maximum EMV decision rule. However, some decision makers are risk avoiders or "risk averse", and others look for risk or are "risk seekers". The utility functions typically associated with these three types of decision makers are shown in Figure 11.5. For convenience the utilities are represented on a scale from 0 to 1, where 0 represents the least value and 1 represents the most. Figure 11.5 illustrates how the same monetary payoff might produce different levels of utility for three different decision makers. The "risk neutral" decision maker who follows the EMV decision rule has a constant marginal utility for increased payoffs. That is, every additional pound in payoff results in the same amount of increase in utility. A "risk averse" decision maker assigns the largest relative utility to any payoff but has a diminishing marginal utility for increased payoffs in that every additional pound in payoff results in smaller increases in utility. The "risk seeking" decision maker assigns the smallest utility to any payoff but has an increasing marginal utility for increased payoffs. That is, every additional pound in payoff results in larger increases in utility.

11.7.4 Decision Trees

Decision trees enable decision problems to be represented graphically. Figure 11.6 shows a decision problem. The problem is described here.

Example project

As a result of growing air traffic in the UK, the government has given consideration to constructing a new airport to the east of London. Two possible locations for the new airport have been identified, but as a result of a lengthy public inquiry, a final decision on the new location is not expected to be made for another year. The Lawrence hotel chain intends to build a new facility near the new airport once its site is determined. The chief executive faces a difficult decision about where to buy land. Currently land values around the two possible sites for the new airport are increasing as investors speculate that property values will increase significantly in the vicinity of the airport. The hotel chain has identified a site for a hotel close to each airport. It has determined the current price of each parcel of land and estimated the present value of the future cash flows that a hotel would generate at each site if the airport is ultimately located at the site. In addition it has determined the resale value it believes it can obtain if the airport is not built at the site.

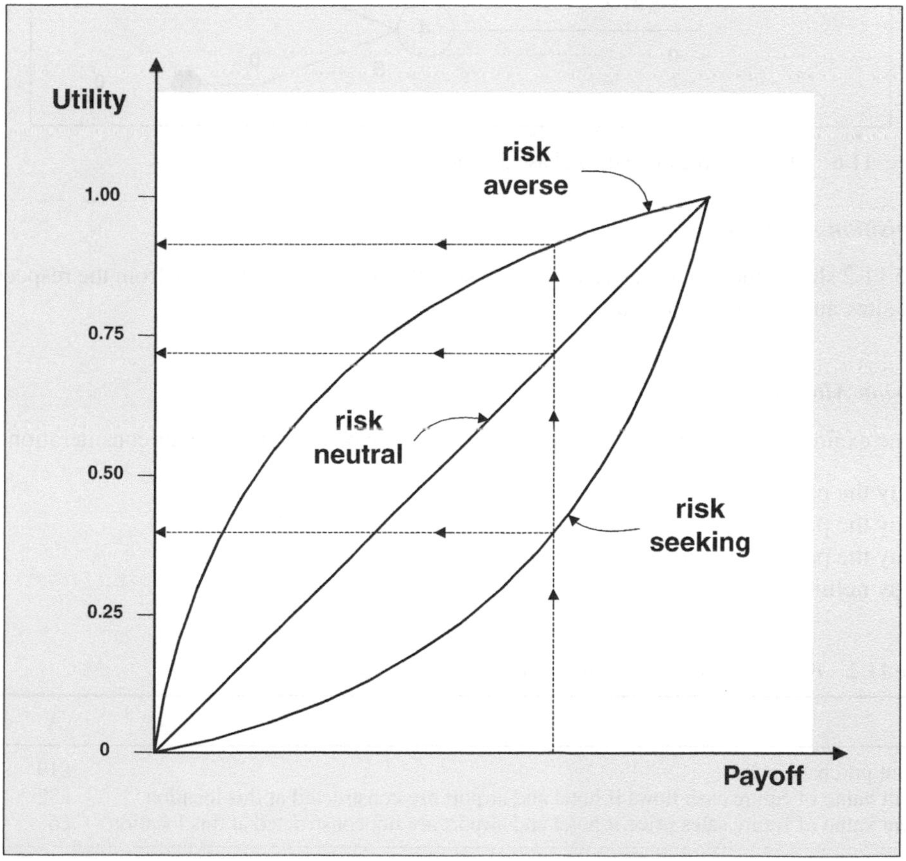

Figure 11.5 Utility functions

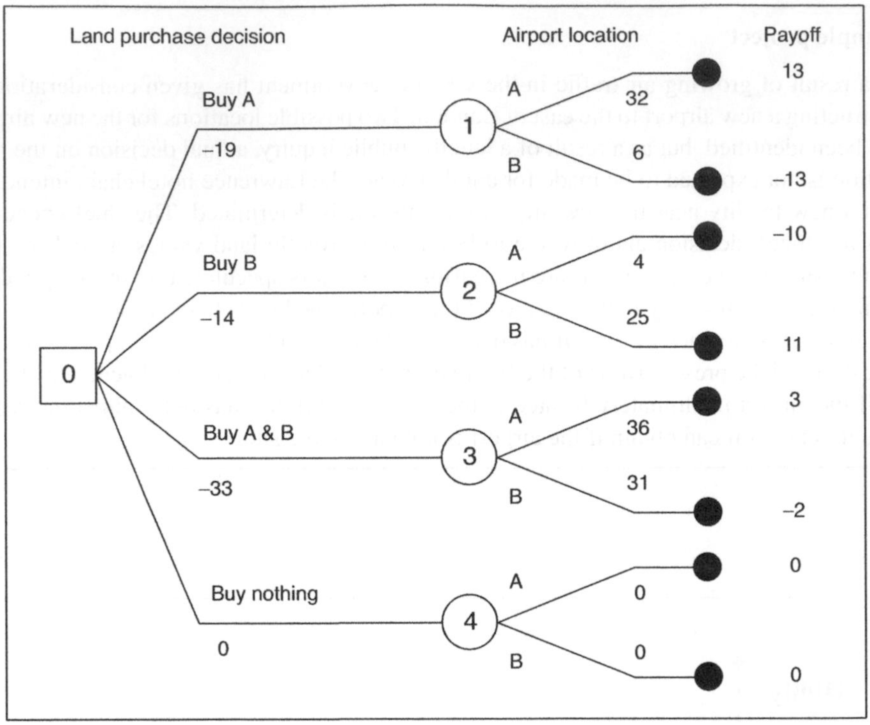

Figure 11.6 Decision tree of land purchase decision

Acquisition Analysis

Table 11.2 shows the purchase price for each site, the projected cash flows from the respective hotel sites and the site resale values.

Decision Alternatives

For the example project described above, the following options were under consideration:

1. Buy the parcel at location A.
2. Buy the parcel at location B.
3. Buy the parcels at locations A and B.
4. Buy nothing.

Table 11.2 Purchase price, projected cash flows and site resale values

	A	B
Current purchase price	£19	£14
Present value of future cash flows if hotel and airport are constructed at this location	£32	£25
Present value of future sales price if hotel and airport are not constructed at this location	£6	£4

All figures are in millions of pounds.

Decision Tree Construction

As shown in Figure 11.6, a decision tree is composed of a collection of *nodes* (represented by circles and squares) interconnected by *branches* (represented by lines). A square node is called a *decision node* because it represents a decision. Branches emanating from a decision node represent the different alternatives for a particular decision. In Figure 11.6, a single decision node (node 0) represents the decision faced by Lawrence Hotels about where to buy land. The four branches coming out of the decision node represent the four alternatives under consideration. The cash flow associated with each alternative is also listed. For example, the value –19 below the alternative labelled "Buy A" indicates that if the company purchases the parcel at location A, it must pay £19 million. The circular nodes in a decision tree are called *event nodes* because they represent uncertain events. The branches emanating from event nodes (called *event branches*) correspond to the possible states of nature or the possible outcomes of an uncertain event. Figure 11.6 shows that each decision alternative emanating from node 0 is followed by an uncertain event represented by the event nodes 1, 2, 3 and 4. The branches from each event node represent a possible location of the new airport. In each case the airport can be built at location A or B. The value next to each branch from the event nodes indicates the cash flow that will occur for that decision–event combination. For example, at node 1 the value 32 next to the first event branch indicates that if the company buys the parcel at location A and the airport is built at that location, a cash flow of £32 million will occur. The various branches in a decision tree end at the small black dots called *leaves*. This is because each leaf corresponds to one way in which the decision problem can terminate. Leaves are also referred to as *terminal nodes*. The payoff occurring at each leaf is computed by summing the cash flows along the branches leading to each leaf. For example, following the uppermost branches through the tree, a payoff of £13 million results if the decision to buy the parcel at location A is followed by the new airport being built at this location (–19 + 32 = 13). As a guide the cash flow values should be verified on each branch and at each leaf, before constructing the whole tree.

Rolling Back a Decision Tree

After calculating the payoffs at each leaf, the EMV decision rule can be implemented – that is, to identify the decision with the largest EMV. A process known as rolling back can be applied to a decision tree to determine the decision with the largest EMV. Figure 11.7 illustrates this process for the example examined here. As the EMV decision rule is a probabilistic method, Figure 11.7 indicates the probabilities associated with each event branch emanating from each event node (i.e. a 0.4 probability exists that the new airport will be built at location A and a 0.6 probability exists that it will be built at location B). To roll back this decision tree, the decision tree is worked through from right to left, commencing with the payoffs, computing the expected values for each node. For example, the event represented by node 1 has a 0.4 probability of resulting in a payoff of £13 million, and a 0.6 probability of resulting in a loss of £13 million. Thus the EMV at node 1 is calculated as:

$$\text{EMV at node 1} = 0.4 \times 13 + 0.6 \times (-13) = -2.6.$$

The expected value calculations for the remaining event nodes in Figure 11.7 are:

$$\text{EMV at node 2} = 0.4 \times (-10) + 0.6 \times 11 = 2.6,$$
$$\text{EMV at node 3} = 0.4 \times 3 + 0.6 \times (-2) = 0,$$
$$\text{EMV at node 4} = 0.4 \times 0 + 0.6 \times 0 = 0.$$

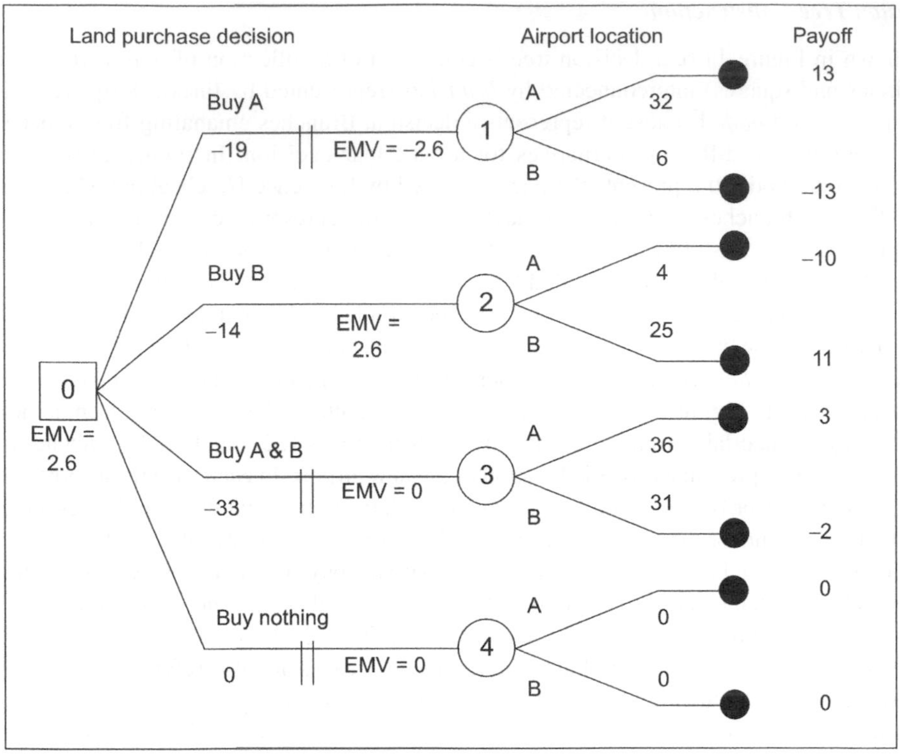

Figure 11.7 Decision tree rolled back

The EMV for a decision node is calculated in a different way. For example, at node 0 a selection has to be made from four alternatives that lead to events with expected values of –2.6, 2.6, 0 and 0, respectively. At a decision node, the alternative that leads to the best EMV is selected. Hence the EMV at node 0 is 2.6, which corresponds to the EMV resulting from the decision to buy land at location B. The optimal alternative at a decision node is sometimes indicated by "pruning" the suboptimal branches. The pruned branches in Figure 11.7 are indicated by the vertical lines (||) shown on the suboptimal alternatives emanating from node 0.

11.7.5 Markov Chain

A Markov chain combines the ideas of probability with those of matrix algebra.[1] The Markov chain concept assumes that probabilities remain fixed over time, but the system that is being

[1] A matrix is a rectangular array of numbers arranged in rows and columns and is characterised by its size (or order), written as: number of rows × number of columns. The whole matrix is usually referred to by a capital letter, while individual numbers, or elements, within the matrix are referred to by lower-case letters, usually with a suffix to identify in which row and in which column they appear. Note that a matrix does not have a numerical value, it is merely a convenient way of representing an array of numbers.

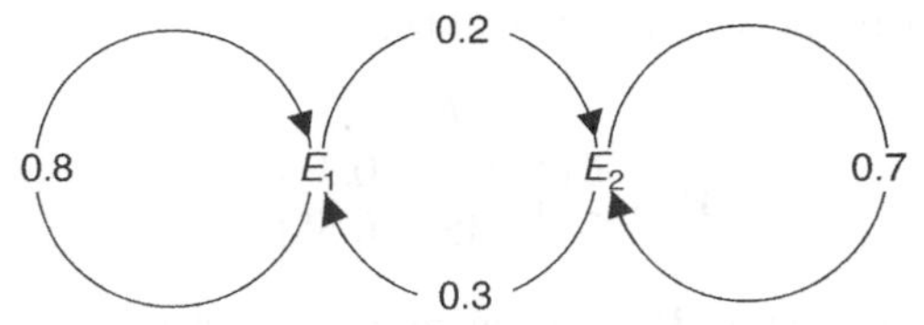

Figure 11.8 Directed diagram

modelled is able to change from one state to another, using these fixed variables as transition probabilities. Consider the following transition matrix:

$$\mathbf{P} = \begin{array}{c|cc|} & E_1 & E_2 \\ E_1 & 0.8 & 0.2 \\ E_2 & 0.3 & 0.7 \end{array}.$$

This means that if the system is in some state labelled E_1 the probability of going to E_2 is 0.2. If the system is already at E_2 then the probability of going to E_1 is 0.3, and the probability of remaining at E_2 is 0.7. This transition matrix could be represented by the directed diagram in Figure 11.8.

If we consider the movement from one state to another to happen at the end of some specific period, and look at the passage of two of these periods, we have the situation in Figure 11.9. The probability of ending E_1 after two periods if the system started at E_1 will be:

$$\mathrm{P}(E_1 \rightarrow E_1 \rightarrow E_1) + \mathrm{P}(E_1 \rightarrow E_2 \rightarrow E_1) = (0.8 \times 0.8) + (0.2 \times 0.3) = 0.7.$$

Starting at E_1 and ending at E_2:

$$\mathrm{P}(E_1 \rightarrow E_1 \rightarrow E2) + \mathrm{P}(E_1 \rightarrow E_2 \rightarrow E_2) = (0.8 \times 0.2) + (0.2 \times 0.7) = 0.3.$$

Starting at E_2 and ending at E_1:

$$\mathrm{P}(E_2 \rightarrow E_1 \rightarrow E_1) + \mathrm{P}(E_2 \rightarrow E_2 \rightarrow E_1) = (0.3 \times 0.8) + (0.7 \times 0.3) = 0.45.$$

Starting at E_2 and ending at E_2:

$$\mathrm{P}(E_2 \rightarrow E_2 \rightarrow E_2) + \mathrm{P}(E_2 \rightarrow E_1 \rightarrow E_2) = (0.7 \times 0.7) + (0.3 \times 0.2) = 0.55.$$

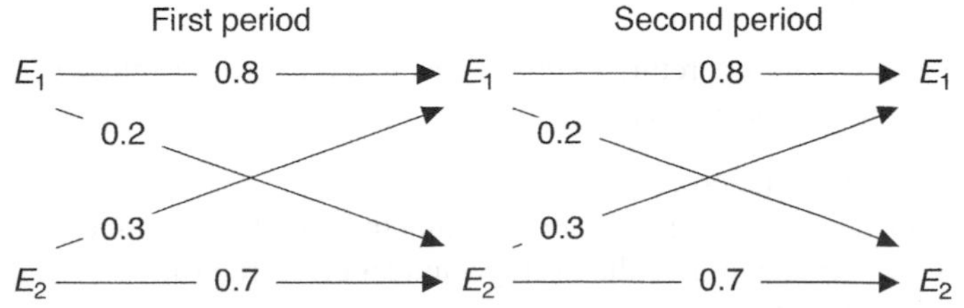

Figure 11.9 Probability over two periods

Hence the transition matrix for two periods will be:

$$\mathbf{P} = \begin{array}{c} \\ E_1 \\ E_2 \end{array} \begin{array}{c} \begin{array}{cc} E_1 & E_2 \end{array} \\ \left|\begin{array}{cc} 0.7 & 0.3 \\ 0.45 & 0.55 \end{array}\right| \end{array}.$$

However, note that this is equal to $\mathbf{P}^2$, the square of the transition matrix for one period. To find the transition matrix for the four periods, we would find $\mathbf{P}^4$ and so on.

The states of the system at a given instant could be an item working or not working, a company being profitable or making a loss, an individual being given a promotion or failing at an interview. In all transition matrices, the movement over time is from the state on the left to the state above the particular column and thus since something must happen, the sum of any row must be equal to 1.

11.7.6 Investment Appraisal

Definition

Investment appraisal is the assessment of different large-scale capital projects to see both what is affordable and what is possible. Businesses should consider both the risks and the benefits involved. Investment appraisal also involves looking at different projects and choosing the one that offers the best chance of success.

Evaluation

If a business wishes to expand, move or increase its productivity, then it needs to make capital investment decisions. Such decisions are made in the context of both cost and market uncertainty – what the market is about to do can only be predicted. Businesses try to minimise risk by accurately predicting costs from a known base. The known base must include the initial cost of the investment at current prices. Predictions involve the known amounts of cash that will flow from the project once it is on stream and the estimated lifespan of the project. Research shows that there are four main techniques used in the UK for screening investment proposals. They are:

- average rate of return, also known as accounting rate of return (ARR);
- payback period (PP);
- net present value (NPV); and
- internal rate of return (IRR).

Average Rate of Return

The ARR is the average annual return expressed as a percentage of initial cost. To calculate the average rate of return a business needs to:

- estimate the lifespan of the project;
- calculate the total profit over the lifespan by taking the initial cost from the net cash inflows;
- divide this by the lifespan to give the average annual profit/return;
- divide this by initial cost; and
- multiply by 100.

Example

For this example the average rate of return (ARR) is 20%
Calculation:

$$\begin{aligned}
\text{total inflows} &= 50 + 50 + 100 + 160 = £360 \\
\text{initial cost} &= £200 \\
\text{overall return} &= \text{net inflows} - \text{initial cost} \\
&= £360 - £200 = £160 \\
\text{average return per year} &= \text{overall return} \div \text{number of years} = £160 \div 4 = £40 \text{ p.a.} \\
\text{average rate of return} &= (\text{average return per year} \div \text{initial cost}) \times 100 \\
&= (40 \div 200) \times 100 = 20\%
\end{aligned}$$

A business will have already established a criterion for implementation, say 5% above current interest rates, and can use this information to make a decision.

Payback Period

Payback refers to the length of time it takes for a business to recoup its original outlay. An ARR of 10% a year would mean that the initial cost would be covered in a ten-year period. The longer the PP, the higher the level of risk involved in the project and the less likely figures will be accurate. It is much easier to predict two or three years ahead than ten or 12. While NPV is theoretically the optimum method as it includes the time value of money, businesses like to use the payback technique as it is straightforward to understand and tells you simply how long you have to wait to get your initial investment back.

Net Present Value

The NPV measures how the value of money decreases over time. If a business earns £1000 today, its worth will be less than £1000 in one year's time and even less in five years' time. Future returns are therefore discounted according to the business' estimate of the future value of money. This is called discounting the cash flow. The discount rate is based on current rates of interest and on inflation estimates. If the project is a high-risk investment, then this will also affect discount rates as will international currency fluctuations, if the business is, for example, buying materials from abroad.

The business will be asking the question "what will £1000 be worth in one, three or five years' time?". The calculation is important, as it can be measured against a risk-free investment such as banking the money. Remember that the opportunity cost (the cost of the next best alternative) of any decision must always be taken into account. If investors are deprived of their money for a year, then they could equally well be deprived of its use by placing it on deposit in a bank or building society. In this case, at the end of the year they could have their money back and have interest as well. So, unless the opportunity to invest offers similar or better returns, they will be incurring an opportunity cost. An opportunity cost occurs where one course of action deprives investors of the opportunity to derive some benefit from

Table 11.3 Results for a business for its first four years

Year	£	Discount factors	Calculation	Present value £
0	(200)	1	(200) × 1	(200)
1	50	0.909	50 × 0.909	45.45
2	50	0.826	50 × 0.826	41.3
3	100	0.751	100 × 0.751	75.1
4	160	0.683	160 × 0.683	109.28

an alternative action. Any investment opportunity must, if it is to make investors wealthier, do better than the returns, which are available from the next best opportunity.

The NPV is positive if the project earns a return above the cost of capital or negative if the project fails to produce a significant return. There are some problems with discounted cash flows and care is needed not to produce misleading results:

- *Equity and entity*. The cash flows do not include the cost of debt and the cost of capital needs to be a weighted cost of capital. Plus currently the cost of equity capital is not subject to a reduction due to tax.
- *Risk*. Many companies use a simple hurdle or "risk adjusted" rate to assess projects. While it is logical to add a margin for risk, this will tend to penalise longer-term projects more excessively. If managers know that the hurdle is 20% then only projects of a certain nature will be put forward for approval.

Using the results from the simple example included in Table 11.3, the present value of the expected inflows is £45.45 + £41.3 + £75.1 + £109.28 = £271.13. This is what the firm believes the future earnings of the project are worth in today's terms:

Calculation:

$$\begin{aligned}\text{the initial outflow} &= £200\\ \text{NPV} &= \text{discounted inflows} - \text{initial outflow}\\ &= £271.13 - £200\\ &= £71.13.\end{aligned}$$

In today's terms the project is expected to be worth £71.13 more than it would cost; therefore it would be worth investing. The higher the NPV, the more the project is worth compared to its cost.

Why NPV is Superior to ARR and PP

NPV is a better method of appraising investment opportunities than either ARR or PP because it fully addresses each of the following:

- *The timing of the cash flows*. By discounting the various cash flows associated with each project according to when they are expected to arise, NPV takes account of the time value of money. The discount factor is based on the opportunity cost of finance (i.e. the return which the next best alternative opportunity would generate) and so the net benefit after financing costs have been met is identified (as the NPV).

- *The whole of the relevant cash flows*. NPV includes all of the relevant cash flows irrespective of when they are expected to occur. It treats them differently according to their date of occurrence, but they are taken into account.
- *The objectives of the business*. The output of the NPV analysis has a direct bearing on the wealth of the shareholders of a business. (Positive NPVs enhance wealth, negative ones reduce it.) Since we assume that private sector businesses seek to maximise shareholder wealth, NPV is superior to the methods previously discussed.

It was discussed earlier that a business should take on all projects with positive NPVs, when they are discounted at the opportunity cost of finance. When a choice has to be made among projects, a business should select the one with the largest NPV.

Internal Rate of Return

The IRR is the last of the four major methods of investment appraisal that are found in practice. This method also uses discounted cash flows and is really a subset of NPV. Instead of using interest rates to work out NPV, it involves a calculation whereby the rate of interest is the unknown factor. The business looks for the point where NPV is zero and reads off the interest rate where this occurs. This can then be compared with current or future predicted rates. The main advantage of this method is that it can be used to compare two capital projects with different initial outlays.

Cost–Benefit Analysis

CBA is used to support decision making in assessing the benefits of a proposed project, programme or policy or to choose between several alternatives. It involves comparing the total expected costs of each option against the total expected benefits, to establish the option with the best cost–benefit ratio (i.e. the project where the benefits outweigh the costs by the greatest margin). Among the first evidence of the application of cost–benefit techniques to projects was the development of highway and motorway investments in the US and UK in the 1950s and 1960s. Over the last 50 years, cost–benefit techniques have gradually developed to the extent that substantial guidance now exists on transport project appraisals in many countries around the world. Whatever the problem under investigation, there are four main stages in the development of a CBA. These are illustrated in Figure 11.10.

The first stage is to identify all of the relevant costs and benefits arising out of the project, programme or policy under examination. On the surface this may sound a relatively simple task. In reality it is not. There are particular problems when it comes to establishing external costs and benefits. These are often controversial, not easily defined in a discrete way and there is the added difficulty that it is not always possible to draw a line in terms of a physical or geographical cut-off.

The second stage involves putting a monetary value on the various costs and benefits. This is relatively straightforward where market prices are available. For instance, in the case of constructing a new motorway the construction and operating costs can be ascertained as well as a the monetary value of the creation of new jobs and lower maintenance on existing roads. However, there will be other costs and benefits associated with the project, sometimes referred to as the externalities, which are difficult to put a value on. These may include the dis-benefits of noise and air pollution, congestion on local roads, loss of recreational spaces, loss of areas

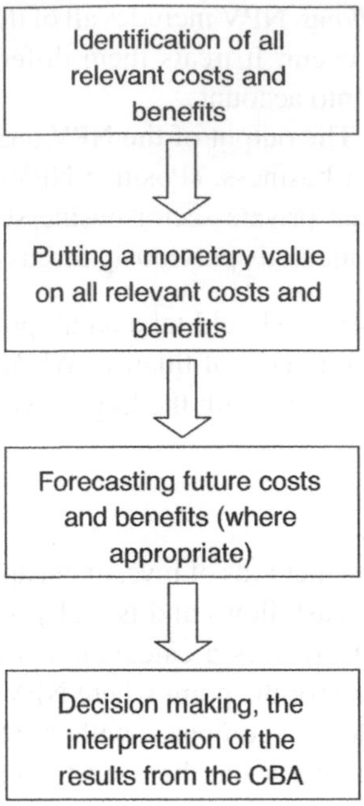

Figure 11.10 Stages in cost–benefit analysis

of outstanding natural beauty, loss of habitat for rare species and/or loss of sites of historical importance. There may also be benefits of shorter journey times, fewer accidents, greater access for tourists and rejuvenation of declining industrial areas. Associated with each of the costs and benefits would be risks and opportunities. For some negative risks, particularly those involving risks to human life or damage to the environment, the ALARP principle may be applied. This divides risks into three regions: a level above which negative risks are intolerable and should not be taken except in extraordinary circumstances; a level below which risks are negligible and need only to be monitored to ensure they remain low; and a central band where risks are made as low as reasonably practicable (ALARP).

The third stage of the CBA applies where projects have long-term implications which stretch well into the future. This would involve projects with a long construction period, a finance payback period of, say, 25 years and a benefit return period of even longer. In such circumstances it is normally necessary to discount the costs and benefits to bring them into "today's money" so that a valid comparison can be obtained. All costs and benefits are expressed as a present value. The present value of all costs and all benefits to all stakeholders can be combined to produce an NPV. A positive NPV implies that the action is beneficial.

The fourth and final stage is where the results of the earlier stages are drawn together so that the outcome can be presented in a clear manner in order to aid decision making. The important principle to recognise is that the value of benefits exceeds the value of the costs, then the project or programme is worthwhile since it provides an overall net benefit to the community.

In summary, CBA can be a very imprecise procedure. In addition, the outcomes of CBAs should be treated with caution because they may be highly inaccurate. Not surprisingly, inaccurate CBAs are likely to lead to inefficient decisions. These estimates have the following problems:

- They depend crucially upon the rate of discount of future costs and benefits used.
- They rely heavily on similar past projects for cost information (often differing markedly in function, size, location, market, contracting strategy and likely contractor skill levels).
- They rely heavily on the project's team members to identify (remember from their collective past experiences) the significant cost drivers.
- They rely heavily on the project's team members' ability (and, most importantly, willingness) to spend time thoroughly identifying and assessing the risks and opportunities to the project.
- They rely on very crude heuristics to estimate the money cost of the intangible external elements due to the absence of reliable or recent data.
- They may include unconscious (or conscious) biases of the team members (who often have a vested interest in a project going ahead).

11.8 PROCESS ACTIVITIES

11.8.1 Basic Concepts of Probability

Chance and the assessment of risk play a major part in a large number of business activities. Hence, probability has found a wide range of business applications such as in investment appraisals which require an assessment of risk and a measure of expected outcomes. Many of the process activities examined here require an understanding of the concepts of probability. Probability represents a new set of conceptual tools. Rather than looking at the world as consisting of *deterministic* situations, where everything is known with certainty, we can now consider a range of outcomes to every situation. More than this, by treating the world as *stochastic*, it is possible to assess the chance of particular outcomes happening in a given situation. Hence, it is important to consider the range of outcomes possible from a situation, so that recognition is given to even the remote (unlikely) outcomes.

There are a number of axioms which may be used to describe probability, which may be paraphrased as follows:

1. The probability of an event lies in the interval $0 \leq \mathrm{P}(E) \leq 1$ (where P represents probability and E represents the event under examination), and no other values are possible.
2. If something is certain to occur, then it has a probability of 1.
3. If two or more different outcomes of a trial or experiment cannot happen at the same time, then the probability of one or other of these outcomes occurring is the sum of the individual probabilities. For example, if $\mathrm{P}(E_1) = \frac{1}{4}$ and $\mathrm{P}(E_2) = \frac{1}{2}$, then $\mathrm{P}(E_1 \text{ or } E_2) = \frac{1}{4} + \frac{1}{2} = \frac{3}{4}$.

In business situations we are concerned about the chances of two or more events happening at the same time, or being mutually exclusive. Hence, it is important to be able to classify events.

Mutually Exclusive Events

A mutually exclusive event is where we add the probabilities together to find the probability that one or the other events occur. For example, if a group of companies consists of 20 public limited companies, 40 private limited companies, ten sole traders and 30 partnerships, then selecting a company at random we have:

$$P(\text{public limited company}) = 20/100$$
$$P(\text{private limited company}) = 40/100$$
$$P(\text{limited liability companies}) = 20/100 + 40/100 = 60/100 = 0.6.$$

Non-Mutually Exclusive Events

Where one outcome has, or can have, more than one characteristic, then these outcomes are said to be *non-mutually exclusive*. In this case it will not be possible to simply add the probabilities together as this would involve counting some outcomes twice. For example, if a group of companies contains both limited and unlimited liability companies that have either a London or Manchester office, then the probability of selecting a company that has limited liability or has a London office will be

P(public limited company or London office) = P(public limited company) + P(London office) – P(public limited company and London office).

Suppose, for example, there were 20 public limited companies with a Manchester office, 30 private limited companies with a Manchester office, 10 public limited companies with a London office, and 40 private limited companies with a London office. As "London office" appears on either side of the equals sign, we need to subtract the probability of this group from our required probability. If, for example, there were 30 public limited companies, of whom are in London, and 70 private limited companies, of whom 40 were in London, then

$$P(\text{public limited company}) = 30/100$$
$$P(\text{London office}) = 50/100$$
$$P(\text{public limited company and London office}) = 10/100.$$

Therefore,

$$P(\text{public limited company or London office}) = 30/100 + 50/100 - 10/100 = 0.7.$$

11.8.2 Sensitivity Analysis

Sensitivity analysis is a useful technique to employ when evaluating the profitability of an investment proposal for a particular project. The technique involves taking a *single* variable (e.g. volume of sales) and examining the effect of changes in the selected variable on the likely performance of the business. By examining the change that occurs, it is possible to arrive at some assessment of how sensitive changes are for the projected outcomes. When the investment appraisal is positive, each input value can be examined to see how much the estimated figure could be changed before the project becomes unprofitable for that reason alone.

Although only one variable is examined at a time, a number of variables, considered to be important to the performance of a business, may be examined consecutively. One form of sensitivity analysis is to pose a series of "what-if?" questions. If we take sales, for example, we might ask the following questions:

- What if the sales volume is 10% higher than expected?
- What if the sales volume is 5% lower than expected?

- What if the sales price is reduced by 20%?
- What if the sales price is increased by 5%?

In answering these questions, it is possible to develop a better "feel" for the effect of forecast inaccuracies on the final outcomes. However, this technique does not assign probabilities to each possible change, nor does it consider the effect on projected outcomes of more than one variable at a time.

11.8.3 Scenario Analysis

Another approach to help managers gain a feel for the effect of forecast inaccuracies is to prepare projected financial statements according to different possible "states of the world". For example, managers may wish to examine projected financial statements prepared on the following basis:

- an optimistic view of likely events;
- a pessimistic view of likely future events (or worse case);
- a most likely view of future events.

This approach is open to criticism because it does not indicate the likelihood of each scenario occurring, nor does it identify the other possible scenarios that might occur. Nevertheless, the portrayal of optimistic and pessimistic scenarios may be useful in providing managers with some feel for the "downside" risk and "upside" potential associated with a project or activity. Scenario analysis is unlike sensitivity analysis in that it will involve changing a number of variables simultaneously in order to portray a possible outcome.

11.8.4 Simulation

Simulation is a technique that is helpful in analysing financial or programme (time) models, where the values of the input data, the variables, may be uncertain. Variables in this context refer to risks, opportunities, costs or durations. Simulation is possible with the aid of commercially available spreadsheet or statistical software. The objective of simulation is to obtain a distribution (and its associated characteristics) for the bottom-line performance figure (measure) of a decision derived from considering the input variables in combination. The thinking behind simulation is similar to the idea of carrying out multiple manual what-if scenarios. (In what-if analysis, a manager changes the values of selected input variables in a model to see what happens to the bottom-line performance figure/measure.) The difference between simulation and manual what-if analysis is that the process of assigning values to variables within the cells in a spreadsheet is automated so that (1) the values are assigned in a non-biased way and (2) the spreadsheet user is relieved of the burden of determining those values. With simulation we repeatedly and randomly generate sample values for each uncertain input variable included in the model and then calculate the resulting value of the bottom-line performance measure. The sample values of the performance measure can be used to determine a cumulative frequency curve to estimate the range of values over which the performance value might vary, to estimate its mean and variance and to estimate the probability that the actual value of the performance measure will be greater than or less than a particular value. All these measures provide greater insight into the risk associated with a given decision than a single value calculation based on the expected values for the uncertain variables. While commercially available software can calculate the output percentiles to several decimal places, to include with a report, say, the 75th percentile to two decimal places (rather than just whole numbers) would be to give the reader a false sense of accuracy, as the input distributions would no doubt have been based on subjective assessments of ranges of financial impact. Alfred North Whitehead (1861–1947), the distinguished logician and philosopher, used to warn against the "danger of the false concreteness". He stated that a measurement does not become more "accurate" by being worked out to the sixth decimal, when the phenomenon is only capable of being verified within a range of 50–70%. This is false "concreteness" and misleading.

11.8.5 Monte Carlo Simulation

Monte Carlo simulation provides a means of evaluating the effect of uncertainty on a planned activity in a wide range of situations. It is typically used to evaluate cost, duration, demand or throughput. Monte Carlo sampling refers to the traditional technique for using random numbers to sample from a probability distribution. The term "Monte Carlo" was introduced during World War II as a code name for simulation problems associated with development of the atomic bomb (Palisade Corporation 2010).

How Monte Carlo Simulation Works

In summary, Monte Carlo looks at a large number of what-if scenarios for, say, the financial outcome of a business activity or project by accounting for a large number of possible values that each variable could take and weighting each value by the probability of occurrence. In operation Monte Carlo simulation generates a number at random for each risk item within the constraints of the probability distribution assigned to it (commonly triangular, PERT, normal, log-normal or rectangular/uniform) and weights this number in accordance with the probability of the risk occurring. (So, for example, if a risk was assigned a uniform distribution with upper and lower limits of £10 and £5 respectively, and a probability of occurrence of 50%, then every other iteration would include the risk and the value assigned to the risk would be between £10 and £5.) This value is then stored. These weighted random numbers are then aggregated to create one model simulation (iteration, trial or scenario), which is one possible value for the business activity or project. We are concerned that the model will reproduce the distributions that have been included in the model. The only way that this can be achieved is by generating a large number of iterations. The process is then repeated, commonly 5000 times, to give 5000 realistic possible outcomes for the project. The statistical data describing each iteration is then aggregated and represented graphically by a histogram to show the range of possible outcomes, a probability distribution to illustrate possible skew and a cumulative frequency curve to show the likelihood (as a percentage) of exceeding the business objective (typically a finance limit).

Percentiles

A simulation will provide a series of values (or a number of possible outcomes) for a business activity or project. These results can be divided into equal parts. A series of values can be subdivided into two equal parts around the median. The median is the middle value of an ordered set of data (listing output values by size). The concept of dividing the data into two equal parts can be extended to dividing the data into four equal parts, separated by quartiles. The quartiles of an ordered set of data are such that 25% of the observations are less than or equal to the first quartile (Q_1), 50% are less than or equal to the second quartile or median (Q_2), and 75% are less than or equal to the third quartile (Q_3). A series of values may also be subdivided into a greater number of equal parts, say 10 or 100, separated by deciles or percentiles, respectively. While percentiles can be calculated manually, they are commonly generated by Monte Carlo simulation tools, to provide confidence levels in, say, the likely outturn cost of an investment. Hence, an 80% confidence figure represents an 80% chance that the cost of the investment will be at this figure or less.

Correlations

During a simulation analysis it is important to account for correlation between input variables. Correlation occurs when the results of sampling from two or more input distributions are related – for example, when the sampling of one input distribution returns a relatively "high" value, it may be that sampling a second input should also return a relatively high value. A good example is the case of one input named "material costs" and a second input called "tender prices". There may be a distribution for each of these input variables, but the sampling of them should be related to avoid nonsensical results. For example, when a high value of material costs is sampled, tender prices should also be sampled as relatively high. Conversely, you would expect that when material costs are low, tender prices would also be low. A record of correlation included in a simulation model should be made so as to facilitate updates when the variables are amended.

Benefits

Monte Carlo simulation offers a number of benefits:

- Models are relatively simple to develop and can be extended as the need arises.
- Readily available proprietary software can be used to automate the tasks involved in simulation and provide outputs suitable for direct transfer into reports.
- Computers can be used to calculate the activity or project outcome distribution quickly.
- The method can accommodate a great number of distributions without difficulty.
- Correlations and other interdependencies can be modelled.
- Sensitivity analysis can be assessed to ascertain the dominant influences in the model.
- The level of mathematics required to perform Monte Carlo simulation is quite basic.
- Greater levels of precision can be achieved by simply increasing the number of iterations.
- As with all spreadsheet-based models, changes can be made very quickly and the results compared with other models/spreadsheets.
- Complex spreadsheet functions can be included (such as MAX, OR and nested IF functions).
- The results obtained can be investigated with ease.
- Monte Carlo simulation is regularly recognised as a valid technique so its results are widely accepted.

Box 11.1 describes one company's use of the application of Monte Carlo simulation to support the analysis of bringing new drugs to market.

Box 11.1 Simulation

Pharmaceutical company Merck & Co. Inc. invests about $2 billion in research and development and capital expenditure annually. Most of the investment goes into long-term risky projects that are impossible to evaluate using traditional cash flow analysis. The reason is that the uncertainties are so wide as to make single point estimates of the various uncertain parameters nonsensical. Instead Merck have developed sophisticated risk analysis models based on Monte Carlo simulation. These models assign probability distributions to the various input parameters and produce a range of possible outcomes in a probabilistic form. Bringing new drugs to market is a very long-term and unpredictable process. Merck have analysed that only 1 in 10 000 explored chemicals reaches the market and becomes a prescribed drug. In 1983 Merck commenced the development of their Research Planning Model, and by 1989 it was being used to evaluate all significant research and development projects over a 20-year horizon. The major inputs to the model are probability distributions for research and development, manufacturing and marketing variables. The model takes account of a number of medical and technological constraints as well as macroeconomic assumptions. It then uses simulation to compute probability distributions of the cash flow and the return on investment from specific projects.

Source: Vlahos (1997).

11.8.6 Latin Hypercube

Latin hypercube sampling is a more recent sampling technology than Monte Carlo. It is designed to accurately re-create the probability distributions specified by distribution functions in fewer iterations than Monte Carlo sampling. Latin hypercube sampling creates a cumulative probability distribution curve for each variable. The significant difference between Latin hypercube and Monte Carlo is that Latin hypercube adopts stratified sampling of the input probability distributions. The process of stratification divides the cumulative frequency curve of each input into equal intervals on the cumulative probability scale of 0 to 1. A sample is then taken from each interval or "stratification" of the input distribution. Sampling is forced to represent values in each interval and hence is forced to re-create the input probability distribution. The number of stratifications of the cumulative distribution is equal to the number of iterations performed. A sample is taken from each stratification. However, once a sample is taken from a stratification, this stratification is not sampled again. During sampling, when a stratification is chosen for sampling, then a value is randomly selected from within that stratification. The sampling technique maintains the independence of the variables (where this is desired) by, in any given iteration, sampling (say) variable 1 from stratification 27 and variable 2 from stratification 5 and so on. This preserves the randomness and independence of the variables and avoids unwanted correlation between variables. If low values of distributions were aggregated together or conversely high values of distributions aggregated together, this would produce unrealistic upper and lower limits of the overall model of results.

11.8.7 Probability Distributions Defined from Expert Opinion

Risk analysis models almost invariably involve some element of subjective estimation. Data commonly does not exist to determine, with any degree of precision, the probability distribution of variables within a model for a number of reasons, as explained by Vose (1996):

- The data has simply never been collected before.
- The data is too expensive to obtain.
- Past data is no longer relevant (new technology, changes in the political or commercial environment etc.).
- The data is sparse, requiring expert opinion to "fill in the holes".
- The area being modelled is new.

When constructing models where probability distributions have been based on subjective estimates, the analyst needs to be aware of the reliance that can be placed on the input data in terms of the knowledge and experience of the subject experts, the degree of involvement of the experts in the process, the culture of the organisation and personal and departmental agendas.

- Were senior project personnel involved in the risk assessment, or were colleagues (with insufficient project knowledge) sent by senior management to deputise?
- Did the experts have sufficient understanding of the nature and condition of the project to be able to safely make an informed and well-reasoned assessment of the risks, based on experience and professional judgement?
- Was sufficient time given to assess the risks and opportunities?
- Was a "sanity" check carried out to review the influence of low-probability, high-impact risks on the overall results?
- Were the differences between discrete and continuous probabilities understood?

11.9 SUMMARY

This chapter has examined a series of mechanisms to aid the risk evaluation process together with a description of the activities that would be undertaken during the course of this process, such as the application of simulation techniques and scenario or sensitivity analysis. The evaluation process is critical to understanding the combined effect of a group of risks and opportunities, and commonly the uncertainty spreads around a series of cost line items. The process is not an end in itself but simply an aid to decision making. The value of the evaluation process will be largely dependent on the quality of the information that formed the inputs. Once a simulation technique has been applied, the results should always be the subject of a sense check to ensure that they appear in the right order (appropriate magnitude). If the results do not appear as expected the probabilities and assessments should be revisited for data entry mistakes. While computer software is now sophisticated and computers enable models to be assessed very quickly, entering the information into spreadsheets and constructing formulas is always subject to human error. Information may have been omitted by mistake, cross-referencing of cells may be wrong or inappropriate functions used. If logical or conditional functions have been used, particularly if they have been nested, they may warrant closer scrutiny. The list of potential errors is virtually endless. If no entry errors are discovered, the appropriateness of the probabilities and assessments should be reviewed. In addition, the risk dependencies should be examined to see if the key relationships have been correctly portrayed and, if so, whether the strength of the correlation reflects the situation under examination. Where the results will be used for investment decisions and the sums involved are considerable, it may be appropriate to have the model independently reviewed.

11.10 REFERENCES

Moore, P.G. and Thomas, H. (1988) *The Anatomy of Decisions*, 2nd edition. Penguin, London.

Palisade Corporation (2010) *Guide to Using @RISK: Risk Analysis and Simulation Add-In for Microsoft®* Excel, Version 5.7. Palisade Corporation, Thaca, NY.

Vlahos, K. (1997) Taking the risk out of uncertainty. In T. Dickson and G. Bickerstaffe (eds), *Financial Times Mastering Management*. Financial Times/Prentice Hall, London.

Vose, D. (1996) *Quantitative Risk Analysis: A Guide to Monte Carlo Simulation Modelling*. John Wiley & Sons, Ltd, Chichester.

12

Risk Treatment: Stage 5

The definition of insanity is to keep doing the same things and expecting a different result. (Stephen Covey)

The previous chapter examined the risk evaluation stage, which entailed combining the risks and opportunities together to determine their net effect. The treatment stage uses all of the preceding risk management effort to produce responses and specific action plans to address the risks and opportunities identified to secure the business objectives. Ensuring that these plans are prepared, considered, refined and implemented is the purpose of this stage. If risk management is to be effective this stage is essential. To spend considerable time, effort and energy in identifying and assessing the potential risks and opportunities and not to plan responses to them would be a poor use of resources. This is where competitive advantage is borne rather than just being envisaged. The structure of this chapter is described in Figure 12.1. The next chapter describes the monitor and review process, which is concerned with monitoring the actual progress of the risk and opportunity actions.

12.1 PROCESS

As described in the preceding chapters, adopting the philosophy of process mapping, each process exists to make a contribution to one or more business enterprise goals. Hence, each process should be measured against specific process goals that reflect the contribution that the process is expected to make to the overall enterprise goals. Processes are simpler to comprehend when they have a primary goal and subgoals. Hence risk planning is described here as having a primary process goal which is accomplished by a series of subgoals, as described in Section 12.2. Any one process is accomplished within a context and might be considered to have two perspectives. The *external view* examines the process inputs, outputs, controls and mechanisms. The *internal view* examines the process activities that transform inputs to outputs using the mechanisms.

12.2 PROCESS GOAL AND SUBGOALS

The primary process goal of risk treatment is to plan specific management responses to both the threats and opportunities identified. The risk treatment process will be sufficient when it has satisfied these subgoals:

- The aim of the treatment process was made explicit.
- Sufficient time was allocated to the treatment process.
- The limitations (if any) of the treatment process were recorded and stated alongside the results.
- A response – retain, remove, reduce or reassign (transfer) – was selected for each of the risks and opportunities.
- The risk appetite of the organisation was made explicit, captured and documented.

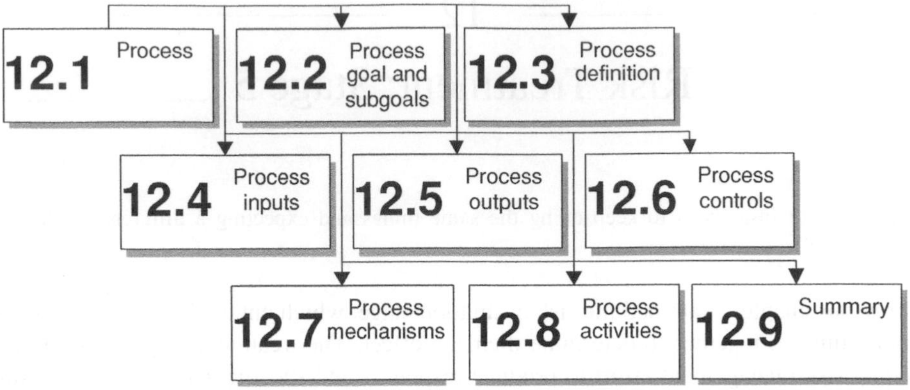

Figure 12.1 Structure of Chapter 12

- Specific actions were decided upon and recorded with an "implementation by" date as appropriate.
- The response actions were created by the appropriate business specialist.
- Consideration was given to which stages of the business process the identified risks and opportunities related to, so that actions can be prioritised.
- It was recognised that secondary risks may arise from risk response actions.
- There was recognition that the magnitude and hence the potential impact of identified risks or opportunities would not remain static as a result of stimuli in the business environment.

12.3 PROCESS DEFINITION

The risk treatment process is described by an IDEFO diagram (see Figure 12.2). The diagram describes a process with inputs entering on the left of the box, outputs leaving on the right of the box, controls entering from above and mechanisms or enablers entering from below.

12.4 PROCESS INPUTS

A risk register, details of existing insurance policies, description of the business risk appetite and industry betas are inputs to the risk treatment process:

- *Risk register*. The risk register is an output from all of the preceding processes as is incrementally developed throughout the risk management process.

12.5 PROCESS OUTPUTS

- *Risk responses*. Risk responses contained within a schedule or recorded individually with each response on a separate page of a document will contain as a minimum the risk ID, risk description, impact in terms of time and cost, the risk response category (i.e. remove, reduce or transfer) actions to respond to the risk or opportunity, the owner, manager, actionee, the date by which the actions will be implemented, the anticipated cost of the response and

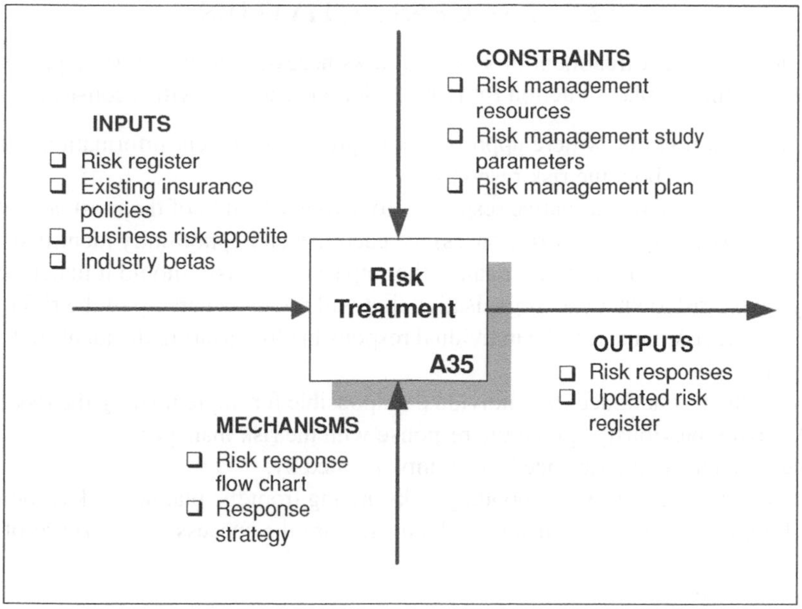

Figure 12.2 Risk treatment process

any secondary risks that may arise from the risk. The schedule or document will also contain a description of the interrelationship with other risks and possibly the strength of the correlation.

12.6 PROCESS CONTROLS (CONSTRAINTS)

The business risk management culture, risk management resources, risk management study and risk management plan (where one exists) regulate the risk process. These constraints were discussed previously in Section 8.6.

12.7 PROCESS MECHANISMS

Mechanisms can be methods, tools, techniques or other aids that provide structure to the process activities. There are two primary process mechanisms:

- resolution strategy (techniques/tools); and
- risk response flow chart.

A resolution strategy is a pre-defined plan designed to respond to a particular reoccurring risk.

A risk response flow chart illustrates the decision options that are made to arrive at the desired risk response category. It is a decision-making aid in deciding whether it is more appropriate to (say) transfer a risk rather than attempt to remove it.

12.8 PROCESS ACTIVITIES

The activities of the risk treatment process are tasks necessary to transform a prioritised list of risks into a concrete plan of action for risk resolution. These activities consist of:

- conducting risk research, where appropriate, to provide sufficient information to make an informed decision about the risk response;
- developing appropriate alternative responses to permit selection of the most advantageous;
- developing a risk response (or responses) for each risk and opportunity identified;
- assessing the cost of the response against the impact of the risk, should it materialise;
- identifying the risk owner (the organisation that will retain ownership of the risk);
- identifying the risk manager (the individual responsible for ensuring the identified response is implemented);
- identifying the risk actionee (the individual responsible for implementing the risk response action), having previously agreed the response with the risk manager;
- deciding when the responses need to be implemented by;
- considering the emergence of secondary risks arising from the planned risk response;
- establishing early warning indicators which measure the success or otherwise of the risk response;
- defining the business's risk appetite.

12.9 RISK APPETITE

Risk appetite (also referred to as risk preference, attitude, tolerance or capacity) can be defined as the amount of risk a business is prepared to tolerate (be exposed to) at any point in time. A business's tolerance will be a reflection of its capacity to absorb risk. While businesses can benchmark their own tolerance with other businesses in their market or industry (where information is available), each business's risk tolerance is unique. A business's appetite for risk will vary in accordance with its objectives, culture, as well as evolving conditions in the overall business environment. Within the insurance industry a board defines and communicates the company's risk appetite or risk tolerance via a three-stage process (PricewaterhouseCoopers 2004). First, objectives for shareholder value creation are defined based on a combination of issues, including the market, business processes or regulatory requirements. Second, the company establishes a tolerance for earnings variance based on its stated objectives. Third, business units are required to bid for an allocation of the company's overall risk tolerance in pursuit of their business plans. The risk tolerance can be expressed as capital, as earnings variance, as liquidity and balance sheet activities, and as guidelines for investment. Attitude to risk can be categorised as risk averse, risk neutral or risk seeking. A business's risk appetite will vary according to such issues as the perceived financial exposure of particular risks, the current success of the business, trends in the economy and the attitude of individual board members.[1] A business's view may additionally be coloured by what other initiatives have already been taken, of which the outcome is not yet known, whether the company as a whole would be affected

[1] The presence of the "novelty gene" in a person is considered to influence that person's attitude to risk. Recent research by Dr Richard Ebstein and his team at Sarah Hezog Hospital in Jerusalem and Dr Robert Coninger of the Washington School of Medicine in St Louis has led to the discovery of a D4 dopamine receptor gene (abbreviated to D4DR) which may give rise to the trait of seeking excitement through change and novelty. It is nicknamed the "novelty gene". Dr Ebstein describes the disposition of those with the D4DR gene as being characterised by a recurring desire for sensation. These people are impulsive, excitable risk takers who need constant stimulation to satisfy their exaggerated appetite for arousal (Stuttaford 1999).

if the outcome were not favourable or whether the company's reputation would be irreparably tarnished. Once an organisation has developed its tolerance levels, the business risk culture can then be used to inform senior management in their decisions about risk tolerance levels for individual projects and programmes in their application for approval. Boards will hope to make informed decisions within their risk tolerance; however, there are several factors which may erode the quality of the information they are presented with, as demonstrated by the checklist below. Boards will have to decide how reliable the information is that they are presented with. They will have to consider for instance the experience of the analyst, the quality of the information upon which the analysis was based, whether risk exposure has been deliberately suppressed to gain project sanction or how effective risk management activity will be.

Checklist: risk evaluation and assessment of the organisation's willingness to take on risk

- Is the timing and level of risk management planned, agreed and implemented appropriate to the different acquisition life cycle stages, risk appetite, decision complexity and level of risk exposure?
- Is the organisation's risk appetite clear? Is there an understanding and commitment as to what level of risk is acceptable for a project and the ability to communicate this? Does this reflect the potential for increasing organisational performance?
- Is a consistent approach and degree of effort being adopted throughout an analysis to assess the potential impact and probability of identified threats?
- Is there a good understanding of the relationship between the likely potential impact and the probability of the risk occurring (such as very high impact but extremely low probability)?
- Is the risk information required being communicated effectively to support the necessary decision-making process, in a timely, clear and cost-effective manner?
- Is there a clear understanding of the difference between the resolution of a known problem or issue and responding to risks, and is there an appropriate mechanism for moving an issue to the risk register and vice versa?
- Is a consistent approach being taken regarding the identification and prioritisation of the risks in the risk management process and in any associated issue management process?
- Are the required skills available to carry out the analysis?
- Are the risks being understated or overstated when assessed and evaluated for commercial, political or individual reasons?
- Is there buy-in at all levels of the organisation to the process of assessing and evaluating the threats? How was this established, and is the process embedded?
- Can risk management processes be implemented sufficiently quickly enough to be able to support rapid change? For example, e-commerce developments increasingly require IT developers, business relationship managers, human resources, facility management etc., to gear up to deliver a solution to the market within very tight timescales.
- Is there a demonstrable correlation between the planned risk management activities (including assessment) and the level of risk exposure?

Source: Office of Government Commerce (2002).

Example

An example of a risk seeker, a company looking to capitalise on opportunities, is Boeing, one of the largest manufacturing concerns in the world. Boeing has a corporate tradition of risking the company on breakthrough aviation products every couple of decades. In the 1930s Boeing gambled on a new bomber that became the B-17 and is considered by many to be a major contributor to winning World War II. In the 1950s Boeing took the gamble to build the first all-jet commercial passenger plane in the US (the 707) on speculation without having a single customer in hand. Douglas Aircraft, a competitor, was so focused on filling all its orders for the propeller-powered DC-7 that it failed to move quickly enough into the development of jet engines and was subsequently acquired by Boeing. In 1968 Boeing built the first jumbo jet, the 747, without enough customer orders to guarantee it could break even. If any of those projects had failed it is reasonable to assume that Boeing would probably have gone out of business.

12.10 RISK RESPONSE STRATEGIES

12.10.1 Risk Reduction

The risk response called *reduction* is also known as treatment or mitigation. One form of risk reduction is risk diversification – the reduction of risk by distribution through, say, investment in multiple stocks rather than a single stock. Diversification is the strategy adopted by those who do not want to "put all their eggs in one basket". Wilkinson (2003), in his focus on the treatment of hazardous materials in a manufacturing context to prevent personal injury, describes two general approaches that may be taken to reduce risk: reducing the likelihood of a risk occurring and limiting the loss should the risk materialise. Wilkinson describes methods to reduce the likelihood of occurrence of risks through protection, controls and maintenance and methods of risk reduction through the act of risk spreading such as dispersing chemical storage. The petrochemical industry, while not being able to remove the threat of adverse weather conditions, designs rigs to withstand high winds. Contractors, while not being able to remove the threat of plant failure, regularly maintain their plant and keep critical spares close to hand. Credit card companies, while not being able to remove default risk, reduce the impact by setting interest rates at a level which compensates and outsource debt recovery. Many companies lose critical personnel. A business cannot prevent it happening but, as Brent Callinicos of Microsoft explains, it is possible to examine companies that have suffered a sudden departure (McCarthy and Flynn 2004). In this way it can be seen how a given company responded, establish what the public reaction was, examine how the market moved and use this information to inform an appropriate response. You cannot stop a tsunami happening, but it sure is helpful if you can have advance warning.

12.10.2 Risk Removal

The risk response called *removal* is also known as avoidance, elimination, exclusion and termination. Risk removal is the strategy adopted to eliminate a risk altogether when a negative outcome is anticipated. The greatest opportunity to remove a risk is at the commencement of any business activity or project embarked upon to accomplish a business improvement. As discussed in Chapter 24, while underdeveloped parts of the world represent very attractive untapped markets, the political uncertainty associated with the host nation(s) may be so high that the risk of doing business may be too great to make the opportunity worthwhile. When risks that were previously accepted (as a result of failure to properly examine and hence appreciate their true potential impact) materialise, they either significantly reduce the benefit of a project or totally erode its business case. On realisation of a risk's true potential impact after a project has started, the act of abandonment of the project or even postponement (to await more favourable circumstances) can be very expensive.

In the case of risk removal, three tests must be applied:

1. *Opportunity*. On removal of the risk, is a significant opportunity being lost as a result of the risk–opportunity balance being incorrectly assessed?
2. *Business objective*. Having removed a risk or risks by selecting an alternative course of action, is the activity or project outturn now going to satisfy the original business objective?
3. *Cost*. Does the cost of removal of the risk outweigh the impact should it materialise? The true cost of removal may not be immediately apparent if removal is in incremental steps rather than as the result of a single action.

12.10.3 Risk Reassignment or Transfer

The risk response called *reassignment* is also known as transfer and deflection. Risk reassignment is the strategy adopted to move a risk onto another entity, business or organisation. Contracts and financial agreements are the principal way in which risks are transferred. Transferring a risk does not reduce its likely severity; it just removes it to another party. In some cases transfer can significantly increase the impact of the risk, as the party to whom it is being transferred is unaware that it is being required to absorb it. The commonest form of risk transfer is by means of insurance. However, transfer through insurance rarely totally transfers a risk, as policies typically include excesses, as with motor insurance. The responsibility for initiating this form of risk response lies with the business that owns the risk in the first instance.

Considering the merit of transferring a risk requires a business to consider both its and the other parties' objectives, the relative abilities of the parties to assume the risk, the degree of control over the context of the risk and the potential gain or loss incentive (Perry 1986). Hence, in the case of risk transfer four tests must be applied:

1. *Objectives of the parties*. What is a party's motivation for transferring or accepting the risk, and is it transparent?
2. *Ability to manage*. Transfer can only ever be effective if the party that assumes the risk, the recipient, has the ability to manage the risk. That is, it has the ability to implement an action or actions, which can directly either reduce or remove the risk.
3. *Risk context*. The ability of a business to manage a risk will be determined not only by its ability to take direct action, but also by the context of the risk – that is, how static or volatile the source of the risk is, and hence the degree of fluctuation in the likely impact and probability of the risk.
4. *Cost effectiveness*. It is usual for a premium to be charged by the party accepting the transferred risk. The issue is whether or not the premium to be paid is less than the likely cost of absorbing the financial impact of the risk, should it materialise. An example of risk transfer is where businesses seeking cost certainty in the procurement of new industrial premises, pass on the risk of poor ground conditions to the contractor. This type of risk

transfer is not cost effective as the risk may or may not materialise and the contractor will have made some provision for this risk in his price. Whether the contractor reflects the full cost of the risk in his price will depend on whether he is in competitive tender, the quality and extent of the soil survey, the competitiveness of the market place, his order book and his knowledge of the area.

Even when a business believes it has transferred a risk, it is usually not totally immune from impact. If a risk is transferred to a contractor (say), and the contractor fails to manage it, resulting in the project being late, even though the contractor may be subject to a penalty clause for late delivery, there is no escape from having a late project.

12.10.4 Risk Retention

The risk response called *retention* is also known as acceptance, absorption or tolerance. Risk retention is the strategy adopted when *either* it is more economic to do so *or* there is no alternative, as the option to transfer, reduce or remove the threat is not available. In the case of risk retention, three tests must be applied:

1. *Options*. If the decision has been taken to retain the risk as it is considered there is no alternative, is it clear that all possible options for removal, reduction or transfer have been examined?
2. *Timing*. Even when it appears that a risk will have to be retained as it is thought there is no alternative, the situation should not be accepted as a *fait accompli*. The business environment never remains static and options may arise even in the short term, in terms of, for example, insurance, contract terms, outsourcing or pursuing alternative markets. Hence, it will be important to monitor the context of the risk through regular risk reviews and understand when a decision has to be made. Proactive risk management will be required to ensure alternative courses of action are not missed.
3. *Ability to absorb*. If the conscious decision has been taken to retain a risk as it is considered more economic to do so, is it clear *either* what the impact would be should it materialise *or* what the likelihood of its occurrence is? Is the risk considered to consist of one isolated event, or could it be a series of events? Would there be a ripple effect if the risk materialised? Will the impact be purely financial or will it also affect, say, staff turnover, reputation, share price or market share?

12.11 SUMMARY

Risk treatment is an essential element of the overall process of risk management. It builds on the preceding activities. You cannot respond to a risk that has not been identified, and you are all at sea if you do not know what risks will hurt you the most, when they are likely to occur and how much it will cost to keep them within manageable limits. The speed of change is so great, and so to capitalise on identification and analysis, response planning must quickly follow. The four risk responses labelled reduction, removal, reassignment and retention have been examined. Once the risk response category has been identified for a particular risk, specific actions must then be devised to reflect that response. Commonly it is not possible to remove a risk in its entirety (unless a different business activity is adopted altogether). Hence, where a risk can be reduced to within acceptable limits and the cost of the risk response actions does not outweigh the benefits, it is possible to retain a risk while carrying out reduction strategies.

Another example of risk reduction is the response of the ferry companies to the opening of the Channel Tunnel between Dover and Calais. They could not remove the risk, but rather than accept defeat they reduced the threat by lowering their prices. As a consequence they did not lose custom to the extent they had feared and the Channel Tunnel did not see its sales figures reach projections, meaning that their debt was not repaid at the rate they had hoped.

12.12 REFERENCES

McCarthy, M.P. and Flynn, T.P. (2004) *Risk from the CEO and Board Perspective*, McGraw-Hill, New York.

Office of Government Commerce (2002) *Management of Risk: Guidance for Practitioners*. The Stationery Office, London.

PricewaterhouseCoopers International Limited (2004) Enterprise-wide risk management for the insurance industry. Global Study. http://www.pwc.com

Perry, J.G. (1986) Risk management – an approach for project managers. *Project Management*, 4(4), 215.

Stuttaford, T. (1999) Addicted to risk? It's your novelty gene. *The Times*, 14 October, p. 38.

Wilkinson, S. (2003) *Risk Control*. Witherby, London.

13 Monitoring and Review: Stage 6

Risk is like fire: If controlled it will help you; if uncontrolled it will rise up and destroy you.
(Theodore Roosevelt)

The previous chapter described the risk treatment stage. This chapter describes what is commonly understood to be a key stage within the overall risk management process known as the monitoring and review stage. It is worth reiterating that the individual stages within the practice of risk management as a whole are iterative in that it is frequently necessary to revisit earlier stages when more information becomes available or circumstances change. Each stage relies on inputs from earlier stages. Stage 6 is critical to the successful implementation of the risk management process as a whole. All risk management process maps state a need to ensure risk responses to identified risks are implemented and that implementation is proactively managed. Risk monitoring and review requires undertaking five key activities:

1. *Reacting* to early warning indicators to forewarn managers of the need to make risk management interventions.
2. *Registering* changes in the details of the risk and opportunities already captured on the risk register.
3. *Recording* emerging risks and opportunities, lessons learnt and changes in the internal and external context.
4. *Reviewing* whether the risk managers and actionees are implementing the responses for which they are responsible.
5. *Reporting* on the success or otherwise of the risk and opportunity management actions implemented to date, the need for additional response actions and the changes in the overall risk exposure profile of the business.

The structure of this chapter is described in Figure 13.1.

13.1 PROCESS

As described in the preceding chapters, adopting the philosophy of process mapping, each process exists to make a contribution to one or more business enterprise goals. Hence each process should be measured against specific process goals that reflect the contribution that the process is expected to make to the overall enterprise goals. Processes are simpler to comprehend when they have primary goals and subgoals. Hence, risk management is described here as having a primary process goal which is accomplished by a series of subgoals, as described in Section 13.2. Any one process is accomplished within a context and might be considered to have two perspectives. The *external view* examines the process inputs, outputs, controls and mechanisms. The *internal view* examines the process activities that transform inputs to outputs using the mechanisms.

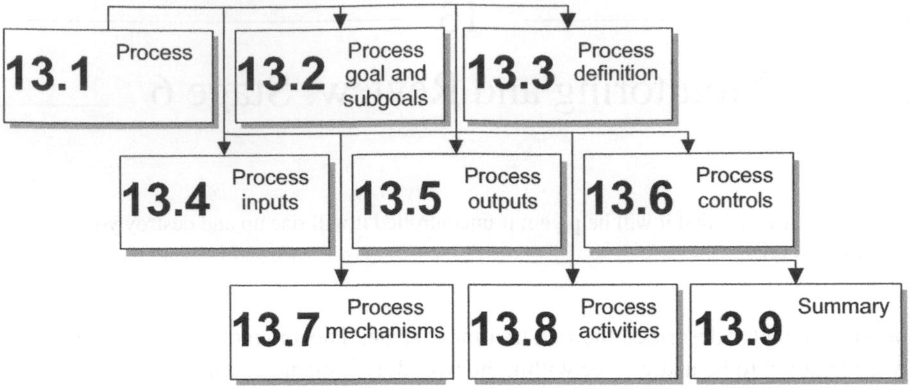

Figure 13.1 Structure of Chapter 13

13.2 PROCESS GOAL AND SUBGOALS

The primary process goal of monitoring and review is to monitor the performance of risk response actions to inform the need for proactive risk management intervention. The monitoring and review process will be sufficient when it has satisfied these subgoals:

- Early warning indicators have been developed and used to forewarn managers of the need to make risk management interventions.
- The internal and external contexts are monitored to establish if the current analysis of opportunities and risks needs to be revisited or in extreme cases whether the risk management framework or policy require revision.
- Risk actionees and managers are implementing the risk and opportunity responses for which they are responsible in a timely manner.
- Risk registers are regularly updated whereby risk events that have either materialised or are time expired are removed and newly identified risks and opportunities are added. Additionally, the register is modified to reflect any changes in the assessment of a risk in terms of its likelihood of occurrence or its potential impact, managers or owners, categories of risk, degree of completion of risk management actions, secondary risks or the cost of risk management action.
- Reports are issued on a regular cycle, providing visibility of the progress made in the success or otherwise of the risk management actions.
- Contingencies are revised to reflect the current risks, opportunities and their assessment.

13.3 PROCESS DEFINITION

The monitoring and review process is described by an IDEFO diagram (see Figure 13.2). The diagram describes a process with inputs entering on the left of the box, outputs leaving on the right of the box, controls entering from above and mechanisms or enablers entering from below.

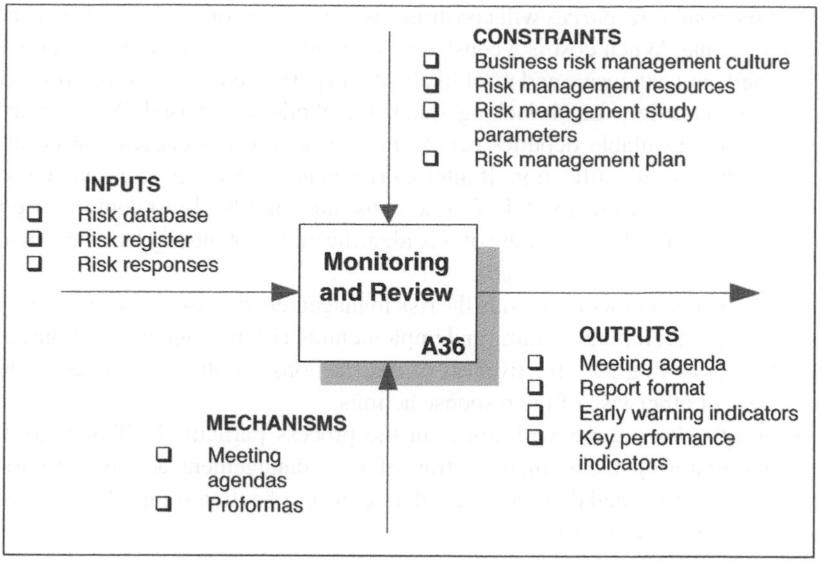

Figure 13.2 Risk monitoring and review process

13.4 PROCESS INPUTS

The risk register is an input into the monitoring and review process. The risk register has been incrementally developed throughout the preceding process. The risk register can capture the risk responses, or each risk can have its own dedicated record, which captures the planned risk response(s). Additionally, where a risk database has been populated during earlier processes, it would also be an input to this process.

13.5 PROCESS OUTPUTS

Regular updates of the risk register and reports on the effectiveness of the risk response actions are the output from the monitoring and review process. Each report will provide a risk and opportunity status, recording the progress (or lack of it) made against actions assigned to each risk and opportunity. Key performance indicators (KPIs) can be used as a way of tracking business sensitive issues, so that should certain levels be reached corrective action will be triggered. These KPIs could be measuring staff turnover, liquidity, absenteeism, sickness, sales, stock levels, fluctuations in the share price, fleet vehicles involved in accidents, loss of customers, vehicle breakdowns, customer complaints, supplier default, late payments and so on, depending on the nature of the business.

13.6 PROCESS CONTROLS (CONSTRAINTS)

From Figure 13.2, the business risk management culture, resources, study and plan (where one exists) regulate/constrain the risk identification process.

- The *business risk management culture* will constrain the risk identification process in terms of the degree of importance, commitment and enthusiasm attached to the process and the extent of support provided when the risk management process is initiated.

- The *risk management resources* will constrain risk identification in terms of cost and hence resources and time. When cost is a constraint, particularly when external support is being commissioned, less expensive and most likely less experienced staff will be allocated to the assignment. When solely time is lacking, quicker methods can be used. When departmental managers are not available deputies can be invited or fewer attendees can be invited to participate in the risk identification. If internal risk management resources are constrained, the process can be accelerated. All of these constraints are likely to compromise process effectiveness, particularly the breadth of risk identification, potentially leaving "blind spots" (areas of risk exposure not addressed).
- *Actionees and managers* will constrain the risk management process if they have insufficient time to devote to developing, refining and implementing risk management actions, pursuing opportunities, assessing the effectiveness of their actions, or attending meetings to report on the success or otherwise of the response actions.
- *Infrequency of risk meetings* will constrain the process particularly if they are the sole vehicle for monitoring the implementation of risk management actions, discussing the emergence of new risks and debating secondary courses of action if initial risk management actions are not totally effective.

13.7 PROCESS MECHANISMS

Mechanisms can be methods, tools, techniques or other aids that provide structure to the process activities. There are two primary process mechanisms for this stage:

- meeting agendas; and
- proformas.

13.8 PROCESS ACTIVITIES

The activities of the monitoring and review process are tasks necessary to ensure that monitoring and review is a proactive process which executes responses, monitors effectiveness and then intervenes to implement corrective action. Hence these activities consist of:

- *executing* risk response actions;
- *monitoring* the effectiveness of risk management actions;
- *controlling*, that is, intervening when events do not go according to plan.

13.8.1 Executing

The time, effort and energy expended in the earlier process of risk treatment, in terms of developing planned actions to respond to the risks and opportunities identified, will largely be wasted, unless they are executed. The what, when and who of execution will have been agreed and recorded in the treatment process, described in the previous chapter.

13.8.2 Monitoring

Having executed the activities agreed in the treatment stage, it is necessary to monitor progress to discern the movement in risk exposure. While monitoring is valuable, it must be recognised for what it is. It is a process of observation. It is neutral. It does not alter events. For monitoring to be effective it must be embedded into a business. It must be part of the culture. It must focus not only on the success (or otherwise) of the planned responses to previously identified risk and opportunities, but also changes in the business that might signal new emerging risks. The market in which businesses operate never stands still. In a high-risk environment, the one thing that can be expected is that not everything will happen according to plan. A fall in sales, increased absenteeism in the workforce, late delivery by suppliers, an increase in the number of returned products or a decrease in margins may all be a signal of emerging risks. A system of early warning indicators (EWIs) – predetermined trigger points – is required to draw the manager's attention to either the lack of effectiveness of risk management actions or changes in essential measures. The first question to answer on deciding on the measures to adopt is "What do we measure?" not "How do we measure?".

The selection of the measures will be made by determining which sets of events are singled out as being important. Events considered important should not only include measurable (quantifiable) events but also unquantifiable events. The measurable results are things that happened, they are in the past. There are no facts about the future. Improvements made by competitors to their products are only measureable when the ability to influence has passed by. Therefore, a balance between the measurable and the non-measurable is a central problem for management. Monitoring, which does not look ahead (only backward) in terms of boundaries and restraints, will possibly misdirect and misinform. Additionally, the greater the energy focused on previously identified risks, the greater the danger that what looks like effective monitoring will actually mean less effective risk management and control. One of the objectives of monitoring is the collection of information on risks for later use. Lessons learnt during the management process can be used to improve future risk management processes.

Monitoring activities should include an understanding of whether:

- actionees and managers are working together successfully;
- new risks and opportunities are being identified across all business sectors;
- the emergence of changes in legislation and compliance are giving rise to new risks to the business;
- risks that have not materialised or have been overtaken by events (such as changes in the market) have been closed out;
- the risk register is regularly updated;
- previous insurance arrangements are still valid;
- hedging opportunities have changed;
- funding opportunities have changed;
- previous market analysis is still valid;
- risk management is providing the contribution anticipated;
- the models and methods of capital adequacy are still appropriate;
- cash flow stress testing is undertaken to assess liquidity risk;
- activities and functions which attract reputational risk are represented.

13.8.3 Controlling

Unlike monitoring, controlling is not a neutral activity. Controlling requires intervention. Control activities concentrate on using the information gathered from monitoring to inform decision making. Controlling means understanding who needs what information for what purpose and when. To give a manager control, controls must satisfy seven specifications (Drucker 1979):

- They must be economical.
- They must be meaningful.
- They must be appropriate.
- They must be congruent.
- They must be timely.
- They must be simple.
- They must be operational.

Control is a Principle of Economy

The less effort needed to gain control of the process, the better the control design. The fewer controls required, the more effective they will be. Usually adding more controls does not give better control. All it does is create confusion. The first question that needs to be asked when designing a system of control is "What is the minimum information I need to know to support control?". The answer will vary depending on the type of business and the risk management framework established. The ability of proprietary computer software or an in-house database to spew out masses of data does not make for better controls. On the contrary, what gives control is asking the question "What is the smallest number of reports and statistics needed to understand a phenomenon and to be able to anticipate it?".

Controls Must be Meaningful

This means that the events to be measured must either be significant in themselves or must be symptoms of at least potentially significant developments (such as emerging new risks e.g. Apple facing competition to their iPad from rival products). Controls should be related to the specific objectives of the risk management process and relate to such questions as "Have the risk responses been implemented, were they effective and what residual risk remains?". Additionally, the controls should be focused on those risks that will have the greatest impact on the business.

Controls Have to be Appropriate to the Character and Nature of the Phenomenon Measured

The controls must give the right information for effective action. It is of little benefit just to report that a risk response action has not been completed by the due date. There must be a succinct description of what this means for the business. If a risk previously identified is now considered to be a greater threat, how large is the threat now, what is its probability, is the risk static or fluctuating and has the ability to manage it diminished or increased? If a new opportunity has emerged, what is the window for capitalising on the potential benefits, how likely is it that competitors will have identified the same benefit, are competitors better placed to respond and are what the likely rewards? If a new risk has been identified what are its characteristics?

Measurements Have to be Congruent with the Events Measured

It is particularly important in the field of risk management for managers to think through what kind of measurement is appropriate to the event being measured. They have to know when "approximate" is more accurate than a firm-looking figure worked out in great detail. They have to know when a range is more accurate than even an approximate figure. They have to know that "larger" and "smaller", "earlier" and "later", "up" and "down" are quantitative terms and often more robust and accurate than any specific figures or range of figures. It can be an important piece of information that an event cannot be measured with precision but can only be described within a range or as a magnitude. To say "we have 33% of the market" sounds reassuringly precise. However, it is likely to be so inaccurate a statement as to be virtually meaningless. It might have been relevant for one point in time. What it really means is that "we are not a dominant player in the market, but we are not marginal either".

Controls Have to be Timely

Frequent measurements and very rapid "reporting back" do not automatically give better control. Indeed, they may frustrate control. The time dimension of control has to correspond to the time span of the event measured. There is often reference to "real time" controls. These are controls that inform instantaneously and continuously. There are events where "real time" controls are desirable, such as in the pharmaceutical industry during drug production. However, few events need such controls. Most risks have an "event window" when they are likely to occur. Controls can be tailored around these windows.

Controls Need to be Simple

Complicated controls do not work. They deter the participants, at worse they confuse. They misdirect attention away from what is be controlled towards the mechanics of control. The what, when, how and why need to be transparent. Controls should not be an automatic copy of what was applied elsewhere. They need to suit the particular circumstances of the moment. They need to be revisited on a regular basis to discern if they are still effective.

Controls Must be Operational

They must be focused on action. Action rather than information is their purpose: "have the planned risk response actions been implemented?" or "have the potential opportunities been examined and results obtained?". The results must always reach the person or persons who are capable of taking controlling actions. Controls must not be constrained by predetermined meeting dates. Controls must be flexible enough to suit the circumstances.

13.9 SUMMARY

There is clearly little point in identifying risks if they are not going to be responded to. The previous chapter looked at response planning. This chapter has described control. Control is all about being proactive, and for risk management this means managing the response process to ensure responses are implemented and their effectiveness is monitored. Management time and energy must be focused on the issues that matter. The activities carried out during the analysis and evaluation processes will inform the priority issues. Management intervention must also be timely to be effective, and, with the rate of change in the market place, timing is all. For actions to be productive they need to be simple (as far as they can be), so that they are readily understood by those charged with carrying them out.

13.10 REFERENCE

Drucker, P.F. (1979) *Management, an Abridged and Revised Version of Management: Tasks, Responsibilities, Practices*. Pan Books, London.

14
Communication and Consultation: Stage 7

The single biggest problem in communication is the illusion it has taken place.
(George Bernard Shaw)

The most important thing in communication is hearing what isn't said.
(Peter Drucker)

The previous chapter examined the monitoring and review stage, which described the need to proactively manage the implementation of responses to identified risks and opportunities. It described realisation of risk management through five key activities: *reacting* to early warning indicators, *registering* changes to existing risks and opportunities, *recording* emerging risk exposure and changes in the internal and external context, *reviewing* response implementation and *reporting* on the success or otherwise of the risk and opportunity management actions. The communication and consultation process refers to the dialogue that takes place across all of the risk management stages to support their effective implementation (ISO 2009). Risk management is a tool rather than an end in itself, and its role is to support decision making. Hence, the value of the tool will be in how effectively the risk management outputs are communicated and understood by those making decisions for the benefit of the business. Clearly leading up to the global financial crisis there was a breakdown in communication between the risk function and company boards to enable executives to appropriately assess and exercise judgement over the risks being faced. The structure of this chapter is described in Figure 14.1. The next chapter describes financial risk management and in particular the seven most significant financial risks to face any business. Their significance to any one business will vary and will depend on a vast number of variables, some of which are described.

14.1 PROCESS

As described in the preceding chapters, adopting the philosophy of process mapping, each process exists to make a contribution to one or more business enterprise goals. Hence, each process should be measured against specific process goals that reflect the contribution that the process is expected to make to the overall enterprise goals. Processes are simpler to comprehend when they have primary goals and subgoals. Hence the communication and consultation process is described here as having a primary process goal which is accomplished by a series of subgoals, as described in Section 14.2. Any one process is accomplished within a context and might be considered to have two perspectives. The *external view* examines the process inputs, outputs, controls and mechanisms. The *internal view* examines the process activities that transform inputs to outputs using the mechanisms.

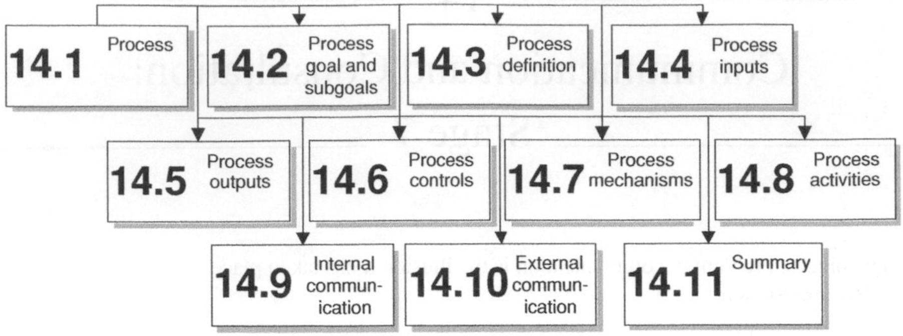

Figure 14.1 Structure of Chapter 14

14.2 PROCESS GOAL AND SUBGOALS

The primary process goal of risk communication and consultation is to develop plans for communication and consultation at the outset. The risk communication and consultation process will be sufficient when it has satisfied these subgoals:

- The aim of the communication and consultation process was made explicit.
- Sufficient time was allocated to map out the communication and consultation process.
- Communication was truthful, relevant, accurate and understandable.
- A risk management policy and framework were prepared and issued.
- Stakeholder needs regarding risk management were established, recorded, disseminated and acted upon.
- Risks were adequately identified as far as they were reasonably foreseeable.
- Different areas of expertise were involved in the risk identification and assessment process.
- Consensus was reached on the scales of impact and probability to be adopted.
- Agreement was secured on the risk responses.
- Agreement was secured on the level of contingency to be established.
- Agreement was secured on the risk software to be adopted for both data capture and reporting.
- The output of the evaluation modelling was reviewed and assessed against expectations and subsequently debated.
- Risk management was integrated with business change management.
- Risk reports were prepared on a regular basis advising of changes to the risk exposure and the degree of success being realised by risk response activities.
- Identified risks over a predetermined threshold were elevated to senior management when identified.

14.3 PROCESS DEFINITION

The risk communication and consultation process is described by an IDEFO diagram (see Figure 14.2). The diagram describes a process with inputs entering on the left of the box, outputs leaving on the right of the box, controls entering from above and mechanisms or enablers entering from below.

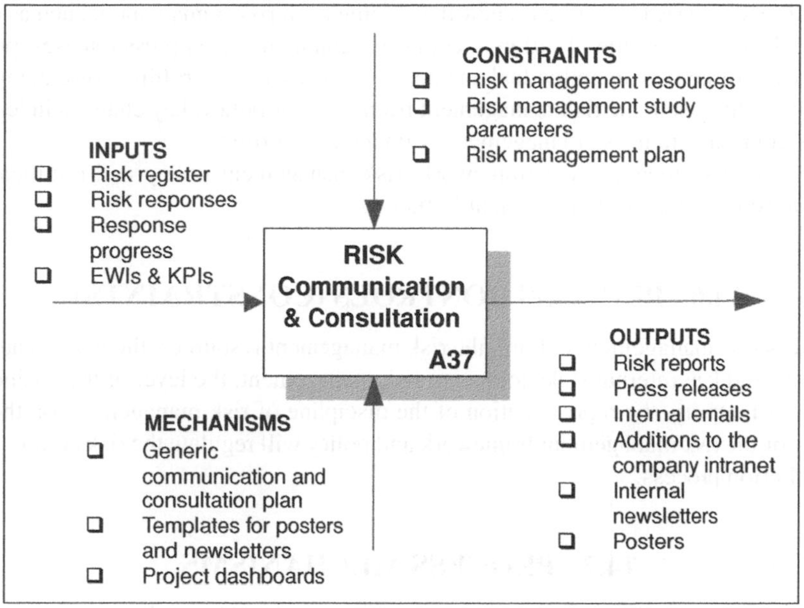

Figure 14.2 Communication and consultation process

14.4 PROCESS INPUTS

The process inputs are naturally the inputs from the proceeding stage together with the inputs from all of the proceeding stages, as communication and consultation occurs throughout the risk management process:

- *Risk register*. The risk register is an output from all of the preceding processes as it is incrementally developed throughout the overall risk management process.
- *Risk responses*. The treatment stage is responsible for defining risk response actions.
- *Response progress*. The monitoring and review stage is responsible for executing and monitoring the effectiveness of risk management actions and intervening where responses are not as effective as envisaged and further action is required.
- *EWIs and KPIs*. These reporting metrics provide a key reporting tool to aid the communication of business health.

14.5 PROCESS OUTPUTS

- *Risk reports*. Board reports, programme and project reports, contributions to annual reports.
- *Press releases*. Reporting on responses to risk/business continuity incidents.
- *Internal e-mails*. Specific announcements on risk management relating to where the discipline has added value to the business.
- *Company internet site*. Documents to be added to the existing internet site risk management framework, policy, process, committees and their composition, procedures, terms and definitions, templates, names and contact details of risk management representatives.

- *Internal newsletters*. Reporting on new developments in risk management such as changes in the political risk profile of different countries, commenting on press releases, providing updates on responses to individual incidents, changes in or profiling risk management personnel, changes in the risk management framework or policy, key changes in legislation such as environmental risk management or financial reporting.
- *Posters*. The risk management framework, risk management policy and ramifications of non-compliance and risk management initiatives.

14.6 PROCESS CONTROLS (CONSTRAINTS)

The business risk management culture, the risk management resources, the general attitude of departments and key internal stakeholders to risk management, the level of the business' risk management maturity, the representation of the discipline of risk management on the board, the quality of the risk management framework and policy will regulate the risk communication and consultation process.

14.7 PROCESS MECHANISMS

Mechanisms can be methods, tools, techniques, templates or other aids that provide support to the process activities. There are a number of process mechanisms for this stage:

- generic communication and consultation plan;
- templates for posters and newsletters;
- project dashboards.

14.8 PROCESS ACTIVITIES

The activities of the communication and consultation process are the tasks undertaken to strive to ensure that the risk management process overall is effective. These activities consist of:

- creating a risk management framework to ensure that risk information derived from the risk management process is adequately reported and used as a basis for decision making and accountability;
- creating a risk management policy to communicate the objectives for managing risk, the commitment to make the necessary resources available to manage risk, how risk management performance will be measured and reported, and methods for implementing continuous improvement;
- creating risk management performance indicators for assessing and communicating risk management maturity;
- communicating the benefits of risk management to stakeholders;
- communicating accountabilities and responsibilities for managing risk throughout the organisation;
- communicating how risk management will be embedded in all of the organisation's practices and processes;
- establishing internal communication mechanisms;
- establishing external communication mechanisms.

14.9 INTERNAL COMMUNICATION

A business should establish internal communication and reporting mechanisms in order to support and encourage accountability and ownership of risk and opportunity management. These mechanisms should ensure that:

- the framework and any subsequent updates are appropriately communicated throughout the business;
- the policy and any subsequent updates are appropriately communicated throughout the business;
- templates devised to develop consistency in risk capture, management and reporting are communicated, including detailed explanation of their purpose and application and the ramifications arising from lack of use;
- risk information is stored in a way it is readily accessible to the business as a whole;
- there is adequate and appropriate dialogue with key internal stakeholders who will influence the success of embedding risk management, with the purpose of
  - sharing the objectives of risk management,
  - explaining the framework, policy and process,
  - describing the expected benefits of applying the discipline,
  - the intended method of implementation,
  - expected participation,
  - reporting requirements,
  - describing the current level of risk management maturity, the desired level of maturity and the planned route for realising the maturity goal;
- there is an open channel to maintain a dialogue with key stakeholders and others to aid the implementation of risk management.

14.10 EXTERNAL COMMUNICATION

A business should establish external communication and reporting mechanisms in order to deliver open and honest information on the risks that the business faces and how it is responding. These mechanisms should ensure that:

- external reporting to comply with legal, regulatory and governance requirements is timely, accurate and complete (in terms of not omitting information which would be critical to shareholder decision making);
- appropriate external stakeholders are engaged and an effective exchange of information is ensured;
- answers to questions on the risk management framework, policy, process, procedures or reporting are given promptly;
- where possible, risk management information is used to provide a positive message to build confidence in the organisation;
- the business is equipped to quickly communicate with stakeholders in the event of a crisis with current accurate information in a consistent way.

14.11 SUMMARY

This chapter has described the goal and subgoals of communication and consultation, together with the process inputs and outputs and the process activities. Risk communication and consultation form a link between all of the stages of the process and as a result are another essential element of the overall process of risk management. The aims of risk management must be clearly communicated to all employees for it to be effective. Risk management cannot be conducted by an individual tucked away in a corner office working in isolation. Risk management affects everyone in the business as each individual can contribute to risk identification, assessment and management. Conversely, each employee can be involved in an activity which causes a breach of business ethics, attracts adverse media coverage, contravenes government legislation, causes reputational damage, diminishes the share price and/or adversely affects shareholder relations.

14.12 REFERENCE

ISO (2009) *ISO 31000:2009 Risk Management – Principles and Guidelines*. ISO, Geneva.

Part III

Internal Influences – Micro Factors

This part of the book examines the way in which internal *micro* influences impact business performance. These micro sources of risk are distinct from the *macro* factors discussed in Part IV, in that micro risk factors are to a large degree generated internally and hence within the sphere of influence of any one business, whereas macro factors are predominantly outside the control of individual businesses. Micro factors include financial, operational, technological, project, ethical, and health and safety risks. The chapters contained in this part of the book describe the internal sources of risk included in the risk taxonomy described in Chapter 9. The sequence in which the following chapters appear mirrors the sequence in which the subjects are recorded in the taxonomy.

Figure P3.1 Structure of Part III

15
Financial Risk Management

The core of the international financial crisis is that we have global financial markets and no global rule of law.
(Jacques Attali)

This chapter examines financial risk management, the first of six classes of risk exposure that businesses face, relating to what I have termed "internal processes" within the risk taxonomy proposed in Chapter 9. The other five classes of risk within these internal processes are discussed in Chapters 16–20. All six classes of risk are considered to be controllable by businesses to a large degree. This chapter focuses on what are considered to be the seven most significant financial risks to face any business. Their significance to any one business will vary and will depend on a vast number of variables, such as the strength of the relationship with their key customers, the volatility of the markets they operate in, the number of competitors, the length of time the business has been in existence, the common level of overseas trade, dependency on suppliers, market share and so on. There is a clear overlap between financial and economic risk, and, reflecting the theme of Section 1.1, these classes of risk cannot be looked at in isolation, but must be viewed together. The decisions taken by a business in managing these risks will determine its performance, position and longevity. Through sound financial management, businesses can evaluate business strategies that are appropriate to their risk appetite, market and exposure profile. It will be seen that financial sources of risk have the potential to be "fatal" in that they can bring about the demise of a business and hence require a clear management strategy. The structure of this chapter is illustrated in Figure 15.1.

15.1 DEFINITION OF FINANCIAL RISK

What is financial risk? Financial risk is the exposure to adverse events that erode profitability and in extreme circumstances bring about business collapse. It can include the failure of financial systems, regulatory non-conformances or compliance issues, as well as bad debt, adverse changes in exchange rates, overdependence on a single supplier, loss of a key customer, loss of overseas investments and poor hedging decisions. Additionally, it can include poor investment decisions concerning plant, machinery and buildings. The essential feature of investment decisions is time. Investment in a new manufacturing plant, railway or ship, for instance, involves making an economic outlay at one point in time which is expected to yield economic benefits to the investor at some other point in time. The risks associated with such investments are that they cost more to procure than at first envisaged, and that when complete the market into which the investor wishes to compete has changed, diminished or even disappeared.

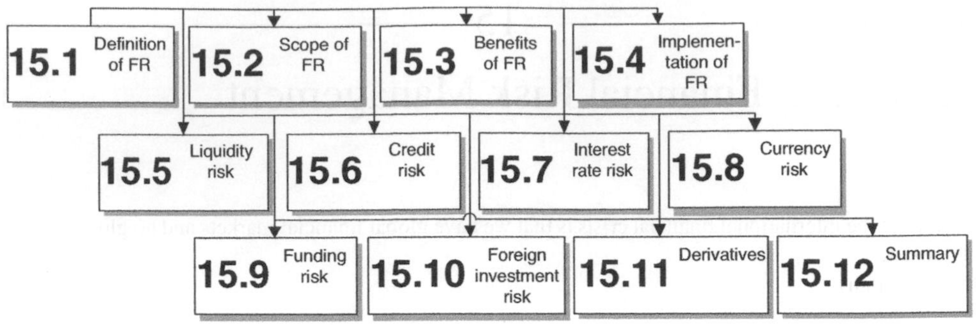

Figure 15.1 Structure of Chapter 15

15.2 SCOPE OF FINANCIAL RISK

The sources of risk considered to be embraced within the term "financial risk" are very considerable. These may be considered to include:

- liquidity risk arising from a short-term inability to meet financial obligations such as payment of suppliers, the premise's landlord or staff;
- credit risk, stemming from lack of payment of goods supplied to customers;
- interest rate risk which affects consumer's disposable income, resulting in a deterioration of trade for, say, retailers, house builders and manufacturers;
- inflation on, say, investment projects in terms of cash flows and the discount rate over the life of the project;
- currency risk in terms of expected cash flows from overseas investments being adversely affected by fluctuations in exchange rates;
- funding risk for borrowers, in relation to being unable to meet capital repayment requirements (and interest) and having to pay fixed charges on the company assets;
- foreign investment risk such as restrictions on the right to repatriate funds, high levels of taxation on profits remitted oversees, the temporary freezing of bank account balances or the expropriation of assets;
- derivatives risk arising from speculation in the market or hedging by, say, buying forward with the aim of acquiring a commodity at a price lower than the price prevailing at the time;
- systems risk such as loss as a result of failure caused by the breakdown of business procedures, processes or systems and controls;
- outsourcing risk arising from the default of a counterparty that has gone into liquidation, failed to deliver goods by a due date or breached contract conditions.

15.3 BENEFITS OF FINANCIAL RISK MANAGEMENT

Financial risk management affords a business benefits – for example, it:

- improves financial planning and management, which sits at the heart of corporate governance;
- facilitates more robust investment decisions;
- informs hedging decisions;

- encourages the development of constant monitoring of markets and the economy to inform decision making;
- encourages the practice of due diligence when outsourcing and engaging with counterparties.

15.4 IMPLEMENTATION OF FINANCIAL RISK MANAGEMENT

The development of a sound system of financial risk management will depend on a number of issues such as:

- the development of robust financial systems and internal controls;
- the development of concise, lucid reporting tools;
- the preparation of a cash budget plan, to diminish the likelihood of the threat of liquidity risk;
- securing credit insurance to cover non-payment of goods or services/bad debt;
- carrying out comprehensive due diligence on counterparties whose default could seriously harm the business;
- monitoring predicted changes in interest rates, so that business activity can be modified to diminish its effect;
- carrying out a robust assessment of planned investments, using tried and tested techniques.

15.5 LIQUIDITY RISK

Liquidity risk is the risk that a business will be unable to obtain funds to meet its obligations as they fall due either by increasing liabilities or by converting assets into money without loss of value. The more liquid an asset, the more easily it can be converted into money. *Near money* is an example of an asset that can be quickly converted into a medium of exchange at little cost. In the UK the most obvious type of near monies is *time deposits* with banks and building societies. They pay higher rates of interest than current accounts. Depositors need to give notice if they wish to withdraw from the account (hence the term "time" deposit). Extreme liquidity results in bankruptcy. Hence, liquidity risk can be a "fatal" risk. However, such extreme conditions are commonly the outcome of other risks. For instance, significant losses due to the default of a key customer can raise liquidity issues and doubts as to the future of the business. All companies will only stay solvent by ensuring that all cash obligations (salaries, rents, tax, etc.) can be met by a combination of investment liquidity, funding sources and contingent liabilities (liabilities that can be terminated quickly).

15.5.1 Current and Quick Ratios

One crude way to measure liquidity is to consider the relationship between the current assets and the current liabilities. *Current assets* are those assets which are either in the form of cash or can reasonably be expected to be turned into cash within one year from the date of the balance sheet. Current assets might consist of cash and bank balances, debtors ("accounts receivable" in the US) and stocks ("inventories"). The debtors figure is usually net of an allowance (or provision) for doubtful debts. *Current liabilities* are those liabilities which are expected to have to be paid within one year from the date of the balance sheet. Liabilities commonly consist of creditors ("accounts payable" in the US), taxation owing, and dividends payable and short-term borrowing. The relationship between the current assets and the current liabilities is known as the *current ratio* and is defined as follows:

$$\text{current ratio} = \frac{\text{current assets}}{\text{current liabilities}}.$$

The financial statements in Box 15.1 relate to a fictitious firm, Lawrence plc, which owns a small chain of wholesale/retail carpet stores. For the year ended 31 March Yr X, the current ratio of Lawrence plc is:

$$\text{current ratio} = \frac{\pounds 574.3}{\pounds 321.8} = 1.8.$$

The ratio reveals that the current assets cover the current liabilities 1.8 times. In some texts, the notion of an "ideal" ratio (usually 2) is suggested for businesses. However, as Atrill (2000) argues, this fails to take into account the fact that different types of business require different current ratios. For example, a manufacturing business will often have a relatively high current ratio because it is necessary to hold stocks of finished goods, raw materials and work-in-progress. It will normally sell goods on credit, thereby incurring debtors. A supermarket chain, on the other hand, will have a relatively low ratio, as it will hold only fast-moving stocks of finished goods and will generate mostly cash sales. The higher the ratio, the more liquid the business is considered to be. As liquidity is vital to the survival of a business, a higher current ratio is normally preferred to a lower one. However, if a business has a very high ratio this may suggest that funds are being tied up in cash or other liquid assets and are not being used as productively as they might otherwise be.

Box 15.1 Current ratio calculations

	Yr X		Yr Y	
Balance sheets as at 31 March	£000	£000	£000	£000
Fixed assets				
Freehold land and building at cost	451.2		451.2	
Less accumulated depreciation	70.0	381.2	75.0	376.2
Fixtures and fittings at cost	129.0		160.4	
Less accumulated depreciation	64.4	64.6	97.2	63.2
		445.8		439.4
Current assets				
Stock at cost	300.0		370.8	
Trade debtors	240.8		210.2	
Bank	33.5		41.0	
	574.3		622.0	
Creditors: amounts due within one year				
Trade creditors	(221.4)		(228.8)	
Dividends proposed	(40.2)		(60.0)	
Corporation tax due	(60.2)		(76.0)	
	(321.8)	252.5	(364.8)	257.2
		698.3		696.6

A more immediate measure of liquidity can be found by excluding stocks from the numerator (the number above the line). The resulting ratio is known as the *quick ratio* or *acid-test ratio*:

$$\text{quick ratio} = \frac{\text{current assets} - \text{stocks}}{\text{current liabilities}}.$$

15.5.2 Mitigation of Liquidity Risk

Mitigation is a key aspect of liquidity and is defined as the payment of debts when they fall due. To prevent the situation developing where debts cannot be met, a company should prepare a cash budget, that is, a plan of future cash receipts and payments based on specified assumptions about such things as sales growth, credit terms, issue of shares and expansion of plant. The purpose of drawing up cash budgets is to ensure that the company neither runs out of cash nor keeps cash idle when it could be profitably invested.

15.6 CREDIT RISK

Credit risk is the oldest and perhaps the most important of all risks in terms of the size of potential losses. Credit risk may be defined as the financial loss suffered due to the default of a borrower or counterparty under a contract. Banks describe credit risk as the risk that customers default, that is, fail to comply with their obligations to service debt. For HSBC credit risk arises principally from direct lending, trade finance and its leasing business, together with off-balance sheet products such as guarantees and credit derivatives.[1] Credit risk for professional consultancies such as solicitors, architects or town planners is where customers default on payment of invoices. Manufacturers who sell goods on credit face the risk that the customer may not ultimately pay. Default by a small number of large customers can generate large losses, which can lead to insolvency. The "quantity" of the risk is the outstanding balance lent to the borrower. The "quality" of risk results both from the chance that the default occurs and from the guarantees that reduce the loss in the event of default. The amount at risk, the outstanding balance at the date of default, differs from the total potential loss in the event of default, due to the potential recoveries. The recoveries depend upon any credit risk mitigations, such as guarantees (collateral or third party), the capability of negotiating with the borrower and the funds available (if any) to repay the debt after repayment of other lenders. Finally, potential recoveries from default cannot be predicted in advance. As a consequence, credit risk may be said to have three main components: default, exposure and recovery. These are discussed in turn.

15.6.1 Default Risk

Default risk is the probability of the event of default. There are several definitions of "default": missing a payment obligation, breaking a covenant, entering into an illegal procedure, or economic default. *Payment default* is declared when a scheduled payment has not been made for a minimum period, such as three months after the due date. *Breaking a covenant* occurs when

[1] HSBC Holdings plc, Capital and Risk Management, Pillar 3 Disclosures as at 31 December 2008.

fixed upper and lower bounds of a financial ratio are not adhered to, and this is recognised as a technical default. Such a default usually triggers legal proceedings commenced by negotiation. The default may also be purely *economic*, without being associated with any specific event. An economic default occurs when the economic value of assets goes below the value of the outstanding debts. The economic value of assets is the value of future expected cash flows discounted to the present day. If the market value of assets drops below that of liabilities, it means that the current expectations of future cash flows are such that the debt cannot be repaid. Default risk is measured by the probability that default occurs during a given period of time. Default depends upon the credit standing of a borrower. Such credit standing depends on several factors such as market outlook, the size of the company, its competitive context, the quality of management and the shareholders. Default probability cannot be measured directly. Historical statistics can and are used by such organisations as banks. From the statistical records of observed defaults the ratio of defaults in a given period over the total sample of borrowers can be derived. It is a default rate which often serves as a historical proxy for default probability.

15.6.2 Exposure Risk

Exposure risk relates to the uncertainty surrounding the payment of future amounts. For all lines of credit where there is a repayment schedule, the exposure risk is considered small. This is not true for all lines of credit. Committed lines of credit allow the borrower to draw on those lines whenever he/she wants to, depending on needs and subject to a limit fixed by the bank. Project financing implies uncertainty in the scheduling of outflows and repayments. Other exposure risks arise with derivatives. Here the source of uncertainty is not the borrower's behaviour but lies in market movements. The liquidation value of the derivatives depends upon such moves and changes constantly. Whenever the liquidation value is positive, there is a credit risk for the bank, since it loses money if the counterparty defaults.

15.6.3 Recovery Risk

Recovery risk relates to the uncertainty over the likely recovery of outstanding amounts due. Recovery risk depends upon the type of default. A payment default does not mean that the borrower will never pay, but it is the catalyst for various types of actions from renegotiation up to the obligation to repay all outstanding balances. If no corrective action can be considered it is common for legal procedures to take over. In such circumstances, all commitments of the borrower will be suspended until some legal conclusion is reached. At best, recoveries are delayed until the end of the legal procedure. At worst, there are no recoveries at all as the company is resold or liquidated and no excess funds are available to repay an unsecured debt. The credit loss of any transaction can always be described as the product of three terms:

$$\text{loss} = \text{exposure} \times \text{default} \times \text{severity}.$$

Loss is the actual economic loss to the company as the result of the default or downgrade of a counterparty – that is, as the result of a *credit event*.

15.6.4 Credit Insurance

Credit insurance is a risk mitigation action for credit risk. Competitively priced credit insurance covering payment for the sale of goods and services is available in many countries and, according to Holliwell (1998), can offer:

- protection against bad debts, usually up to a maximum of two years – the level of the cover will be subject to negotiation, but 90% is common, and the party taking out the insurance can usually decide the risk that they want to keep and the percentage to insure;
- cover for all or only a selection of the buyers of your products or services;
- insurance for either domestic or international trade, or both;
- cover for "country" risk, including delays in transferring money from the buyer's country, the actions of governments which prevent delivery or payment, including those countries through which goods or monies have to pass, and war;
- international debt recovery services (the insurer may be prepared to contribute towards the costs of recovery and will have access to specialist lawyers and debt collectors in different countries);
- the benefit of the credit insurer's skills and experience, based upon their exposure in the markets;
- "pre-credit risk" insurance for the costs incurred during the manufacture period before shipment;
- cover for the seller's costs and expenses and for contractual interest due from the buyer;
- the opportunity to win business by offering attractive terms because credit risk is no longer a significant factor;
- the ability to argue for cheaper finance from the bank, as the potential negative impact of any bad debts on the business has been reduced (and hence the ability to use the cost savings from the cheaper finance to pay for the credit insurance); and
- cover for the losses in meeting forward exchange commitments, where the buyer has defaulted.

Offer of Cover from an Insurer

Insurers providing credit risk, in reaching their decision, will principally take into account:

- industry sector;
- country risk (if appropriate);
- the types of goods or services that are being sold;
- terms of trade; and
- the track record of existing buyers of the goods or services.

Conditions Upon Which Insurance Claims are Settled

The conditions upon which insurers will settle will vary and depend on their policies, but there are common terms, which are described by Holliwell as follows:

- the goods or services have been delivered or otherwise provided;
- the debt is valid and that the buyer actually exists (the insured party would have to satisfy itself that the buyer was genuine);

- the buyer is not disputing payment (the insurance policy will state settlement terms if the buyer disputes – that is, whether the insurer with pay anything, a reduced amount or whether the goods have to be resold before the insurer will make any contribution to the loss);
- credit limits have been respected (it is anticipated that businesses will set discretionary limits for individual buyers); and
- insurance premiums have been paid – these normally comprise a basic fee plus a premium based on the level of activity.

15.6.5 Counterparty Risk

Default risk (Section 15.6.1) occurs when other organisations that a business trades with may not honour their obligations in terms of failing to pay for or deliver goods or services, or to repay a borrowing. On the assumption that the business has fulfilled its obligations under the transaction, default on the part of the counterparty may arise as they:

- have become insolvent;
- have themselves been let down by a trading partner;
- cannot obtain the resources (plant, labour and or materials) necessary to complete the transaction;
- are prohibited from meeting their obligations through national trading controls.

When dealing with a counterparty, take time to understand the risks and implement risk response actions to limit exposure on the assumption that not all risk can be removed in its entirety. Sensible actions include:

- undertaking an appropriate level of due diligence;
- recognising that circumstances change and that the circumstances of long-term trading partners may change;
- establishing the background to a potential counterparty in terms of their legal form (limited liability company, partnership or sole trader), their activity (primary, related to natural resources; secondary, processing of materials such as manufacturing; or tertiary, services such as insurance and banking), and size (turnover, assets, number of employees and capitalisation);
- not committing to a single customer or supplier;
- knowing the extent of your exposure at all times;
- acting immediately in the event of default, or its likelihood.

15.6.6 Due Diligence

For a business considering an undertaking such as entering into a major contract, committing to a joint venture, acquiring a business or lending money to a third party, it will need to undertake due diligence as part of the evaluation process. The extent to which due diligence will be appropriate will depend upon individual circumstances but will primarily be judged on what damage could be done if the activity went "sour" and it had an adverse impact on the business. Holliwell (1998) offers a checklist of those issues that may need to be considered as part of due diligence in Box 15.2.

Box 15.2 Due diligence checklist

- Strategic plans and vision
- Nature and diversification of products and services
- Market potential and industry risks
- Product life cycles
- Technological risk
- Research and development programme and costs
- Market shares and the order book
- Competitors, positionings and differentiations
- Spread of client base and dependencies/continuity
- Customer care policy and practice
- Public relations
- Trends such as turnover and costs
- Benchmarking of key factors
- Whether growth has been generic or by acquisition
- Changes in the nature of the business
- Intra-group trading and exposures
- Management style
- Ethics and culture
- Historic and management accounts
- Financial strengths and weaknesses
- Key financial ratios
- Basis of funding and terms of borrowing
- Accounting, depreciation and dividend policies
- Off-balance sheet transactions
- Control of treasury functions (are the exposures and risks understood?)
- Economic risks
- Terms of trade
- Debtors and creditors, their spread and collection/payment periods
- Currency exchange rate exposures
- Interest rate exposures
- Lease and hire purchase agreements
- Security given and available
- Borrowing covenants
- Bases of valuations, including property, stock and intellectual assets
- Insurance, including assets, key man and loss of profits
- Human resources, including spread and depth of skills and experience, continuity and succession planning
- Board and organisational structures
- Executives' contracts, remuneration and benefits
- Subcontracting and outsourcing
- Resources, including dependencies and threats to suppliers
- Stock levels
- Operation and IT risks
- System costs and useful life
- Property, plant and equipment
- Environmental issues
- Licences, goodwill, intellectual assets, franchises
- Capital and contractual commitments
- Contingent liabilities, including warranties and guarantees
- Pension, health and welfare commitments
- Group structure
- Equity structure and holders, including warrants, options and conversion rights
- Legal entity and jurisdiction of business
- Legal issues, including ownership of assets
- Litigation
- Regulatory issues
- Management information systems and knowledge management
- Country risks
- Political risks
- Risk management culture, policy and risk aversion
- Sensitivity analysis
- Disaster scenarios

Source: Reproduced with permission from *The Financial Risk Manual*, J. Holliwell, Pearson Education Limited © 1998.

Shim and Siegel (2001) provide signals for both quantitative and qualitative corporate failure which can be incorporated in prompt lists, questionnaires and risk registers to "interrogate" the board and department heads and establish the risks facing a company at a particular point in time (see Box 15.3).

Box 15.3 Quantitative and qualitative factors

Quantitative factors in predicting corporate failure

- Low cash flow to total liabilities
- High debt-to-equity ratio and high debt to total assets
- Low return on investment
- Low profit margin
- Low retained earnings to total assets
- Low working capital to total assets and low working capital to sales
- Low fixed assets to non-current liabilities
- Inadequate interest/coverage ratio
- Instability in earnings
- Small-size company measured in sales and/or total assets
- Sharp decline in price of stock, bond price and earnings
- A significant increase in beta (beta is the variability in the price of the company's stock relative to a market index)
- Market price per share is significantly less than book value per share
- A significant rise in the company's weighted-average cost of capital
- High fixed cost to total cost structure (high operating leverage)
- Failure to maintain capital assets. An example is a decline in the ratio of repairs to fixed assets

Qualitative factors in predicting failure

- New company
- Declining industry
- Inability to obtain adequate financing, and when obtained there are significant loan restrictions
- A lack in management quality
- Moving into new areas in which management lack expertise
- Failure of the company to keep up to date, especially in a technologically oriented business
- High business risk (e.g. positive correlation in the product line; susceptibility to strikes)
- Inadequate insurance coverage
- Fraudulent actions (e.g. misstating inventories to stave off impending bankruptcy)
- Cyclicality in business operations
- Inability to adjust production to meet consumption needs
- Susceptibility of the business to stringent governmental regulation (e.g. companies in the real estate industry)
- Susceptibility to energy shortages
- Susceptibility to unreliable suppliers

- Renegotiation of debt and/or lease agreements
- Deficient accounting and financial reporting systems

Source: Shim and Siegel (2001).

15.7 BORROWING

If a company is borrowing money, it will want to know on what basis the interest rate is determined, what the interest rate will be on commencement of the borrowing, whether the interest rate is fixed or variable, and when the interest will be payable. The rate of interest paid will depend on a combination of some or all of the following:

- *Amount*. The rate of interest often varies according to the amount of money involved (known as the "principal" or "capital"). Larger amounts usually attract preferential rates as the overhead and control costs may be proportionally lower.
- *Term*. This is the length of time the monies are to be borrowed. The longer the term, the greater the opportunity for something to happen that could prevent the borrower from paying all or some of the borrowings. The credit risk of this may be small if repayment is on demand or due within a short period of time, but the situation may change radically in, say, three years.
- *Forecasts*. If market interest rates are projected to either increase or decrease, then that will be taken account of in fixed rates for medium- and longer-term loans or deposits.
- *Inflation*. A provider of funds hopes to earn a rate of interest at least equal to the rate of inflation over the term of the loan, otherwise the amount the provider obtains at the outturn will, in current terms, be less than the original principal (or capital).
- *Risk*. The greater the concern of the lender that they may be unable to recover all or some of their money, for whatever reason, the higher the reward they will want for putting their funds at risk.
- *Opportunity cost*. The rate of interest may be influenced by the fund provider forgoing other transactions to commit to this transaction.
- *Market*. Interest rate charges may be influenced by regulatory requirements, international competition and the publishing of available rates.

The subject of interest rates is discussed in Chapter 21.

15.8 CURRENCY RISK

There is always a risk that the expected cash flows from overseas investments will be adversely affected by fluctuations in exchange rates. As a result, the value of a foreign currency receivable or payable when translated into the currency of the country where the business is located may be more or less than expected. For example, a UK-based business may receive less than expected from a transaction with a French business because of a rise in the euro against sterling. The kinds of business operations that will expose a business to exchange rate risk include the import or export of goods or services, investing in overseas assets such as factories, and raising finance from overseas sources. Where a business is engaged in overseas transactions involving large sums, an adverse movement in exchange rates can be catastrophic and so it will usually adopt some form of "hedging" to minimise the risk. There are various ways in which hedging can be carried out, and the most important of these are discussed in Chapter 21.

15.9 FUNDING RISK

Most companies rely to a greater or lesser extent on the issue of loan capital to finance their operations. Lenders will normally enter into a contract with a company which will clearly set out the rates of interest to be paid on the loan and the due dates for interest payments and capital repayments. Businesses that have secured loans may have had to offer security to the lender by pledging a *fixed charge* on assets held by the company, or a *floating charge* which "hovers" over the whole of the company's assets. The riskiness of loan capital (for lenders) can be measured in terms of default risk. A number of credit-rating agencies, including the Standard & Poor's Corporation and Moody's Investor Services (both contactable on the web), attempt to place loan capital issued by companies into categories according to the level of default risk. The lower the risk of default (by the borrower) the higher the rating category that will be assigned to the debt. The ratings used by the two leading agencies mentioned above are very similar. Where a company is rated by both of these agencies, the risk category assigned to it is usually the same. Where a difference arises it is normally only a difference of one risk category.

To arrive at an appropriate debt rating, an agency will rely solely on published information. In the case of Standard & Poor's, credit ratings are based on current information furnished by the obligors or obtained by Standard & Poor's from other sources it considers reliable; it does not perform an audit in connection with any credit rating and may, on occasion, rely on unaudited financial information. Credit ratings may be changed, suspended or withdrawn as a result of changes in, or unavailability of, such information, or based on other circumstances. The rating classification assigned to the debt will be derived from an assessment of all of the relevant information, and as it is not formula-driven some subjective assessment comes into play. Once a debt has been assigned to a particular category, it will tend to remain in that category unless there is a significant change in circumstances.

The analytical framework and methodology of determining the credit estimate is similar to that of traditional credit ratings, with credit estimates being determined by a rating committee utilising a global rating scale (see Box 15.4). Issue credit ratings can be either long-term or short-term. Short-term ratings are generally assigned to those obligations considered short-term in the relevant market. In the US, for example, that means obligations with an original maturity of no more than 365 days. Long-term issue credit ratings are based, in varying degrees, on the following considerations: likelihood of payment, and the capacity and willingness of the obligor to meet its financial commitment on an obligation in accordance with the terms of the obligation. It takes into consideration the creditworthiness of guarantors, insurers, or other forms of credit enhancement on the obligation, and the currency in which the obligation is denominated. The issue credit rating is not a recommendation to purchase, sell or hold a financial obligation, inasmuch as it does not comment on the market price or suitability for a particular investor.

Box 15.4 Standard & Poor's credit rating definitions

The issue rating definitions are expressed in terms of default risk.

AAA	An obligation rated "AAA" has the highest rating assigned by Standard & Poor's. The obligor's capacity to meet its financial commitment on the obligation is extremely strong.
AA	An obligation rated "AA" differs from the highest-rated obligations only to a small degree. The obligor's capacity to meet its financial commitment on the obligation is very strong.
A	An obligation rated "A" is somewhat more susceptible to the adverse effects of changes in circumstances and economic conditions than obligations in higher-rated categories. However, the obligor's capacity to meet its financial commitment on the obligation is still strong.
BBB	An obligation rated "BBB" exhibits adequate protection parameters. However, adverse economic conditions or changing circumstances are more likely to lead to a weakened capacity of the obligor to meet its financial commitment on the obligation.
BB, B, CCC, CC and C	Obligations rated "BB", "B", "CCC", "CC" and "C" are regarded as having significant speculative characteristics. "BB" indicates the least degree of speculation and "C" the highest. While such obligations will likely have some quality and protective characteristics, these may be outweighed by large uncertainties or major exposures to adverse conditions.
BB	An obligation rated "BB" is less vulnerable to nonpayment than other speculative issues. However, it faces major ongoing uncertainties or exposure to adverse business, financial or economic conditions which could lead to the obligor's inadequate capacity to meet its financial commitment on the obligation.
B	An obligation rated "B" is more vulnerable to nonpayment than obligations rated "BB", but the obligor currently has the capacity to meet its financial commitment on the obligation. Adverse business, financial or economic conditions will likely impair the obligor's capacity or willingness to meet its financial commitment on the obligation.
CCC	An obligation rated "CCC" is currently vulnerable to nonpayment, and is dependent upon favorable business, financial and economic conditions for the obligor to meet its financial commitment on the obligation. In the event of adverse business, financial or economic conditions, the obligor is not likely to have the capacity to meet its financial commitment on the obligation.
CC	An obligation rated "CC" is currently highly vulnerable to nonpayment.
C	A "C" rating is assigned to obligations that are currently highly vulnerable to nonpayment, . . . or obligations of an issuer that is the subject of a bankruptcy petition or similar action which have not experienced a payment default. Among others, the "C" rating may be assigned to subordinated debt, preferred stock or other obligations on which cash payments have been suspended in accordance with the instrument's terms or when preferred stock is the subject of a distressed exchange offer, whereby some or all of the issue is either repurchased for an amount of cash or replaced by other instruments having a total value that is less than par.

D	An obligation rated "D" is in payment default. The "D" rating category is used when payments on an obligation . . . are not made on the date due even if the applicable grace period has not expired, unless Standard & Poor's believes that such payments will be made during such grace period. The "D" rating also will be used upon the filing of a bankruptcy petition or the taking of a similar action if payments on an obligation are jeopardized. . . .

Note: The ratings from "AA" to "CCC" may be modified by the addition of a plus (+) or minus (–) sign to show relative standing within the major rating categories.

Source: http://standardandpoors.com/ratings/definitions-and-faqs/en/us

15.10 FOREIGN INVESTMENT RISK

When considering investment opportunities, a business may wish to examine opportunities abroad. These kinds of opportunities, however, attract additional risks to those associated with domestic market opportunities. The first step in the management of corporate foreign investment risk is to acknowledge that such risk does exist and that managing it is in the interests of the business and the stakeholders. The next step is much more difficult: the identification of the individual risks, an understanding of their magnitude and what steps might be taken to address them.

15.10.1 Country Risk

Risks arise from the geographical distance of the market, which can increase both the cost and time associated with debt collection. While legal action can be taken against a customer within a foreign jurisdiction that does not pay, the costs of recovery are not guaranteed. Most governments encourage investment from overseas business because of the beneficial economic effects; however, some governments adopt policies that are discouraging. These policies may include restrictions on the right to repatriate funds, high levels of taxation on profits remitted overseas, the temporary freezing of bank account balances and the expropriation of assets. Therefore, when making an overseas investment decision, there must be some assessment of the risks. There are specialist agencies that produce indices of "country risk" and which may be of assistance. However, while it is possible to devise country risk indices (with weighting applied to each criterion) the criteria employed may not always be relevant to the investment under review.

If the risks described are considered significant, the investing business has the option to abandon the proposal or try to develop strategies which attempt to minimise or overcome them. Such strategies might include a joint venture with the overseas government or with a local business in the host country, ensuring local labour, plant and materials are used wherever possible, or by becoming a "good corporate citizen" through charitable donations. In some instances it may be possible to transfer (at a price) certain risks such as the expropriation of assets by taking out insurance with a credit insurance business. This will mean the risk is being passed to the insurer.

15.10.2 Environment Risk

There may be insufficient experience of the business environment within the overseas market within which it is intended to invest. There will most probably be different laws, working practices, cultural and ethical norms and taxation regimes, which may have a profound effect on the viability of an investment proposition.

15.11 DERIVATIVES

The options and futures markets come under the general heading of markets in derivative products or derivatives. Their main function is the redistribution of risk. The customers who use these markets fall into two main categories: those who want to hedge (guard against) a risk to which they are exposed in the normal course of their business and those who are prepared to accept a high risk in return for the possibility of large rewards, the traders and speculators. The term "derivative" stems from the simple fact that they are financial products derived from some other existing product. Shares, currencies, bonds and commodities such as zinc and cocoa are all products. People are familiar with what these are. Derivatives are based on these existing products. Derivatives are contracts between two parties (the "buyer" and the "seller"), which are known as the counterparties. They fall into three principal categories: *options*, *futures* and *swaps*. Derivatives are available to cover many types of exposure, including:

- interest rates;
- foreign currency exchange rates;
- commodities, such as energy (e.g. oil and gas), bullion (e.g. gold and silver), base metals (e.g. copper and nickel) and agriculture (e.g. sugar); and
- equities.

The gain or loss under a financial derivative depends on (or "derives from") movements in the market price of the asset or index to which the contract relates (known as the "underlying"). A derivative contract where one counterparty has to pay a premium (e.g. an *option*) also has the right but not the obligation to exercise the contract. Derivative contracts where no premium is paid at the outset (e.g. *futures* and *swaps*) commit both counterparties to complete the transaction at settlement dates(s), which normally means that one of them will have to make payment to the other. Derivatives can be either "exchange traded" or "over the counter" (OTC).

The risks associated with exchange traded and OTC derivatives include:

- credit risk;
- settlement risk;
- aggregation risk;
- operational risk;
- liquidity risk;
- legal risk;
- reputational risk; and
- concentration risk.

15.11.1 Exchange Traded Derivatives

Exchange traded derivatives are bought and sold on recognised exchanges throughout the world, among the best known being the Chicago Board of Trade (CBOT) and the London

International Financial Futures Exchange (LIFFE). Trading on most exchanges is conducted by a combination of "open outcry" and computer-based dealing. LIFFE has been progressively transferring its business to screen-based dealing. Once a deal has been agreed, a clearing house associated with the exchange (and which may be owned by the exchange itself or by banks or other financial institutions) steps in. From that point onwards, the clearing house acts as the counterparty to both the buyer and the seller of the contract. Each counterparty is therefore taking a risk on the clearing house and not on the other counterparty. The price of an exchange traded derivative is whatever it fetches in the market. There are, however, restrictions on the minimum amounts by which prices for each of the standard derivative contracts can move. These minimum movements are known as "ticks" and differ from one type of derivative to another.

15.11.2 Over-the-Counter Derivatives

OTC derivatives are contracts written to meet the specific needs of individual clients, such as businesses, banks or governments. They are usually provided by banks or other financial institutions and cannot be traded on any exchange. The actual contract document is likely to be based on the standard terms and conditions of an organisation such as the International Swaps and Derivatives Association (ISDA). The most common OTC derivatives are *options* and *swaps*. With an OTC contract, the pricing of the derivative is negotiated between the counterparties, normally between a bank and a client.

15.12 SUMMARY

In this chapter various aspects of financial risk have been examined which have to be managed to maintain a business as a going concern.

Liquidity risk has to be managed to ensure a business has sufficient funds to meet its obligations, and two ratios to measure liquidity – the current ratio and the quick ratio – were discussed. Cash budgets are cited as a proactive mitigation action to minimise the risk. However, this response cannot eradicate the risk. Credit risk was defined as the economic loss suffered due to default by a borrower or counterparty. For a bank this means default by a customer on a loan, for a consultancy it means the non-payment of fees, and for a manufacturer it means non-payment of goods. Credit risk was broken down into default, exposure and recovery risk. A mitigation action, credit insurance, was described.

Interest rate risk can have both an upside and a downside. Lower interest rates mean more borrowing and consumer spending, and lowering of the exchange rate makes our products cheaper abroad. Higher interest rates lead to less consumer spending and greater saving. Additionally the risk exists that the expected cash flows from overseas investments will be adversely affected by fluctuations in exchange rates. As a result, the value of a foreign currency receivable or payable when translated into the currency of the country where the business is located may be more or less than expected. It was discussed that where a business is engaged in overseas transactions involving large sums, an adverse movement in exchange rates can be catastrophic, so that businesses usually adopt some form of hedging to minimise the risk.

Funding risk for businesses was seen to be an inability to repay loans resulting in a fixed or floating charge on their assets. For lenders, other than imposing charges, the other mitigation action at their disposal was the use of a credit rating agency such as Standard & Poor's. Businesses examining opportunities in overseas markets are exposed to a harsher risk environment than businesses pursuing domestic market opportunities. This additional risk is based on the geographical distance of the market, which can increase both the cost and time associated with debt collection. Derivatives were described as a way of guarding against risk and also as a source of speculation for those wishing to trade on the markets. Three primary reasons for wanting to manage these risks are to preserve the business, protect it against financial losses, and ensure that a business (where listed) is able to maintain dividend payments within pre-determined limits. Investors are normally unlikely to welcome "surprises" in dividend policy and may react by selling their shares and investing in a business which has a more stable and predictable dividend policy. This behaviour will lower the value of a business's shares and will increase the cost of capital.

15.13 REFERENCES

Atrill, P. (2000) *Financial Management for Non-specialists*, 2nd edition. Pearson Education, Harlow.
Holliwell, J. (1998) *The Financial Risk Manual, a Systematic Guide to Identifying and Managing Financial Risk*. Pearson Education, UK.
Shim, J.K. and Siegel, J.G. (2001) *Handbook of Financial Analysis, Forecasting and Modeling*, 2nd edition. Prentice Hall Press, Paramus, NJ.

16

Operational Risk Management

Organisations that choose to remain blissfully ignorant of operational risk will continue to operate under a false sense of security. They will remain "under-controlled" in areas where they have the most risk and "over-controlled" in areas where they have the least risk. So without addressing op risk head on, recognising and understanding it and acknowledging the crucial role that it plays, we face the prospect of another global financial crisis in the not too distant future.
(Samad Khan)

The previous chapter examined financial risk management as the first of six classes of risk exposure that businesses face relating to internal processes within the risk taxonomy proposed in Chapter 9. This chapter examines the second of the internal processes, called *operational risk*. A business cannot claim that it has an enterprise risk management process if it does not address operational risk. According to the FSA, "operational risk is present in all firms and can affect a firm's solvency, the fair treatment of its customers and the incidence of financial crime" (Financial Services Authority 2002). The perceived significance of operational risk is also illustrated by the results of a study (cited by Carey and Turnbull 2001) where financial managers, in consideration of the risks facing their company and assessing their relative importance, considered the principal risks to be generally operational and strategic. The effectiveness of operational risk will depend on how comprehensive the identification process is. The structure of this chapter is based on the risk taxonomy included in Chapter 9, which is used as a vehicle to examine the elements, attributes and features of operational risk and describe an appropriate response strategy.

It is clear that there has been an increasing interest and development in the application of operational risk (Peccia 2001; Financial Services Authority 2002, 2005; Maxant 2004[1]), particularly among financial institutions, as a result of six drivers:

- as a consequence of the widely reported operational risk losses that led to losses at Société Générale, Allied Irish and the destruction of Barings Bank;
- from the trend of managing risk under a RAROC[2] framework leading to a shift from the control (minimisation) of risk to the management of risk (balancing the need for risk control with the cost of control);

[1] Deloitte's 2004 Global Risk Management Survey was based on interviews with senior executives from the world's top 162 global financial institutions. The survey was intended as a global benchmark for the state of risk management in the financial sector. According to the survey, operational risk management (ORM) continued to be considered a challenging and relatively new field compared to more established risk management disciplines. However, the survey did show an increase over 2002 in the number of firms that had established ORM programmes even though the majority of respondents indicated that at least some improvement in functionality was still required.

[2] In the financial sector in particular there is no performance level without a price to pay in terms of risk. Hence, the risk–reward combination is meaningful. When risk is omitted from performance evaluation it is not possible to compare the performance of transactions or business units, evaluate the risk to be transferred to counterparties or subdivide the perceived total business risk between business units or individual transactions. Risk-adjusted profitability addresses these issues. One of the main solutions to defining risk-adjusted profitability is known as *risk-adjusted return on capital* (RAROC). This adjusts the return for risks, for instance by calculating margins net of statistical defaults. RAROC is expressed as a ratio and adjusts the earnings by the expected loss and uses risk-based capital as a measure of unexpected loss: RAROC = expected return/economic capital.

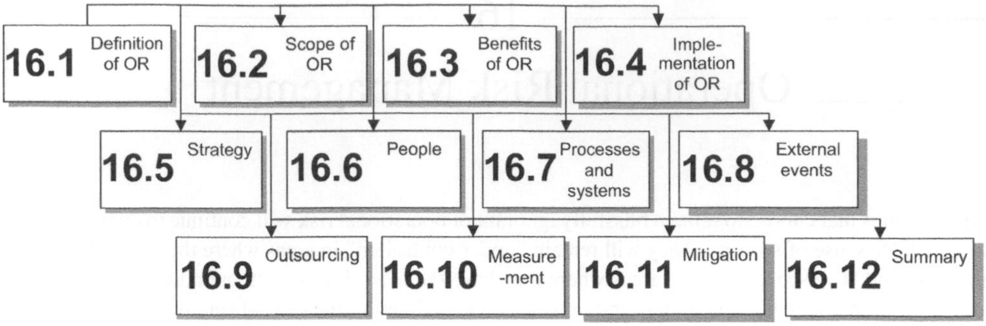

Figure 16.1 Structure of Chapter 16

- from the fact that operational risk now forms a substantial part of the current risks assumed by businesses and the expectation that this will increase;
- as a result of the realisation by financial institutions that risk management is not an add-on activity, but is a core competency which can be used to competitive advantage;
- the 2010 Basel Accord, which describes the supervisory regulations governing the capital adequacy of internationally active banks and calls for the measurement and mitigation of operational risk;[3] and
- as a result of FSA initiatives responding to the Basel Accord.

The increasing interest in operational risk led to the formation of the Institute of Operational Risk in January 2004 (in accordance with the requirements stipulated by the UK Secretary of State in regard to the formation of an institute). It was formed as a professional body in response to a need to promote and maintain standards of professional competency in the discipline of operational risk management.

The structure of this chapter is illustrated in Figure 16.1. The next chapter examines the third of the internal processes, technological risk.

16.1 DEFINITION OF OPERATIONAL RISK

What is operational risk? It is difficult to flaw the logic of Peccia (2001) who states that, at a rudimentary level, running a business is concerned with employing people to carry out processes aided by technology and external dependencies to achieve defined business objectives. Hence, he argues, it makes sense to define operational risk as "the potential for loss due to failures of people, processes, technology and external dependencies". The same "ingredients" are included in the definition provided by the Basel Committee[4] and adopted here due to its broader acceptance: "Operational risk is the risk of loss resulting from inadequate or

---

[3] Basel III "is a comprehensive set of reform measures, developed by the Basel Committee on Banking Supervision, to strengthen the regulation, supervision and risk management of the banking sector".

[4] The Basel Committee, established by the central-bank governors of the Group of Ten countries at the end of 1974, meets four times a year. It has four main working groups which also meet regularly. The Committee provides a forum for regular cooperation on banking supervisory matters. Its objective is to enhance understanding of key supervisory issues and improve the quality of banking supervision worldwide. The Committee's members come from Argentina, Australia, Belgium, Brazil, Canada, China, France, Germany, Hong Kong SAR, India, Indonesia, Italy, Japan, Korea, Luxembourg, Mexico, the Netherlands, Russia, Saudi Arabia, Singapore, South Africa, Spain, Sweden, Switzerland, Turkey, the United Kingdom and the United States. It usually meets at the Bank for International Settlements in Basel, where its permanent Secretariat is located.

failed internal processes, people and systems or from external events" (Basel Committee on Banking Supervision 2004). While this definition has been produced for the banking sector, it has a broad application. The FSA (2002) correctly states that "ultimately firms need to decide for themselves what operational risk means to them" and any firm needs to "consider a more specific definition of operational risk that is appropriate to the range and nature of its business activities and its operating environment". Hence, businesses need to define operational risk in terms of their end product and the resources and processes engaged to produce that product.

16.2 SCOPE OF OPERATIONAL RISK

The sources of risk considered to be embraced within the term "operational risk" are very considerable. The FSA Integrated Prudential Sourcebook consultation paper (Financial Services Authority 2001) describes operational risk as covering the following:

- Business risk, which includes adverse changes to a firm's market, customers or products, changes to the economic and political environments in which the firm operates and strategic risk which a firm faces if business plans, supporting systems and the implementation of these plans adversely affect the firm.
- Crime risk including potential theft, fraud and computer hacking.
- Disaster risk such as fires, floods and other natural disasters and terrorist activity.
- Information technology risk, including unauthorised access and disclosure, and data corruption.
- Legal risk including loss arising from legal action against it and from inadequate, incomplete or otherwise unsound legal documentation and practices.
- Regulatory risk relating to the lack of observance of rules set by a regulatory body.
- Reputational risk from negative publicity about its business practices or internal controls.
- Systems risk loss as a result of failure caused by the breakdown of business procedures, processes or systems and controls.
- Outsourcing.

FSA (2002) later describes some specific areas of concern relating to operational risk, which include:

- highly automated and integrated technology that has the potential to transform risks from minor manual processing errors to major systematic failures;
- the growth in e-commerce that brings with it some new and potentially significant operational risks for both consumers and firms (e.g. fraud and system security issues);
- firms that outsource their activities may suffer some loss of control over them, which could affect the quality and availability of their products; and
- insourcing, where firms take on the operational risks of third parties.

The FSA definition of "business risk" given earlier includes market, economic and political risk. These risk sources are treated as separate classes of risk (within the risk taxonomy proposed in Chapter 9). Crime and information technology risk are not discussed in this chapter (but see Chapter 17). Additionally legal risk[5] (included within the Basel Committee's definition of operational risk) is treated as a separate class of risk within the risk taxonomy

[5] Legal risk includes, but is not limited to, exposure to fines, penalties or punitive damages resulting from supervisory actions as well as private settlements.

proposed here – the rationale being that the FSA and Basel initiatives are aimed at the financial institutions, whereas the taxonomy proposed here is for a broader audience. This greater subdivision enables a more balanced risk breakdown structure to be created, with the attendant benefits of description, resourcing and management. Where a particular class of risk is included within a taxonomy is of minor importance. The important issue is that risks are comprehensively identified, robustly assessed and proactively managed. If there is a relationship between different risk sources, that relationship will exist regardless of where the sources are placed within a taxonomy.

16.3 BENEFITS OF OPERATIONAL RISK

Operational risk management affords a business benefits by:

- improving the ability to achieve its business objectives;
- providing management the opportunity to focus on revenue generating activities rather than fire-fighting one crisis after another;
- minimising day-to-day losses;
- providing a more robust enterprise risk management system;
- contributing to the establishment of a system which enables the correlation of different classes of risk to be understood and, where appropriate, modelled.

16.4 IMPLEMENTATION OF OPERATIONAL RISK

The development of a sound system of operational risk management will depend on a number of issues:

- The risk management system must not overly constrain risk taking, slow down decision-making processes or limit the volume of business undertaken.
- The implementers of the risk management framework must be separate individuals from the managers of the individual business units.
- Risks must be managed at an appropriate level in the organisation.
- A culture must be developed which rewards the disclosure of risks when they exist, rather than encouraging managers to hide them.

16.5 STRATEGY

We now turn to the various elements of operational risk. Figure 16.2 illustrates a possible taxonomy for *strategy risk*. This element is illustrated here as having seven attributes, each of which is described in turn.

16.5.1 Definition of Strategy Risk

A business's strategy is its overall approach to achieving its objectives. Objectives are the results required within a particular timeframe, and results are the measure of performance. Strategy is a description of what the business will do and the rationale behind it. For example, Virgin Mail Order's early strategy for music record sales was to compete in the market place by means of mail order (as its company name suggests), undercutting prices offered by the existing well-established high street retailers. Adopting the wrong business strategy, failing to

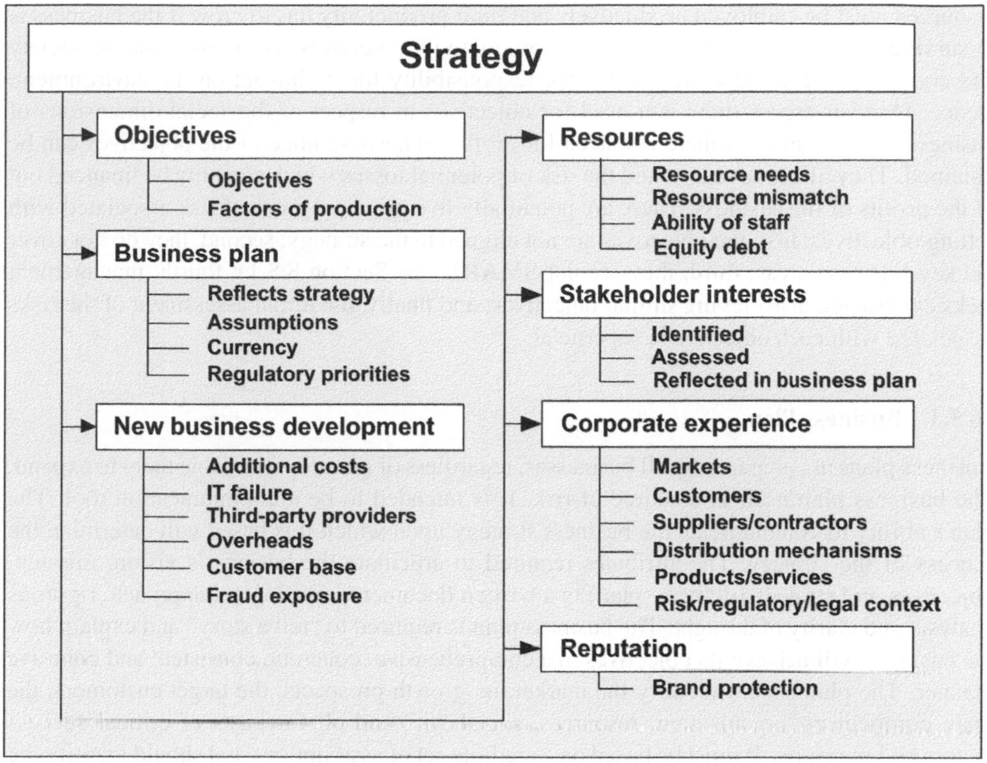

Figure 16.2 Taxonomy of strategy risk

execute a well-thought-out strategy or not modifying a successful strategy over time to reflect changes in the business environment are forms of operational risk. Strategic risk, then, may be defined as the risk associated with initial strategy selection, execution or modification over time, resulting in a lack of achievement of overall objectives.

16.5.2 Objectives

For a strategy to succeed, the objectives must be clearly stated and understood. Objectives are the basis for work and the assignment of work. They determine the structure of a business, the key activities that must be undertaken and the allocation of people to tasks. So objectives are the foundation for designing the organisational structure and processes of a business as well as the work of individual business units and managers. Drucker (1979) argues that objectives are always needed in eight key areas: marketing, innovation, human organisation, financial resources, physical resources, productivity, social responsibility and profit requirements. He explains the need for these objectives by assessing the operation of a business. Drucker states that a business must first be able to create a customer. Hence, there is a need for a marketing objective. Businesses must be able to innovate, or else their competitors will make them obsolete. Hence, there is need for an innovation objective. All businesses depend on the three factors of production – human resources, capital resources and physical resources. There must be objectives for their supply, their employment and their development, respectively. The resources must be employed productively and their productivity has to grow if the business is to survive. There is a need, therefore, for productivity objectives. Business exists in society and community, and therefore has to take responsibility for its impact on the environment. Hence, Drucker argues there is a need for objectives in respect of the social dimensions of business. Finally, he says there is a need for profit – otherwise none of the objectives can be obtained. They all attract a cost and the risk of potential losses which can only be financed out of the profits of the business. There are potentially five principal areas of risk associated with setting objectives: first, the objectives are not aligned to the strategy; second, they do not cover the key business areas; third, they are not SMART (see Section 8.8.1); fourth, management lacks experience in achieving similar objectives; and finally, the initial assessment of the risks associated with each objective is superficial.

16.5.3 Business Plan

Business plans are prepared by all businesses, regardless of maturity, to enable them to expand. The business plan itself is a source of risk. It is intended to be a communication tool. The plan's ability to communicate the business strategy upon which it is based will determine the success of the strategy. The attributes required to articulate the business's vision, mission, objectives and strategy (delivery plan) in a written document are a logical approach, rigorous analysis and clarity of thought. The business plan is required to "tell a story" and explain how the business will achieve its objectives in a comprehensive, coherent, consistent and cohesive manner. The plan should identify the market, its growth prospects, the target customers, the main competitors, organisation, resources, social "fit" and all forecasts of critical success factors and measures. It must be based on a credible set of assumptions and should identify the assumptions to which the success of the business is most sensitive. These assumptions will be about the future size of the market, the economy, potential changes to the market, competitor behaviour and the ability of the business to deliver. These assumptions are potential risks. Hence, the plan should record the risks facing the business arising from an understanding of the assumptions, how these will be addressed and the anticipated degree of success.

The plan will also record the risk management process for the risks identified against the objectives, as discussed in Section 16.5.1. As the plan will be referred to not only for the initial business idea, but also for successive business decisions, it will require regular updating. For this reason the plan will have a long-term influence on the attainment of the business idea. There are therefore a series of potential areas of risk associated with creating and implementing a business plan. These include the plan not readily articulating the strategy; not explaining how the objectives will be accomplished, particularly at start-up through short-term detailed operating plans, not making the assumptions explicit; not identifying the risks (and their responses) associated with the assumptions, not taking account of regulatory priorities (if applicable); and not updating the risks to the objectives or being regularly updated.

16.5.4 New Business Development

This risk category refers to the risks associated with plans for entering new business areas, expansion through mergers and acquisitions, providing new services and enhancing infrastructure (e.g. physical plant and equipment, as well as information technology and networking). While competition places additional pressure on businesses to protect profitability through the development of new products and services, such activity attracts additional risks from: the lack of recovery of additional research and development or marketing costs; the failure of new information technology and the associated loss of reputation; poor performance by third-party providers; the overheads and staff costs from high-volume, low-margin services outweighing the corresponding increase in profits; new services not attracting new customers or the new services attracting higher losses due to fraud or theft. The financial sector, in particular, is exposed to these types of risk. For example, financial sector businesses expanding internet banking services to include electronic bill payment services and increasing existing bank card issuing programmes significantly increase their risk exposure. Larger financial institutions often specialise in specific retail payments and invest in the resources and expertise to support high-volume transaction processing applications. Smaller financial sector businesses also compete in some retail payment segments through the use of advanced distributed information technology platforms and third-party service providers. Many retail payment system services are transaction-intensive and priced competitively, based on volume. Financial sector businesses wishing to compete in high-volume, transaction-intensive retail services are required to make significant investments in information technology. Strategic plans should reflect these investments and link business-line goals and objectives with planned information technology enhancements. To mitigate strategic risk, management should have a strategic planning process that addresses its retail payment business goals and objectives, including supporting information technology components. As financial institutions often rely on third-party service providers for retail payment system products and services, the strategic plan should include a comprehensive vendor management programme. In summary, strategic plans should demonstrate that management has assessed the risks and documented the business's programme to mitigate them.

16.5.5 Resources

As discussed in Section 16.5.2, an objective for a business is to be innovative in order to differentiate itself from its competitors. Hence, part of the business strategy will be to achieve, sustain and enhance competitive advantage. Some businesses are more successful than others because they have resources that are inherently different from those of their competitors, who may not be able to acquire or replicate similar resources. Businesses should therefore acquire or develop in house such unique resources in order to attain competitive advantage. Resources will include capital, energy, raw materials, people, buildings, land and machinery. Additionally, similar businesses may have identical resources but one business is able to outperform the other due to productivity. The risks to resources relate to: a lack of comprehension of resource needs to meet objectives; a mismatch between objectives and existing resources; a mismatch between production/sales projections and procurement of planned resources; experience, qualifications and technical ability of staff; and a mismatch between equity-debt and spend profile. Human resources are discussed more fully in Section 16.6.

16.5.6 Stakeholder Interests

Stakeholders are those individuals or organisations who can affect or be affected by the business. They include shareholders, lenders, employees, suppliers, business partners, customers, analysts and in some instances society at large. There should be a clear understanding of stakeholders' interests. Stakeholder analysis should be undertaken which identifies the primary and possibly conflicting expectations of the stakeholders and their power of influence.

Stakeholders are likely to have conflicting interests or different priorities. Hence, they should be recorded, disseminated, openly debated (when appropriate) and where possible aligned through negotiation. The agreed course(s) of action should be recorded and forwarded to the stakeholders, acknowledging where stakeholders' wishes have not been taken on board. Stakeholder requirements should be reflected in the business plan; if omitted, they could be the source of problems in the long term.

16.5.7 Corporate Experience

A business's corporate experience will reflect on the risk exposure profile of the business's strategy. Issues reflecting corporate experience will include knowledge of markets, customers, suppliers, contractors, distribution mechanisms, products and services, and the legal and regulatory constraints (compliance requirements) of the industry. This is not an exhaustive list and will vary depending on the service provided, market segment and the industry.

16.5.8 Reputation

One of the most valuable assets a business can have is its reputation. One significant measure of a business's reputation is its brand value. As discussed in Chapter 25, branding has value because it is a market growth tool. Successful branding can secure long-term competitive advantage. Customers can be persuaded that a product is different from its rivals to such a degree that they believe the rival product(s) to be inferior (regardless of whether this is the case or not). While brand value can be protected by trademark legislation, any action a business takes or any statement it makes can add to or eradicate brand value. Schmitt (2001) considers five interrelated aspects of effective reputation management. First, reputation management needs to be broadly conceived – the concept of branding having expanded from single products to the organisation as a whole; examples given are the Guggenheim Museum in New York and well-known leading figures such as Virgin Group's Richard Branson (see Branson 2002). Second, brand reputation is an ongoing undertaking and should not be confused with short-term crisis management. Third, the corporate brand has been discovered as an essential new marketing initiative. Businesses that traditionally focused on the branding of their individual products are now focusing on the organisation as a whole. Fourth, organisations need to take a unified approach to reputation management across the whole business by instilling the brand into its employees so they become familiar with it and live it in their day-to-day activities. On every face of the organisation to the outside world, whether it is trade fairs, news conferences, or communication with the public, the message needs to be consistent. Lastly, as a result of the internet, brand protection almost needs to be real-time, to cope with the new form of brand scrutiny. This requires effective management of the corporate website, links to other sites, selective presence on other websites and fast and adequate response to electronic queries. Schmitt also refers to the emergence of websites such as epinions.com which capture and organise public concerns and allow customers to express their opinions, make suggestions and post their complaints online. The web enables customers to quickly and easily convey their concerns to a very broad audience in minutes.

Reputation erosion from single or multiple events can present a serious risk to a business. A poor reputation can impede the sale of goods or services, harm the recruitment of high-calibre staff, deter desirable business partners, and/or make debt more expensive to obtain. Core value statements, which talk about trust and honesty, need to be shared and put into practice throughout an organisation. More and more consumers are interested in anything related to a brand, from the ingredients of its products to an organisation's behaviour on environmental issues and beyond to a wide range of other economic, social and political issues.

16.6 PEOPLE

A business must establish appropriate systems and controls for the management of *people risk* that may result from the actions of employees or the business itself. These systems and controls must be implemented throughout the duration of an employee's tenure with the organisation, beginning with recruitment and ending with the employee's resignation, retirement or termination. On the upside, people can differentiate one organisation from another and are an important source of competitive advantage. Figure 16.3 describes a possible taxonomy for people risk composed of seven attributes, each with their respective features.

16.6.1 Definition of People Risk

People risk may be described as a combination of the detrimental impact of employee behaviour (which may occur anywhere on the continuum between profit erosion and business failure) and employer behaviour (which impairs employee efficiency, health and safety or loyalty). In

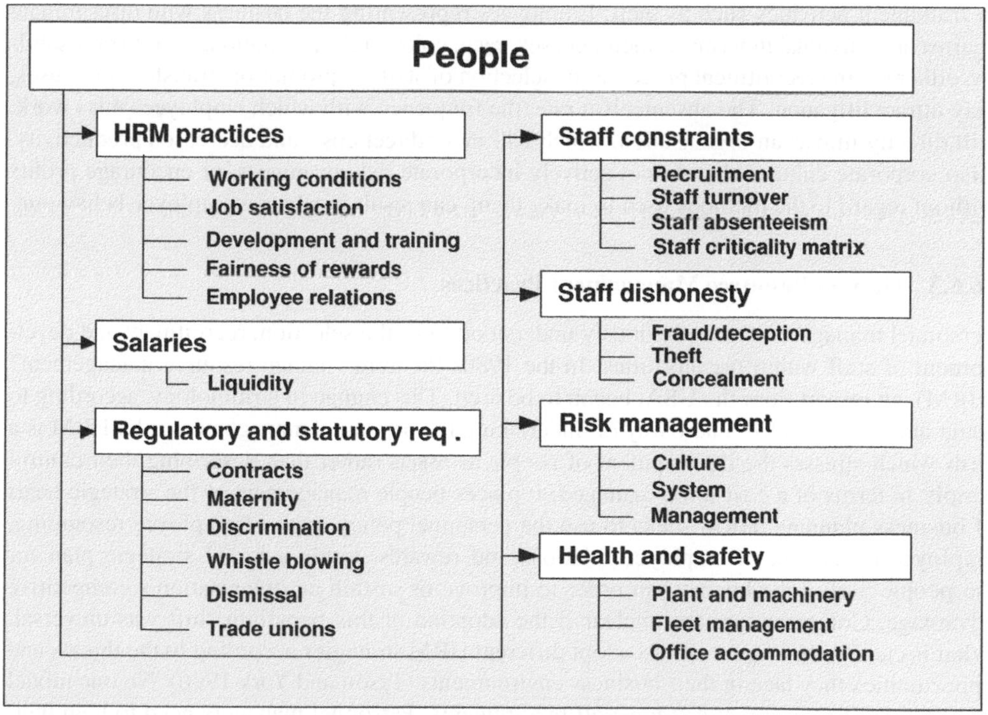

Figure 16.3 Taxonomy of people risk

simple terms, the impact of people risk can be described as having three levels of severity. At the lowest level people risk may be defined as events which, should they materialise, would erode profitability, share value and/or credit rating in the short term. At an intermediate level, people risk events may have a detrimental impact on the business's wealth or reputation in the long term. At the highest level, people risk events may bring about the eventual collapse of a business, as a result of the conscious or unconscious actions of one or more employees.

16.6.2 Types of People Risk

As employees have a high impact on business profitability, it is important to manage them effectively. The effectiveness of human resource management can be measured by absenteeism rates (staff constraint), labour turnover (staff constraint), accident rates (health and safety), productivity (management), quality of finished goods (management) and customer satisfaction (management). However, people risks are broader than solely efficient utilisation of employees, and typically result from human resource management practices, late payment of wages/salaries, lack of compliance with statutory or regulatory requirements, staff constraints, dishonesty, corporate culture (which does not cultivate risk awareness), risk management or poor health and safety. Staff constraints occur when *either* companies cannot fill new critical positions (because of shortages in particular trades or professions, and because compensation and other incentives are not sufficiently attractive to new candidates) *or* when staff retention is poor, leading to business disruption. Incompetence becomes an issue when employees lack the level of skills and knowledge to do their jobs correctly. Lack of professional training and development would further compound human errors. Dishonesty within a company can lead to fraudulent activities such as theft. Employees representing the business who discriminate against an individual in terms of their age, sex, race, colour, religion, national origin or disability, either during recruitment processes or selection of staff for promotion, transfer or bonuses, may attract litigation. The absenteeism rate, the frequency with which employees miss work, will directly impact an organisation, which will incur direct costs and decreased productivity. Also corporate cultures that do not actively incorporate risk awareness, or encourage profits without regard to the methods used to make them, can result in adverse employee behaviour.

16.6.3 Human Resource Management Practices

Personnel management was previously understood to be the selection, recruitment and development of staff within organisations. In the 1980s the term "human resource management" (HRM), an import from the USA, began to be used. This change in terminology, according to some authors, signalled a new way of undertaking the personnel management role: HRM is a term which stresses the development of people as assets rather than describing their control simply in terms of a cost to be managed; it places people management at the strategic heart of business planning. HRM seeks to use the personnel policy areas of employee resourcing, employee development, employee relations and rewards within a broad strategic plan for the people part of the business in order to improve or sustain an organisation's competitive advantage. Commentators are unclear if the adoption of this paradigm shift was universal. What is clear is that organisations adopt different HRM strategies according to the threats and opportunities they face in their business environments (Tyson and York 1996). No one model of personnel management can meet all requirements. Personnel managers need to help their organisations to adapt the existing HRM practices to the changing environment so that they can contribute to their organisation's goals. HRM can contribute to the improvement of working conditions, the creation of job satisfaction, the development and training of employees, the maintenance of harmonious relationships and the fairness of rewards. It can also assist with the threats to organisations, such as low productivity, unfair dismissals, absenteeism, accidents and social abuses (such as bullying), stress induced by unrealistic workloads, and sexual and racial discrimination.

16.6.4 Ability to Pay Salaries

Consideration of employee remuneration requires an assessment of the adequacy of the risk indicators and their ability to provide accurate and timely information upon which management can act. Payment of salaries is clearly a liquidity issue. The salary burden must be managed against the current and anticipated income, by balancing staff numbers accordingly.

16.6.5 Regulatory and Statutory Requirements

Contracts

For a contract of employment to exist one person must employ another to perform a particular task as part of their business in a manner that they dictate. The ordinary principles of the law of contract apply. So in a contract of employment there must be an offer and an acceptance, which is in effect the agreement. There must be: an intention to create legal relations, consideration, capacity, consent of the parties, and no mistake, misinterpretation, duress or undue influence. In addition, the contract must not be illegal. It must accord with the legislation included in Box 16.1. Certain written particulars have to be given to the employees to accord with the Employment Rights Act 1996 (ERA). Employers must provide all employees (full- or part-time) with written particulars unless the employee has entered into a written contract with the employer containing all of the relevant terms.

Box 16.1 Employment legislation

Individual/issue	Examples of protective legislation
Age	Code of Practice on Age Diversity in Employment
Agency workers	Working Time Regulations 1998, National Minimum Wage Regulations 1999
Carers	Maternity and Parental Leave etc. Regulations 1999
Colour	Race Relations Act 1976 (RRA)
Disability	Disability Discrimination Act 1996
Employee representative status	Employment Rights Act 1996 (ERA)
Employees exercising a statutory right	ERA
Employees on fixed-term contracts	Fixed–Term Employees (Prevention of Less Favourable Treatment) Regulations 2002
Ethnic group	RRA
Gender	Sex Discrimination Act 1975 (SDA)
Gender reassignment	SDA
Health and safety representative	ERA
Marital status	ERA

Individual/issue	Examples of protective legislation
Maternity	ERA and Maternity and Parental Leave etc. Regulations 1999
Nationality	RRA
Parental status	Maternity and Parental Leave etc. Regulations 1999
Part-time employees	Part-Time Workers (Prevention of Less Favourable Treatment) Regulations 2000
Pregnancy	ERA and SDA
Race	RRA
Religion	Planned legislation
Reservists	Reserve Forces (Safeguard of Employment) Act 1985
Sexual orientation	Planned legislation
Shop workers who refuse to work on Sundays	ERA
Trade union membership	Trade Union and Labour Relations (Consolidation) Act 1992
Trustees of occupational pension schemes	ERA
Unfair dismissal	ERA
Whistleblowing	Public Interest Disclosure Act 1998

Source: Osman (2003).

Maternity

Under the ERA a pregnant employee who has, on the advice of her doctor, midwife or health visitor, made an appointment to obtain antenatal care must have the time off to keep it and must also be paid. An employer, who, acting unreasonably, does not give the employee these rights can be taken to a tribunal by the employee, but this must normally be during the first three months following the employer's refusal. Employees with two or more years' service are entitled to maternity leave from the 11th week before the birth, with the right to return up to 29 weeks after the birth, with statutory maternity pay payable for 18 weeks plus paid time for antenatal care.

Discrimination

The number of employers subject to litigation arising from perceived or actual contravention of employment legislation governing discrimination has risen sharply in recent years. Discrimination is prohibited in relation to those protected regarding recruitment, pay and benefits, promotion, training, terms and conditions, transfers, dismissal, action short of dismissal and any other detriment.

Whistleblowing

Employers have to take account of the relatively new form of influence on management, whistleblowing, which describes the practice of an employee metaphorically blowing a whistle to draw the attention of those outside of the business to some form of unethical practice inside the business. Previously this was done by individuals taking a personal risk with their employment. After the enactment of the Public Interest Disclosure Act 1998, workers who "blow the whistle" about any wrongdoing within their employer's organisation are protected (as far as the "umbrella" of the Act extends). The provisions within the Act protect workers from being subjected to what it calls "a detriment" by their employer. The DTI guidance[6] states that "detriment may take many forms such as denial of promotion, facilities or training opportunities which the employer would otherwise have offered". Additionally, employees who are "protected" by the provisions may make a claim for unfair dismissal if they are dismissed for making a protected disclosure. A qualifying disclosure will be a protected disclosure where it is made to the worker's employer or to a person whom the worker reasonably believes to be solely or mainly responsible for the relevant failure. Particular kinds of disclosures qualify for protection which collectively are termed "qualifying disclosures". Qualifying disclosures are disclosures of information which the worker reasonably believes show that one or more of the following issues is currently taking place, previously took place or is likely to happen in the future: a criminal offence; the breach of a legal obligation; a miscarriage of justice; a danger to health or safety of any individual; damage to the environment; or the deliberate concealment of information tending to show any of the issues just referred to.

Dismissal

Dismissal of a member of staff is never just about managing one individual. Methods of dismissal become part of a business's culture, can modify the behaviour of remaining staff and, in the wider context, if a trend emerges, can increase or decrease the attractiveness of the business to potential employees. The risk for any employer in releasing a member of staff is not adhering to the prevailing legislation relating to wrongful and unfair dismissal. An employee can claim that they were wrongfully dismissed in a common law action for breach of contract in the civil courts or in an industrial tribunal. This is relevant where the employee claims that the employer did not dismiss him or her in accordance with their contract. An example of this type of claim is where the employer had failed to give proper notice as recorded in the contract. The amount of the compensation or damages awarded would normally aim at placing the employee in the financial position in which they would have been, had the wrongful dismissal not taken place. Additionally, businesses have to be mindful of not being guilty of unfair dismissal. Since the Trade Union and Labour Relations (Consolidation) Act 1992, as amended by the Employment Rights (Dispute Resolution) Act 1998, employees have the right not to be unfairly dismissed. The grounds for classifying dismissal as being unfair as defined by the Act include: employees taking or seeking maternity, paternity or adoption leave; requesting flexible working arrangements; seeking to assert a statutory employment protection right; taking or proposing certain types of action on health and safety grounds; and performing or proposing to perform duties relevant to his or her role as an occupational pension scheme trustee.

Trade Unions

The change in how businesses deal with unions has been brought about by the changing political, economic and industrial context in which businesses trade and not by a shift in management ideology (Pinnington and Edwards 2000). The Conservative government, elected

[6] http://www.dti.gov.uk/er/individual/pidguide-pl502.htm

in 1979 with Margaret Thatcher as Prime Minister, was committed to reducing union power which it saw as one of the primary causes of Britain's poor performance internationally. Consequently, through numerous pieces of legislation between 1979 and 1997, the government restricted the ability of the unions to take industrial action and regulated their internal affairs in a way that the law had not previously done. The government also reduced the influence of the unions in the public sector, particularly at its intelligence centre known as GCHQ.

16.6.6 Staff Constraints

While it might seem self-evident that businesses have to attract and retain staff for business development, as Peter Drucker has said:

> Any experienced executive knows companies or industries [are] bound for extinction because they cannot attract or hold able people. Every experienced executive also knows that this is a more important fact about a company or an industry than last year's profit statement.

Drucker went on to say that any manager who considered that staff retention "was a 'nonproblem', would be quickly – and correctly – dismissed as an ass" (Drucker 1979).

> All organisations now say routinely, "People are our greatest asset." Yet few practice what they preach, let alone truly believe it. Most still believe, though perhaps subconsciously, what nineteenth-century employers believed: people need us more than we need them. But, in fact, organisations have to market membership as much as they market products and services – and perhaps more. They have to attract people, hold people, recognise and reward people, motivate people, and serve and satisfy people. (Drucker 1992)

Recruitment

The recruitment process will have a direct impact on the quality of the staff employed, the future retention of existing valued client relationships (and associated repeat business) and staff retention. There is a direct correlation between recruitment and the success of a business. The key aspects of recruitment are: recruiters, job analysis, job descriptions, interviews, selection, induction and integration.

Recruiters. The effectiveness of the recruitment process will be influenced by the behavioural characteristics of the recruiter in terms of whether they are personable, enthusiastic and competent. Other desirable qualities of the interviewer include humility, maturity, the ability to think objectively, listening (not overtalking) and freedom from extreme opinions. The risk exists that a new person in the human resources department or a line manager may be given a recruitment assignment, before that individual has been given interview training. In addition, novice recruiters may not be aware of the mission, goals, structure and services/products of the organisation, the job requirements and the corresponding experience required. The business will be exposed to poor selection if the interviewer does not establish an interview plan, maintain rapport, listen actively, pay attention to non-verbal clues, provide honest and full answers to questions, use questions effectively, split facts from inference, avoid overselling the position, control the interview, consistently ask the same questions of all candidates, and avoid biases and stereotypes. One typical bias is for interviewers to consider candidates who have interests, experiences and backgrounds similar to their own to be more acceptable. Stereotyping involves forming generalised opinions of how people of a given gender, race or ethnic background think, feel and act. A further trap is the influence of "beautyism": discrimination against unattractive persons and favouritism towards attractive individuals is a persistent and pervasive form of employment discrimination.

Job analysis. Gaining a clear understanding of the full extent of the role that an organisation wishes to fill is critical to describing the job description to match that role. When analysing a job through *job analysis*, an organisation needs to understand: the *roles* the job holder has to fill; the *tasks* the job holder will have to do; and the *skills* the job holder will need to carry out to complete the tasks allocated to the role. The procedure of job analysis involves undertaking a systematic investigation of jobs by following a number of predetermined steps specified in advance of the study. When completed, job analysis results in a written report summarising the information contained in the analysis of somewhere between 20 and 30 activities. Human resource managers will use this data to develop job descriptions. The ultimate purpose of job analysis is to improve organisational performance and productivity. Job data may be obtained in several ways. The more common methods of analysing jobs are through interviews, questionnaires, observation and diaries. Several different job analysis approaches are used, each with specific advantages and disadvantages. Three of the more popular methods are functional job analysis, the position analysis questionnaire system and the critical incident method (Sherman *et al.* 1998).

Job descriptions. Job descriptions are a valuable tool in performing human resource functions. To be effective, job descriptions must use statements that are terse, direct and simply worded. Several problems are frequently associated with these documents, including:

- being poorly written, using vague rather than specific terms;
- providing little guidance to the job holder;
- not being updated as the job duties change;
- violating the law by containing aspects not related to job success;
- limiting the scope of the activities of the job holder through the provision of unnecessary constraints; and
- being economic with the truth and only portraying the favourable aspects of a position.

On this last point, Sherman *et al.* consider that organisations may be able to increase the effectiveness of their recruitment efforts by providing job applicants with *realistic job previews* (RJPs). An RJP informs applicants about all aspects of the job, including both its desirable and undesirable facets. Proponents of the RJP believe that applicants who are given realistic information regarding a position are more likely to remain in the position and be successful, because there will be fewer unpleasant surprises (see Box 16.2). When there are skills shortages the recruitment process takes on even greater significance.

Box 16.2 Realistic job descriptions

The importance of the management of expectations is explained in an article by Chris Wyche, who describes British Airways' approach to their recruiting difficulties in the early 1990s. At the time British Airways (BA) had the largest centralised commercial recruitment operation in the UK, recruiting nearly 5000 people each year. For some time BA had been finding it increasingly difficult to find skilled recruits in areas such as information technology, finance and engineering. In addition, there had been a clear downturn in the supply of skilled young people. All of these trends were occurring side by side and concurrently with a demand for skilled labour driven by business growth. The widening gap between supply and demand led BA to create a recruitment marketing team. In addition, BA established quality standards for the marketing team, developed different training programmes for line managers to help increase their understanding of the market place and engaged an advertising company to assist the recruitment drive. In their efforts to promote BA as the first choice among employers, the recruitment department made special efforts to maintain the delicate balance between projecting the genuine opportunities of working for a company of the size and diversity of BA, and the tendency to paint too rosy a picture of the realities of working within a large organisation. This was seen as especially important since retaining talented employees in a diminishing labour market was perhaps more important than attracting them in the first place.

Source: Wyche (1990).

Interviews. Interviews remain the mainstay of the selection process. However, they can be plagued by problems of subjectivity and personal bias.

Selection. Selection must contain a screening process to establish the fitness and propriety of employees, including their honesty, integrity, reputation, competence, capability and financial soundness. Screening should be more stringent for employees who will occupy positions of high personal trust (e.g. security administration, payment and settlement functions); and people occupying positions requiring significant technical competence (e.g. geologists and pharmaceutical research scientists).

Induction of employees. A risk to any business is a poor induction process. It seldom receives the very careful attention it deserves. It makes good sense to help new recruits to integrate into their new surroundings and hence become efficient and effective in their work as quickly as possible. Failure to do so can, at the very least, lead to erratic progress and, at the other extreme, the possible loss of customers. Inadequate induction is also recognised as a significant contributor to turnover during the first year of employment. The problems of social adjustment that newcomers have to face are simply not appreciated or sympathetically handled. Newcomers experience loneliness and a sense of disorientation when coping with unfamiliar surroundings. The format and content of an induction programme will vary according to the size and type of organisation and the existing knowledge, experience and seniority of the recruit. It often consists of two stages: organisation-wide induction followed by business unit induction. The purpose of induction is to ensure that new employees (Institute of Management 1999):

- are integrated into their working environment as quickly as possible;
- learn the relevant aspects of the organisation's mission, culture, policies, procedures and methods of working;
- become productive and well motivated;
- become aware of the skills and knowledge needed for the job; and
- understand their responsibilities.

Induction of directors. A more significant risk to a business is the poor induction of its directors. Newly appointed directors need to be inducted efficiently so that they can quickly familiarise themselves with the company's activities and begin to apply their skills and experience, for which they have been appointed, for the benefit of the company and its shareholders.

The need for a proper induction process for this purpose has long been recognised: the Cadbury Report made it clear in 1992 that newly appointed board members are entitled to expect proper induction into the company's affairs. Despite this recognition, a telephone survey carried out in 2002 for the Higgs Review found that less than a quarter of non-executive directors received a formal briefing or induction after appointment (Higgs 2003). Commenting that the current position is not acceptable, the Higgs Review concluded that companies must set aside adequate resources and ensure that sufficient time is available for a thorough induction of new directors. It is recommended that the chairman should take the lead in providing a properly constructed induction programme, which should be facilitated by the company secretary.

The Higgs Review suggested that each company should develop its own comprehensive formal induction programme tailored to the needs of the company and individual directors. It recommended that a combination of selected written information should be made available together with presentations, meetings and site visits to provide new appointees with a balanced real-life overview of the company. The Review also suggested that the induction process should contain three main elements:

- Build an understanding of the nature of the company, its business and the markets within which it operates covering:
  - the company's products or services;
  - group structure including subsidiaries and joint ventures;
  - the company's constitution, board procedures and matters reserved for the board;
  - summary details of the company's major risks and management strategy;
  - key performance indicators; and
  - regulatory constraints.
- Build a link with the company's people including:
  - meetings with senior management;
  - visits to company sites other than headquarters, to learn about production or services and meet employees in an informal setting; and
  - participating in board strategy development.
- Build an understanding of the company's main relationships, including meeting the auditors and developing a knowledge of in particular:
  - who are the major customers;
  - who are the major suppliers; and
  - who are the major shareholders and what is the shareholder relations policy – participation in meetings with shareholders can help give a first-hand feel as well as letting shareholders know who the non-executive directors are.

The Institute of Chartered Secretaries and Administrators (ICSA) worked closely with the Higgs Review team on the creation of that checklist, in order to keep it brief and to the point. The ICSA has produced and undertaken to maintain on its website (http://www.icsa.org.uk) a guidance note detailing a full list of appropriate induction material. The guidance note describes "essential information to be provided immediately" which is subdivided under three headings: "directors' duties", "the company's business" and "board issues". It also describes "additional material to be provided during the first few months" and "additional information which the company secretary might consider making the director aware of".

Staff Turnover

Regardless of technological change, the risk of staff turnover will always haunt enterprises. The loss of key personnel within, for example, research and development departments of pharmaceutical businesses, information technology and software houses, high-profile design consultancies, prominent advertising agencies, and leading edge technology manufacturers can be particularly debilitating. The loss of key personnel has led to the abandonment of the development of projects and new products as the holder of the knowledge to bring these products to the market place has walked out of the door. Many personnel experts believe that the majority of employee resignations occur for common, easily recognisable and avoidable reasons. While there is no specific list that can help an employer in any one industry, there are common reasons why people leave a place of employment. Managing staff turnover must be part of managing a business to secure its continued growth and wealth creation. The widely recognised benefits of managing turnover include:

- retention of top performers critical to the success of the business;
- retention and development of the organisation's knowledge base;
- retention of information which would be useful to a competitor;
- a reduction in advertising and recruitment agency costs;
- improved employee morale;
- a reduction in the time managers are diverted away from activities directly contributing to business creation;
- a reduction in the amount of time existing staff spend in inducting and training new staff; and
- a reduction in the time that staff are not fully productive in terms of those working their notice or new arrivals.

The most cited reasons for turnover include alternative employment opportunities, pay and compensation, the lack of perceived fairness of the distribution and the equity of rewards such as bonuses, lack of professionalism, inadequate communication of the responsibilities of the role, the culture of the organisation in terms of overall management, degree of office politics, poor peer group relations, lack of promotion opportunities, inadequate working environment and facilities, and harassment. This is not an exhaustive list. In addition, an individual normally leaves an organisation for a combination of reasons and usually not just, say, an increase in salary.

By taking a simple systems perspective (see Figure 16.4), it is easier to understand the source of the catalysts of staff turnover and how they may combine to lead to a member of staff to leave and join a new employer, join an agency as a freelance consultant or start their own business (Chapman 2002).

Ostergen (1989) carried out a study of "key persons" in "knowledge-intensive firms". She defined key persons as people with technical and managerial roles and she chose computer businesses as her subject firms. Ostergen interviewed a number of key persons in a variety of firms and categorised the different types of direct and indirect consequences suffered when key persons leave these "knowledge-intensive firms" (Box 16.3).

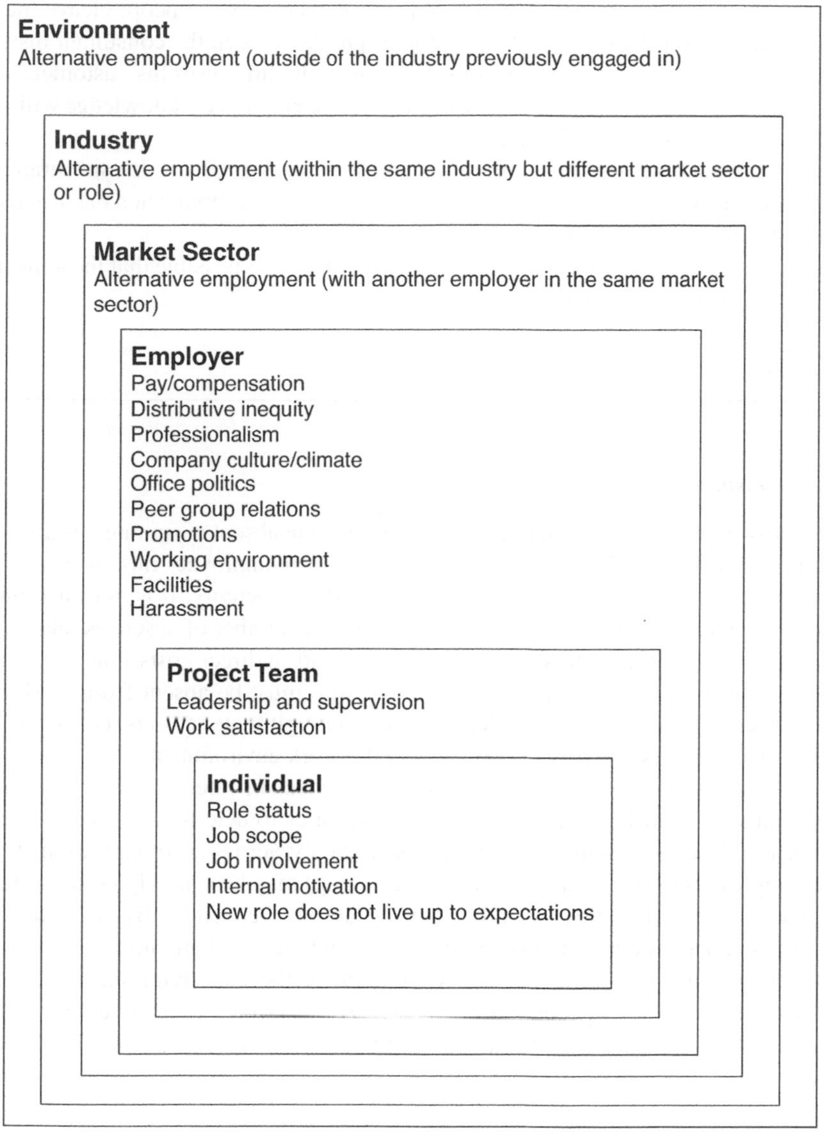

Figure 16.4 Systems perspective of sources of turnover

Box 16.3 Loss of key personnel

1. The first loss is that of human beings, we can call it "mass-escape". When some of the key persons leave the firm to start something of their own, a lot of people feel that it's safer to go with them than to stay at the disintegrating firm. Another reason why a great number of people follow these key persons is because they feel the pressure from the group.

2. Besides "mass-escape" the firm will lose markets when several people leave the firm, let's call it "market-escape". Often a customer prefers to keep the consultant instead of hiring a special firm. Hence, if the consultant quits, the firm loses his customer.
3. A third kind of escape is "technology-escape", i.e. the employee's knowledge will vanish and also future progress in this area.
4. We can also mention "legitimacy-escape". The firm loses goodwill. It's not strange that rumours circulate when staff leave. This can lead to problems both when the firm recruits staff and when it sells its services.
5. Last we can discuss "culture-escape". When staff leave, it's possible that the atmosphere will change and result in a new culture.

Source: Ostergen (1989).

Staff Absenteeism

How frequently employees are absent from their work – the absenteeism rate – is also directly related to human resource planning and recruitment. When employees miss work, the organisation incurs direct costs of lost wages and decreased productivity. It is not uncommon for organisations to hire extra workers just to make up for the number of absences totalled across all employees. In addition to those direct costs, there are indirect costs that may underline excessive absenteeism. There will always be some who must be absent from work because of sickness, accidents, serious family problems or other legitimate reasons. However, chronic absenteeism may signal some deeper problems in the work environment.

While an employer may find that the overall absenteeism rate and costs are within an acceptable range, it is still advisable to study the statistics to determine if there are patterns in the data. Rarely does absenteeism spread itself evenly across an organisation. It is very likely that employees in one area (or occupational group) may have nearly perfect attendance records, while others in a different area may be absent frequently. By monitoring these differential attendance records, managers can assess where problems might exist and more importantly begin planning ways to resolve or improve the underlying causes. Employers look at "stick and carrot" regimes where progressive discipline procedures are applied to employees having a recurring record of absenteeism whereas incentives are provided for perfect attendance.

Staff Criticality Matrix

A staff criticality matrix can be developed which will show visually which individual is carrying out a critical task during any one phase of an activity and hence whose unplanned or unexpected departure could threaten the completion date of the overall activity. In this way consideration can be given to the proactive measures to be taken to compensate for the loss of a key member of staff at a critical moment in the completion of an activity.

The example below in Box 16.4 describes an activity that is composed of four phases and requires four different individuals to carry out separate interlinked tasks to complete the activity within the required timeframe. The tasks range from "Comparing" through to "Synthesising" and they have been ranked in order of importance to completing the activity, where 1 is the least important and 6 is the most important.

Box 16.4 Staff criticality matrix

Task	Importance ranking
Synthesising	6
Coordinating	5
Analysing	4
Compiling	3
Copying	2
Comparing	1

	Phase A	Phase B	Phase C	Phase D
Employee A	6	4		
Employee B	5	5	5	5
Employee C	1	1	2	3
Employee D	1	1	2	3

16.6.7 Staff Dishonesty

One of the most prominent cases of dishonesty was the fraudulent activity of Nick Leeson, who through his actions brought about the demise of the 233-year-old Barings Investment Bank. Leeson was the chief derivatives trader at Barings, managing an operation in futures markets on the Singapore Monetary Exchange buying and selling derivatives futures pegged to the Nikkei 225 stock index. He ran up liabilities of $1.38 billion hidden in his "88888" account, more than the entire capital reserves of the bank. Investors saw their savings wiped out and some 1200 employees lost their jobs. The Dutch bank ING agreed to assume nearly all of Barings' debt and acquired the bank for the princely sum of £1. In December 1995 a court in Singapore sentenced him to $6\frac{1}{2}$ years in prison after he had pleaded guilty to two counts of deceiving the bank's auditors and cheating the Singapore exchange. The aftershocks of Barings' fall prompted a complete overhaul of financial regulation in the City of London and a re-examination of risk controls by investment banks. However, outside the UK, rogue trading events continue to emerge. In 2002 US currency trader John Rusnak was charged with covering up $691 million (£474 million) of trading losses so that he could boost his own earnings. He was indicted by a federal grand jury on charges of bank fraud, false entry in bank records and aiding and abetting. The indictment followed a four-month investigation into trading at Allfirst Financial, a subsidiary of Allied Irish Bank (AIB), from 1997 to 2001. In 2010 rogue trader Jérôme Kerviel of Société Générale was found guilty of forgery, breach of trust and unauthorised computer use and was sentenced to three years in prison, banned from trading for life and ordered to repay the bank's losses of €4.9 billion. The trial centred on the allegation that Jérôme had bet €50 billion of Société Générale's money without the bank's knowledge.

16.6.8 Risk Management

The effectiveness of operational risk management and its contribution to securing a business's objectives will be directly proportional to the way it was initially established and how it has been subsequently maintained.

Risk Management Culture

Business culture, expressed in simple terms, is a business's accepted way of doing things. Risk management *culture* is a subset of business culture. It is the pervasive business-wide view of how risk management should be implemented. It is all about beliefs, attitude, judgement, approach and outlook, which manifests itself through employee behaviour. It is an all-embracing term which covers an array of issues such as board promotion of risk management, working methods, appetite for risk, lines of reporting, the delegation of responsibility and the frequency of reporting. Culture will dictate whether risk management is considered an administrative burden, a task to be undertaken to satisfy regulatory requirements or a process to improve the chance of securing the business's objectives. Most importantly, a business's risk management culture encompasses the general awareness, attitude and behaviour of its employees to risk and the management of risk within the organisation. Hence, an organisation's risk management culture can be considered a measurement of how successfully risk management has been embedded within that organisation. Indicators of a prevalent risk management culture are leadership, sponsorship, risk management system, allocation of responsibilities, reporting, training, assessment of effectiveness and approach to continuous improvement. The answers to the following (and similar) questions will determine the nature of an organisation's risk management culture. This is a very small sample of review questions the author has prepared for evaluating cultures.

- Which executive is responsible for operational risk?
- What is the role of the audit function in overseeing operational risk?
- What is the role of non-executive directors in the execution of operational risk?
- What are the objectives of the risk management process?
- Is a training and education programme established for each level in the business, which has a duty to identify, assess and report risks to management?
- Has a risk management system been established?
- Is the framework applied at all levels of the organisation?
- Is there a database of losses to inform ongoing risk identification and assessments?
- Is guidance provided on how to assess potential operational risks against expected profits, when making significant business decisions?
- Is there guidance on the retention of records of risk identification, assessment, evaluations, response planning and management?
- Is the risk culture maintained?
- Is a format for reports described to ensure the results are easy to understand and quick to assimilate?
- Have senior management acceptable levels of exposure to specific risk types?
- How is the operational risk process benchmarked?
- What process is adopted to secure continuous improvement?
- Is there consistency between the reward structure for staff and the risk culture?

System

The level of maturity or sophistication of a business's risk management *system* will influence the effectiveness of risk management within the business. But what do we mean by risk management system? The term "system" is used here as the collective noun for the group of action plans consisting of the *strategy*, *framework*, *policy*, *process* and *profile*. While the literature on operational risk refers to these terms, descriptions are brief and/or inconsistent. The meaning adopted here is as follows:

- *Risk strategy* is a description of the overall objective of the risk management process, normally expressed in terms of its contribution to the business objectives.
- *Risk framework* is the overall plan of implementation of risk management, which includes a combination of the policy, profile, process and exposure. This description is consistent with the Basel II description of a framework which states: "The framework should cover the bank's appetite and tolerance for operational risk, as specified through the policies for managing this risk, including the extent and manner in which operational risk is transferred outside the bank. It should also include policies outlining the bank's approach to identifying, analysing, monitoring and controlling/mitigating the risk" (Basel Committee on Banking Supervision 2004, item 737, p. 161). It is also consistent with the M_o_R definition of a framework, which spans issues such as policy and process – where process is understood to consist of identification, analysis, ownership, reporting, responding and management (Office of Government Commerce 2002). Frameworks are not static documents and must be updated to reflect significant changes to a business's organisational structure, outsourcing arrangements, products or services, operating environment, acquisitions or mergers, geographical area of operation, funding arrangements and customer base. This should cover issues such as:
  - Are the objectives of the framework clearly stated?
  - Does the framework state how operational risk contributes to corporate governance?
  - Does the framework contribute to securing the business objectives?
  - Does the framework influence decision making?
  - Is the vocabulary adopted within the framework explained?
  - Is the framework comprehensive, describing the risk scope, risk policy, risk profile and risk process?
  - Is the framework sufficiently clear, and is the vocabulary adopted used consistently?
  - Has the framework been explained to the board, business department heads and line management, and is it understood?
  - Does the framework state the risk taxonomy adopted and hence the elements, attributes and features being addressed?
  - Is an individual or a committee assigned the responsibility of maintaining the framework?
  - Does the framework state how risks are quantified?
  - Does the framework (depending on the business of the firm) state how the regulatory capital to be set aside for risk losses is to be calculated?
  - Has an independent review of the framework been undertaken to evaluate it against best practice?
- *Risk policy* is a statement of the operational risks that the firm is prepared to accept and those that it is not prepared to accept, and, where possible, the acceptance threshold (the firm's risk appetite/tolerance). The policy should describe the business's appetite for all classes of risk and their corresponding elements, otherwise the policy will be too broad and difficult to administer.
- *Risk profile* describes the types (classes and elements) of operational risk that are faced by the firm and its clients and its exposure to those risks:
  - Has a risk taxonomy been prepared which has been tailored to the business's operating context?

- How frequently is the profile reviewed and updated?
- Is the profile used in the risk identification process?
- How is competition in the market place monitored?
- How is regulatory compliance monitored?
- How is consumer behaviour monitored?
- Are sales of the business's products or services immune from changes in the economy or do sales mirror economic upturns/downturns?

- *Risk process* consists of how the firm intends to identify, analyse and evaluate its operational risks; implement risk response planning and management, allocate risk owners, managers and actionees; determine the method of deriving internal capital allocation from operational risk assessments, the allocation of responsibilities for managing and implementing the process, the establishment of thresholds for particular operational risks (based on predetermined risk appetite/tolerances) which when reached trigger a response (review and management action); guidance on how boundary risks would be treated, incident management and escalation requirements, reporting requirements, definition of the terms used and quality assurance requirements.
  - Is it clear who has responsibility for risk identification, analysis, evaluation, response planning and management?
  - Are new services and products critically evaluated for their risk exposure?
  - Are contracts with customers and third parties evaluated for risk ownership?
  - Are risk management models and assessment tools independently checked?
- *Risk exposure* relates to the extent of operational risk faced by a firm and is usually expressed in terms of either the likelihood and impact of a particular type of operational loss such as fraud, or the aggregated impact of all of the risks identified.

Section 16.6.8 should be read in conjunction with Section 1.9.4 headed "Risk Management Framework", which describes the generic outline of a framework, and Figure 1.2, which shows the relationship between the framework and other RM documents.

Management

The success or otherwise of the operational risk *management* system will be influenced by the way in which the following series of issues are addressed: creation of an operational risk framework, integration of the framework with corporate governance of the firm as a whole, and management through the allocation of responsibilities. Having an operational risk framework is not an end in itself. The answers to the following (and similar) questions will determine the effectiveness of the management processes. This is a very small sample of review questions the author has prepared for evaluating operational risk frameworks.

- *Risk process*. The degree of robustness of the risk management process will also directly influence the effectiveness of an operational risk system. Robustness will depend on whether the process represents a logical sequence of activities involving: comprehensive identification (with no blind spots); realistic analysis (which resembles reality as closely as possible); appropriate evaluation methods (which remove double-counting, reflect correlation between risks, take account of interrelationships and assign appropriate probability distributions to risks); evaluate appropriate responses and actively drive through risk response actions. Risk identification needs to be vigorous. The business's end-to-end operating cycle needs to be examined for the sources of risk or, where construction projects have been commissioned, the project life cycle. The data used for identification and analysis needs to be examined for its integrity (honesty), relevancy and sufficiency. Robustness will be influenced by the timeliness of the process – how frequently risk registers are updated, risk assessments are revalidated, and the success or otherwise of risk actions is monitored. Companies that proudly announce in their annual reviews that they review their operational risks once a year have obviously lulled themselves into a false sense of security or their risk management aspirations do not go beyond responding to regulatory pressures and obtaining a "tick in the box".

- *Enforcement*. Human resource procedures and processes can be used in a direct way to embed a risk management system. It can be enforced by a series of measures such as employees' contracts of employment, induction, risk management training, appraisals and performance reviews. Contracts of employment can make explicit the statutory and regulatory content of the firm and its risk profile. They can include reference to those actions of an individual that would lead to either disciplinary action or termination of the contract. These could be actions that would be harmful to the business in terms of solvency, reputation or customer relations, attracting legal proceedings and/or invoking regulatory sanctions.
  - Are risk management responsibilities (in line with the framework) included in job descriptions?
  - Are the ramifications of a lack of adherence to risk management responsibilities included in terms of employment?
  - Are employees (including new board members and non-executive directors) introduced to the operational risk framework during induction?
- *Risk training*. Risk training answers the questions why, how, when and who. Training increases comprehension of the risk management system by providing: an understanding of the objectives and benefits; comprehension of the terminology; familiarity with the process; and the ramifications of inactivity. Knowledge can increase the willingness to participate, apply the process and take the initiative to improve understanding and efficacy of the process.

Training and Development for Non-Executives

Boards face the risk that non-executive directors are ill equipped and have insufficient expertise to fulfil their obligations as members of a unitary board. The survey completed by the Higgs Review revealed that two-thirds of non-executive directors and chairmen had not received any training or development. The Review explains that in this context the word *training* means continued professional development. The Review stated that for existing directors, knowledge of issues such as strategy, management of human and financial resources, audit and remuneration can often usefully be updated and expanded with updates on legal and regulatory obligations being helpful. The survey also found that some non-executive directors were concerned about the increasing amount of knowledge necessary in order to fulfil their roles on board committees. As mitigation, following the recommendation of the Review, non-executive directors should regularly appraise their individual skills, knowledge and expertise and determine if tailored professional development would help them develop their expertise and meet their board obligations.

- Are employees informed of corporate governance requirements, the Combined Code and the Listing Rules?

- Is the training refreshed at regular intervals?
- Does the training cover the entire system, including identification, assessment, evaluation, response planning and management?
- Are staff appraised of in-house quantification methods (where developed)?
- Is the risk management framework readily accessible on the company intranet?
- Are the risk indicators and the risk indicator templates saved on the intranet?

16.6.9 Health and Safety

Businesses commonly are faced with a multifaceted health and safety task as a result of a number of different workplace "environments" relating to (for example) plant and machinery, fleet management and office accommodation. The Health and Safety Commission (HSC) and the Health and Safety Executive (HSE) are responsible for the regulation of almost all the risks to health and safety arising from work activity in Britain. Their collective mission is to protect employees' and the public's health and safety by ensuring risks in the workplace are properly controlled. The HSE and HSC look after the health and safety in nuclear installations, mines, factories, farms, hospitals, schools, offshore oil and gas installations, the national gas grid, the movement of dangerous substances, railway safety and other aspects of the protection of both employees and the public. Local authorities are responsible to the HSC for the enforcement of health and safety in offices, shops and other parts of the service sector. Health and safety management, while being a statutory requirement for businesses, can be used to increase operational efficiency, enhance working environments and improve financial performance. The HSE statistics show that there were 235 fatal injuries and 159 809 other recorded injuries in 2003/2004.[7] Additionally, in the same period, 30 million working days were lost due to work-related ill health from 2.2 million people suffering from an illness which they believed was caused or made worse as a result of their current or past work. There are obvious benefits for both employees and employers if injuries and ill health are reduced. Effective health and safety management clearly helps reduce the unacceptable toll of suffering, anguish and disabilities that accidents and ill health bring. For businesses, compliance with legislation and the implementation of robust health and safety practices minimise the likelihood of prosecution and penalties, reduce disruption to operations, afford the opportunity of achieving a cost-effective and efficient use of resources and reduce the incidence of litigation.

16.7 PROCESSES AND SYSTEMS

A business should establish and maintain appropriate systems and controls for the management of operational risks that specifically arise from failures or inadequacies in management processes and systems. These systems and controls commonly span a business's end-to-end operating cycle. Hence, they need to reflect the sequence of activities undertaken, be integrated, reflect customer requirements and be regularly updated to reflect changes in the market place. Figure 16.5 describes the structure of Section 16.7 and a possible taxonomy of the process and systems element of operational risk.

[7] http://www.hse.gov.uk

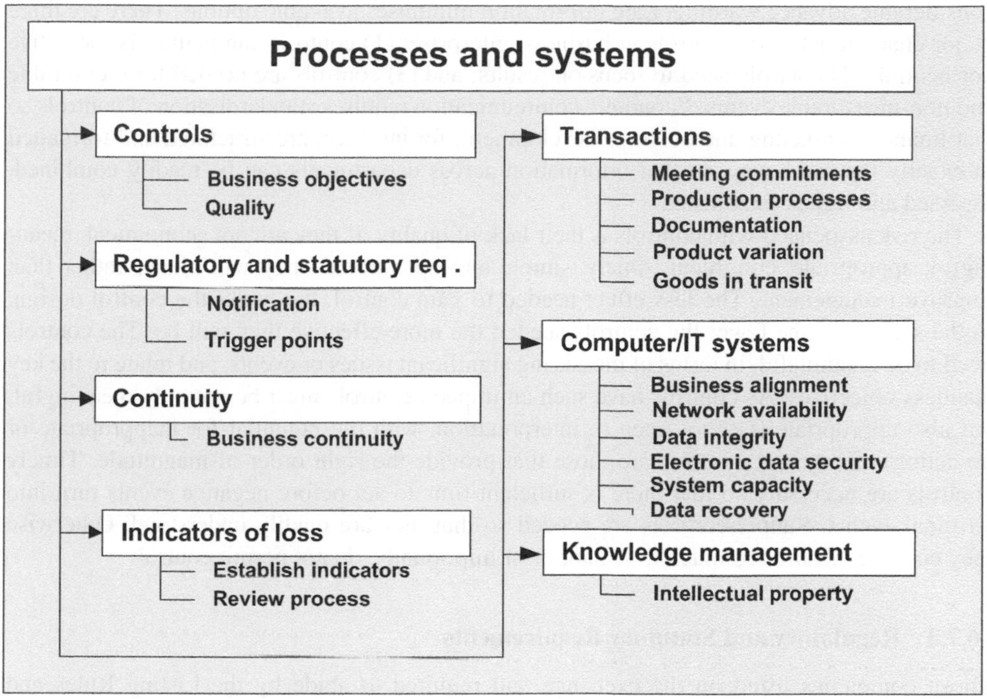

Figure 16.5 Taxonomy of processes and systems risk

16.7.1 Definition of Processes and Systems Risk

Processes and systems risk may be defined as the failure of processes or systems due to their poor design, complexity or non-performance, giving rise to operational losses. As a result a business may experience a wide range of problems, including an inability to meet orders, poor quality control, settlement processing errors, fraud and information security failure. As the result of the increasing reliance on company-wide computer systems, IT now has the potential to transform risks from minor manual processing errors to major disruptions, simultaneously stopping hundreds of individuals from working.

16.7.2 Controls

While it is not made explicit in dictionary definitions, "controls" in a business sense is not the plural of "control". Perhaps the best way to describe the distinction between the two is that controls pertain to means and control to an end. Not only do more controls not necessarily give more control, but the two words have different meanings altogether. The synonyms for controls are measurement and information. The synonyms for control are direction and constraint. Controls relate to recording events in the past. Control deals with providing strategic direction, future events. Hence, financial controls relate to recording past financial events to provide the controllers with information to decide about future direction. Accordingly, controls have an important time element. Like a ship, a change of direction or stopping an activity requires considerable advance warning. Late information minimises available options. There are three major characteristics of controls in business enterprise: (1) controls can neither be objective nor neutral; (2) controls need to focus on results; and (3) controls are needed for measurable and non-measurable events. Economic communication requires standardisation of controls so that finance, marketing and business development, for instance, are structured and formatted in exactly the same way, so that information across departments can be readily combined, assessed and trends established.

The risk associated with controls is their lack of quality. If they are not economical, meaningful, appropriate, congruent, timely, simple and operational, they will hinder rather than improve management. The less effect needed to gain control, the better the control design. In the same way the fewer the controls needed the more effective they will be. The controls need to be meaningful, in terms of measuring significant issues or events, and relate to the key business objectives. As controls have such an impact, controls must be not only meaningful, but also appropriate (i.e. not open to interpretation, with the potential for inappropriate or no action). Congruent controls are those that provide the right order of magnitude. Timely controls are necessary so that there is sufficient time to act before negative events turn into terminal events. Simple controls are needed so that they are readily understood. Otherwise they confuse, absorb too much time and, most importantly, do not permit control.

16.7.3 Regulatory and Statutory Requirements

Those companies listed on the exchange and required to abide by the Listing Rules and Handbook and are required to notify the FSA of any operational risk that may have a significant regulatory impact. This requirement includes notification of a significant failure in systems and controls, a significant operational loss or the intention to enter or significantly change a material outsourcing arrangement. Hence, just having controls is not enough. It must be possible to discern when they are not being effective. Pre-agreed trigger points are required so that a notification is issued to the FSA when certain parameters are exceeded.

Businesses within regulated industries have to ensure that they comply with the restraints imposed. An example is the pharmaceutical industry, which is subject to a series of balances and checks. Prices charged and profits made by branded drug companies are modulated by the Pharmaceutical Price Regulation Scheme (PPRS). The PPRS has three objectives: to secure safe and effective medicines for the National Health Service at reasonable prices; to promote a strong and profitable pharmaceutical industry with sustained research and development capacity; and to encourage the competitive and efficient development and supply of medicines. The pharmaceutical industry also has to cope with pressures from other sources. In 2005 the Office of Fair Trade announced a probe into pricing practices by drug companies, pressure was exerted by HM Treasury to reduce waste in public health spending and the National Institute for Clinical Excellence (a government advisory body on the efficacy and cost of medicines) began to flex its muscles.

16.7.4 Continuity

A business must make arrangements for the continuity of its operations in the event that a significant process or system becomes unavailable or destroyed. This risk relates to IT service outages and unreliability causing disruption to the business. Actively managing this attribute of risk requires capabilities in incident and problem management, IT service management, business continuity and disaster recovery. In many businesses temporary loss of primary computer systems can cause significant disruption of business processes, while permanent damage to critical servers and/or software applications can suspend business operations for days. Business continuity is discussed further in Section 16.8.2.

16.7.5 Indicators of Loss

The risk indicators are used to facilitate regular quantitative assessment and monitoring of risk exposures and mitigating responses. There is only ever any value in establishing indicators if data is collected and reviewed on a regular cycle and specific response plans are put in place to address the findings. During the first reviews it may be discovered that more information needs to be collected so that more meaningful responses can be implemented. Indicators are tailored to suit a business's specific services and operating context. Typical indicators are:

- bank borrowing against credit limits;
- cost of raw materials;
- sales revenue;
- third-party defaults;
- shareholder complaints;
- lawsuits;
- business continuity events (see Section 16.8.2);
- customer complaints;
- contracts secured and contracts lost.

16.7.6 Transactions

Meeting Commitments

A common process risk for any business relates to the processing of transactions. Transaction risk relates typically (but not exclusively) to the manufacturing sector. The risks to a business emanate predominantly from not honouring commitments to a customer in terms of time, quality and quantity. This includes the potential for errors in any stage of a business transaction including pricing, design, manufacture, sales, confirmation, documentation and fulfilment. At any stage of the transaction process, a company is faced with risks that can cause financial, reputation and/or customer loss. For instance, a pricing error at the time of entering into contract can cause lower profitability or loss, while a fulfilment problem can result in a customer ceasing to do business with the firm.

Production Processes

Another common process risk is product defects. Control charts have been used in manufacturing (Eppen 2001) to measure variability, and the goal was to continue to reduce the variability in processes until zero defects became an obtainable goal. Japanese companies led the world in adopting and perfecting process improvement (risk reduction), and the quality of their products has been testament to their philosophy on variability management. Eppen describes total quality management (TQM) starting to make an important impact on manufacturing in the US in the second half of the 1970s. US electronics group Motorola made process improvement a central part of its corporate strategy. The Motorola approach became known as the Six Sigma system and became the guideline for all Motorola processes, not just manufacturing. The basic idea was to improve processes so that the probability of a defect was effectively zero. The goal of a TQM company is to take the variability (risk) out of the entire process of producing a good or service.

Documentation Risk

Documentation risk might be considered a subset of transaction risk, as it is one step in the overall process of a transaction. Decisions made as the result of information contained in documentation which is incomplete, incorrect, inconsistent, open to interpretation, time expired and so on may result in inappropriate or misguided business activity. It may lead to poor decisions regarding oil exploration, drug development or company acquisition. Contracts, a form of documentation (Lam 2003), are a significant source of documentation risk as evidenced by the volume of case law. Any company is likely to have a large number of contracts with third parties in existence at any one time. They are the potential source of dispute and disagreement, which may lead to legal action. Any dispute can deflect senior management away from their core duties and disputes that have to be resolved through the courts can consume a significant amount of management time.

Product Variation Risk

Variability risk is the risk experienced by manufacturers where a customer requires: (1) a specific feature of their product that when added distinguishes the product from otherwise identical products; and (2) the modified product to be sent to a specific destination, when that product is typically distributed among several locations (Eppen 2001). The risk relates to a process burden and how successfully it is responded to. Eppen cites the sale of printers in western Europe by electronics manufacturer Hewlett-Packard as an example of variable risk, as follows. In general, a different power source and set of instructions are required for the printers for each country. In the original design, the power source was an integral part of the printer. Printers were assembled in Vancouver, Washington, packaged along with appropriate instruction documents and shipped to European warehouses. The process was redesigned based on the principle of delay. The printer itself now serves all countries. The power source is built into the power cord. Printers can be assembled in the US and shipped in bulk to warehouses in Europe. When an order arrives, the printer and its appropriate power cord and instructions are packaged and dispatched to the customer. This change substantially reduced the level of inventory needed to satisfy uncertain demand – Eppen explains that this is very important for a product that measures obsolescence in months.

Goods-in-Transit Risk

Reddaway (2001), from his experience as insurance manager within group risk at Glaxo Wellcome, states that one facet of operational risk management responsibility for large global organisations is goods in transit, which he describes being composed of three key areas: theft, accumulation risks and recoveries. Global companies face the serious problem of hijacking and theft in such areas as eastern Europe and Latin America. He cites accumulation risks arising when shipping goods and batches of goods or containers come together in one place – for example, while waiting for transportation, in a vessel or at a customs warehouse. Reddaway also refers to the situation where international regulations can stipulate that compensation should be made according to the weight of the cargo rather than the real commercial value of the goods. The mitigation action is to seek immediate legal and technical advice to demonstrate negligence of the carrier and press for settlement based on actual commercial value as opposed to weight.

16.7.7 Computer/IT Systems

IT systems include the computer systems and information technology infrastructure required for the automation of processes and systems, such as application software, operating system software, network infrastructure, and desktop and server hardware. There is a strong overlap between operational risk and technology risk. This section should be read in conjunction with Section 16.8.2 and Chapter 17, as the subjects overlap. As computer technology becomes increasingly necessary in more and more areas of business, operational risk events due to computer failures have become an increasing concern. Computer system risks include business alignment, network availability, data security, system capacity, unauthorised access/use and data recovery.

Business Alignment

The board must be assured that its IT system reflects its business needs and that the organisation is exploiting the system to full advantage. Hence, boards commonly have a series of questions for the IT department such as those included here. The risk for any business is that it has not invested wisely and its IT system is not adding to business performance and, in particular, competitive advantage. Quite distinct differences in strategy will be evident in businesses where IT is the business and where IT has a minor role. When IT is the business (e.g. for companies such as Yahoo! and Google), or the key driver of operational effectiveness and service delivery (e.g. for companies such as airlines, banks, insurance companies, hotels and, increasingly, estate agents), every IT project is able to have a direct impact on the performance of the business and thus on its worth in the market place. A good example of where IT is the business and its systems differentiate it in the market place is Dell Computers. Dell's capacity to deliver a customer-specified configuration PC in a short timeframe clearly sets it apart from its competitors in the same industry. Dell drives its business through IT, as evidenced by its use to analyse price sensitivities to incremented system improvements and to forewarn key suppliers of likely trends.

- Does the IT system reflect the business strategy and hence the business plan?
- Does the pace of development of the company's technological infrastructure match the pace of development of its business?
- How much benefit is being received from its IT?
- Is there sufficient in-house support provided for the IT users?
- Are ongoing systems acquisition, development and maintenance appropriate to the current business requirements?
- Are the risks inherent in the existing system and any planned changes understood?
- Can planned changes be implemented predictably?

Network Availability

The automation of processes and systems may reduce a business's susceptibility to people risks such as human error, but will increase a business's dependency on the reliability of its IT systems. If centralised systems go down then all hosted services go down, information processing ceases and all users are impacted. A temporary loss of a network can prevent access to files, the intranet, the World Wide Web, customer details, personal calendars, e-mail, personal contact details, transaction records and so on. In larger organisations where the IT system has evolved and grown over time across a number of sites and buildings, and there have been changes in the IT staff, knowledge of the network is incomplete and mapping of IT services to infrastructure is not as good as it might be. When one server fails its root cause may not be immediately apparent. Network redundancy and resilience – alternative routes for traffic to follow – are the key design principles that are now adopted to reduce the risk of individual link outages.

Data Integrity

Corrupted or degraded data is as valueless as data that is completely lost, when the extent of corruption or degradation is unknown. When discovered, corrupted information cannot be used until it has been checked, double-checked and confidence in its use has been restored. All the time and expense involved in correcting the wake left behind the use of invalid data, finding and identifying errors, rebuilding and then validating the data, could have been used profitably. However, there may always be that nagging doubt that the data is a risk to the business as corrupted elements remain undiscovered. The situation is obviously far worse when corrupted or degraded data is assumed to be reliable and is treated as a valuable asset. In the instance staff and customers rely on and make decisions on the data, the consequences could be very significant, particularly where personal safety is put at risk. Those intent on fraud may affect data integrity by changing prices, delivery instructions, status of the goods (new or damaged) or payment details. Degraded assets can destroy an organisation. If a bank's customer transaction records have been systematically distorted, it can destroy the bank. Unintentional corruption or loss typically arises from the following sources:

- A simple power failure at a critical moment can lead to information being lost (not saved) and the extent of the loss is not discernible or the software generates errors.
- Computer software errors such as bugs in the software allowing data to be overwritten, data to be written to one file and not another, prices to be changed without records being kept, deliveries to be authorised twice, calculations that generate incorrect values and so on.
- A hardware malfunction or a telecommunication failure leading to information not being saved, or loss of receipt of information from external sources.
- User error through the disregard of software warning messages resulting in data corruption or loss.

Electronic Data Security

Security of electronically held data is a serious risk, which, because of the internet, now has a global perspective. For example, during January 2000[8] a United Nations agency was

[8] BBC News (2000) Hacker hits UN website. 20 January. http://news.bbc.co.uk

forced to close down part of its website after it was hacked into by a "cyber-vandal". A link on the site of the World Intellectual Property Organisation (WIPO),[9] instead of showing a collection of laws for electronic access, displayed lyrics of a Bruce Springsteen song under the caption "Children of the Darkstar". The information held in electronic form within a business's information systems needs to be protected against unauthorised access, which may result in theft, corruption, corporate espionage and/or disclosure. The risk can originate from inappropriate access to or use of technology from inside the business, but more commonly from outside it. While financial crime is nothing new, the ways in which financial crime is being committed are changing. Criminals are increasingly using information technology to commit crime, as it is quick with low risk but potentially high rewards. Hackers see the infection of systems with viruses or data corruption as a sport. Viruses or data corruption can prevent important information from being used by the business.

Of the responding firms to independent surveys conducted by the National Hi-Tech Crime Unit and the DTI, 83% and 94% respectively reported that they had been the subject of system intrusions. These took the form of virus attacks, denial of service attacks (causing websites to crash), financial fraud and system penetration. Of particular concern is "phishing", which is aimed at identity theft. "Phishing" attacks are where criminals send hoax or spoof e-mails misrepresenting corporate identity to deceive individuals into disclosing their personal financial data such as account numbers and passwords. They create websites that mimic the brands of "high street" financial firms. Gartner Research reported in May 2004 that "phishing" attacks cost US banks and credit card companies $1.2 billion in 2003 (Financial Services Authority 2004). A study conducted by Information Risk Management of 18 online banking companies showed that 72% were vulnerable to long-term hacking attacks.[10]

Three recent highly publicised examples of system breach occurred during 2004 and 2005. The first relates to a Californian man, Nicholas Lee Jacobsen, 21, who was arrested in October 2004 for hacking into T-Mobile's network and taking names and social security numbers of 400 customers.[11] T-Mobile is a subsidiary company of Deutsche Telekom and has about 16.3 million subscribers in the US. It is thought his hacking campaign took place over at least seven months, during which time he read e-mails and personal computer files, but failed to obtain customer credit card numbers, which were stored on a separate computer system. The arrest came a year after T-Mobile uncovered the unauthorised access. The US Secret Service had been investigating the case. It is interesting that the Associated Press agency reported that the hacker also read personal files on the Secret Service agent who was apparently investigating the case. A Los Angeles grand jury indicted Mr Jacobsen with intentionally accessing a computer system without authorisation and with the unauthorised impairment of a "protected" computer between March and October 2004.

The second, more serious incident, which occurred in 2004, relates to the arrest of 53 people by federal police in Brazil for allegedly stealing $30 million (£16 million) from Brazilians through internet fraud.[12] They said the arrests were made across four states in the north of Brazil. The group secured the money by sending e-mail attachments infected with a virus which was able to store details of people's internet bank accounts, police said. Computer fraud experts at the time declared Brazil as the global capital of hacking and internet fraud. Cristiano

[9] WIPO is an intergovernmental organisation, one of 16 specialised agencies affiliated to the UN. It oversees the administration of various multilateral treaties regarding intellectual property such as trademarks and copyright.

[10] Seven in ten banks at risk from cyber-crime, *Metro*, 15 April 2005.

[11] BBC News (2005) US hacker breaks into T-Mobile. 13 January 2005. http://news.bbc.co.uk

[12] BBC News (2004) Brazil holds $30 m fraud hackers. 21 October 2004.http://news.bbc.co.uk

Barbosa, who coordinated the operation which resulted in 53 arrests, told the *Jornal do Brasil* newspaper: "They diverted money from e-mail accounts across the country – mainly in the south or south-east where people use e-mail more frequently." At a conference in the capital, Brasilia, in the previous September, federal police said the country was home to eight out of ten of the world's hackers. The amount of money lost in internet financial fraud in Brazil outstripped that lost through bank robberies, the conference was told. And security experts from other countries said that some 96 000 hacking attacks were launched from Brazil last year – six times more than any other country. The explosion in hacking was blamed, in part, on weak legislation. Police have to prove fraud has taken place in order to prosecute, as hacking itself is not a crime in that country.

Lastly, on 17 March 2005, the police in London announced they had foiled one of the biggest attempted bank thefts in Britain.[13] It would appear that the plan had been to steal £220 million from the London offices of the Japanese bank Sumitomo Mitsui. Computer experts are believed to have tried to transfer the money electronically after hacking into the bank's systems. Yeron Bolondi, 32, was seized in Israel, after the UK National Hi-Tech Crime Unit uncovered an attempt to transfer £13.9 million into an account there. Unit members worked closely with Israeli police. The investigation was started in October 2004 after it was discovered that computer hackers had gained access to Sumitomo Mitsui's computer system in London. They managed to infiltrate the system with keylogging software that would have enabled them to track every button pressed on computer keyboards. From that they could have learned account numbers, passwords and other sensitive information. Bolondi was initially charged with money laundering and deception, but the police said at the time that their investigation was continuing. They issued a warning to banks and businesses to watch out for cyber-criminals. The National Hi-Tech Crime Unit was launched in April 2001 with responsibility for tracking down the growing range of criminals who operate in cyberspace. Takashi Morita, head of communications at Sumitomo Mitsui in Tokyo, said at the time the incident made the news the company had not suffered any financial loss as a consequence of the robbery attempt. But they would, wouldn't they!

In 2004 the SANS[14] Institute (a non-profit group which trains and certifies computer security professionals) published a list of hackers' favourite top 20 security breaches to help organisations find out if they were closing the most commonly exploited vulnerabilities.[15] With more than 2500 software vulnerabilities found every year, the list was published to assist organisations in prioritising their risk response actions. The list includes loopholes found in both Windows and Unix/Linux software.

The report which accompanies the top 20 fleshes out individual vulnerabilities and what organisations can do to close these holes. Almost 60% of the loopholes listed in the 2004 report were in the 2003 top 20 list. The institute advises this was because only half of all organisations patched their systems and had not yet reached the point of fixing their vulnerabilities automatically.

[13] BBC News (2005) UK police foil massive bank theft. 17 March. http://news.bbc.co.uk

[14] SANS claims to be the most trusted and the largest source for information security training and certification in the world. It develops and maintains the largest collection of research documents about various aspects of information security, and it operates the internet's early warning system – Internet Storm Centre. The SANS (SysAdmin, Audit, Network, Security) Institute was established in 1989 as a cooperative research and education organisation. It is located at 8120 Woodmont Avenue, Suite 205, Bethesda, MD 20814, USA (http://www.sans.org).

[15] BBC News (2004) Top 20 computer threats unveiled. 9 October. http://news.bbc.co.uk

Top 10 Windows	Top 10 Unix/Linux
1. Web servers and services	1. Bind domain name system
2. Workstation service	2. Web server
3. Windows remote access services	3. Authentication
4. Microsoft SQL server	4. Version control systems
5. Windows authentication	5. Mail transport services
6. Web browsers	6. Simple Network Management Protocol
7. File-sharing applications	7. Open secure sockets layer (SSL)
8. LSAS	8. Misconfiguration of enterprise services
9. E-mail programs	9. Databases
10. Instant messaging	10. Kernel

System Capacity

Basically system capacity relates to the amount of memory on a server which impacts all businesses. However, with the growth of the internet and e-commerce, system capacity can relate to the number of individuals that can access a system at any one time. There have been instances when airlines have made special offers and potential customers have been unable to access the airlines' websites due to demand.

Data Recovery/Loss

Back-ups have been used since the earliest beginnings of IT, when the inherent unreliability of the hardware necessitated rigorous copying of data, and the frequent use of the backed-up data. Now that reliability has been dramatically improved, businesses are less likely to use data restoration facilities, explaining the common reality that restoration and recovery are not completely effective.

16.7.8 Knowledge Management

Information is increasingly forming the lifeblood of any organisation. A company's knowledge or intellectual property is usually the basis of the company's current success. The security of that information is key to competitive advantage. Hence, a key component of knowledge management is keeping commercially sensitive information, which would be useful to competitors, secure. How successful a company is in this task depends on a company's information culture. This can be defined as the values, attitudes and behaviour that influence the way employees collect, organise, save, discuss, process, communicate and use the information.

Intellectual Property

National and cultural standards regarding intellectual property differ significantly around the world. The rules for intellectual property protection between different countries are neither uniform nor equally enforced. The tracking of online dissemination and use of intellectual property, as well as the capability to collect payment, is not well developed in interactive networks. This is especially true of the internet, which on inception was never intended to be used for commercial purposes. Companies can also be exposed and vulnerable when working in collaboration with other companies on large projects, when under normal circumstances these companies would be competitors. The intellectual property protection issues arising out of collaboration are discussed in the *Financial Times* article in Box 16.5.

Box 16.5 Intellectual property

Companies are struggling to protect their intellectual property [IP] when they embark on collaborative development projects and could risk serious financial consequences as a result, according to a government-funded study being launched today. Part of the problem lies with senior management, which has been slow to draw up comprehensive IP policies, and ensure that these are understood by the likes of information technology managers and engineers, as well as in house lawyers and patent attorneys, it says. "Any corporate IP policy needs to sit at the interface of IT strategy, commercial and contractual policies and engineering and design practices", says the study. But, it continues: "those involved in corporate strategy mainly concern themselves with managing tangible and financial assets, and IP management seems to be treated as a specialist task that does not readily find its way into the boardroom". The report points out that the risk of IP losses can be exacerbated by the growing use of electronic data networks, which are often a feature of collaborative projects, although it notes that technology can be employed positively to track the use of IP assets. It cites, for example, research conducted by the UK National High Tech Crime Unit last year. This found that 12% of businesses had experienced instances of data theft through the internet, causing losses of about £7 billion. "Anecdotes abound of engineers that are only too happy to share proprietary and commercially sensitive technical details with peers in other companies. Although some of these anecdotes involve instances in which such exchanges were not facilitated by electronic networks, concerns remain about what would happen when the digital systems for collaboration are in place that could allow a loquacious engineer to send reams of technical information across to project partners with the click of a button." The report stems from research by academics at the University of Sussex for the UK Economic and Social Research Council and the Ministry of Defence, and will be launched at a conference of government procurement officials, lawyers and executives at the Institute of Advanced Legal Studies later today. According to Dr Puay Tang, one of the authors, the research was based on more than 60 interviews with people involved in collaborative development. The initial focus was on defence companies and subcontractors, but Dr Tang claims that the conclusions are broader. "The lessons are generic. Our results are pretty transferable," she said yesterday. The report suggests that clear and specific contractual conditions can help in IP management.

Source: Tait (2005).

16.7.9 Project Management

Organising any activity, which has predetermined constraints, requires the discipline of project management. However, how best to organise the efforts of individuals to achieve a desired outcome has been one of the world's most important, difficult and repeatedly controversial problems. The organisation of relief for tsunami, hurricane and earthquake victims in 2005, for instance, is evidence of the difficulties of project management. Project management is about planning, controlling and coordinating a project from inception to completion on behalf of a sponsor (or sponsors), where the sponsor(s) may be within or outside of the business. It is concerned with the identification of the sponsor's objectives in terms of function, time, cost and quality. It involves the integration, control and monitoring of the contributions to the project to achieve these desired outcomes. A fundamental aspect of project management is people management – working through others and relying on their timely collaboration to achieve the project objectives. Each project will have its own unique risk profile and will require the discipline of risk management to provide greater certainty of achieving its objectives. Large projects, when they fail or do not achieve their objectives, may have a debilitating effect on a business or may in extreme cases bring about their downfall. Railtrack plc in the UK and Boeing in the US, for instance, have both experienced serious project problems that have shaken them to the core. It is reported that Railtrack was placed into administration due predominantly to the lack of performance of the West Coast Main Line project.

16.8 EXTERNAL EVENTS

External events (see Figure 16.6) are events that occur outside the business, which may require a response in the form of change management (such as organisational change) or the instigation of contingency events to cope with, say, a natural disaster.

16.8.1 Change Management

The introduction of change needs to be handled with care to avoid a series of common risks. Where radical IT change is planned it is often prudent to consider the implementation of a small pilot study initially to understand the implications of the change and how they may be best addressed. Time must be taken to define as precisely as possible the requirements of the new solution. If cross-company deployment of new IT facilities is planned, then it is important that no one user or interest group steers the specification to solely suit their own needs, but a solution is defined that will meet the needs of the whole enterprise. As a solution appears as if it will become accepted and integrated into the IT system, its design must make for a long-lasting and durable solution that can be cost-effectively maintained, enhanced and upgraded. A significant change management project usually warrants an integration plan to raise awareness of the change in advance of the deployment of the solution and to ease its introduction with the aid of familiarisation and training sessions appropriate to each of the disparate user groups. Deployment of the change into the business environment should follow established channels. Existing user champions, system experts and user group members would be the natural and

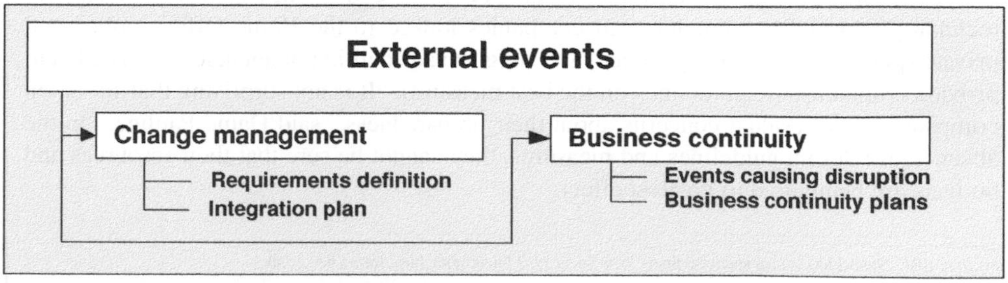

Figure 16.6 Taxonomy of external events risk

logical choice for educating the users of the new system to support the integration and use of the system at the earliest opportunity.

16.8.2 Business Continuity

Business continuity management is a holistic management process that identifies potential impacts that threaten an organisation and provides a framework for building resilience and the capability for an effective response, which safeguards the interests of its key stakeholders, reputation, brand and value creating activities (Business Continuity Institute 2002).

Events Causing Disruption

There are a number of issues that can disrupt businesses, or in severe circumstances put them out of business. The issues of attacks by animal rights activists, computer hackers and terrorists are addressed by a BBC report included in Box 16.6.

Box 16.6 Corporate security

Firms afraid of attacks from animal rights activists, computer hackers or terrorists, have spent £1m on extra security this year, says a new survey. The Confederation of British Industry (CBI) survey of 100 top companies suggested most had shaken up their security arrangements. And two-thirds now employ a chief security officer. The CBI says the government could boost confidence by being more open about its contingency plans. The organisation said businesses were spending significantly more on security than five years earlier and some of the newly appointed security chiefs even had places on firms' board of directors. Despite the extra security measures, 60% of those questioned in the Mori telephone poll said they were concerned about their firm's level of preparedness for an attack. CBI director general Digby Jones said: "Business Britain understands the meaning of risk and is working hard to calibrate the additional risks posed by security. But the risk assessment process in a business would be much improved if there was greater transparency from government and other key agencies. Business needs to have more confidence that it is getting its contingency planning right." Mr Jones also complained policing to deal with some threats, including those posed by animal rights activists, had been inadequate. "In a democratic society, companies and everyone that works for them must be protected from those who seek to prevent lawful business activity," he said. The problem of business security is on the agenda in the opening session of the CBI national conference in Birmingham on Monday. Dame Pauline Neville Jones, chairman of security technology specialists QinetiQ, said companies looked to the Home Office, police and private specialists for security advice. She suggested a coalition of agencies was needed to provide commonsense guidelines on the best measures. "It is not surprising that six in ten companies have residual concerns about their preparedness," said Dame Pauline. "In the absence of relevant guidelines and measures, they cannot be sure that their resources and budgets are being used to greatest effect."

Source: BBC News (2004) £1m spent on firms' new security. 7 November. http://news.bbc.co.uk

However, a more comprehensive assessment of the issues that can derail a business was captured in a survey led by the Chartered Management Institute (CMI). In January 2004, the CMI, supported by COLT Telecom Group plc, Nortel Networks and the Business Continuity Institute, undertook a survey of managers to understand the main events which disrupted their organisation in the last year. The results obtained were derived from postal questionnaires sent to individual CMI members. A total of 461 responses were obtained. The results are illustrated in Figure 16.7. Twenty-five per cent of respondents suffered a loss of IT capacity, whereas 1% of respondents suffered terrorist damage.

Business Continuity Plans

The survey also uncovered that 53% of organisations had no business continuity plan. It was found that the production of business continuity plans was more prevalent in companies with an annual turnover of £11 million. The survey concludes that:

> Businesses that require an uninterruptible power supply, particularly those protecting sensitive data on computer systems, require a back-up UPS battery supply designed to cut-in, during a mains black out. These systems must be capable of supplying large amounts of power with a high degree of reliability. They are commonly composed of lead acid batteries as they have a low self-discharge rate, which reduces the cost of recharging.

16.9 OUTSOURCING

Businesses frequently decide to outsource aspects of their operations to independent third parties for reasons including cost, efficiency and/or risk transfer. Outsourcing provides the opportunity to bring significant benefits to a business; however, it may alter the risk *profile* of a business from a number of aspects. The boundaries of outsourcing risk need to be clearly understood, due to the relationship with business continuity, information security, regulatory risk and so on. Included here is a non-exhaustive list of potential risks:

- The business (in a regulated industry) does not notify the regulator of the intention to outsource a significant element of operations.
- Failure on the part of the third party to adhere to contractual arrangements as a result of significant changes to their people, processes and systems.
- Reduced control over the people, processes and systems deployed by the third party.
- Business (working in a regulated industry) does not require the third party, with whom it has entered into a contract, to pre-agree any changes in the service provision, which would contravene its operating agreement.
- Third-party competency, financial standing or expertise is not adequately assessed prior to contract.
- Third-party performance is not monitored on a regular basis during the life of the contract.
- Business does not agree rights of termination of the contract (with the third-party supplier) in the event that the supplier becomes insolvent or goes into liquidation or receivership.
- The business does not prepare a business continuity plan for the instance where the supplier significantly underperforms or goes into liquidation.
- The third party's business continuity plan is not reviewed by the business prior to contract.
- The third party is not required to either agree in advance outsourcing of services to a subcontractor or notify the business when outsourcing has been undertaken.

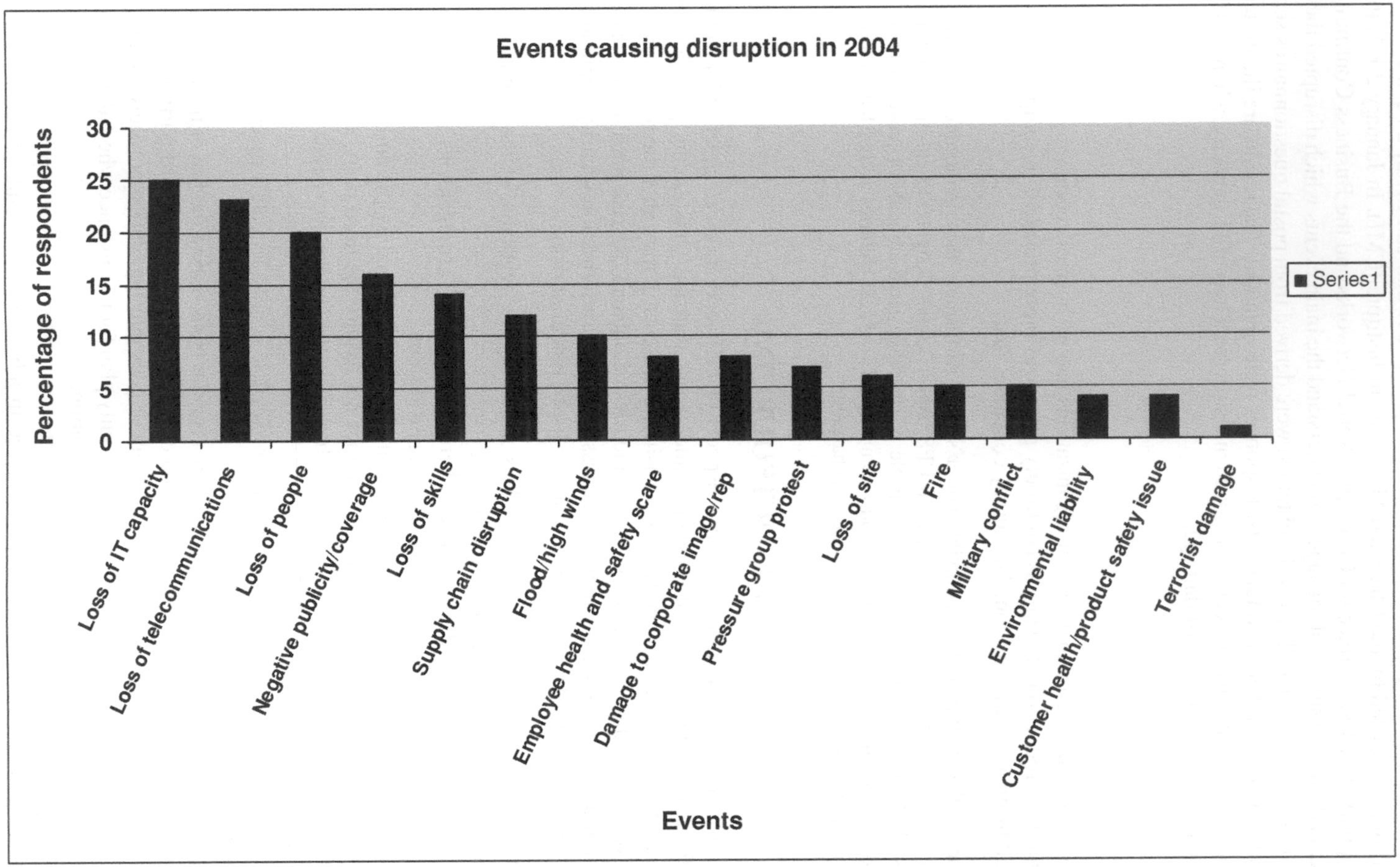

Figure 16.7 Events causing disruption to organisations in 2004

- Information security breach arising from outsourcing to a third-party provider who has inadequate security protocols.

16.10 MEASUREMENT

A key aspect of operational risk management is control. However, to be able to prioritise management action and focus on those issues likely to have the greatest detrimental effect on the operation of the business, it is necessary to measure their likely impact. To accomplish this measurement it is necessary to have both historical data and high-powered analytical tools. As identified by the FSA, many organisations are unable to measure a number of operational risks quantitatively yet, due to the absence of these facilities (Financial Services Authority 2002). Measurement is important as it enables businesses to set aside monies to cope with possible adverse events and to know the extent of insurance required, if the decision is taken to transfer the risk of potential losses. Transfer is always at a cost and it is never absolute.

16.11 MITIGATION

The success of mitigation will depend on a number of things. It will depend for instance on: the degree to which risk management is embedded in the business and championed from the top; the robustness of the risk identification process; the willingness and the desire of senior management to distil from the assessment process the priority issues and take the time to prepare specific tailored risk response actions; the experience of senior management of the risk exposure issues being addressed; and the comprehensiveness of the approach. The quality of the response actions themselves will depend on the amount of time taken to research the problem and examine alternative responses. A fall in sales may be from a number of interrelated issues.

16.12 SUMMARY

This chapter has examined the elements of operational risk, namely strategy, people, processes and systems, external events, and outsourcing. There is a growing interest in this area of enterprise risk management due to its ability to address everyday management issues which, if not properly controlled, will at least be debilitating and at worst affect survival. Operational risk not only affects systems and processes but also emanates from what might be described as first principles, that is, the setting of the strategy, the development of the business plan and the allocation of resources to that plan. Businesses must establish a series of systems and controls to manage people risk – the behaviour of the business to its people, and the behaviour of its people within the business. Every step in the recruitment, induction and ongoing management of staff impacts a business's operational risk exposure. Processes and systems are a significant area of operational risk, spanning such subjects as business continuity, transaction risk, IT and information security. Operational risk impacts businesses from within and from outside. External events such as flooding, power failure and terrorist damage can all derail a business. And last but not least, outsourcing, while frequently a CEO favourite for cutting costs, can bring a business to its knees as fast as any other serious operational risk event. The dependence on others for critical suppliers or operational activities will always be a major source of risk.

16.13 REFERENCES

Branson, R. (2002) *Losing My Virginity: The Autobiography*, updated edition, p. 257. Virgin Books, London.

Basil Committee on Banking Supervision (2004) *International Convergence of Capital Measurement and Capital Standards: A Revised Framework*. Bank for International Settlements, Basel.

Business Continuity Institute (2002) *Business Continuity Management: Good Practice Guidelines*, Version BCI DJS 1.0, 01/11/02 (ed. D.J. Smith). http://www.thebci.org

Carey, A. and Turnbull, N. (2001) The boardroom imperative on internal control. In J. Pickford (ed.), *Mastering Risk Volume 1: Concepts*. Financial Times, Harlow.

Chapman, R.J. (2002) *Retaining Design Team Members: A Risk Management Approach*. RIBA Enterprises, London.

Drucker, P.F. (1979) *Management*. Pan Books, London.

Drucker, P.F. (1992) The new society of organisations. *Harvard Business Review*, September–October, 100.

Eppen, G.D. (2001) Charting a course through the perils of production. In J. Pickford (ed.), *Mastering Risk Volume 1: Concepts*. Financial Times, Harlow.

Financial Services Authority (2001) *Consultation Paper 97a, Annex C: Draft Rules and Guidance*, June. FSA, London.

Financial Services Authority (2002) *Consultation Paper (CP) 142: Operational Risk Systems and Controls*. FSA, London.

Financial Services Authority (2004) *Countering Financial Crime in Information Security*. FSA, London.

Financial Services Authority (2005) *Operational Risk Management Practices, Feedback from a Thematic Review*, February. FSA, London.

Higgs, D. (2003) *Review of the Role and Effectiveness of Non-executive Directors*. Department of Trade and Industry, The Stationery Office, London.

Institute of Management (1999) *People Management*. Hodder and Stoughton Educational, London.

Lam, J. (2003) *Enterprise Risk Management, from Incentives to Controls*. John Wiley & Sons, Inc., Hoboken, NJ.

Maxant, R. (2004) *Global Risk Management Study* (fourth biannual global risk management survey of financial institutions). Deloitte & Touche LLP.

Office of Government Commerce (2002) *Management of Risk: Guidance for Practitioners*. The Stationery Office, London.

Osman, C. (2003) Part 8.2, Employment Practices. In A. Jolly (ed.), *Managing Business Risk: A Practical Guide to Protecting Your Business*. Kogan Page, London.

Ostergen, K. (1989) *Man as a Critical Resource – A Study of Patterns in Keypersons Behaviour*. Department of Business Administration, Umeå Universitet, Sweden.

Peccia, T. (2001) Designing an Operational Risk Framework from a Bottom-up Perspective. In C. Alexander (ed.), *Mastering Risk Volume 2: Applications*. Financial Times Prentice Hall, Harlow.

Pinnington, A. and Edwards, T. (2000) *Introduction to Human Resource Management*. Oxford University Press, Oxford.

Reddaway, R. (2001) The devil in the details: Attaining coverage for the global corporation. In J. Pickford (ed.), *Mastering Risk Volume 1: Concepts*. Financial Times, Harlow.

Schmitt, B. (2001) Branding puts a high value on reputation management. In J. Pickford (ed.), *Mastering Risk Volume 1: Concepts*. Financial Times, Harlow.

Sherman, A., Bohlander, G. and Snell, S. (1998), *Managing Human Resources*, 11th edition, p. 93. South Western College Publishing, Cincinnati, OH.

Tait, N. (2005) Business warned to protect creative jobs, *Financial Times*, 4 March.

Tyson, S. and York, A. (1996) *Human Resource Management*, 3rd edition. Butterworth-Heinemann, Oxford.

Wyche, C. (1990) British Airways flies the marketing flag, *Personnel Management*, October, 125–127.

17
Technological Risk Management

The first rule of any technology used in a business is that automation applied to an efficient operation will magnify the efficiency. The second is that automation applied to an inefficient operation will magnify the inefficiency.
(Bill Gates)

The previous chapter examined operational risk management as the second of six classes of risk exposure that businesses face relating to internal processes within the risk taxonomy proposed in Chapter 9. This chapter examines the third of the internal processes, called *technological risk*. The ubiquitous technologies of today are information (IT), communication and controls. These technologies can raise productivity, lower costs and drive growth. Internet technology has become critical to companies such as airlines, banks, insurers, retailers and hotels. While previously IT was used to remove administrative and bureaucratic burdens, so that staff could spend more time doing their "proper jobs", it is now an enabler of the business plan and growth targets. In today's highly competitive business environment, the effective use of technology has the ability to transform enterprises and contribute to enhanced and sustainable stakeholder value. This improvement in value is commonly being driven by technology investment, optimisation of resources and diligent maintenance, to preserve reliability. However, rapid technology changes and the convergence of technologies (such as computer and telecommunications) are constantly redefining "industrial boundaries" so that "old" industrial sectors become barely recognisable. Even as far back as 1999 Microsoft felt the pressure (Gates 1999), illustrated by Bill Gates' statement that "in three years every product we make will be obsolete. The only question is whether we'll make them obsolete or somebody else will." The pace of technological change is illustrated by an advert which appeared in 2011 for Rouen Business School in Normandy, France, which said that "The ten most sought-after jobs in 2010 did not exist in 2004". The structure of this chapter is based on the risk taxonomy included in Chapter 9 and describes important areas of focus for enterprises.

So changes in technology are both an opportunity and a threat in terms of market share and market development. In addition, the introduction of technology within a business can open the door to a series of debilitating risks, which may seriously erode profitability, reputation and competitive advantage, or at worst lead to business failure. Disasters occur even at the frontiers of technology, often with no quick recovery and debilitating effects on the business. An example is the Fukushima Dai-ichi nuclear plant in Japan. Hence, CEO and board-level direction and oversight are essential. This requires technology governance structures with the right level of executive involvement and commitment, and clarity, about the business's risk appetite. In addition, clearly defined levels of responsibility for risk management are required, including, as a minimum, identification, measurement, management and monitoring.

Technology is an all-embracing term. While there are numerous technologies, the common technologies considered to be important to business and considered here are information, communications and control. This chapter seeks to provide a definition of technology risk

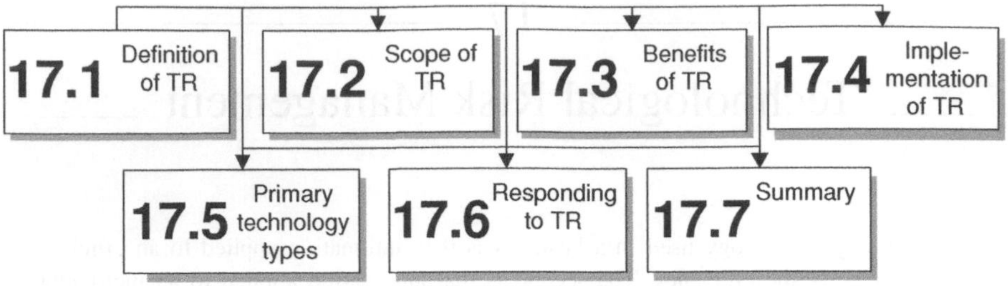

Figure 17.1 Structure of Chapter 17

management, examine the primary types of technology of interest to businesses, sources of risk and possible responses. The structure of the chapter is shown in Figure 17.1.

17.1 DEFINITION OF TECHNOLOGY RISK

Before being able to define technology risk, it is important to understand what we mean by "technology" in a business context. The economists' perspective of technology is that it is a component of capital goods (one of the factors of production) used by businesses to produce commodities consumed by society. Technology is thus a subset of production, the process that transforms inputs into a set of outputs. Traditionally, this means turning raw materials and component parts into finished goods. Examples of this transformation process that we are all familiar with from television advertising are car assembly plants, and television and computer production lines. While technology is important to the process of production, its application is broader. The *Oxford Everyday Dictionary* definition is as follows: "technology is the scientific study of mechanical arts and applied sciences". Mechanical arts and the applied sciences are continually adopted for product design. Hence, the commodities produced by the manufacturing process exhibit incremental advances in technology and their functionality, quality and reliability influence market creation, market share and market growth. In addition, now that we are part of the Information Age, technology is not only about manufacturing and product design. It is also about the integration of mechanical processes and information, such as the use of software to drive mechanised air cargo warehousing at British Airways, optimise production in steel plants at the Anglo-Dutch steelmaker Corus, and produce aircraft components at Boeing. In addition, it is about "digital nervous systems", a term coined by Bill Gates to describe a well-integrated flow of information to the right parts of the organisation at the right time (Gates 1999). Hence, technology risk may be defined as events that would lead to insufficient, inappropriate or mismanagement of investment in technology, in terms of manufacturing processes, product design and/or information management. Mismanagement may include poor business continuity planning, security and or protection of intellectual property. The most significant effect of this risk would be erosion of market share.

17.2 SCOPE OF TECHNOLOGY RISK

A sample of the sources of risk that are considered to be embraced within the term "technology risk" are recorded below. The potential list is considerable. Any examination of the sources of risk needs to be tailored to the specific activities of a business.

- Lack of investment in technology and the resultant erosion of ability to compete.
- Inadequate technology governance and in particular IT governance.
- Inadequate management of outsourcing.
- Lack of alignment of IT to the business objectives.
- Inadequate protection against viruses, hacking and loss of confidentiality of information.
- Inadequate flexibility of production to be able to economically produce small production runs.

17.3 BENEFITS OF TECHNOLOGY RISK MANAGEMENT

Technology risk management affords a business benefits as it:

- Improves the quality of information for decision making. Business leaders who succeed will take advantage of a new way of doing business based on the increasing velocity of information and building advanced processes and products faster than the competition.
- Sets out the risks to investment in technology and promotes a proactive approach to managing technology projects.
- Maps the threats to existing business practices from emerging business-to-customer relationships. Gates (1999) claims that "Today US businesses are ahead of businesses in other countries in the adoption of digital technologies. The many reasons include an openness to risk taking, individual empowerment and labour mobility."
- Draws attention to exposure to the loss of market share arising from a competitor's improvement in product design.
- Forces a continuous review of developments in technology within manufacturing processes (technology advances can improve productivity).
- Provides insights into the disbenefits of not aligning technology to strategy and business operations. To get the full benefit of technology, business leaders will need to streamline and modernise their process and their organisation. The goal is to make business reflex nearly instantaneous and to make strategic thought an ongoing, iterative process – not something done every 12–18 months, separate from the daily flow of the business (Gates 1999).

17.4 IMPLEMENTATION OF TECHNOLOGY RISK MANAGEMENT

The development of a sound system of technology risk management will depend on whether attention is being paid to a number of issues, including but limited to the following:

- managing investment in technology to secure the business objectives and optimise investment benefits;
- ensuring that the right information reaches the right people at the right time through a combination of management information systems, intranets and e-mail;
- understanding the risks of outsourcing and to manage them;
- monitoring competitors to avoid being "outmanoeuvred" by the introduction of new technologies that shift industrial boundaries;
- embracing new developments in e-commerce;
- implementing information security.

A key aspect of technology risk management is not being outwitted by the competition and as a minimum keeping pace with their developments – the ideal goal being to set the pace.

17.5 PRIMARY TECHNOLOGY TYPES

17.5.1 Information Technology

Information technology is the collection, storage, processing and communication of information by electronic means. There are various types of (IT) tools:

- Software applications include spreadsheets, databases, word processing, graphics packages, drawing packages, desktop publishing, presentation packages and expert systems. Spreadsheets save considerable time, make repeated calculations simple, aid accuracy, allow managers to set up mathematical models, investigate the effects of different strategies such as asking "what-if?" questions and provide many tools for analysis. For example, they provide graph and chart facilities together with compound interest, depreciation, optimisation and goal-seek functions. Databases are sets of files organised to provide easy access to their content. Expert systems cover a particular area of expertise and draw conclusions from computer-stored knowledge obtained from specialists with domain knowledge. Their purpose is to capture the expertise of key people and make their knowledge available to users of the program.
- Management information systems are systems designed by organisations to collect and report information on projects and programmes which allow managers to plan, monitor and evaluate their performance.
- Intranets are computer networks based on the same technical standards as the internet but designed for use within a single organisation. Intranets are cheaper and simpler to install than proprietary networks, and companies are increasingly using them to circulate internal information such as phone directories, job openings, training, marketing and publicity material.
- "Telematics" is the term given to the technology that enables remote access to vehicle data over a wireless network.
- "Information assets", increasingly the lifeblood of any business, covers subjects like customer contacts, manufacturing process innovation, product design and IT development.

Software Applications

This risk type deals with failures in IT applications. Applications are typically proprietary off-the-shelf software packages, customised proprietary software, bespoke software commissioned from a vendor or software developed in-house. Certain applications, such as those which are "job-specific" and used for accounting, marketing, project management and human resources, will be the domain of the departments of the same name, whereas there will be other packages such as word-processing and spreadsheets which will be used right across the business. The impact of any one application failing to perform as expected can range from minor irritation to major downtime during which employees are idle or are unable to tackle priority tasks. The degree of impact will also be dependent on whether the application is department-specific or is used company-wide, and whether the application is loaded on a server or on individual PCs. Customised bespoke software developed in-house can be the most problematical. For example, applications that are not easily maintained and changed over time (to reflect changing needs) may form a constraint to introducing further change. Applications that are poorly documented or not well structured may be difficult to fault-rectify with confidence. In addition, significant defects can be introduced unwittingly by software developers when only minor changes are made, as they were not the original authors and do not appreciate the structure of the application. These types of risk require software engineering capabilities particularly for maintenance, enhancement, integration, testing and release management and subsequent change management, system administration, monitoring and problem management.

Management Information Systems

Management information systems for projects include scope definition, work breakdown structures, organisational breakdown structures, programming, budgeting, change control, value management, earned value analysis, risk management and contingency planning. The risk associated with these systems relates to the lack of implementation or their poor execution in terms of the accuracy of the data they contain, the completeness, currency (whether they are kept up to date), revision control or lack of the creation of a baseline from which to measure progress.

Intranets

Intranets are touching everyone's lives, from the US Marine Corps (who have adopted a situation awareness application) to physicians in southern Virginia and North Carolina (who can access patients' records remotely over the web) to school children in Reading, England (who can access the school intranet remotely). Intranets can offer a business considerable time savings if they contain information which is readily accessible by a significant percentage of the employee population. The downside risk is that, should an intranet be unavailable for any length of time, that same employee group would be unable to perform some or many of their routine tasks.

Telematics

The word "telematics" can be used to describe any integrated communication and computer system, but it is now used almost exclusively within the context of vehicles. Telematics adapts information and communications technologies (ICT) to create vehicle management systems, thereby helping to resolve many logistical problems. Telematics provider Thales Telematics combines the technologies of precise positioning, using the satellite global positioning system (GPS), and data and voice communication, using the GSM and GPRS digital mobile network. According to Thales, the benefits of telematics include auto geofencing, which immobilises unattended vehicles; proof of delivery, preventing disputes over arrival/departure times of delivered goods; and exact positioning, enabling quick localisation in case of vehicle breakdown. In America the most popular telematics system is General Motors OnStar, which provides features such as stolen vehicle tracking, remote diagnostics and remote door unlocking made possible through integration of an in-car computer, GPS and mobile phone.

Telematics is used for monitoring, safety and convenience. A telematics system can notify an operator when the "check engine" light goes on or when an airbag deploys. When a safety measure is detected in the car, an operator calls the car to make sure the passengers are not in difficulty – and if they are, the operator sends help. GPS tells the operator where to send the police and ambulance services. Telematics is considered to have considerable market potential and hence, rather than posing a risk, is a business opportunity area. Similarly, the UK Department for Transport initiative to place microchips in car number plates in order to monitor vehicle movements is considered to be another business opportunity. In 2006 the Department for Transport successfully trialled the use of radio frequency identification in conjunction with the electronic vehicle identification (EVI), considered a consistent method of identifying vehicles. The trial placed electronic tags on vehicles to explore the merit of placing microchips in number plates as a tool to detect stolen/cloned vehicles, vehicle excise duty evasion and vehicles involved in criminal activity. The trial was very successful. There are similar initiatives in Europe and America which see the "electronic number plate" also being used for road pricing and parking management. However, this is just one of three technology projects under way in the UK specifically to support the delivery of road safety, congestion management and other policy objectives. Other than EVI, a White Paper on *The Future of Transport* (which highlights a commitment to encourage and enable greater adoption of technology) also talks about an intelligent transport systems technical framework and cooperative vehicle highway systems (Department for Transport 2004).

Information Assets

Information is the energy of enterprises generating new products and services and enabling new ways of working. Money, men, machines and materials were the business resources of the past. Today we have a fifth resource – information. As described by Earl (1997), businesses have to be able to manage information as an asset, both as a lever for business development and as a process for managing organisations. Examples of information assets can be found in the airline and retail industries. Competition in commercial airlines is substantially based on commanding electronic channels of distribution by way of reservation systems and on aggressive sales and yield management by analysis of customer databases. Within the retail sector loyalty cards (offered by organisations such as Tesco and Boots) not only provide frequent and volume purchasing discounts to customers, but also capture individual customer behaviour allowing offerings to be customised and the development of newly targeted services. Businesses are building an information-driven market place through the amalgamation of media, computing and telecommunications. Earl describes the latest visible sign of this convergence as the creation of the "infotainment" sector. New sectors are emerging and conventional sector boundaries are being eroded. All of these businesses are seen as information businesses.

In the information age business strategy cannot be formulated as a whole without considering information as an asset. Signs of this approach Earl describes as being "exemplified by the current wave of mergers and acquisitions around ownership of both information content and information distribution", hence "information intensive businesses have to consider threats such as entrants from other sectors – insurers offering banking, retailers offering insurance, software companies offering money transmission and so on – and intermediaries being taken out by direct electronic traders". It is no longer the case that information strategies support business strategy; they are interwoven. IT strategy and business strategy are one. It is now seen that a business strategy is not complete if the information resource is ignored. Information and IT can create or destroy business.

If information is a fundamental component of business strategy then its protection is key. This type of risk relates specifically to damage, loss or exploitation of information assets held within IT systems. For many businesses, this risk commences with not recognising the information assets actually held. The risk of information asset risk can vary significantly. For example, customer information such as credit card details may be copied and used for fraudulent purposes or the "theft" published, which is damaging to the reputation of the business. Similarly, information may be obtained by a competitor, removing competitor advantage. Core business processes reliant on critical information may be severely degraded – as when an account enquiry function contains out-of-date or inaccurate information, thereby becoming ineffective.

17.5.2 Communications Technology

- Conference calls enable several people to hold a single telephone conversation.
- With the use of the internet, websites can be used to advertise the firm and offer the facility to buy direct over the net; e-commerce is the fastest growing business sector.
- Broadband[1] provides a more efficient quicker data communication channel. It enables workers to work from home as efficiently as in the office, cutting down on the firm's costs.
- Video conferencing allows people to "meet" without leaving their offices. This cuts down on time and on travel costs.
- E-mail allows instant communication and the use of attachments enables the exchange of text, drawings, spreadsheets and diagrams in a format that is readily usable by the recipient.
- Network systems in which computers are linked to one another over a network; this may be used for sharing software or files.

Broadband

The term "broadband" is used to describe any high-speed connection to the internet. However, there are in fact several different types of broadband connection available in the UK through different internet service providers. These include satellite, cable, wireless and ADSL. Satellite access is available anywhere in the UK, so for some businesses in remote areas it will be the only option. It is, however, expensive to install and planning permission may be required for the dish. Cable requires a special cable connection to be installed into business premises. There are few wireless service providers in the UK and a receiver on the outside of the premises is required to send the signals through to connection points inside. These in turn send the data through to your computer(s). ADSL[2] (Asymmetric Digital Subscriber Line) is a technology that allows normal telephone lines to carry more data. ADSL is one variation of the DSL (Digital Subscriber Line) broadband technology, designed to bring large amounts of information through ordinary telephone lines.

However, as broadband provides an always-on connection, there is a greater risk that businesses could receive spam, viruses or be vulnerable to theft or hackers. Unsolicited commercial e-mail or unsolicited bulk e-mail, generally known as "spam", is a nuisance for internet users, particularly when received in large numbers. In most cases, spam will appear merely as unwanted commercial e-mail – junk e-mail advertising – although at times

[1] The main difference between broadband and traditional telephone communications is that broadband is based on digital technology, and the telephone is based on analogue technology. While analogue technology is fine for voice communications, it has limitations when it comes to transmitting data. Digital technology is much more precise and accurate, and enables both higher speeds and greater reliability.

[2] British Telecom estimates that 99% of its fixed telephone lines are capable of supporting ADSL broadband at download speeds of 256Kbps or more and virtually all telephone exchanges are broadband enabled (Ofcom 2010).

advertising some fairly distasteful products. These might be get-rich-quick schemes, which attempt to defraud recipients, pornography and offers to sell prescription drugs. A computer virus (a program designed to alter the way a computer operates, without the knowledge or consent of the user) is a major concern to users as it can erase information.[3] The authors of viruses attempt to infect computers with a virus that becomes a source of a great deal more virus-laden e-mail, sent to every e-mail address stored on a computer. Viruses can also attempt to compromise the user's personal information to third parties – several of them have appeared of late to steal credit card information. There have been a number of viruses that have been designed to turn the systems they infect into "zombies", to be controlled by the virus authors, typically to send out more spam. One of the most publicised risks to information systems is that of unauthorised access, often referred to as hacking. The DTI Information Security Breaches Survey 2002 found that data theft accounted for about 6% of security incidents. A further risk exposure is systems failure, the temporary loss of availability of broadband.

Business can take some basic steps to reduce this risk exposure by: installing firewall software to help prevent outsiders penetrating their system; installing anti-virus software to filter destructive data; keeping the operating system, firewall, virus protection and other software up to date; turning off unnecessary permissions and applications such as FTP (File Transfer Protocol), mail servers and file and print sharing; looking at the sources of system failure; making sure technical support is available, whether this be in-house or by engaging specialist support contractors; and ensuring employees understand the importance of security by making IT security a part of employees' contracts and drawing up and enforcing policies for the use of e-mail and the internet. In addition, businesses should not keep information on individual computers as this creates a security risk. Create a robust back-up system. Set up a procedure for making regular complete and partial back-ups. Back-ups should be stored off-site and kept away from heat, moisture and magnetic fields.

Video Conferencing

The concept of video conferencing[4] was first developed by the US firm AT&T. Video conferencing systems were first offered commercially in Japan in 1984. Its selling point is that it is a tool that offers competitive advantage through improved communication. Some argue that video conferencing equipment will soon become as common as the telephone as businesses will not want to face the risk of being as outdated as companies without fax machines were

[3] A business's internet service provider may use Network Address Translation (NAT), a protocol that ADSL providers like since a NAT application allows several computers using private IP addresses to use a single IP address to access the internet. Using NAT can make the connection more secure, as a casual intruder will access a router and not the business's internal network, although NAT is not resilient to all types of attacks. NAT configurations are not suitable for hosting web servers, video conferencing, setting up virtual private networks and various other internet applications. Routed IP overcomes these usage problems but is more vulnerable and requires firewall and virus scanning back-up. Also different services have inherent security concerns. Most cable providers have responded to this sharing problem by upgrading networks and equipment to a new standard called Data Over Cable Service Interface Specification (DOCSIS). DOCSIS-compliant modems encrypt traffic from the providers' communication centres and local hubs to the end user's PC.

[4] The key advantage of video conferencing is that it saves considerable travelling time and cost for geographically dispersed groups that wish to communicate within a business, business-to-supplier, business-to-customer or business-to-business. It removes the risk to individuals in travelling to remote locations. It enables employees to build closer relationships with colleagues despite geographical distance. Employees benefit personally, as it eliminates the stress or tiredness that can be brought about by long-haul travel. It increases the proportional productivity and efficiency of employees as they spend less time travelling and more time working. It improves communications in organisations. It enables more personal contact between employees and colleagues at remote locations. It facilitates organisation-wide improvement in sharing best practice, performing tasks and managing projects. Vast geographical distances can be bridged in no time at all. It also provides the opportunity for more individuals to be involved in decision making.

in the late 1980s. It provides a means of communication that is obviously richer, enabling more intense human interaction than a telephone conversation, though perhaps not as good as a face-to-face meeting.

The initial problem with video conferencing was that as it generated such large amounts of data, it consumed incredible amounts of bandwidth.[5] The arrival of ISDN[6] and its gradual implementation by the telephony companies together with the compression of bandwidth gave a lifeline to video conferencing. Unfortunately ISDN is an imperfect network technology as there is a need for multiple ISDN channels to support the high bandwidth needed (even after compression), line drops, different delays between channels, calls are expensive and ISDN lines must be installed into all premises that require a link. ISDN has formed the basis of group video conferencing systems. These are usually located in a specific room within a business that has to be reserved in order to schedule when they are used. This is restrictive and takes away the spontaneity of using video conferencing. All information such as audio, data and video is transmitted in digital form at high speed over the public switched telephone network (PSTN). Normally conferences are between two participants (known as point-to-point, a conference with two groups and a single information path between each group). Protocol H.320[7] is the established standard using dedicated ISDN lines and relates to point-to-point. Technology now allows several groups of people to participate. A multipoint control unit connects and manages all the ISDN lines. A multipoint conference is one with three or more groups involved and multiple information paths between the groups. These two types of video conferencing, point-to-point and multipoint, have their own protocol.[8]

However, with the rise of communication platforms based on the Internet Protocol (IP) there has been a migration of video conferencing from ISDN to IP. This has arisen because systems are less expensive, IP networks do not have to pay per minute usage fees (they are free of charge), there is significantly higher reliability (as ISDN uses a combination of data channels), there is improved convenience (easier to install) and enhanced audio and visual quality (through far greater bandwidth). IP forms the basis of desktop video conferencing systems, which are PC based. The popular transport media used for desktop video conferencing are: local area networks (LANs), wide area network (WAN) internet, ADSL and virtual private networks. They all have strengths and weaknesses. With a LAN the bandwidth is significantly more than ISDN, and the video quality is much higher, approaching that of television. Like LANs, the internet, virtual private networks and ADSL are other forms of TCP[9]/IP networks and hence can be used as transport media in desktop conferencing systems. Protocol H.323[10] is a newer standard using IP to carry information for point-to-point and many-to-many communications.

---

[5] Bandwidth is the data capacity of a service, measured in thousands or millions of bits per second. In video conferencing systems a larger bandwidth is used to spread or "dither" the signal in order to prevent interference.

[6] ISDN stands for "Integrated Services Digital Network". An ISDN line is similar to a standard telephone line and provides a digital connection to a communication network at a speed of about 128 kilobytes per second.

[7] H.320 is an ITU (International Telecommunications Union) umbrella standard that is used to describe visual and audio communications systems over "narrowband". In video conference terms, it is used to denote video conferencing over an ISDN link/line.

[8] A protocol is the standard set of rules that enables video conferencing units to "talk" to each other.

[9] Transport Control Protocol (TCP) is a reliable protocol designed for transmitting alphanumeric data. It can stop and correct itself when data is lost. There are other standards-based transport protocols used with conferencing such as User Datagram Protocol (UDP) and Real-Time Protocol (RTP). Generally each configures data into packets, with each packet having a "header" that identifies its contents. Each offers a different degree of reliability.

[10] H.323 is an International Telecommunications Union (ITU) umbrella standard that is used to describe multimedia communications systems over IP-based network infrastructures. In video conference terms it is used to describe video conferencing over IP networks. The ITU is a UN agency.

There are a series of risks associated with video conferencing, and they predominantly relate to technology rather than human interaction issues. Systems that want to talk to each other must be compatible. For instance, without pre-planning there is the risk that systems adopted do not conform to the telecommunications standards set by the ITU and/or the Internet Engineering Task Force and hence are not compatible. The bandwidths do not keep up with the pace of change in video conferencing technology. User expectation levels are not aligned to what is realistically possible from the available systems.

E-commerce

Electronic commerce or e-commerce is the buying and selling of goods on the internet. It is doing business electronically. Other terms that are used when talking about e-commerce are B2B and B2C, shorthand for business-to-business, where companies do business with each other, and business-to-customer, where companies do business with customers using the internet. These are considered to be the main forms of e-commerce. The biggest volume of trade on the internet is B2B. Technology companies such as Cisco and Oracle were among the first to transfer their purchasing and indeed most of their sales to the web.

Online retailing is often referred to as "e-tailing". Possibly the best-known example of an e-tailer is Amazon, whose name has become synonymous with e-commerce. "Bricks-and-mortar" refers to companies with traditional outlets. "Bricks-and-clicks" refers to companies that use a mixture of offline and online channels. The internet is radical as it makes a difference to a whole range of managers' day-to-day activities such as locating a new supplier, collecting customer data, obtaining news updates and exploring new markets. The changes that the internet brings are simply more pervasive and varied than anything that has gone before.

In the summer of 1999 Larry Ellison, CEO of Oracle, announced that the company would cut $1 billion from its global corporate expenses of $7 billion. "I've been in business for more than 30 years, and I think this is far the biggest productivity advancement I've seen in my life", commented Jeff Henley, Oracle's chief financial officer (Percival-Straunik 2001). If customers can submit electronic orders, it can save a business considerable resources normally required to implement the process. In addition, customers can track the progress of their orders for themselves. Businesses that are out there early are not just getting ahead of the learning curve. They are redefining business boundaries. For instance, since its foundation 17 years ago, Amazon.com has transformed itself from a simple bookseller into one of the world's biggest online retailers, selling everything from DVDs to kitchen appliances. Search engine advertising in particular has proved to be very lucrative. In January 2010, Google reported that its total revenue had increased by 17% from a year earlier to $6.67 billion. Nearly 97% of its 2009 revenue of $22.889 billion was generated by advertising, and the majority of that through AdWords – the sponsored listings that appear next to search results. Google has diversified considerably from its origins and continues to expand through acquisition. One of its first and most well-known acquisitions was Keyhole Incorporated, changing the name of its product Earth Viewer (which gave a 3-D view of the Earth) to Google Earth. Later, Google purchased the online video site YouTube and GrandCentral whose name was subsequently changed to Google Voice.

E-mail

E-mail has become an indispensable tool in our daily lives. Communication is now almost instantaneous. E-mail is now taken for granted just the same way PCs are. Coca-Cola was one of the first companies to establish worldwide communication using its own custom e-mail system in the 1980s. Now e-mail is ubiquitous. It flattens the hierarchical structure of an organisation. It encourages people to communicate and express their opinions. Senior managers have become more accessible. E-mail and telephone communication are changing the requirement for employees to have to live close to work (although not until the integration of the phone and PC, so that you can see who you are talking to, will it overcome the solitary nature of home working and make remote interaction with colleagues and customers possible). Internal communication within businesses can be facilitated by attaching links to pages in the intranet to e-mails sent to staff. While e-mail is invaluable, the dependence placed upon it causes considerable disruption when its availability is temporarily removed. The e-mail server must have the same power back-up and support services that are provided for other major business systems. E-mail must be protected from viruses, spam or "junk mail" and various strains of virus, which were discussed under the heading "Broadband".

17.5.3 Control Technology

Specific computer-based production control systems include the following:

- Computer-aided design (CAD) is the use of computers to assist in the production of designs, drawings and data for use in the manufacturing process. It increases flexibility and speed.
- Computer-aided manufacture (CAM) is the use of computers to support the manufacturing process. It improves speed and quality and reduces costs.
- Flexible manufacturing systems (FMSs) are an important concept made possible by computer numerically controlled (CNC) machines, affecting the integration of manufacturing cells, productivity and quality, in a wide variety of strategic industries.
- Mechatronics is the combination of microelectronics, mechanical engineering and computer technology.
- Computer-integrated manufacture uses IT to integrate one or more parts of the manufacturing process – for example, design and production (CAD and CAM).
- Manufacturing resource planning (MRP) can be described as both a computerised planning system and an inventory management model.
- Operational research (OR), also known as management science (OR/MS), looks at an organisation's operations and uses mathematical or computer models, or other analytical approaches, to find better ways of doing them. Operational research consultants use analytical tools to aid decision making such as critical path analysis, production, scheduling, Markov chains, queuing theory, replacement, simulation, stock control, dynamic programming, decision theory and game theory.

Computer-Aided Design

CAD has traditionally been a computer-aided system for drafting, creating and communicating a two-dimensional design, or a three-dimensional model, for a product or components of a product. Heavily used in the electronics, mechanical engineering and construction industries, CAD seeks to visualise a design (a concept) before manufacture. This design can then be tested and evaluated (e.g. physical shape, size and volume, aesthetic attributes, fluid dynamics, material suitability and conductivity) prior to manufacture. The key concept of CAD is the representation of an idea or concept using the most appropriate ICT tools for the product that is being designed. CAD is delivered through the use of software. The risks associated with CAD relate predominantly to the use of the software rather than the software itself. The better-known products have been available for many years now, have been constantly developed and hence can be considered robust.

The hardware used to support CAD must have sufficient memory and processor speed. The largest risk associated with CAD is that the benefits that should be delivered are not realised. There are a host of reasons why this might arise, among them inadequate user training, the choice of the wrong software product for the task in hand, and the way the users work together.

A key characteristic of design in all industries, without exception, is the degree of specialisation that is taking place, which means that designs are produced by several discipline specialists. The integration of the designs of the specialists holds the key (for the most part) to the quality of the design produced. In advance of the preparation of the design, the layering conventions have to be agreed. A single drawing will be made up of multiple layers and each layer, for example, may contain the design of a single discipline (for building design there may be architectural, structural, mechanical, electrical, drainage and acoustic layers). The coordination of the information contained on the layers will reflect directly on the design, and reworking may be required to correct deficiencies. Additionally, information must be produced in a particular sequence and when the information is not available from one discipline to suit the program the whole design is delayed. If one designer omits information, which has to be added later, each of the other disciplines may have to modify their design to coordinate with the change.

Computer-Aided Manufacture

CAM is the computer control of manufacturing production machines, ranging from CNC[11] machines to high-performance programmable industrial robots, which can perform a variety of industrial tasks. CAM is commonly linked to CAD systems. The resulting integrated CAD/CAM system then takes the computer-generated design drawings and converts them directly into instructions for the production machines within the overall manufacturing system, optimising consistency between design and finished product. The development of CAD had little effect on CNC initially due to the different capabilities and file formats used by drawing and machining programs, but as CAD applications such as SolidWorks and AutoCad began to incorporate CAM intelligence, and as CAM applications such as MasterCam adopted sophisticated CAD tools, design and manufacture became more integrated.

[11] CNC machines existed as far back as the 1950s. They used the existing technology of the day – paper tapes with regularly spaced holes punched in them to feed numbers into controller machines that were wired to the motors positioning the work on machine tools. The electromechanical nature of the controllers allowed digital technologies to be readily incorporated as they were developed. By the late 1960s numerically controlled (NC) machining centres were commercially available, incorporating a variety of machining processes and automatic tool changing. Such tools were capable of doing work on multiple surfaces of a workpiece, moving the workpiece to positions programmed in advance and using a variety of tools – all automatically. In addition, the same work could be carried out over and over again with extraordinary precision and very little additional human input. NC tools immediately raised automation of manufacturing to a new level once feedback loops were incorporated (the tool tells the computer where it is, while the computer tells it where it should be). What finally made NC technology enormously successful was the development of the universal NC programming language called APT (Automatically Programmed Tools). Announced at MIT in 1962, APT allowed programmers to develop postprocessors specific to each type of NC tool so that the output from the APT program could be shared among different parties with different manufacturing capabilities. In a production environment, a series of CNC machines may now be combined into one station, commonly called a "cell", to progressively machine a part requiring several operations. CNC machines commonly control activities such as welding, soldering and milling machines. CNC machines represent a special segment of industrial robot systems, as they are programmable to perform many kinds of machining operations (within their designed physical limits, like other robotic systems). Source: "What is CAD/CAM?", Harvard Design School, *Computer Resources Manual*, http://www.gsd.harvard.edu/inside/cadcam/whatis.htm

The commonly used examples of CAM often refer to engineering processes, such as car manufacture. An advantage of CAM is that it can be used to facilitate mass customisation: the process of creating small batches of products that are custom designed to suit each particular client. Without CAM and the CAD process that precedes it, customisation would be a time-consuming manual and costly process. However, CAD software allows for easy customisation and rapid design changes, while the automatic controls of the CAM system make it possible to adjust the machinery automatically for each different order. Box 17.1 describes an example of investment by a US car manufacturer in robots to secure this ability to quickly change product lines to respond to the loss of competitive advantage. The key risks associated with CAM are the lack of full integration with CAD, a lack of investment in CAD/CAM, insufficient investment in research and development, losing sight of the developments in CAD/CAM and how they may improve productivity and/or create competitive advantage, not reflecting customer preferences in product lines and not benchmarking output against competitors. Laid over these risks are the typical operational risks of outsourcing, maintenance regimes, maintenance contracts, business continuity, security and protection of intellectual property.

Box 17.1 CAM and the use of robotics: Flexibility in production

It was recently announced that DaimlerChrysler was borrowing a page from Toyota's book and retooling its factories with robots that can switch from building one model to another model in 42 seconds. These new robots would allow Chrysler to build two or three models on the same assembly line. The first plant to be retooled with the flexible robotic system is a Neon factory in Belvidere, Illinois, at a cost of around $415 million. Vice president of manufacturing Frank Ewasyshyn was open with the press in disclosing that the inspiration for the retooling had come from their competitor Toyota. The more flexible plants will allow Chrysler to respond more quickly to shifting market demands. Pilot models can be more easily assembled and existing models can be modified to meet niche markets. Increased flexibility is critical in the current highly fragmented automobile market. Traditional product lines had all but disappeared as consumers demand a wider range of choices. Most automakers only produce between 70 000 and 100 000 units of a model every year and some niche models, like Chrysler's Viper, have annual runs of 5000 vehicles. This flexibility in production is set to continue. Dr Ken Young, Chairman of the Warwick University-based British Automation and Robot Association (BARA), has advised that advances to give robots "eyes" will mean that they could soon do much more. The new "vision systems" would equip robots with a camera, which would then be used to take a picture for analysis by the robot. The new technology – expected to come into use in three or four years – would have a massive impact, said Dr Young, as they would soon be far more flexible in what they could and could not do. "In the past robots have only been able to do exactly the same thing, but now they can adapt what they are doing to whatever is in front of them", he said. It would have enormous benefits to businesses. Robotics are seen as key to business development. The BBC reported that in the third quarter of 2004, 507 robots had recently been bought. Their use being key to the car industry was demonstrated by the fact that of the 507 sold, 400 had been bought between two car manufacturers: Toyota based in Derbyshire, and BMW based in Oxford.

Source: Based on Yahoo! News (2005) DaimlerChrysler borrows page from Toyota's manufacturing book, 3 August; BBC News (2004) Sales of robots hit record high, 18 October.

Flexible Manufacturing Systems

FMSs aim to integrate the use of flexible automation equipment such as NC machine tools and industrial robots and achieve improved efficiency in the small-scale production of a large number of products. The serious efforts made to shorten the time between design and production by enabling the data prepared by CAD to be used by CAM are symbolised by the term CAD/CAM.

Mechatronics

Mechatronics is a term, coined in Japan, which describes the combination of microelectronics and mechanical engineering (Technova 1983). Mechatronics was first used in terms of the computer control of electric motors by an engineer at Japan's Yaskawa Electric Company in the 1960s (Ashley 1997). While innumerable definitions of mechatronics are offered, the favoured definition adopted here is that offered by Loughborough University (UK), which states: "Mechatronics is a design philosophy that utilises a synergistic integration of Mechanics, Electronics and Computer Technology (or IT) to produce enhanced products, processes or systems." The UK Institution of Mechanical Engineers stresses that mechatronics is not just the combination of disciplines but a fundamental way of looking at doing things, a total synergy. On its website the institution states:

> First and foremost Mechatronics should be seen to represent technology integration and not merely a combination of the primary disciplines. In fact the "fusion" of mechanical, electronic and computer based structures into a complete Mechatronics "product" can only achieve its desired functionality through a process of systematic integration of all inherent disciplines involved right through from the conceptual stages.[12]

The use of mechatronics by Japan (a country with scarce natural resources whose economy is dependent on exports) saw the launch of a formidable challenge to the dominance of the US and Europe in the electrical goods market. After massive investment Japan was able to produce large quantities of very precise machinery at low cost with very high quality and reliability. Mechatronics opened up enormous technological possibilities. The developments in mechatronics over the last 40 years has been startling and the pace of change formidable. Current examples of mechatronics are computer disk drives and the increasingly sophisticated and ever smaller camcorders and compact disc players. These would never have been plausible by adopting a traditional single disciplinary or combined approach.

Significant to the development of mechatronics has been CAD/CAM, discussed above, which is described as the core technology of mechatronics. The risk for manufacturers immersed in producing products which embrace mechatronics is the rate of change. The introduction of new products by competitors can make some products obsolete almost overnight. The home entertainment market is a prime example. Significant sums have to be invested in research and development, and new products may have a very short life before more advanced products are placed on the market by competitors.

[12] http://www.imeche.org/knowledge/industries/mechatronics-informatics-and-control/about-the-group/mechatronics-forum/what-is-mechatronics

Manufacturing Resource Planning

MRP is essentially a spreadsheet tool that converts sales forecasts into purchasing requirements for materials and components, and plans actual production. The techniques of MRP can be utilised in the manufacturing of goods involving a number of stages in the production process. The stock levels of raw materials, single components, subassemblies and finished goods are analysed in the MRP approach. In essence, much of the demand can be accurately predicted when the demand for the finished product is known. For instance, an order for a specific customised model of a sports car received by BMW results in the exact requirements of the subassembled components such as body panels, which determine the raw materials requirements such as steel and types of paint. The control of stock where the demand for one item depends on the demand for a higher-order item can be achieved using MRP. The system would automatically generate orders for given items based on the demand for finished goods.

Operational Research

Perhaps the base way to describe OR is to use case studies. The studies cited here are drawn from the website of the Operational Research Society.[13] The first relates to improving the speed of car body production at PSA Peugeot Citroën. To set the study in context, PSA Peugeot Citroën is the seventh largest car maker in the world and the second largest in Europe.

> In 1998, to meet its new CEO's ambitious targets for growth, innovation, and profitability, PSA decided to focus on bottlenecks in the car body shops in its plants. The shops' single flow architecture limited PSA's ability to handle diverse models, and ingrained beliefs and practices in production line design were causing inefficiency. The car body production line needed a new architecture that could handle model diversity and new car launches easily and quickly, and a method of sustaining quick innovation without overinvestment. Solving these problems would require operations research expertise in simulation, Markov chains, and sophisticated analytical methods.

The PSA OR team used a multi-method approach to evaluate performance, developing an iterative three-step design process that took advantage of the speed of analytical methods and the accuracy of simulation.

The OR tools improved throughput with minimal capital investment and no compromise in quality – contributing $130 million to the bottom line in 2001 alone. PSA personnel, initially sceptical about OR, had the opportunity to compare the results that the analytical methods predicted with the actual outcomes. Persuaded by the accuracy of the forecasts, they and other PSA divisions adopted the tools and initiated further OR projects. Christophe de Baynast, Director of Car Structure Entity at PSA, declared "We expect $130 million revenue gain just from this improvement. Other benefits due to this work include aid in decision making and forecasting, more accurate and faster shops' design, and better knowledge of manufacturing systems in our staff and our suppliers."

The second study relates to Air New Zealand's need to improve the way it scheduled tours of duty and rosters. The challenge consisted of two problems: a tours-of-duty planning problem to generate minimum-cost tours of duty (sequences of duty periods and rest periods) to cover all scheduled flights, and a rostering problem to assign tours of duty to individual

[13] http://www.theorsociety.com

crew members. Solving these problems required operational research expertise in complex scheduling and routeing. Air New Zealand staff and consultants, in collaboration with the University of Auckland, developed eight optimisation-based computer systems to solve all aspects of the tours-of-duty planning and rostering processes for the airline's national and international operations. As of 2000, these systems had saved NZ$15.655 million per year while providing crew rosters that better respect crew members' preferences.

17.6 RESPONDING TO TECHNOLOGY RISK

17.6.1 IT Governance

According to Thomas (2005), the idea of IT governance has come about as a way of imposing order on chaos. While chaos may be too strong a word, there have been a number of spectacular IT failures, particularly in the public sector. There is conviction that careful management of risk has a direct relationship with the success of any business, as expressed by Paul Beach (head of corporate banking at Atos Consulting): "If shareholder value is coming out of your IT, then good corporate governance of that is absolutely vital because a risk poorly managed could drive that value away" (Thomas 2005). Without a doubt, the pervasive use of IT by businesses has created a critical dependency for business operations. In addition, an increasing number of businesses rely on their IT for competitive advantage or hope to successfully leverage IT to achieve competitor advantage. Hence, boards need to extend governance to include IT. Business IT governance is all about boards exercising a set of responsibilities and practices with the view to providing strategic direction, ensuring the IT objectives are achieved, ascertaining that the risks are managed and verifying that the businesses IT resources are appropriately deployed.

The definition of IT governance used by the Control Objectives for Information and Related Technology (COBIT)[14] standard, which is issued by the IT Governance Institute (ITGI), is "a structure of relationships and processes to direct and control the enterprise in order to achieve the enterprise's goals by adding value while balancing risk versus return over IT and its processes". To accomplish IT governance the development of an IT framework is required which describes the benefits and opportunities sought, the current IT capability and any shortfall and corresponding investment needs, accountability, funding, rules on the use of the system, risk, control processes and effective systems. Some believe the primary focus of IT governance is risk management. For Jay Heiser (research director at Gartner Research Group), "Governance is that whole area of controlling unwanted activity". He argues that IT governance is largely about managing and reducing risk – making sure that systems are secure from viruses or unauthorised access, for example, or making sure that regulatory requirements are not being breached (Thomas 2005). But the benefits of risk management are much broader than this. The IT Governance Institute states that successful enterprises understand and manage the risks and constraints of IT. Managing IT risk involves looking at the risks to business performance arising from IT systems. Approaches that businesses may use to ensure that this risk is minimised and appropriately assigned include the following:

[14] COBIT, now in its fourth edition (with the release of the fifth edition planned for 2012), is internationally accepted as good practice for control over information, IT and related risks. COBIT is used to implement governance over IT and improve controls. The IT Governance Institute (http://www.itgi.org/) was established in 1998 to advance international thinking and standards, in directing and controlling an enterprise's IT. ITGI offers symposia, original research and case studies to assist enterprise leaders and boards of directors in their IT governance responsibilities.

- Setting the IT risk appetite for the business which will then inform decisions on such matters as outsourcing, system integration, insurance, security and business continuity.
- Maintaining a risk register, proactively managing the risk response actions and regularly reviewing the high-level risks at board meetings.
- Assigning clear risk accountability. For enterprises this means being aware that final responsibility for risk management resides with the board. While the board may wish to delegate responsibility for aspects of risk management, it is responsible for overseeing its successful implementation. The board cannot abrogate its responsibilities.
- Being aware of the fact that a transparent and proactive approach to risk management can create competitive advantage.
- Insisting that risk management is embedded in the management structure so that the business is able to respond quickly to changes in risk profiles. Managers must know how to escalate risk information through the organisation, what to report, when to report, to whom and in what format (how much detail).
- Understanding whether IT is aligned to the business objectives and how critical IT is to growing the business.
- Understanding the IT fit. No business can stand still and survive. There must be an understanding of how IT systems must change to reflect and be aligned to organisational change.
- Understanding whether IT projects overrun, exceed their budgets, fail to satisfy end-user needs or require expensive ongoing maintenance.
- Understanding how robust return-on-investment studies inform IT expenditure.
- Conducting lessons learnt workshops on completed IT projects and ensuring managers of subsequent projects familiarise themselves with the content of the lessons learnt reports.
- Ensuring the board is aware of the latest developments in IT and whether a lack of investment would erode market share or their business.
- Being clear about both the risks and opportunities associated with new technology, outsourcing, integrating legacy IT systems within acquired companies with existing IT systems, adoption of new business models and e-commerce.
- Avoiding an overdependence on a single vendor by analysing the IT industry[15] and identifying alternative suppliers.
- Ensuring IT risk is aggregated so that risk management effort can be prioritised and there is an awareness of the consequences of risk mitigation action. That is, the reduction in one risk does not increase risk exposure elsewhere in the business to the extent that the overall business IT risk is increased.
- Ensuring comprehensive two-way communication. IT practitioner language can be an obstacle to communication. IT managers must be adept communicators and hence, where

[15] As described by Gates (1999): "The realignment of the computer industry from vertically integrated vendors to horizontally integrated, customer-driven solutions has brought prices down dramatically and offered more choice. In the old vertically integrated computer industry, a customer would buy almost all of the elements of a solution from a single company – the chips, the computer systems built on the chips, the operating system, the network, hardware and service. Every vendor – IBM, Fujitsu, HP, Digital, NCR and others – had its own vertical solution. Sales volumes were low, and prices were high. Integration among vendors was difficult and expensive. Switching costs for customers were very high since every piece of the solution would have to change. These vertically integrated vendor solutions are being displaced by the PC approach in which specialised companies give customers a choice in each of the infrastructure layers: chips, computer systems, system software, business applications, networking, systems integration and service."

necessary, must sharpen their language and presentation skills to ensure the full business implications of IT are communicated. Decision making by the board will be dependent on the quality of the information it is supplied to make those decisions and the board's capacity to make such judgements. The board needs to know how much IT risk the organisation is taking.

17.6.2 Investment

There are still widespread concerns that IT projects are not delivering against targets. Research by Accenture, based on interviews with 300 executives in the UK and Ireland, identified two main concerns over IT development projects. Andrew Morlet, head of the strategic IT effectiveness group at Accenture, explains: "First there was a strong consensus that IT spending is going up and the alignment between business and IT is improving. But second there is still a belief that IT is not delivering against investment" (Manchester 2005). The essential features of investment decisions, regardless of who the decision maker is, are time and cost. Investment involves making an outlay of something of economic value, usually cash, at one point in time which is to yield economic benefits to the investor at some other point in time. Typically the outlay precedes the benefits. In addition, the outlay is typically a large single amount and the benefits arrive in a stream of smaller amounts over a fairly protracted period. However, it is common for a chief executive to be charged with accomplishing certain improvements in bottom-line performance within a given timeframe.[16] Hence, right from the outset there may be pressure on a project to show quick improvements in performance and/or costs. IT projects are commonly evaluated from a risk and benefit perspective, where risk is associated with the degree of certainty over accomplishing the time, cost and quality parameters. The investor seeks to understand the predictability in returns in terms of the business benefits, the life of the asset and the maintenance costs of the asset during its useful life.

It is possible to see the investment process as a sequence of six key stages (see Figure 17.2). Each of the stages must be given proper consideration by managers. There are a number of inherent risks in any investment process. The common risks include lack of stakeholder involvement, lack of adequate end-user consultation, insufficient project definition, poor assessment criteria, inadequate time devoted to the selection process, lack of adequate assessment of the implementation and in-use risks and insufficient input from IT specialists.

Determine the Investment Funds Available

The board commonly determines the amount of funds available for IT investment within a business. Usually the limit on available capital for any one period will mean that the funds available will not be sufficient to finance all of the projects identified. When this occurs some form of "capital rationing" has to be undertaken. This means that managers are faced with the task of deciding on the most profitable use of the investment funds available. The projects may consist of new IT projects to support new processes or projects to replace existing hardware and software to reduce operating costs and improve overall business bottom-line performance.

[16] It is interesting to note that when Coca-Cola's CIO, Bill Herald, conducted the company's first ever information strategy review in 1997 to align with the company's business strategy the company came to realise that, despite its earlier investments, it too often treated information technology as an expense to control rather than as an enabler of better business. As a result of this realisation the thinking at Coca-Cola moved from a preoccupation with savings to sharing information globally so that individuals did not reinvent the wheel, spending time re-creating information already available in the company (Gates 1999).

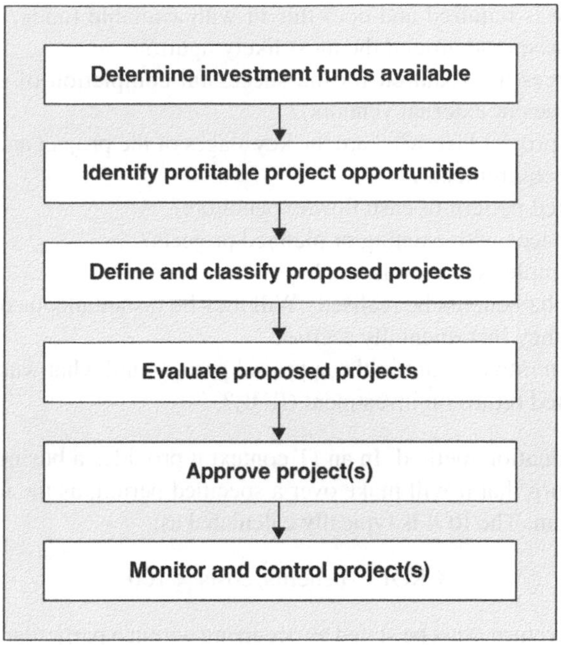

Figure 17.2 The investment decision-making process

Identify Profitable Project Opportunities

An important part of the investment process is the search for profitable investment opportunities. However, management may have submitted so many proposals, and are clamouring for funding, that this task is quickly executed. It may be simply a gathering process.

Defining and Classifying Proposed Projects

This stage of the process aims to convert promising ideas into full-blown proposals. This means that further information will be required, much of it of a detailed nature. The first stage will involve collecting enough information to allow a preliminary screening. Many proposals fall at this first hurdle because it soon becomes clear that they are unprofitable or unacceptable for other reasons. Proposals that are considered worthy of further investigation continue to the next stage.

Evaluate Proposed Projects

Once a project has passed the preliminary screening stage and has been fully developed, a detailed evaluation is usually carried out. For projects of any size, this will involve providing answers to a number of key questions including the following:

- What is the nature and purpose of the project?
- What are the perceived benefits and how were they determined?
- Does the project align with the overall objectives of the business?

- How much finance is required and does this fit with available funds? If this is not a hard number, what is the spread around the most likely figure?
- What other resources are required for the successful completion of the project (such as number of employees or external vendors)?
- How long will the project last, what are the key stages of the project and does the timescale reflect the CEO's requirements?
- What is the expected pattern of cash flow/expenditure?
- What are the interfaces with existing or planned projects?
- What are the key implementation issues?
- How quickly will the benefits be realised? Will they be instantaneous once the new system is in place or will they incrementally accrue?
- Has risk been taken into account in the appraisal process and what was the outcome?
- What is the expected return on investment (ROI)?

The ROI is a key evaluation method. In an IT context it provides a business with an estimate of the percentage return that it will make over a specified period, as the result of investing in a new computer system. The ROI is typically calculated as:

$$\%\,\text{ROI} = \text{benefits}/\text{costs} \times 100.$$

To be meaningful the return must be stated as occurring within a particular timeframe/payback period. The opportunity cost must also be considered. This is the financial gain that would have been made if the sum of money assigned to the IT project were invested elsewhere (e.g. in a bank). For reliance to be placed on the ROI calculation the two key components, current costs and anticipated cost, must be fully calculated, capturing all of the contributory costs.

An example of an ROI estimate for a replacement system, looking at a three-year period, is as follows:

$$\%\text{ROI} = \text{benefits/costs} \times 100$$

$$\%\text{ROI} = \frac{\begin{array}{l}(\text{current annual costs})*3 - ((\text{new estimated annual costs}*3 + \text{risk})\\ +(\text{investment costs} + \text{risk}))\end{array}}{(\text{current annual cost})*3} \times 100.$$

The current annual costs must take account of the existing software licence costs, software support, hardware support, employee costs, overheads (rent and/or building running costs) and any other maintenance costs. The investment costs will include the hardware and software purchase costs, installation costs, the cost of removal of any redundant hardware/software and any consultancy fees. The new estimated annual costs will comprise the software licence costs, software vendor support, hardware vendor support, employee operators and overheads such as rent and building running costs.

Jordan and Silcock (2005) describe what they call "investment traps whereby projects do not live up to expectations and the business benefits envisaged at the outset of a project are not realised":

- project overspend increases the initial outlay;
- the project runs late and the achievement of business benefits is deferred;
- business take-up and usage is less than planned, diminishing business benefits;

- a merger or acquisition requires significant redevelopment effort to accommodate the unforeseen needs;
- the system is more costly to support, maintain and enhance than planned increasing costs and lowering net benefits;
- business requirements change or system requires replacement sooner than planned, reducing the asset's useful life.

Approve Project(s)

Once the managers responsible for authorising the investments are satisfied that the project should be undertaken, formal approval can be given. However, a decision on a project may be postponed if more information is required from those proposing the project or if revisions are required to the proposal. In some cases the project may be rejected if it does not provide a return on investment.

Monitor and Control Project(s)

Making a decision to invest in new IT support does not automatically cause the investment to be progressed and be implemented without problems. Managers will need to actively manage the project through to completion. Managers should receive progress reports on the project at regular intervals. The reports should provide information on the status of the project in terms of actual expenditure against planned expenditure, actual progress against planned progress, the effectiveness of risk management actions and any changes to the brief. In extreme cases managers may abandon the project if circumstances appear to have changed dramatically for the worse.

17.6.3 Projects

There is abundant evidence that IT projects have a poor track record of delivery against their objectives. The delivery objectives, as with projects across other industries, are typically time, cost and quality, including functionality (Jordan and Silcock 2005). Project risk management as a discipline is discussed in Chapter 18.

When a project is late and allowed to slip, the business is deprived of the planned benefits for longer. Projects that are late but are not permitted to miss their completion date attract inherent risks in terms of their functionality, operability and maintainability. They do not deliver all that they were planned to deliver and there are operational problems that require rectification. A functionality and quality gap will amount to a loss of benefits. When the project is solely for internal use, operational problems reduce efficiency. Depending on the scale of the project, this shortfall may be a minor performance issue or a loss of automation resulting in a far lower reduction in headcount than had been hoped for, with significant cost implications. When projects are outward facing and are at the interface with the general public, business partners or suppliers, the ramifications of poor performance can damage a business' reputation. Projects that are late or struggling from a combination of inexperienced personnel often request more project funds. Each time more funds are awarded the initial benefits become diluted. The cost of slippage is the sum of the direct costs of in-house resources retained longer, the costs of consultancy support (if the reason for delay lay with the sponsor), the indirect costs of lost business benefits for the period of the slippage and the impact on any dependent projects. When a high proportion of the original budget has already been spent, which may already amount to several millions, and an application is made for more funding which puts the project on the limit of acceptability (and there is no certainty on the final outturn cost), the decision to abandon the project can be very difficult. To overcome poor project experiences there is a school of thought that "big and ugly" projects should be abandoned for a set of interlinked "small and beautiful" projects. However, this introduces the problem of managing a programme of interdependent projects with a different set of risks which may represent a significant challenge to manage.

When a project fails there is a "lost opportunity" risk where the original objectives of the project have not been met and the previously identified shortcomings have not been addressed. In addition, if the money invested in the project had simply been invested in a bank it would have earned interest. Alternatively, the money could have been invested in another project. Jordan and Silcock (2005) describe a further negative outcome of failed projects as "collateral damage" risk. This risk they describe as referring to project team members who suffer a crisis of confidence and lose the trust of the planned recipients of the project.

17.7 SUMMARY

This chapter has examined three technologies collectively described as "primary technology types". These are information, communications and control technology. Risks are identified against all three technologies; however, all types present businesses with very significant opportunities and the potential to achieve competitive advantage. Hence, one of the most serious risks to a business does not emanate from the technology itself but from a lack of investment in IT. Communication is now predominantly via broadband and e-mail. While this communication method provides an exchange of information at a speed hitherto unattainable, it was highlighted it has inherent risks. As broadband provides an always-on connection, there is a greater risk that businesses could receive unsolicited commercial e-mail (spam), viruses or be vulnerable to theft or hackers. While video conferencing is now more feasible through greater bandwidths, there are a series of risks that can undermine its performance such as lack of compatibility of systems, bandwidths not keeping pace with video technology and user difficulties.

The experience of Oracle and others is that e-commerce has cut millions of dollars from their global corporate expenses, and it has been described as the biggest productivity advancement for 30 years. The risk for businesses in not taking advantage of this technology is that competitors will steal on them, and late-comers will have considerable difficulty in playing "catch-up". Likewise, CAD/CAM linking design with manufacture combined with MRP and operational research (to create optimum processes and configurations of plant) provides businesses with the most sophisticated production capabilities available. Car manufacturers have found that without the very latest robots they cannot achieve the versatility in production runs that competitors can achieve. Hence, the risk again is being left behind due to a lack of investment. IT investment, the common factor across all businesses, while having the potential to drive business improvement, can be a business's Achilles' heel. Hence the continuing attention being given to IT governance. A number of high-profile IT project failures clearly demonstrate that IT projects are exposed to considerable risk, and the discipline of project management has not yet had broad-based success.

17.8 REFERENCES

Ashley, S. (1997) Getting a hold on mechanatronics. *Mechanical Engineering Magazine*, The American Society of Mechanical Engineers, found on www.memagazine.org

Department for Transport (2004) *The Future of Transport: A Network for 2030*, Cm 6234. Department for Transport, London.

Earl, M. (1997) Managing in the information era. In T. Dickson and G. Bickerstaffe (eds), *Financial Times Mastering Management*. Financial Times/Prentice Hall, London.

Gates, W.H. (1999) *Business @ the Speed of Thought: Succeeding in the Digital Economy*. Penguin, London.

Jordan, E. and Silcock, L. (2005) *Beating IT Risks*. John Wiley & Sons, Ltd, Chichester.

Ofcom (2010) Joint UK submission from the Department for Business, Innovation & Skills and Ofcom to the European Commission's Consultation on Universal Service Principles in e-Communications, May 2010.

Manchester, P. (2005) A change in attitude helps prevent project failure. *Financial Times*, 17 June.

Percival-Straunik, L. (2001) *E-Commerce*. The Economist, in association with Profile Books, London.

McLean, M. (ed.) (1993) *Mechatronics: Developments in Japan and Europe*. Frances Printer, London.

Thomas, K. (2005) A new era of accountability. *Financial Times*, 17 June.

18

Project Risk Management

One of the key implementation issues that must be addressed is how to overcome a corporate culture that is lacking or even negative toward risk management.
(Edmund Conrow)

The previous chapter examined technological risk, the third of six classes of risk exposure that businesses face relating to internal processes within the risk taxonomy proposed in Chapter 9. Technological risk overlaps with the fourth of the internal processes, *project risk management* (the subject of this chapter), as clearly technology improvements are introduced as projects. Businesses, in order to grow and improve bottom-line performance, regularly engage in activities which because of their characteristics are defined as projects. There is a myriad of definitions for the term "project".[1] The definition proposed here is as follows: a project is a unique activity with defined objectives, undertaken in pursuit of the achievement of beneficial change, typically constrained by limited resources. "Unique" means that the product or service is different in some distinguishing way from all similar products and services within the organisation, existing or planned. The "beneficial change" sought is commonly a component or subset of the organisation's strategic goals. "Resources" can refer to finance, personnel and software tools, for example. Critical to a project is that it has a defined start and finish date. Hence, the project life cycle distinguishes projects from non-projects. Examples of projects include developing a new drug, implementing organisational change, opening a new mine, starting up a new passenger airline, introducing new process software, constructing a new production plant, and launching a satellite. However, unless these projects are appropriately managed, rather than contributing to business growth they have the potential to damage the organisation's reputation, erode stakeholder relationships, diminish the share price and critically undermine financial performance. The existence of effective project risk management has the ability to significantly contribute to the realisation of a project's objectives (Davies *et al.* 2009; NAO 2011), while its absence has been cited as a contributor to project failure (NAO 2000, 2004, 2005, 2009a, 2009b). The process of project risk management is now well documented; however, its successful integration into projects still presents a challenge. All too often project risk management occurs in a "parallel universe" where the work of the risk analyst (risk coordinator, risk engineer or risk manager) and the project discipline leads[2] happen side by side and never come close enough to being fully integrated. This chapter explores some of the challenges encountered in embedding risk management within a project. Some of these issues are not for the faint-hearted. The structure of this chapter is shown in Figure 18.1.

[1] A number of project definitions can be found in the Wideman Comparative Glossary of Project Management Terms v3.1; see http://www.maxwideman.com/pmglossary/PMG_P09.htm

[2] The term "project discipline leads" refers to the domain experts – for example, for a construction project this might be the architect, structural engineer and electrical engineer, and for a rail project the signalling, traction power and/or communications engineer.

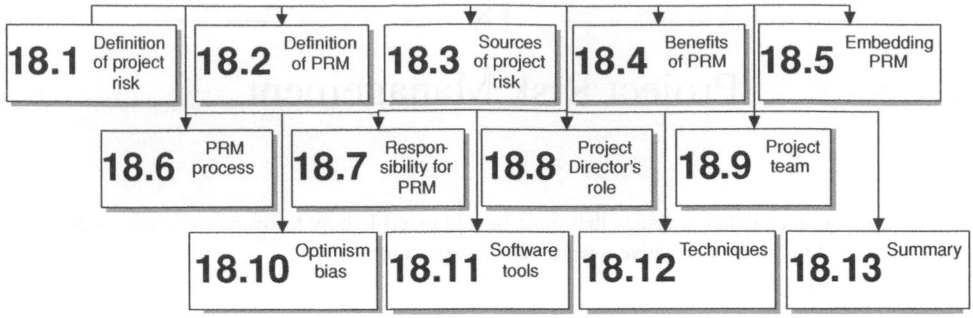

Figure 18.1 Structure of Chapter 18

18.1 DEFINITION OF PROJECT RISK

What is project risk? Risk is defined within the broad-based standard ISO 31000 *Risk Management – Principles and Guidelines* (ISO 2009) as "effect of uncertainty on objectives". A definition of project risk is found in British Standard BS 31100 (BSI 2008): "risk relating to delivery of a product or service, usually with the constraints of time, cost and quality". However, this definition does not articulate the two dimensions of project risk, namely likelihood and impact. When making an assessment of the significance of two or more risks, these two dimensions must be considered. Likelihood may also be described as "probability", and the impact is sometimes referred to as "consequence" or "effect". In addition, the definition of project risk must make reference to the fact that the event may have a negative or positive effect on a project's objectives. These two types of project risk are called threat and opportunity, respectively. Hence, a possible definition of project risk is as follows: project risk is the likelihood of a negative or positive event impacting on a project's objectives.

18.2 DEFINITION OF PROJECT RISK MANAGEMENT

"Risk" refers to risk exposure, whereas "project risk management" (PRM) refers to the management of that exposure in the pursuit of achieving predefined goals. Hence, PRM has two primary functions: a management function (the "what") to drive down risk exposure and exploit opportunities and a goal-seeking function to support the satisfaction of a project's aims or objectives (the "why"). PRM thus requires both a support management process and comprehension of the project objectives. (Typically objectives include cost, time and quality; however, they may also include goals relating to cost-in-use, the environment, maintainability, reliability, reputation, safety, scope and sustainability.) There are a number of descriptions and definitions of PRM but they only focus on one facet, the "what" and not the "why". Here are some examples:

- The US Project Management Institute's *PMBOK*® Guide states: "The objectives of Project Risk Management are to increase the probability and impact of positive events and decrease the probability and impact of negative events in the project" (PMI 2008).
- BS 31100 defines risk management as the "coordinated activities to direct and control an organization with regard to risk".

- Patel and Morris[3] describe PRM as a process but omit reference to the fact that its overarching goal is to secure the achievement of a project's objectives: "The process of identification, assessment, allocation, and management of all project risks".

Reflecting on the primary aim of PRM, a possible definition is as follows: PRM is the management process designed to exploit opportunities and treat risks to secure a project's agreed, defined and disseminated objectives.

18.3 SOURCES OF PROJECT RISK

The sources of risk considered to be embraced by the term "project risk" are very considerable. They will emanate from the external business environment, the industry within which the organisation sits, the sponsor's organisation (or department) and the project itself. The primary categories of project risk might be considered to include the following list, but will be far more extensive and vary depending on the project context such as the industry, location and sector (public or private). There is no "one size fits all".

- Business environment
- "Host" industry
- Sponsor's organisation
- Business case
- Project brief
- User requirements
- Project team
- Design, specification, layout
- Internal approvals
- External approvals
- Project controls (such as scheduling, cost management and change control)
- Procurement
- Implementation
- Testing and commissioning (for physical projects)
- Handover

18.4 BENEFITS OF PROJECT RISK MANAGEMENT

PRM has the "potential" to afford a business a series of benefits, which may include:

- supporting satisfaction of the business case;
- increasing the likelihood of a project achieving its objectives;
- forcing a project's objectives to be made explicit and prioritised;
- ensuring that realistic contingencies are defined;
- ensuring that the procurement route, form of contract and contract conditions reflect the sponsor's risk appetite and project objectives (where third parties are engaged in project delivery);
- forcing the team to think collectively and collaboratively;
- engendering a proactive management culture – "tackling tomorrow's problems today";
- validating the funding requirements;
- enhancing value for money;
- protecting the organisation's reputation and stakeholder confidence;
- improving accountability, decision making and transparency;

[3] Patel, M.B. and P.W.G. Morris. Centre for Research in the Management of Projects (CRMP), University of Manchester, UK, 1999.

- facilitating change to be executed more effectively and efficiently and improving project management;
- providing a better understanding of and compliance with relevant legal and regulatory requirements;
- improving integration of the project into the organisation's operations.

The use of the word "potential" is deliberate: as with all disciplines, success is dependent on a number of factors such as those described in the next section.

18.5 EMBEDDING PROJECT RISK MANAGEMENT

The effectiveness of an organisation's PRM is directly proportional to the maturity of its risk management practices and the degree to which they have been successfully integrated into the projects it implements. Integration here refers to whether risk management activities are well defined and described in the project plans and processes and incorporated into the project life cycle (the incremental stages in the project's life). Integration additionally relates to whether these activities happen routinely, as part of project management processes, or if they have to be "policed" and regularly prompted before they occur. In this context the terms "integration" and "embedding" are treated as if they are synonymous. Hence, well-defined risk management practices will count for little unless they are embedded within the organisation as a whole and are part of the project culture. However, embedding risk management within a project is not straight forward as there are a number of commonly recognised challenges.

18.5.1 Common Challenges in Implementing Project Risk Management

There are a number of challenges to the implementation of PRM that occur time and time again. These include, but are not restricted to:

- lack of clearly defined and disseminated risk management objectives;
- lack of senior executive and project director commitment and support for PRM;
- lack of a risk maturity model to guide the goals for risk management;
- lack of a change process to introduce the discipline (in situations where some form of PRM has not previously been embarked upon);
- no common risk language (terms and definitions);
- lack of articulation of the sponsor's risk appetite (i.e. risks the project will and will not take);
- no definition of risk management roles and responsibilities;
- lack of risk management awareness training to build core competencies;
- no integration of risk management with other project disciplines;
- reticence of project personnel to spend time on risk management;
- risk owners not automatically taking responsibility for the risks assigned to them;
- no clear demonstration of how risk management adds value and contributes to project performance;
- overcomplicated implementation through confusing policies, strategies, frameworks, plans, and verbose and mutually incompatible procedures;
- lack of alignment between the overall business strategy, the project business model and the risk management objectives for projects.

18.5.2 Lack of Clearly Defined and Disseminated Risk Management Objectives

Prior to the commencement of risk management activities on a project it is important for the risk management objectives to be defined to ensure planned activities are directed towards the desired outcomes. If management are particularly concerned about the impact of the project on existing operations (e.g. manufacturing), the project being completed on time (e.g. getting a drug to market prior to a competitor) or customer relations (e.g. initiating internet banking), the risk management objectives must reflect this required focus. Once the objectives are agreed they must be communicated to the team to ensure the risk process both commences and continues toward this focus.

18.5.3 Lack of Senior Executive and Project Director Commitment and Support

Risk management can never be successfully embedded within a project through attempts to drive it from the "bottom up" by risk management personnel. It needs to be driven from the "top down" by the project director and senior project personnel who clearly demonstrate belief in and commitment to PRM, not just because the project sponsor's remit calls for risk management but also because there is a belief that successful project management is dependent on knowing the risks, knowing their potential impact and proactively responding to them.

18.5.4 Lack of a Risk Maturity Model

The benefits of PRM derived by an organisation will depend directly on the level of maturity its risk management practices have attained. In the absence of an organisation-wide knowledge infrastructure, repeatable results are typically entirely dependent on the availability of specific individuals with a proven track record. This rarely provides the basis for long-term success and continual improvement. The starting point for an organisation wishing to embed risk management is an understanding of its own current competencies, its requirements and the planned method of accomplishment.

Risk maturity models can be used in a structured process to understand current risk management competencies and where and how improvement may be achieved. They can be used to answer the questions "What is it we want to accomplish with risk management and why?" and "What resources will it take to realise those benefits?". In general terms, a risk maturity model is a representation of mature practices for appraising an organisation's risk management competency. Models are typically structured as a series of distinct incremental steps which progressively drive greater benefits. The old adage of "If you fail to plan, you plan to fail" is just as true for implementing risk management. With the aid of a maturity model, organisations can set long-term goals for risk management having a clear understanding of the current maturity of their working practices and the areas that require improvement.

A maturity model provides:

- a shared goal;
- a common language;
- a description of incremental steps of improvement;
- a means of identifying the current level of risk management capability;
- a guide for process improvements;
- a tool to support a structured approach to process improvement.

18.5.5 Lack of a Change Process to Implement the Discipline

Where risk management is not already integrated into project management processes it needs to be implemented as part of a formal change process. Organisations that ignore the change management aspects of risk management do so at their peril. Integration is a building process where awareness and capabilities are incrementally developed. It is important not to overestimate the capacity and capability of an organisation to introduce a change. Introduction of the change will involve:

- establishing the objectives of the change;
- establishing the benefits the change will deliver;
- establishing the organisation's willingness and readiness to change;
- establishing the timeline of the change, the cost (if any) and the resources required;
- defining the substance of the change, the sequence of implementation and the method of effecting the change;
- recognising the factors that may influence, constrain or block the implementation of the change;
- defining the contributions that will be required to effect the change;
- understanding the interdependencies between this change and other change initiatives planned or already commenced;
- deciding if it is appropriate to carry out a pilot implementation first, to absorb the lessons learnt.

18.5.6 No Common Risk Language (Terms and Definitions)

Confusion can occur when there is a lack a common understanding of at least the key risk management terms, undermining effective implementation. The easy remedy is the production and dissemination of a schedule of terms and definitions which is regularly reviewed and expanded and/or modified to suit the implementation of projects. Ideally the risk management terms and definitions should be a subset of project–wide terms and definitions which are fully coordinated across all disciplines. For instance, the disciplines of risk management and cost planning should have the same understanding of what a contingency is and how it is defined. Likewise risk management and value management should have the same understanding of what an opportunity is and so on.

18.5.7 Lack of Articulation of the Project Sponsor's Risk Appetite

Projects are the tools to deliver the business strategy, growth and performance targets. Hence, projects must reflect and be subservient to the business strategy and the capital constraints of the organisation. The board and management will normally balance the cost of a project against its benefits (within the environmental context of the organisation) and, importantly, decide on the tipping point – the amount of risk that they are willing to accept before the business case for the project is negated. Without boards articulating this position, decisions are likely to be taken which are inconsistent with the cost–reward balance. In addition, they will be unable to challenge recommendations made by management about the initiation and progress of the project over time. Hence, a project sponsor (a manager with project delivery responsibility) should be given the project's "risk appetite" which will influence the degree of tolerance or latitude the project has in securing an individual project's objectives. However, risk appetite is commonly "defined" by exhibited behaviours rather than by a formal documented process. For the trouble with risk appetite is that it is not the easiest concept to define or communicate even though, for instance, each time an organisation enters into a contract with a third party, it is making a decision about its tolerance for risk. In most cases risk appetite is defined by a mixture of both qualitative and quantitative measures. Qualitative measures are typically intangible, such as customer and stakeholder relationships or reputation, whereas quantitative measures relate typically to potential changes in the capital investment required.

Why is risk appetite important? Without a definition of "risk appetite" an organisation may embark on a project without thinking through the possible levels of failure that may arise and the organisation's ability to withstand them. If the project failed in its entirety, what would that mean? How are procurement routes together with forms and conditions of contract to be objectively chosen unless there is both a recognition of the inherent risk exposure in each and an agreement on the "ceiling limit" to acceptable risk exposure?

18.5.8 No Definition of Roles and Responsibilities

A lack of a clear designation of responsibility for leading or participating in risk management will inhibit its effective implementation. The roles and responsibilities for risk management across a project team need to be clearly established and communicated. This is typically accomplished through a combination of activities, as follows:

- preparing a risk management plan and incorporating within it a RACI chart (or a variant thereof; see Section 18.9.3) which describes the individual risk management activities, those responsible and accountable for implementing them and those who will be consulted or informed in the process;
- presenting the risk management plan to the client and the project team;
- including risk responsibilities within job descriptions;
- reviewing whether the responsibilities included within job descriptions are being followed and conducting staff performance reviews;
- clear allocation of risk response actions to designated risk owners and risk actionees (where an actionee supports the risk owner in risk mitigation implementation);
- clear designation of risk reporting requirements;
- identification of an individual or group that will provide guidance and support in the implementation of risk management.

18.5.9 Lack of Risk Management Awareness Training to Build Core Competencies

Comprehension of, commitment to and implementation of risk management will be directly influenced by project team member's knowledge of the discipline of risk management and how it should be integrated with the other project disciplines. Methods of building awareness include:

- formal risk management training or seminars and refresher training;
- presentations of the risk management plan and procedure;
- the creation of risk management "champions" who are able to provide support to team members;
- the inclusion of plans, procedures, presentations and risk management guidance papers on the company intranet.

18.5.10 Lack of Integration of Risk Management with Other Project Disciplines

For a project to be successful, risk management needs to be integrated with the other project disciplines throughout the project life cycle, from inception through to handover and implementation. Risk management should:

- Be integral to and support project management and hence overall project delivery.
- Support feasibility assessments and the selection of a preferred option.
- Form an integral part of gate reviews.[4] Gate reviews are a safeguard for the sponsor. They discern whether the original business case is being satisfied and whether further funding should be committed to the project. The culmination of a gate review is the recommendation as to whether the project should proceed to the next phase or additional work should be undertaken.
- Be a formal part of the change control process so that there is an understanding of how the planned change(s) will affect the overall risk exposure of the project.
- Be integrated with stakeholder management so that there is an awareness of the degree to which any one stakeholder can influence the outcome of a project so that risk management effort can be prioritised.
- Influence the procurement route, form of contract and contract conditions.
- Be integrated with cost planning and scheduling to provide robust realistic defendable contingencies for cost and time, respectively.
- Be integral with project governance whereby the organisational structure, responsibility matrix and job descriptions appropriately reflect the project scope.
- Be integral with the discipline of value management through an understanding of the client's value system and the project's value chain and recognising that risk exposure is a by-product of the pursuit of value.
- Complement and support earned value analysis and scenario modelling.

18.5.11 Reticence of Project Personnel to Spend Time on Risk Management

A significant obstacle to the implementation of risk management is the reticence of project personnel when it comes to devoting time to its application. Many consider it adds little value, and the pace of projects means that there is always a myriad of pressing issues for project personnel to respond to as opposed to spending time on identifying risks which may or may not happen. Hence, risk management has to be "sold" to projects on the basis that while it is not more important than say design, estimating, scheduling or change management, it is equally important. For risk management is a key tool in delivering project sponsor satisfaction through identifying threats to successful delivery and responding to them. The ability to manage risk is one of the core competencies of any organisation and its employees.

[4] A gate review is a "peer review" carried out at the end of intermediate project stages by an independent individual or team from outside the project, who use their knowledge and experience to examine the progress to date and the likelihood of the project being successful. The review uses a series of interviews, documentation reviews and the team's experience to provide valuable additional perspective on the issues facing the project team, and provides an external challenge as to whether or not the current phase has been completed. Critically, the review will recommend whether or not the project should move to the next project phase, and hence whether further funds should be committed to the project.

18.5.12 Risk Owners not Automatically Taking Responsibility for Assigned Risks

Experience illustrates that project personnel will attend regular risk meetings having not looked at the risk register or the risks allocated to them as owners since the last meeting. As a consequence, any opportunity to progress risk response actions in the interval between the two will have been lost. When this lack of participation is demonstrated by one individual the impact can usually be contained. When all project personnel are exhibiting the same behaviour, risk response planning is largely ineffective and a cultural shift is required. Risk management then needs to be taken back to "first principles". The risk champion must reinforce why the project is engaging in risk management, the benefits sought from its application, the downside if it is not implemented, the ramifications if the project benefits are not realised and the lessons learnt.

18.5.13 No Clear Demonstration of How Risk Management Adds Value and Contributes to Project Performance

The holy grail of PRM is being able to both describe and demonstrate through examples that the disciple adds value and contributes to project performance. To persuade the project team to spend time on risk management when they could be directly contributing to project delivery requires winning their hearts and minds. It requires an explanation that the *raison d'être* of project management is the management of risk from inception to handover. It involves assessing risk at project sanction, during options analysis, at each gate review, as part of feasibility analysis, during procurement route and contract selection, as part of contract clause reviews and tender return reviews and during management of execution. Project sponsors require the project manager they have engaged on a project to manage the risks to their project so that it satisfies its objectives and meets the pre-agreed cost and time constraints.

18.5.14 Overcomplicated Implementation from an Unclear Risk Policy, Strategy, Framework, Plan and Procedure

Embedding risk management will be hindered by a myriad of documents that are intended to provide guidance but have been prepared without a clear understanding of their unique purpose, content or applicability. The documents in common circulation are risk policies, strategies, frameworks, plans and procedures. Many organisations prepare a suite of documents which are supposed to be hierarchical in nature but are not – they do not clearly state their purpose and their content overlaps. Project participants have mountains of documents to digest and should only be provided with those that directly affect the project objectives, their scope of work and the deliverables.

18.5.15 Lack of Alignment between the Business Strategy, Business Model and the Risk Management Objectives

PRM will be fruitless if the business model for the project does not adequately support delivery of the business growth strategy. If the business strategy calls for, say, a freight railway with a particular annual throughput and the scope of the project will not accomplish that throughput, there will be a gap between what the business needs and what the project will deliver. Organisations must have the ability to prepare business cases or commission them and have the ability to assess models that are prepared on their behalf.

18.5.16 Lack of the Integration of Risk Management Activities into the Day-to-Day Activities of Project Managers

Often risk management is seen as a peripheral activity which will make no difference to the project outcome. However, a project manager's role is the management of planned tasks to a predetermined schedule to achieve agreed objectives. The successful completion of these tasks will clearly influence the overall satisfaction of the objectives. Each of these tasks will have challenges, which may be described as threats or risks. For risk management to be embraced the discipline needs to be recognised as the tool to manage these threats (or opportunities) to the planned tasks. While there will always be risks that materialise that a team had not identified, it could be argued that the primary source of risk on a project is a lack of adequate management of known threats. These threats will emanate from both inside and outside the project team's organisation(s). The term "management" does not imply that the threats will be under the total control of the team, but suggests that management processes should include an appreciation of how events may unfold and knowing how to make provision for them.

Progress reporting is in essence describing whether the project is on track, as measured by whatever metrics have been agreed, and describing the events (materialised risks) which have increased the budget and/or schedule, threats which may derail the project in the future or activities implemented to eliminate or reduce known or potential threats.

18.6 PROJECT RISK MANAGEMENT PROCESS

The PRM process should provide a methodical, efficient and effective way of managing risks to the delivery of a project. The overall PRM process to be followed is similar to the seven stages described for enterprise risk management in Part II. Clearly the focus will be different and the potential sources of risk will relate to the project life cycle, the sponsor's organisation and the wider context (such as the economy, legislation and taxation). For consistency the same process map adopted for enterprise risk management is repeated here in Figure 18.2. Readers implementing PRM should also refer to ISO 31000:2009 for guidance. It should be noted that risk management is an iterative process (and not purely sequential), as indicated by the return arrows in the figure. For the risk management process to remain effective the risk identification and analysis stages should be revisited on a regular cycle as new risks will emerge, existing risks will be overtaken by events or their nature will change (in terms of their probabilities and impacts). Changes in these earlier stages will necessitate changes to the evaluation, treatment, and monitoring and review stages.

18.6.1 Establish the Context

This stage involves establishing the context or environment of a project, which has two dimensions, *external* and *internal*. The external context relates to the political, legal, regulatory, market, technological and economic setting, whereas the internal context relates to, for

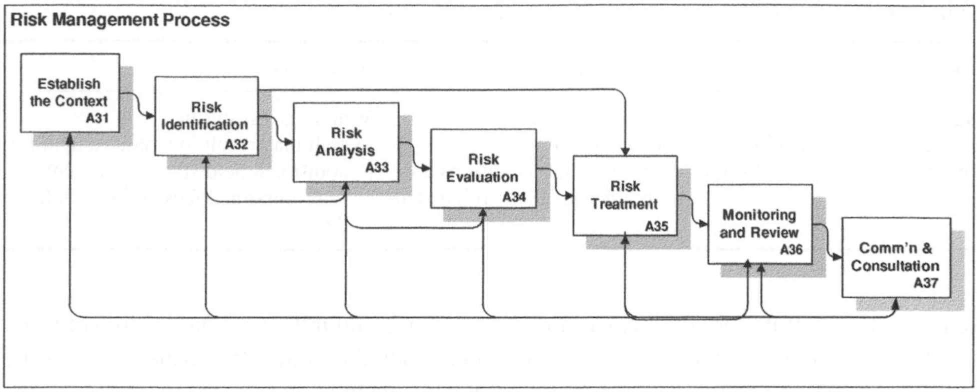

Figure 18.2 Risk management process

instance, the organisation's strategic objectives, its structure, policies, processes, stakeholders, culture, reputation, capabilities (including capital and people) and concurrent projects.

A key aspect of understanding the external context is to establish the legislation that the project will have to adhere to. This may be very significant if the project is related to, say, the oil and gas industry. Depending on the scale of the project, it may involve satisfying extensive environmental and health and safety legislation and sustainability goals, and obtaining approvals will be both expensive and time-consuming.

A key aspect of understanding the internal context is to establish if the business case for the project supports delivery of the organisation's business strategy and in turn whether the project is aligned with the business case. The business case should provide a description of the business needs and the project's contribution to the organisation's business strategy. It should state why the business change is required now, the key benefits that will be derived and the critical success factors. The purpose of the business case is to obtain management commitment and approval for investment in business changes and the resultant project(s) based on a clear rationale for doing so. The business case must provide sufficient information to enable management to discern if the project is desirable, viable and achievable.

The business case will dictate the project objectives. The project team must not only be clear what the project objectives are but also what the prioritised objectives are. The prioritised objectives will be dictated by the sponsor's risk appetite (whether they are risk seeking, risk neutral or risk averse) and the drivers for the project. If a product is being manufactured to reach a market before competitors enter that market, then speed will be of paramount importance. If the business case for a project is only viable if costs are kept below a particular ceiling limit, then cost will be of primary importance. The prioritised objectives will inform the procurement route, the choice of contract and the contract conditions. If speed of completion is the most important objective, then the contract selected will be very different than the one selected for a project where cost certainty is the most important objective.

Project "environments" do not remain static. Once a project has been approved and commenced, progress should be checked against the business case at the gate reviews to ascertain whether the project is still viable and the planned benefits will be realised. If the costs have risen above the cost plan, the project has been delayed and/or the scope is no longer achievable, the benefits may have been eroded to the point where the project should be re-scoped or

Table 18.1 Distinguishing between the cause, risk and impact

Risk Cause	Risk Description	Risk Impact
As a result of...	There is a risk that...	Which may result in...
As a result of changes in environmental legislation	There is a risk that environmental conditions are more onerous than anticipated	Which may result in a requirement for additional field studies, an increase in professional fees and schedule delay

even abandoned. If the project's environment is dynamic and influences on the project cause priorities to alter, the business case may need to be adapted to reflect the changes and ensure the project remains aligned to the business's strategic direction.

18.6.2 Risk Identification

Risk identification is the process of determining which risks may affect the project and establishing their characteristics. Identification is typically achieved through a facilitated workshop or meeting attended by the project personnel, accompanied, where appropriate, by one or more of the following: stakeholders, end users, the project sponsor, external subject-matter experts and risk specialists. Identification is typically supported by stimuli such as checklists, prompt lists, a risk breakdown structure, drawings or flow charts. It is important that risk descriptions are not a confused mixture of causes, risks and impacts. The common problem is for participants in the process to offer risks such as cost overrun or late completion. It is not possible to define response actions (mitigation measures) for these risks without knowing the actual risk event and the causes. To illustrate this problem using a new port project and the subject of approvals as an example, the cause may be "changes in environmental legislation", the risk may be "environmental conditions are more onerous than anticipated" and the impact is "schedule delay". Just stating schedule delay as the risk would not provide any guidance as to where in the project life cycle the problem is likely to occur or how it might be addressed. A way to overcome this problem is to use a prefix to the cause, risk and impact as exemplified in Table 18.1.

Risks have different characteristics depending on their source, and as a result a project team's ability to influence their occurrence spans from no influence at all through to an ability to control, as illustrated in Table 18.2. External threats emanate from outside the business. Internal threats emanate from within the business.

18.6.3 Risk Analysis

Risk analysis involves the identification of the probability and impact (positive or negative) of the identified risks and opportunities. Analysis can be qualitative or quantitative depending on the requirements of the risk process and the information available. Qualitative assessments use labels such as high, medium or low, whereas quantitative assessments provide a percentage likelihood such as 50% (or a range such as 40–60%) and an impact in terms of time, cost and any other measures of project success. Analysis should consider whether the risks are interdependent or independent. In other words, is the occurrence of one risk dependent on the occurrence of another? May one project event give rise to two (or more) risks with different

Table 18.2 The sources of risk and a project team's ability to influence their probability and impact

External Uncontrollable (no influence)	Internal Alterable (able to influence)	Internal Controllable (able to control)
The project is unable to control...	The project is able to influence...	The project is able to control...
Government policy	Sponsor decision making	Clarity of the objectives
The economy	Project objectives	Project processes
Inflation	Choice of procurement route	Project reporting
Taxation	Project duration	Application of software tools
Bank interest rates	Change management process	Discipline integration
Import duties	Project team performance	Team communication

probabilities where only one of the risks can occur? Is there a situation where a number of risks may occur concurrently but only the risk with the largest time impact will affect the schedule?

18.6.4 Risk Evaluation

Risk evaluation typically looks at the combined net effect of the identified risks and opportunities and is commonly accomplished with the aid of proprietary software that has an in-built simulation technique such as Monte Carlo or Latin hypercube. A prerequisite for carrying out evaluation is that a probability and impact have been defined for each risk and opportunity, as described in the risk assessment stage above. The output of the evaluation stage can be used to define the project contingencies for time and cost. Evaluation can be conducted before and after the definition of response actions, and so the risk outputs need to be described by such statements as "if no response action is undertaken the risk exposure will be in the range ..." or "as a result of the defined response actions and assuming that they will be 100% effective (or whatever percentage is appropriate) when implemented, the risk exposure will be in the range ...". Reporting an overly optimistic mitigation success rate early in a project (as has occurred on government projects) can lead to underfunding and interruption of a project's progress.

18.6.5 Risk Treatment

Risk treatment is the action of responding to an identified risk. This is arguably the most important aspect of the overall process. If a risk is left untreated then arguably it has been accepted by the project. Should the risk materialise, the project will have to live with the consequences, whatever they are. There are a number of response options.

18.6.6 Risk Monitoring and Review

Monitoring and review is an ongoing process of implementing and examining the success or otherwise of the planned responses. It entails evaluating the perceived benefit of the response, its attendant costs and the likelihood of new risks being triggered by the response. If the decision is taken to implement the response, there needs to be clarity as to who will do so and when.

18.6.7 Communication and Consultation

Communication and consultation take place at commencement and throughout the risk management process. The activities of the communication and consultation process are the tasks undertaken to strive to ensure that the risk management process is effective. These activities consist of:

- communicating with the sponsor and stakeholders (as appropriate) to establish their risk management expectations;
- clarifying with the project sponsor the project objectives and the ranking/prioritisation of the objectives;
- clarifying the objectives for risk management on the project;
- establishing alignment between the overall business strategy, the project business model and the risk management objectives referred to above;
- having a dialogue with the project sponsor to establish the sponsor's risk appetite;
- agreeing across the project disciplines the risk management terms and definitions – this cannot be completed in isolation from the project as risk interfaces with and crosses over into a number of disciplines;
- holding a seminar to debate how risk management will be integrated with the project management function to ensure risk management becomes part of day-to-day management activities;
- clarifying through dialogue the purpose, format, content and frequency of risk reports;
- debating how risk management will be integrated with procurement, cost planning, change management and any other discipline that will interface with risk management;
- making sure that those allocated risk management responsibilities within the risk management plan understand their duties;
- agreeing how contingencies will be defined either by adopting quantitative risk analysis results or cost plan percentages verified by risk results;
- presenting the risk management plan and explaining how it will be applied;
- engaging with project team members to ensure risk registers and responses are up to date and the team are proactively managing risk.

18.7 RESPONSIBILITY FOR PROJECT RISK MANAGEMENT

Successful PRM cannot be driven from the bottom up but must be championed from the top. Ultimate responsibility for PRM must rest with the project director, who must be instrumental in setting the right culture. The rationale for implementing risk management must be crystal clear and delivered with conviction. Within the team, roles and responsibilities for PRM should be clearly communicated. Project individuals should be held accountable for implementing assigned response plans and actions. Where appropriate, specific risk management duties should be added to project personnel's job descriptions so that there is no ambiguity and team members know from the outset what is expected of them. Project team members must engage in the process, and superficial lip service will lead to the ineffective management of threats and opportunities.

18.8 PROJECT DIRECTOR'S ROLE

The director has overall responsibility for the delivery of a successful project in terms of satisfying the stated objectives. Typically the director is supported by one or more project managers who are in the driving seat in terms of having responsibility for the day-to-day management of aspects of the project, including implementing the risk management process. The role of the project director may include:

- setting the risk management culture of the project;
- promulgating open and honest communication about risk exposure as soon as project team members become aware of a potential problem;
- approving the risk management plan;
- establishing the need to carry out quantitative schedule and cost risk analysis;
- agreeing the level of risk management resources;
- mandating the implementation of risk management practices;
- periodically participating in risk meetings and/or workshops;
- agreeing the level of project contingency with the project sponsor;
- approving responses to the most serious risks;
- chasing risk response actionees to ensure agreed responses are implemented;
- approving expenditure to address the most serious risks;
- overseeing the integration of third parties into the risk management process;
- regularly reporting risk status to key project stakeholders (the board and end users);
- agreeing the release of the risk contingency over the life of the project;
- agreeing with the sponsor the degree of risk transfer to contracting parties.

18.9 PROJECT TEAM

The composition of the project team and how it performs will have a fundamental impact on the realisation of a project's objectives. It rarely receives sufficient attention. Team performance is a major source of potential risk – hence its inclusion here. It generally falls into the category of "alterable (able to influence)" particularly where the team is composed of representatives from either different departments in the same business or different companies. There are a number of issues which can undermine the effectiveness of teams, some of which are recorded below.

18.9.1 Lack of Team Structure

A lack of team structure will lead to confusion over the levels of authority, decision making, lines of communication, reporting and accountability. One of the first activities on a project must be to define the team structure, in terms of the composition and the extent of the team, who reports to and takes instruction from whom, and areas of responsibility. The team structure must be coordinated with the project's resource schedule, indicating the start and completion time for each team member and the role that they will fulfil. While this may sound trivial, on projects lasting several years both resource planning and management of the associated costs are dependent on a resource schedule.

18.9.2 Lack of Definition of Roles

Each team member must have a clear job description which is not just a list of generalities but a clear description of the processes to be followed and the deliverables to be produced. The role descriptions must be coordinated with the project brief, the team structure, the work breakdown structure and the responsibility assignment matrix. Team members need to understand whom they report to and whom they are supervising and, importantly, how they will be required to interact with other team members. The matrix will inform them whom they have to engage with to complete their assigned tasks. In addition, project members need to be issued with a copy of the project schedule so that they are aware when their deliverables need to be produced by.

18.9.3 Lack of Responsibility Assignment Matrix

A responsibility assignment matrix such as a RACI matrix describes the activities to be undertaken and those who are to take part in their completion. RACI is an acronym for "*r*esponsible, *a*ccountable, *c*onsulted and *i*nformed". Against each activity should be listed an individual who is accountable, one who is responsible and, where appropriate, the individuals that have to be consulted and informed. The matrix lists the tasks in the sequence they need to be completed. The matrix is unequivocal in showing "where the buck stops".

18.9.4 Poor Leadership

Poor leadership leads to a lack of project governance. Governance is required to establish a project correctly at the outset and then to continue to manage its delivery. The project scope, objectives, team structure, processes and the tools for tracking and reporting progress must be in place at the beginning. The style of leadership can also be a barrier to successful teams. An autocratic leader will suppress commitment, enthusiasm and initiative. Project managers require the skills to empower project team members. Empowerment gives team members a sense of ownership of and responsibility for their deliverables. They will have a greater feeling of belonging to the project. Poor leadership is also attributed to low morale and poor motivation. Motivation will be enhanced if the project manager continuously recognises the achievements of the team members through praise, encouragement or through formal recognition awards. Poor leadership can exacerbate the management of a sponsor's expectations. Repeated failure to meet pre-agreed deadlines will attract unfavourable feedback from the sponsor and change a "positively charged" team atmosphere into a lacklustre team. Setting realistic goals, adequately resourcing the team and establishing appropriate processes are critical to satisfy schedule milestones.

18.9.5 Poor Team Communication

Communication can fail on a number of fronts. The physical location of team members is an important factor in helping or hindering communication. Due to the complexity of projects they require the participation of a number of disciplines that need to interact. If they are not located in the same building, communication can be slow or ineffective. Due to the size of project teams, their duration and the turnover of staff during the life of a project, document control is critical to successful projects. A clear audit trail of what was issued to whom, when and for what purpose is vital.

18.10 OPTIMISM BIAS

The presence of optimism bias in project appraisals is common. Those undertaking PRM should readily understand the concept and recognise when within the project life cycle it should be considered. It might be described as the frequent tendency of appraisers to be overly optimistic about key project parameters, including capital costs, operating costs, works duration and benefits delivery. It is most prevalent early in a project when project details are scarce and there is insufficient information to carry out quantitative risk analysis to guide investment planning. As will be described below, the UK government recommends the adoption of the adjustment of capital estimates for optimism bias to counter over-optimism in predicted outturn costs (or, depending on your viewpoint, either unfathomable behaviour or deliberate manipulation of project information to secure project sanction).

18.10.1 The Investment Decision

Business managers seek to maximise the value of the firm by investing in opportunities which yield a positive return. These decisions are critical. Businesses thrive or wither depending on which opportunities or projects are pursued from the many competing opportunities identified. Making an investment or capital allocation decision requires estimating the value of each project, which is a function of the size, timing and predictability of future cash flows. Selection of the project(s) to pursue is typically done through the evaluation of an investment appraisal. This is the planning process used to determine the benefits (or otherwise) of long-term investments in, say, new machinery, replacement machinery, new plants, new products, new offices or research development projects. Investment appraisal is typically accomplished through the development of a business case. The business case provides a framework for planning and management of the business change. It is used to obtain management commitment and approval for investment in business change through the delivery of a project or projects. The business case must show the purpose of the project, articulate the need for its implementation (the benefits), demonstrate it is achievable and convey that the proposer fully understands the implications of implementing the business change, specifically in terms of being able to support delivery with adequate project, risk and quality management. In addition, it must show it is affordable – this should include checking that all relevant costs are included (capital, operating and maintenance) and any likelihood of significant cost changes over the life of the project. There should also be confirmation from the sponsor that funds are available.

The public sector uses a similar approach to that of the private sector before significant funds are committed to investment projects to encourage a thorough and analytically robust approach to appraisal and evaluation. However, in the UK and Europe there is the tendency for the true costs and duration of public funded projects to be underestimated and the benefits overstated. In the UK the HM Treasury provides guidance to other public bodies on how proposals should be appraised before the commitment of funds, in the form of the *Green Book* (see Section 7.12 above). The new edition (HM Treasury 2003) includes for the first time an explicit adjustment procedure to redress what it describes as the systematic optimism that historically has bedevilled the appraisal process.

18.10.2 Optimism Bias

Optimism bias is described as the tendency for a project's costs and duration to be underestimated and/or the project benefits to be overestimated. Optimism specific to cost is expressed as the percentage difference between the estimate at the appraisal stage and the final outturn cost. Likewise, optimism specific to time is expressed as the percentage difference between the schedule duration at the appraisal stage and the final outturn duration. Studies have shown that optimism bias is caused by a failure to identify and effectively manage project risk. The *Green Book* (HM Treasury 2003, p. 29) records that "there is a demonstrated, systematic, tendency for project appraisers to be overly optimistic". It advises that to reduce this tendency "appraisers should make explicit adjustments in the form of increasing estimates of the costs and decreasing and delaying the receipt of estimated benefits". HM Treasury recommends that adjustments for optimism bias should be empirically based, adopting data from past projects or similar projects elsewhere which are adjusted for the unique characteristics of the project in hand.

18.10.3 Monitoring

It is assumed that once the sponsor has committed funds to the project the sponsor will monitor its delivery using (as a minimum) the optimism bias risk area contributors to guide data gathering and risk management practices. In addition it is also assumed that this monitoring will run through each gate review and the sponsor will expect to see evidence that the most likely cost estimate (or mean) has been adjusted downwards to reflect the developing experience and improved knowledge of the project.

18.10.4 Using Numerical Indicators in Project Decision Making

The adoption of the methodology to address optimism bias is intended to develop a robust approach to assessment of the capital cost of public infrastructure projects to determine whether the project should be supported. The approach is based on the project team preparing the business case asking the following questions:

- What project information is not available, what assumptions have to be made and what is the potential range of possible outcomes?
- Is the risk exposure profile of the project fully understood?
- Are the major risks identified manageable?
- How will the risk exposure profile change over time?
- Are the boundaries of the project clearly defined?
- Is the scope clearly defined?
- Are the stakeholders and their requirements known?

Proposed changes to the project during its life cycle should be assessed against the original business case to establish if the scheme is more or less attractive and whether or not the business case is undermined.

18.10.5 Causes of Optimism Bias

The very comprehension of the causes of Optimism Bias is important as it enables responses to be prepared to reduce the potential exposure to surprises in, say, cost and duration.

According to Mott MacDonald (2002), the top 11 project risk areas contributing to the recorded capital expenditure optimism bias, in terms of frequency of occurrence, are as follows:

1. Inadequacy of the business case (58%)
2. Environmental impact (19%)
3. Disputes and claims (16%)
4. Economic (13%)
5. Late contractor involvement in design (12%)
6. Complexity of contract structure (11%)
7. Legislation (7%)
8. Degree of innovation (7%)
9. Poor contractor capabilities (6%)
10. Project management team (4%)
11. Poor project intelligence (4%)

With reference to item 1 above, "in most instances, the inadequacy of the business case was stated to be the major cause of project time and cost overruns" (Mott MacDonald 2002, section 2.4.3, p. 17). The study goes onto say: "This fundamentally demonstrates the need to concentrate significant effort and diligence to ensure the business case comprehensively represents the requirements of all of the project stakeholders in terms of the agreed scope and objectives". Furthermore, "most of the traditionally procured projects in the sample were inadequately defined (in terms of requirements and project scope) in the approved business case".

18.10.6 The Distinction between Risk Events and Optimism Bias

The critical issue to avoid in a project appraisal is being overly pessimistic by double-counting risk exposure through the addition of quantitative risk analysis results and optimism bias figures, which are based on the same risks. The UK DfT guide recommends the addition of the mean quantitative risk analysis output with the mitigated optimism bias figure. One possible method is to restrict the risk assessment to purely technical risks and the optimism bias figure to business case issues (such as income and operational, maintenance and renewal costs). It is assumed that the project promoter will confirm acceptance (or otherwise) of the split categories of risk incorporated within the quantitative risk analysis and the optimism bias assessment.

18.11 SOFTWARE TOOLS USED TO SUPPORT PROJECT RISK MANAGEMENT

At the time of writing, the following tools listed in Table 18.3 were commercially available to support the implementation of PRM.[5] This is not an exhaustive list. Clearly, as a result of business failures and acquisitions, products can disappear from the market or be "rebadged"

[5] In a departure from BS 31100, the term "tool" is reserved for software products; techniques are listed in Table 18.2.

Table 18.3 Software tools for project risk management

Tool	Supplier	Cost	Schedule	Other*
@RISK for Excel	Palisade Corporation	✓		
@RISK for Project	Palisade Corporation		✓	
Precision Tree	Palisade Europe			✓
Predict! Risk Controller	Risk Decisions Ltd			
Predict! Risk Controller Lite	Risk Decisions Ltd			✓
Predict! Risk Analyser	Risk Decisions Ltd	✓	✓	
Oracle Crystal Ball	Oracle	✓		✓
Active Risk Manager (ARM)	Active Risk Group	✓	✓	
Goldsim	Goldsim Technology Group	✓		
iDecide	Decisive Tools	✓		
DRAMMS	BMT Reliability Consultants Ltd			✓
Primavera Risk Analysis	Oracle		✓	
Risk Solver Pro	Frontline Systems	✓		
RiskTrak	Risk Trak International (RTI)			✓
CURA Quants	CURA Technologies Limited	✓		✓

*Other here refers to a capability other than cost of schedule risk analysis. The software tool may support a specific risk technique or act as a database.

with different names. In the normal way, any software tool must be evaluated by a prospective purchaser for its suitability to the intended task. Inclusion of a product in this list does not imply an endorsement or recommendation.

18.12 TECHNIQUES USED TO SUPPORT PROJECT RISK MANAGEMENT

There are now a number of techniques available to assist in the implementation of PRM. Some of these techniques are recorded in Table 18.4. This is by no means an exhaustive list. Some of the techniques will support multiple stages in the risk management process. Some have a very specific purpose and only support a single stage. It is advantageous to maintain a library of samples of the techniques and, where necessary, an explanation of how to use them.

18.13 SUMMARY

This chapter has provided a definition of the discipline of project risk management, and has described in outline the potential sources of risk and the benefits of implementing the discipline. It describes the challenges faced in implementing the discipline. The mere fact that the project sponsor or the project director thinks it will be a good idea to implement risk management does not mean that staff will readily implement it. It takes considerable effort to understand the obstacles and develop workarounds. Many consider project risk management pedestrian, banal and mundane. Perhaps *project risk management* is a misnomer for the process of managing the threats and opportunities to a project's objectives. It is not conducive to eliciting constructive support. Perhaps it needs relabelling! This chapter additionally describes the risk management process. A process similar to the one described has been implemented in the UK for some 30 years. Additionally, the issues around the project team are described. Often a reflex action of projects is to think that the major risks all sit outside the project team and

Table 18.4 Examples of project risk management techniques (based on Table B.1 of BS 31100:2008), taken from the Manchester Business School's MBA Study Guide Unit 2, "Risk Management Tools and Techniques", July 2007

Technique	Identification	Assessment	Evaluation	Response
Assumptions analysis	✓			
Brainstorming/"thought shower" events	✓			
Bow tie analysis		✓	✓	
Cause-and-effect diagrams	✓			
Checklists	✓			
Cost–benefit analysis	✓	✓	✓	
Decision trees		✓		
Delphi technique	✓			
Expected monetary value		✓		
Failure mode effects analysis	✓	✓	✓	
Fault and event tree modelling		✓	✓	
Flow charts/process maps	✓			
Gap analysis/Pareto analysis	✓	✓		
Hazard and operability study	✓	✓		
Horizon scanning	✓	✓		✓
Interviews and focus groups	✓			
Latin hypercube			✓	
Layer protection analysis	✓	✓		
Markov analysis	✓			
Monte Carlo analysis			✓	
Nominal group technique	✓			
PESTLE (political, economic, sociological, technological, legislation and environment) analysis	✓	✓		
Probability and impact (consequence) diagrams (PIDs)/probability impact grids (PIGs)/Boston grid	✓	✓		
Probability trees		✓		
Process mapping	✓			
Project profile model	✓			
Prompt lists	✓			
Risk breakdown structure	✓			
Risk indicators	✓			
Risk management workshop	✓	✓		
Risk mapping/risk profiling	✓			
Risk modelling		✓		
Risk questionnaires	✓			
Risk register	✓	✓		✓
Risk taxonomy	✓			
Root cause analysis		✓	✓	
Scenario analysis/scenario planning	✓	✓		✓
Sensitivity analysis		✓		
Stakeholder engagement/matrices	✓			
Stress testing	✓	✓		
SWOT (strengths, weaknesses, opportunities and threats) analysis	✓			
Systems dynamics	✓			
Visualisation techniques: heat maps, RAG status, waterfall charts, profile graphs, 3D charts, radar chart, scatter diagram, spider diagram		✓		✓
"What if?" workshops	✓			

emanate from third parties such as local government, suppliers and contractors. Wrong! A major source of risk is the team itself, and this should not be overlooked. The subject of optimism bias is referred to. Those embarking on projects should whenever possible strive to benefit from lessons learnt from other projects. The research conducted to provide guidance on optimism bias has repeatedly found that large projects (in capital expenditure terms) overspent their capital budgets and were late by significant margins. The implementation of the discipline is considerably enhanced by techniques and software tools, and a sample are listed as an aid.

18.14 REFERENCES

BSI 31100 (2008) *Risk Management – Code of Practice*. British Standards Institution, London.

Davies, A., Dodgson, M. and Gann, D. (2009) From iconic design to lost luggage: Innovation at Heathrow Terminal 5. Paper presented at the Summer Conference 2009 at Copenhagen Business School, 17–19 June.

HM Treasury (2003) *Appraisal and Evaluation in Central Government*. The Stationery Office, Norwich.

ISO (2009) *ISO 31000:2009 Risk Management – Principles and Guidelines*. ISO, Geneva.

Mott MacDonald (2002) *Review of Large Public Procurement in the UK*. Mott MacDonald, Croydon.

NAO (2000) *The Cancellation of the Benefits Payment Card Project*. National Audit Office, August.

NAO (2004) *Managing Risks to Improve Public Services*. National Audit Office, October.

NAO (2005) *Improving Public Services through better construction*. National Audit Office, March.

NAO (2009a) *The National Offender Management Information System*. National Audit Office, March.

NAO (2009b) *Department of Transport: Failure of Metronet*. National Audit Office, 5 June.

NAO (2011) *Managing Risks in Government*. National Audit Office, June.

PMI (2008) *A Guide to the Project Management Body of Knowledge (PMBOK® Guide)*, 4th edition. Project Management Institute, Newtown Square, PA.

19
Business Ethics Management

If ethics is poor at the top, that behaviour is copied down through the organisation.
(Robert Noyle)

This chapter examines business ethics, the fifth of six classes of risk exposure that businesses face, relating to internal processes within the risk taxonomy proposed in Chapter 9. All six classes of risk are considered to be controllable by businesses to a large degree. Business ethics is now a vast subject and warrants close scrutiny. While there is burgeoning legislation to regulate companies, there is a heightened awareness and expectation from customers and the general public about the way companies should behave. The media headlines have brought business ethics to the forefront of people's minds in three distinct waves: first, the prominent and highly publicised corporate scandals of Enron, WorldCom, Tyco and others involved in accounting irregularities; second, the economically and ethically destabilising actions of the banking sector whose short-term vision has made the global economic system unstable, necessitated government fiscal intervention and caused financial suffering for millions of taxpayers, most of whom were not responsible for managing the system; and third, hard on the heels of the global financial crisis, the bribery allegations involving formerly well-respected companies such as Siemens, BAE Systems plc, Shell International Exploration and Production Inc. and ALSTOM. While the profit motive of business is understood and accepted, the public and communities do not accept it as an excuse for ignoring the basic norms, values and standards of business practice. As identified by the Financial Services Authority, the key question for businesses is not "show me where it says we can't...?" but rather "how can we improve our standards and conduct our business with honesty and integrity?" (FSA 2002). Ethics is inextricably linked with reputation, and a breach of ethics commonly leads to one or more of the following: shareholder disquiet, reduced share price, reduced profitability, unfavourable media coverage, fines, additional administration – and, in extreme cases, imprisonment. As with other aspects of risk management, the decisions taken by a business in managing the risks relating to ethical conduct will determine its performance, position and longevity. The words of Anita Roddick, the founder of the Body Shop, have rung true: "being good is good for business". This chapter focuses on what are considered to be the key aspects of business ethics to aid comprehensive risk management. The structure of this chapter is illustrated in Figure 19.1.

19.1 DEFINITION OF BUSINESS ETHICS RISK

Ethics is the branch of philosophy that addresses questions about morality. The word "ethics" is commonly used interchangeably with "morality". Morality is a sense of behavioural conduct that differentiates intentions, decisions and actions between those that are good and evil, and right and wrong. So what is meant by the terms business ethics and ethical risk? Business ethics are ethics that refer to the moral rules and regulations governing the business world. Currently the only unifying moral force for businesses within society is the law. For example, in the UK the Financial Services Authority is the "watchdog" that seeks to reduce financial crime and in the US the Securities and Exchange Commission has a similar role

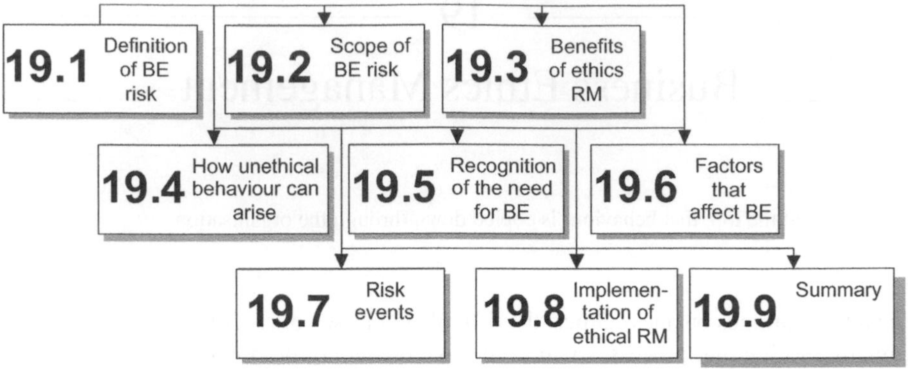

Figure 19.1 Structure of Chapter 19

as an enforcement agency tackling insider trading, accounting fraud and false or misleading information. However, it might be argued that the law merely specifies the lowest common denominator of acceptable behaviour. In addition even in the best legal context, the law will lag behind moral condemnation of certain unscrupulous yet legal business practices. For example, in the past pharmaceutical companies would make exaggerated claims about the miraculous healing properties of their products. Now government regulations prohibit any exaggerated or unproven claims. So prior to the enactment of a law, there will be a period of time when a business practice will be considered immoral, yet the practice will be legal. This will be a continuous problem as changes in technology, products and marketing techniques will soon present new questionable practices that will not be addressed by current legislation. With no widely recognised system of ethics that is external to the law, supra-legal moral obligations in our society appear to be optional. However, encouragingly businesses are seeing the need to assume obligations beyond the law, no doubt in part from the moral expectations of society and the need to manage their stakeholders. Ethical risk refers to exposure to events which may result in criminal prosecution, civil law suits or erosion of reputation. It may also be considered to include an organisation's lack of awareness of society's growing expectations, its receptiveness to change and its ability to devise and implement concrete (not abstract) measures to bring about desired behaviours in both its decision making and everyday business processes. It refers to the absence of openness, integrity and honesty which erodes business reputation and profitability together with public confidence and trust.

19.2 SCOPE OF BUSINESS ETHICS RISK

The sources of risk considered to be embraced within the term "ethical risk" are very considerable and seem to be growing all the time. These may be considered to include, but are not limited to:

- Bribery
- False accounting
- Use of child labour
- Insider trading
- Misleading or exaggerated advertising
- Industrial espionage
- Breach of copyright

- Money laundering (and financing terrorism)
- Tax evasion
- Transgression of industrial relations legislation
- Improper competition practices
- Exploitation
- Counterfeit goods
- Invasion of privacy.

19.3 BENEFITS OF ETHICS RISK MANAGEMENT

Businesses around the world are designing and implementing business ethics risk management programmes to address the legal, ethical, social responsibility and environmental risks they face. By addressing these risks in a systematic way, enterprises can improve their own business performance, expand opportunities for growth and contribute to the development of corporate responsibility capital in their markets. They can realise specific business benefits, such as:

- Enhanced reputation and good will
- Protection from their own employees, subcontractors and agents
- A stronger competitive position
- Expanded access to capital, credit, and foreign investment
- Increased profits
- Sustained long-term growth
- International respect
- Business partner of choice
- Stronger employee performance (improved employee morale and lower turnover)
- Ability to attract high achievers
- Positive media coverage
- Customer loyalty
- Reduced insurance premiums.

Another way of viewing those benefits is looking at what negative events are avoided, such as:

- consumer boycotts;
- prosecution and finds;
- the imposition of harsh government-imposed restrictive laws;
- employee resentment, politicisation of the workforce and the proliferation of restrictive practices;
- reduced ability to attract high-calibre staff;
- unwanted media attention;
- ambiguity in the workforce about the company's culture and acceptable norms of behaviour;
- lack of support of communities or organised opposition to the business and its operations;
- protracted scrutiny by agencies such as the FSA in the UK or the SEC and the Department of Justice in the US; or
- business interruption by environmental organisations.

19.4 HOW UNETHICAL BEHAVIOUR CAN ARISE

There are a number of reasons why unethical behaviour may surface, and the FSA (2002) has identified some of the more common reasons (Box 19.1).

Box 19.1 Reasons for unethical behaviour

Unethical behaviour can arise for a number of different reasons:

- The pressure of short-term gain could be seen to encourage undesirable behaviour. Staff bonus payments may often seem to be geared to pure bottom line success. How risks and tensions can be identified is a constant issue – e.g. truth versus loyalty, one person versus the many? In all of this it is usual for the values and actions of senior management to influence employee levels.
- Some individuals behave unethically because they think it is worth the risk. This may be related to a short-termist agenda, or may be simply personally selfish. People weigh up the pros and cons and take a chance. It is a deliberate risk/reward trade off.
- Others may believe they are behaving ethically but come to operate by a different yardstick to that used by others. They might do something which is deemed unethical, but which seems acceptable from their own perspective.
- Others (and some of these groups are not mutually exclusive) may be unaware of the values embedded in existing regulatory standards. So, they comply (or don't comply!) blindly with the "letter of the law", rather than thinking about the wider effects their behaviour might have.

Source: FSA (2002, p. 11).

19.5 RECOGNITION OF THE NEED FOR BUSINESS ETHICS

It is evident from recent publications and speeches made by representatives of leading organisations on the world stage that, years after the eruption of the final crisis, the business world and the financial sector in particular have a tarnished reputation and that there is still a mountain to climb to address unacceptable business practices. A common call is for "a return to ethics in business". The US Department of Commerce, the Group of Eight (G8)[1] – the world's eight largest industrial market economies – the Organisation for Economic Co-operation and Development, the FSA and the US Department of Justice all describe the need for an improvement in business ethics – in the absence of self-imposed business ethics based on businesses' own "moral" code.

19.5.1 US Department of Commerce

As recognised by the US Department of Commerce, businesses around the world are designing and implementing business ethics programmes to address the legal, ethical, social responsibility and environmental issues they face. The Department considers that by addressing these issues in a systematic way, enterprises can improve their own business performance, expand opportunities for growth and contribute to the development of social capital in their markets. However, the conduct of many businesses must go beyond what has been traditionally expected of them. The Department argues that a fundamental ingredient of any successful market economy is respect

[1] The G8 consists of Canada, France, Germany, Italy, Japan, Russia, the UK and the USA. The European Commission attends as well.

for basic human values: honesty, trust, fairness, reliability and self-discipline. In addition, these values must become an integral part of business culture and practice for markets to remain free and to work effectively. The Department describes the unattractive alternatives to responsible business conduct as being inefficient markets, costly government regulation, a lack of capital, high transaction costs, limited markets and underdevelopment. A well-written guide is available in which the Department articulates its thinking (US Department of Commerce 2004).

19.5.2 The G8 Summit in Italy Pushes for a Return to "Ethics"

Before and during the G8 Summit held in the city of L'Aquila, Italy, in July 2009, there was general recognition that fundamental weaknesses in the world economic system demanded the adoption of guiding norms and principles and a return to ethics in business. At the G8 summit the Italian presidency announced the creation of a Global Standard aimed at developing a set of common principles and standards for propriety, integrity and transparency in international business and finance. The Global Standard draws on a wide range of instruments, established or under development, which share a set of common principles. It classifies them into five categories: corporate governance, market integrity, financial regulation and supervision, tax cooperation, and transparency of macroeconomic policy and data. The Standard will build on initiatives of the Financial Action Task Force on Money Laundering, the Financial Stability Board, the International Monetary Fund, World Bank, OECD and other international organisations. The OECD in particular supports the creation of a Global Standard and Angel Gurría, the OECD Secretary-General, has highlighted business integrity and ethics as being of paramount importance to the world economy.

19.5.3 OECD and Its Approach to Business Ethics

In his 2009 address to the European Business Ethics Forum entitled "Business ethics and OECD principles: What can be done to avoid another crisis?", Angel Gurría described the failure of business ethics as a contributor to the global economic crisis. He believes that the crisis provides the opportunity to build the foundations of a new, more ethical and responsible, business culture. For him, business ethics govern the strength and health of the world economy, and one of the main lessons of the financial crisis is that companies and markets cannot rule themselves. The justification is there for all to see. He argues that financial innovation sacrificed business ethics for the sake of extraordinary profit. He advocates regulation by the application of a set of rules:

> in a world where some corporations rival the power of states, where the excesses of a handful of banks can paralyse the world economy, it is fundamental to have a solid, transparent and updated set of rules that guarantee that business activities can produce their best fruits, avoiding the excesses of market capitalism.

Gurría considers these rules necessary to rebuild trust in banks and the market place. Trust is required for banks to generate the key asset to enable them to discharge their fiduciary responsibility. Trust is borne out of transparency, objectivity, reliability, honesty and prudence – all created from a culture of business ethics. The OECD (2011) has produced the *Guidelines for Multinational Enterprises*, which it considers is the sole instrument covering all areas of business ethics. The *Guidelines* have been developed multilaterally and agreed by governments. They include general principles but also detailed recommendations such as promoting compliance with laws; protecting consumer interests; respecting human rights; caring about employment; industrial relations; and protecting the environment.

19.5.4 UK Financial Services Authority

In his keynote address to the Mansion House conference in October 2010, the Chief Executive of the UK Financial Services Authority (FSA), Hector Sants, explored the role for regulators in facilitating the right culture within firms. Sants said that it is crucial to address the role that culture and ethics play in shaping behaviours and judgements. Addressing how regulators and firms can work together, he said: "It is those who manage the financial institutions, who make the judgements, who should be held responsible for them and for restoring the trust between the financial sector and the public. But there is a legitimate role for a regulator in facilitating the right culture." He said he considered that the starting point was for firms to have a culture which "encourages individuals to make appropriate judgements and deliver the outcomes we are seeking. At all times we want an institution to act with integrity. The regulator's focus should therefore be on what an unacceptable culture looks like and what outcomes that drives. It should not be on defining the culture itself." Focusing on whether it is realistic for regulators to intervene and modify culture, he said that it is important that the regulator focuses on the outcomes that the culture delivers and that it is for a firm to demonstrate that it has a framework for assessing and maintaining that culture. He said: "To be completely clear, a box-ticking approach to regulating culture will not work. The regulator must focus on the actions a firm takes and whether the board has a compelling story to tell about how it ensures it has the right culture; which rings true and is consistent with what the firm does."

19.5.5 US Department of Justice

In his speech to the White House Intellectual Property Theft Summit in December 2010, Attorney General Eric Holder emphasised the need to do more to protect intellectual property rights. He said that "the same technologies that spur growth in the legitimate economy also allow criminals to misappropriate the creativity of our innovators and entrepreneurs and to operate criminal enterprises that profit by selling imitations of legitimate products. In fact, for every quantum leap we have made technologically or commercially, criminals and often entire international criminal syndicates have kept pace. They have developed sophisticated methods for counterfeiting products and trademarks." Holder highlighted that trafficking in counterfeits is not victimless. For these acts threaten the financial stability of firms that sell legitimate goods, suppress ingenuity and destroy jobs. He said: "today, when the theft of a single trade secret can destroy a burgeoning small business, America's entrepreneurs and industry leaders are relying on strong Intellectual Property enforcement". However Holder emphasised it was more than just protecting intellectual property. He described how counterfeit goods could directly affect the public's and public sector employee's health and safety. He gave examples of lives being put at risk through counterfeit drugs or medical devices finding their way into medical care, counterfeit components entering industrial or military supply chains, and substitute low-quality materials being used in the manufacture of bullet-proof vests.

19.6 FACTORS THAT AFFECT BUSINESS ETHICS

Ethical codes that govern businesses often address certain main areas. The following are examples of the subject areas addressed by these codes:

- Honesty
- Objectivity
- Integrity
- Carefulness
- Openness
- Respect for intellectual property
- Confidentiality
- Responsible publication
- Responsible mentoring
- Respect for colleagues
- Social responsibility
- Non-discrimination
- Competence
- Legality

19.7 RISK EVENTS

Every business, even if it strives to comply strictly with the law, runs the risk of crossing the moral boundary through either the independent action of its employees or too imprecise or uncontrolled processes. Examples are provided below where companies have been prosecuted or have suffered reputational damage as a result of the behaviour of employees who have attracted negative media attention.

- *Bribery in the private sector*. Four individuals were given custodial sentences for bribery relating to service contracts between a machinery maintenance firm and Mars, the UK confectionary company.
- *Money laundering*. The UK Serious Fraud Office (SFO) announced in 2010 tthat it was conducting an investigation into the suspected payment of bribes by companies within the ALSTOM group in the UK. It is suspected that bribes had been paid in order to win contracts overseas, and that this had involved associated money laundering and false accounting. The SFO had been working closely with the Office of the Attorney General and Federal Police in Switzerland and a number of police forces in the UK.
- *Improper sales and marketing*. The US Department of Justice announced in 2010 that pharmaceutical manufacturer Allergan Inc., California, had agreed to plead guilty and pay $600 million to resolve its criminal and civil liability arising from the company's unlawful promotion of its biological product, Botox, for uses not approved as safe and effective by the Food and Drug Administration (FDA).

- *Inadequate financial accounting.* In 2010 BAE Systems plc was fined after admitting it had failed to keep adequate accounting records in relation to a defence contract for the supply of an air traffic control system to the Government of Tanzania.
- *Bribery of government contracting officers.* In 2008 the SEC announced that it had charged Siemens AG (Europe's largest engineering conglomerate and the largest electronics company in the world) with engaging in worldwide bribery. Siemens earned more than $1.1 billion in profits on these and several other transactions. Linda Chatman Thomsen, Director of the SEC's Division of Enforcement, said: "This pattern of bribery by Siemens was unprecedented in scale and geographic reach. Siemens made thousands of payments to third parties in ways that obscured the purpose for and the ultimate recipients of the money. Siemens made at least 4,283 payments, totalling approximately $1.4 billion, to bribe government officials in return for business given to Siemens around the world." These payments spanned Siemens' business groups to include transactions as varied as an infrastructure project in Argentina, telecommunications projects in Bangladesh and Nigeria, the installation of electricity lines in China, the construction of power plants in Israel, the design and construction of municipal transit systems in Venezuela and the sale of medical devices in China, Russia and Vietnam. Siemens violated the Exchange Act by failing to have adequate internal controls to detect and prevent the payments and by improperly recording the payments in its books and records. As reported in the press at the time, combined penalties of $800 million were imposed by the US Department of Justice and the SEC. Together with various penalties imposed in Germany, Siemens' aggregate penalties were $1.6 billion. The US penalties are by far the largest monetary sanction ever imposed in a Foreign Corrupt Practices Act (FCPA) case. The Department of Justice and SEC also charged Siemens with books-and-records and internal controls violations related to payments to the Iraqi government in connection with contracts secured by Siemens under the auspices of the United Nations Oil-for-Food Program. The Siemens settlement is notable because of the sheer scope of the improper payments at issue and the size of the penalties that were imposed. According to the government's papers, the improper payments implicated virtually all aspects of Siemens' operations, including its headquarters, subsidiaries and regional operating companies.

In the UK the Serious Fraud Office's investigation into alleged corruption by BAE Systems (a global defence, security and aerospace company) cast a cloud over the group's reputation and potentially jeopardised billions of pounds of defence contracts. BAE had been accused of offering bribes to win the billion-pound al-Yamamah deal with Saudi Arabia. The company denied the accusations and said it had not paid any bribes. A 30-month investigation by the SFO was dropped in December 2006 on the advice of Attorney General Lord Goldsmith. The UK government had intervened, fearing that it would harm Britain's relationship with the Kingdom and jeopardise a new £10 billion defence deal. An SFO appeal against the government decision failed. At the May 2007 annual general meeting BAE Systems defended its business ethics amid increasing calls to reopen a probe into a deal between the UK defence firm and Saudi Arabia. Speaking at the firm's annual general meeting, chief executive Mike Turner said BAE upheld the law and was ethical in "all respects of our business". He added that the long-running inquiry had distorted the view of the company. BAE Chairman Dick Olver also tried to reassure shareholders over the firm's behaviour. "The Attorney General said there was no case to answer, so there can have been no bribes", Mr Olver said. Mr Olver told the shareholders' meeting that BAE had a zero-tolerance policy on corruption and made sure its 88 000 staff were vetted and received ethics training. The contract with Saudi Arabia, which has generated more than £40 billion in sales for BAE since it was agreed by Margaret Thatcher in the mid-1980s, remained under investigation by the US Department of Justice some time after the UK investigation was dropped.

- *Inadequate internal controls*. In 2010 the SEC issued an order and imposed sanctions against Royal Dutch Shell plc (Shell) and Shell International Exploration and Production Inc. (SIEP). The order found that SIEP violated the anti-bribery provisions of the FCPA and Shell violated the record keeping and internal controls provisions of the FCPA.
- *Failure to follow quality standards and procedures*. In 2010 the car manufacturer Toyota recalled vehicles in the US, Europe and China over concerns about accelerator pedals getting stuck on floor mats.
- *Environmental irresponsibility*. In 1989, an Exxon oil tanker called the *Valdez* struck a reef in Alaska's Prince William Sound and created (at that time) the largest crude oil spill in US waters. The tanker trip was part of a routine convoy from Alaska to Long Beach, California, that had been successfully made by other tankers over 8000 times. The captain of the ship, Hazelwood, had assigned the piloting of the vessel to a less experienced officer and had then retired to his quarters. Icebergs were in the path of the ship, which an ineffective radar system failed to detect earlier. The ship was so large that it took a full minute to respond to steering changes. Attempting to navigate around an iceberg, the piloting officer miscalculated and ran the ship into a reef. Oil poured from the ship and, when the weather changed, it ran onto the beaches for hundreds of miles. Initially viewing it as only a public relations problem, Exxon was slow to respond with clean-up efforts, which made the situation worse. The spill had a terrible impact on plant and animal life in the area, which the news media vividly captured in pictures and on television. The clean-up was also expensive; the average cost of rehabilitating a seal was \$80 000. Hazelwood was ultimately fired for not being on the bridge at the time of the disaster. Exxon paid in excess of \$2 billion in the clean-up efforts and, just as significantly, suffered severe loss of reputation as a result of the disaster. According to some reports 40 000 Exxon credit card holders destroyed their cards.
- Another prominent environmental disaster was the BP oil spill in the Gulf of Mexico. As President Barack Obama announced in his address from the oval office in June 2010: "on April 20th, an explosion ripped through BP *Deepwater Horizon* drilling rig, about 40 miles off the coast of Louisiana. Eleven workers lost their lives. Seventeen others were injured. And soon, nearly a mile beneath the surface of the ocean, oil began spewing into the water. Because there has never been a leak of this size at this depth, stopping it has tested the limits of human technology. Already, this oil spill is the worst environmental disaster America has ever faced. And unlike an earthquake or a hurricane, it's not a single event that does its damage in a matter of minutes or days. The millions of gallons of oil that have spilled into the Gulf of Mexico are more like an epidemic, one that we will be fighting for months and even years."
- *Employee claims of sexual harassment*. In 2005 Boeing announced that its board of directors asked for and received the resignation of President and CEO Harry Stonecipher. The board it said had acted following an investigation by internal and external legal counsel of the facts and circumstances surrounding a personal relationship between Stonecipher and a female executive of the company who did not report directly to him. The board determined that his actions were inconsistent with Boeing's Code of Conduct. In a similar case, in 2010 Hewlett-Packard announced that Chairman, CEO and President, Mark Hurd, had decided

with the board of directors to resign his positions effective immediately. Hurd's decision was made following an investigation by outside legal counsel and the General Counsel's Office, overseen by the board, of the facts and circumstances surrounding a claim of sexual harassment against Hurd and HP by a former contractor to HP. The investigation determined that there had been violations of HP's Standards of Business Conduct.

- *Blacklisting of international, national or local organisations*. As reported in the press in December 2010, the World Bank Group barred an Italian engineering company, C. Lotti and Associati Società di Ingegneria SpA, from working on any World Bank-financed projects for the next two years – after the company acknowledged that it had engaged in fraud on a water sector project in Indonesia. The World Bank, United Nations and the European Union all have debarment practices which prohibit companies who have been convicted of fraud, corruption or serious misconduct from receiving funds for sponsored projects. The World Bank especially has a very developed system which allows the Bank to decide to exclude companies by way of an administrative panel.
- *Insider trading*. In 2005 Samuel Waksal, the former CEO of ImClone Systems Inc., and his father, Jack Waksal, agreed to a final resolution of the insider trading case brought against them by the SEC. Sam and Jack Waksal paid more than $5 million in disgorgement and civil penalties from their unlawful trades in ImClone securities. This fine was in addition to a partial settlement of $800 000 paid in 2003 when Sam Waksal was permanently barred from acting as an officer or director of any public company. In its complaint, the SEC charged that in late December 2001, Sam Waksal received disappointing news that the FDA was expected soon to issue a decision rejecting for review ImClone's pending application to market its cancer treatment drug, Erbitux. With that inside information in hand, and before ImClone publicly announced the FDA's decision on 28 December 2001, Sam Waksal unlawfully tried to sell shares of ImClone worth nearly $5 million, directed his daughter, Aliza, to sell all of her ImClone stock, and tipped his father, Jack Waksal, who sold his own ImClone stock as well as the ImClone stock of Patti Waksal, Jack's daughter and Sam's sister.
- *Exploitatation of Third World countries*. Information about the atrocious activities of US tobacco companies over the years is continually being made public. As deceptive and uncaring as they have been in the US, they are even worse in Third World countries. In developed countries, tobacco tar levels have decreased, but in Third World countries they have increased. Almost all developed countries have tobacco legislation, while less than half of Third World countries do, which is partly the result of cigarette companies' heavy lobbying efforts. Without restrictions on cigarette production, advertising and sales, cigarette companies expand the bounds of Third World markets with no thought of the health hazards they create for consumers. In Argentina, for example, tobacco companies buy 20% of all advertising time. Thus, although cigarette smoking is on a decline in developed countries, it is on the rise in Third World countries – and, globally, cigarette consumption is growing faster than population. US tobacco companies create strong incentives for local growers to shift to tobacco production by paying start-up costs to farmers, underwriting loans and guaranteeing purchases. By growing tobacco, less land is available for domestic food production, which is particularly serious for countries with large numbers of people living at subsistence levels. There are also ecological effects of flue-cured tobacco production that requires fire. In Third World countries, wood fires are a main method for curing tobacco, which requires one tree for every 300 cigarettes. This is bad since firewood accounts for 90% of the heating and cooking fuel in developing countries.

- *Health and safety irresponsibility*. In 1984, a gas leak at a Union Carbide chemicals plant in Bhopal, India, released 40 tonnes of poisonous gases (methyl isocyanate) over the city, killing thousands of people and injuring tens of thousands more. The incident was the worst industrial disaster in history. Campaigners say at least 15 000 were killed within days and that the effects of the gas continue to this day. In 1989, Union Carbide paid $470 million in compensation to the Indian government.
- *Invasion of privacy*. There has been a long history of company information leaking from the Hewlett-Packard board to the press. One notable and harmful leak was reported by CNET News on 23 January 2006, in which HP's acquisition, chip supplier, printer development and server plans were all reported. In response to the leaks, HP, led by Chairwoman Patricia Dunn, commenced a rigorous investigation into the leaks by an investigative team composed of a senior counsel in HP's legal department, other employees from HP's global IT and security departments and an external investigative firm. In turn, the investigative firm hired private investigators. From testimony given to the House of Representatives Energy and Commerce Committee in 2006 by the Chief Executive Officer and President, Mark Hurd, and from press articles at the time, it became public knowledge that HP had been involved in unethical practices to address the leaks. The privacy of nine journalists (including reporters for CNET, the *New York Times* and the *Wall Street Journal*), two current HP employees and seven former and current HP board members had been infringed by HP. To varying degrees these individuals were investigated through "pretexting", a technique used by the private investigators to obtain phone records by impersonation. Dunn claimed she did not know beforehand the methods the investigators would use to try to determine the source of the leak. According to press reports, board member George Keyworth was ultimately accused of being the source, and on 12 September 2006 he resigned, although he continued to deny making unauthorised disclosures of confidential information to journalists.

19.8 IMPLEMENTATION OF ETHICAL RISK MANAGEMENT

One way of addressing risk exposure from of a breach of ethics is to devise and implement an ethics system across the organisation as a means of preventive action. That requires clearly thinking through the desired business culture and determining how to make it a reality. The US Department of Commerce (2004) describes how enterprises may design and implement a business ethics programme that meets emerging global standards of responsible business conduct. This section is based on the recommendations of what we will call the *Manual*. Clearly it is only one approach. However, it is a well-considered approach and provides a basis upon which to understand both ethics exposure and potential corrective action.

19.8.1 Areas of Focus

Typically a business ethics programme aims to achieve specific expected outcomes, such as increasing awareness of ethics issues, improving decision making and reducing misconduct. In addition, the effectiveness of a programme will be determined by its areas of focus or orientations. The *Manual* describes four primary orientations as follows, which are not considered to be mutually exclusive:

1. A *compliance-based* approach, which "focuses primarily on preventing, detecting and punishing violations of law".

Figure 19.2 Four levels of a responsible business enterprise

2. A *protecting senior management* approach, which "is introduced in part to protect owners and senior management from blame for ethics failures or legal problems".
3. A *satisfying external stakeholders* approach through which enterprises "hope to maintain or improve their public image and relationships with external stakeholders".
4. A *values-based* approach, which "aims to define organizational values and encourage employee commitment to ethical aspirations".

19.8.2 Levels of Application

The *Manual* describes a responsible business enterprise as being characterised by or having an "identity" based on the extent or degrees of application of ethics management. It describes four levels of a responsible business enterprise – compliance, risk management, reputation enhancement and value added – where a subsequent level builds on the foundation set by the preceding level(s), as illustrated in Figure 19.2. These degrees of application are clearly based on the four areas of orientation described in the preceding subsection.

Compliance Level

The compliance level is the most obvious and rudimentary level of identity. Compliance means meeting all legal requirements through an effective programme to prevent and detect misconduct. This means preparing a formal system, communicating the system, providing training, monitoring its effectiveness and providing corrective action where required. However, at compliance level an organisation will not have reviewed and modified its culture to embrace and embed compliance in its day-to-day activities. This level does not require the business to assess its stakeholder needs or understand how compliance is integrated with the core values of the business.

Risk Management Level

It is a little confusing that the *Manual* describes one of the levels of application as "risk management", as it could be argued a whole ethics programme is a risk management activity. The activity prescribed by the *Manual* at this level is twofold, looking at the business' policies, processes and activities and closely examining the business's culture (to understand both the attitude of staff to compliance and whether there is a history of problems, disharmony and conflict). The *Manual* says that failing to address these risks may lead to one or more of the following: criminal prosecution, regulatory action, debarment from government contracting, civil lawsuits, compromised strategic partnerships, labour–management disagreements, shareholder dissatisfaction and calls from society for more regulation.

Reputation Enhancement Level

Reputation is critical to business survival. The larger an organisation grows, the greater the effort that needs to be expended to ensure that all staff are "on the same page" in terms of both understanding and implementing ethics requirements. The results of not doing that have been illustrated by Siemens and BAE Systems. As many CEOs have described, it takes many years to build a reputation, but it can literally be destroyed over night from revelations of ethical failings. Exposure has increased over time from the increasing number of stakeholders that an organisation deals with (such as charities, non-government organisations, self-styled environmental watchdogs and international institutions). The *Manual* says that enhancing one's reputation among a larger and more diverse group of stakeholders represents a third, more important level of enterprise identity.

Value Added Level

This level refers to a responsible business which, through its considered conduct and behaviour, contributes to the community within which it resides. While it is recognised that businesses have to make a profit, they are expected to take account of their stakeholders. These can include the local community (who may be impacted by the business's treatment of the environment), local suppliers of goods and services (who may be impacted by payment reliability), and employees (who may be impacted by discrimination, harassment, bullying and payment terms). A good example of a company which considered how it would contribute to its community from the outset was the Tokyo Telecommunications Engineering Corporation, which later evolved into the Sony Corporation (see Box 19.2). In its founding prospectus Mr Masaru Ibuka describes how the company intends to contribute to its engineers, households, subcontractors,

Box 19.2 Tokyo Telecommunications founding prospectus

The following are extracts from the founding prospectus of Tokyo Telecommunications Engineering Corporation that Masaru Ibuka drew up in 1946.

Purpose of Incorporation

- To establish an ideal factory that stresses a spirit of freedom and open-mindedness, and where engineers with sincere motivation can exercise their technological skills to the highest level
- To reconstruct Japan and to elevate the nation's culture through dynamic technological and manufacturing activities
- To rapidly commercialize superior technological findings in universities and research institutions that are worthy of application in common households
- To bring radio communications and similar devices into common households and to promote the use of home electric appliances

- To actively participate in the reconstruction of war-damaged communications network by providing needed technology
- To produce high-quality radios and to provide radio services that are appropriate for the coming new era
- To promote the education of science among the general public.

Management Policies

- We shall eliminate any unfair profit-seeking practices, constantly emphasize activities of real substance and seek expansion not only for the sake of size
- We shall focus on highly sophisticated technical products that have great usefulness in **society**, regardless of the quantity involved
- We shall guide and foster sub-contracting factories in ways that will help them become independent, and we shall strive to expand and strengthen mutual cooperation with such factories
- We shall distribute the company's surplus earnings to all employees in an appropriate manner, and we shall assist them in a practical manner to secure a stable life. In return, all employees shall exert their utmost effort into their job.

Source: http://www.sony.net/SonyInfo/CorporateInfo/History/prospectus.html

universities, the general public, the community and the country as a whole. Stakeholders will also include a business's customers who will be directly affected by the quality of its products and services being provided. Although when the subject of business ethics is raised, product quality does not immediately spring to mind, it is nonetheless a component of business ethics. It concerns whether the products are, for instance, safe (from a security, financial, electrocution, allergic reaction or personal injury perspective), reliable or serviceable. Producing low-quality goods can reduce reputation, employee pride and morale, willingness of third parties to form alliances and supplier relationships. Ultimately the manufacture of low-quality goods is not a productive utilisation of the resources consumed and can be detrimental to the overall community within which the business resides. The *Manual* says that for an organisation to be truly responsible, it must fully embrace all four levels of identity.

19.8.3 The System

A system of business ethics may be composed of seven sequential components (as described in Figure 19.3) – which are never complete and hence are shown as forming a circle. Each of the components is described in turn. Once the *evaluate* component has been completed, the business *vision* must be re-examined in the light of both the knowledge accumulated in implementing the system to date and the findings of the evaluation process, to see if the overall goal of the business should be modified.

Vision

The business vision is the starting point for the creation of a business ethics programme. The programme must not conflict with or undermine the intentions of the vision and its underlying objectives. Collins and Porras (1994) argue that the term "vision" has become one of the most

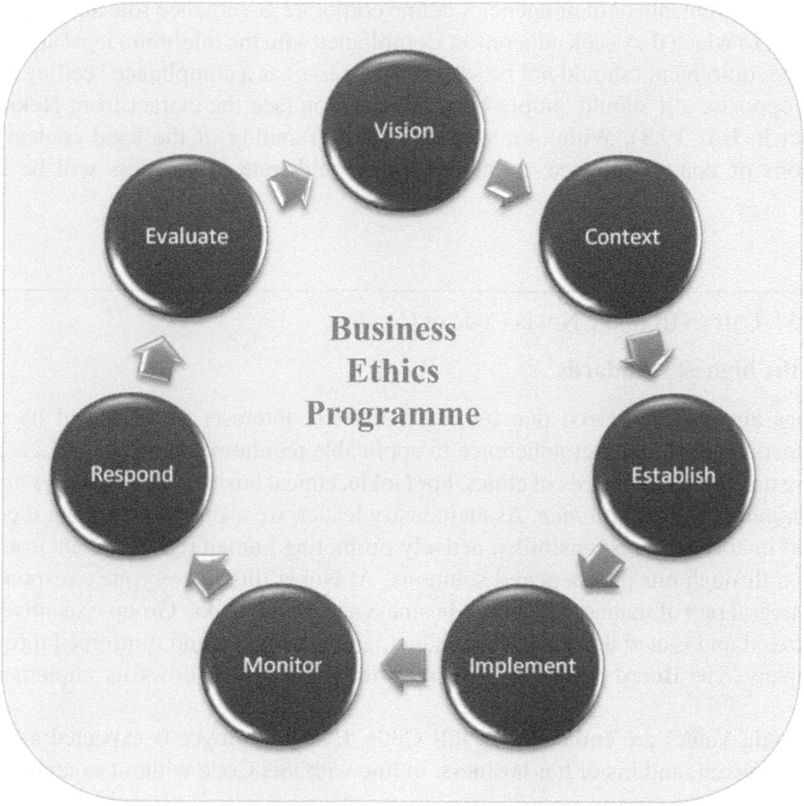

Figure 19.3 Seven-step business ethics programme

overused and least understood words in the English language and has been the subject of a host of different interpretations. They say that a well-conceived vision consists of two major components – a core ideology and an envisioned future. The core ideology defines what a business stands for and why it exists, and the envisioned future defines what the business aspires to become, achieve and create. The core ideology is composed of two components: the core values and tenets (the guiding principles); and a core purpose (the organisation's fundamental reason for existence). The ideology is described as the glue that holds an organisation together as it grows, decentralises, diversifies, expands globally and develops workplace diversity. The envisioned future is something that will require significant change and progress to attain.

Context

Any ethics system is meaningless and most likely to fail without a thorough comprehension of the business's context. A business's context is typically composed of legal, economic, political, environmental, social and technological elements.

Legal. Governments create the legal framework in which market processes and businesses operate, and (to varying degrees) regulate public health, safety, morality and dispute resolution.

In addition, governments or their agencies define corporate governance role and responsibility requirements to which they seek adherence. Compliance with the minimum legal and corporate governance requirements should not be seen by businesses as a compliance "ceiling" to aspire to but the opposite – it should simply be the foundation (see the extract from Nokia's Code of Conduct in Box 19.3). Without a very clear understanding of the legal context and the ramifications of non-compliance, unintentional or deliberate, a business will be left very exposed.

Box 19.3 Extract from the Nokia Code of Conduct

Setting the highest standards

Nokia has always recognized that its own long-term interests and those of its various stakeholders depend on strict adherence to applicable regulation, the Rule of Law and on following the highest standards of ethics. For Nokia, ethical business conduct *does not mean mere minimum legal compliance*. As an industry leader, we aspire to be among the best in the world in corporate responsibility, actively promoting human rights and environmental protection through our products and solutions. At Nokia this is everyone's responsibility and an integral part of managing Nokia's business and brand. Nokia Group Executive Board has approved and issued this Code of Conduct, and it is shared and reinforced throughout the company. The Board periodically reviews this Code and follows its implementation closely.

The Nokia Values are embedded in this Code. Every employee is expected to conduct himself or herself, and his or her business, in line with this Code without exception.

Source: http://www.nokia.com (emphasis added).

Economic. In examining its context, businesses need to understand the nature and extent of government influence in the market, how laws and regulations are introduced, who has any practical ability to steer their direction, and the degree of government influence over the economy.

Political. In appreciating its context a business needs to understand the kind and degree of government influence in the market, the process of the introduction of laws and regulations, planned legislation and the degree of government influence on the economy. Clearly governments do not have total control over their economies, as exhibited by the global financial crisis and struggling economies in the European Union such as Ireland, Portugal, Spain and Greece.

Environmental. The environment constitutes the widest form of context and more recently has attracted considerable media coverage due to the attention given to global warming, rising sea levels and anticipated changes in global climatic conditions. The planet may be described as a series of interdependent ecosystems that have evolved over millions of years but are susceptible to radical change from man's increasingly invasive behaviour. The business ethics programme must be highly sensitive to and informed about man's impact on the environment, international initiatives to reduce harmful practices, current and planned legislation and how society expects businesses to behave.

Social. The behaviour of company employees (and board members) will be influenced by the country within which they originate. Each society has cultural practices that are unique. Employees will bring their beliefs, values and cultural norms to the job. It is widely accepted that people are not born with a culture, but that it is learned. So the term "culture" includes all that we have learned in relation to the norms and values, customs and traditions, beliefs and religions and rituals, and urban fabric (i.e. tangible symbols of a culture such as prominent buildings and districts). Cultures can influence the importance attached to respecting authority, how uncertainty is managed, the importance attached to timekeeping, the methods employed to correct (or criticise) staff, the giving of gifts, privacy, nepotism (hiring friends and relatives), reporting misconduct, gender relations and the display of emotions. Organisational cultures cannot permit staff to make up their own minds as to what constitutes acceptable behaviour (based on their own experiences and upbringing) but must set and drive desired behaviours. However, a business must be sensitive to the culture and norms of the countries within which it has offices, without sacrificing its core beliefs.

Technological. The internet is an international phenomenon. Businesses cannot escape it even if they wanted to. It is the world we live in. The proliferation of social network websites (such as Twitter, Facebook, MySpace, YouTube and Yelp) have all exposed companies to yet another potential route for reputational damage arising from the actions of their staff. There are now over 100 social network websites around the world and membership is enormous. Social network Kaixin001.com in China has a reported membership of 75 million. Research has shown that a high percentage of companies are afraid of what their employees are choosing to share on social networking sites. Office moves, promotions, company reorganisation, the development of new products and the purchase of new technology could be of use to a competitor. Inappropriate discussion on a network site about the behaviour of a client organisation or their staff may bring the business into disrepute and result in a loss of business. Inadequate software or software training may lead to inadequate record keeping, monitoring or auditing. The illegal downloading of software can expose a business to licence infringement. Electronic mail can be a curse in terms of its deliberate use in sexual harassment, bullying, breach of confidentiality, discrimination or pornography.

Establish

Organisation. The system developed for an organisation must reflect its size, complexity, resources, context, culture and stakeholder expectations. Over the past few decades organisations have developed systems to emphasise their strengths and address their recognised weaknesses and, having decided on the functions to be completed, appointed a responsible officer and a business ethics officer, and established a business ethics council. Usually the council and ethics officer have sought guidance on best practice from professional bodies which can provide independent impartial advice on compliance and the implementation of ethics systems. The ethics officer is a part- or full-time role, depending on the size of the organisation and the nature of its business. It is commonly recognised that there is no right or wrong way to develop a business ethics programme; however, there are certain functions that need to be performed. Failure to address these functions will often lead to an inappropriate focus and/or capacity to respond to problems related to ethics and the underlying compliance and social responsibility issues. According to the *Manual*, leading organisations have found that an effective business ethics programme functions at seven levels of responsibility:

1. Overseeing the program at a high level (responsible officer – typically someone reporting to a board member)
2. Performing or coordinating the specific functions of the business ethics program (ethics officer)
3. Advising the responsible officer and business ethics officer and representing the business as a whole (the business ethics council)
4. Advising the responsible officer, business ethics officer and employees and agents about specific professional ethics, compliance and social responsibility issues such as human resources, engineering and/or community issues (the professional ethics 'council' – national professional organisations advising on ethics)
5. Linking the various levels of the business with a central ethics office (where established – otherwise the business ethics officer)
6. Performing related executive and department functions (the chief financial officer, legal counsel, human resources, internal audit, risk management, environment, health, safety, government procurement and investor relations
7. Abiding by standards and procedures and striving to meet responsible stakeholder expectations (every employee and other agent of the business).

Guidance. A well-designed and well-implemented business ethics programme provides employees with both the guidance and the information they need to implement required ethics practices. Guidance, in particular, enables employees to make choices and select courses of action with confidence. Employees need to know what is expected of them, but specifically what they should and (possibly more importantly) what they should not do, to meet the business's ethics objectives. The business needs to develop standards and procedures so that employees are clear who is responsible for which decisions and how individuals will be held accountable for their actions. If standards and procedures are not prepared, it will be difficult to guide employees or hold them accountable for poor judgement or unacceptable behaviour. Criticism levelled at staff when procedures and processes are not in place will undermine employee loyalty, and employee morale will be a serious risk.

Code of conduct. A code of conduct is often the primary means by which management gives guidance to its employees in terms of what is expected of them. A code demonstrates a board's commitment to meeting all applicable legislation and industry specific practices. It is a means by which it can ensure that the organisation conducts its business in an ethical manner to help ensure long-term success for the business, its clients, employees and shareholders. A code of conduct addresses minimum standards of conduct and procedures to reduce the business's risk of liability and damage to its reputation.

Implement

Training. Training is a critical element of ethics risk management implementation to ensure the goals of the ethics programme are communicated, understood and applied. The training programme objectives will depend on the context, desired organisational culture, goals, objectives and the expected outcomes of the ethics programme as a whole. As a minimum, the training sessions should require employees to become familiar with applicable laws and regulations as well as with the business's procedures for reporting and investigating suspected unacceptable conduct.

Monitor

Monitoring business performance is an essential management task. It involves setting performance expectations for individuals, groups, departments and business streams to guide their efforts towards achieving the business's goals and objectives. These performance expectations, to be meaningful and effective, must be SMART (see Section 8.8.1). Clearly business ethics management can be addressed in the same way. Where risks have been identified and assessed for subjects such as conflicts of interest, fraudulent consumer transactions, inaccurate financial records, bribes, corruption and staff mistreatment, a business ethics programme monitors performance on a regular basis. It requires an assessment to be specific against predetermined criteria (compliance requirements), tailored to the nature of the business activities (services or products) and reflect stakeholder interfaces. Key questions need to be raised. Have there been any prosecutions in the last year, and how does this compare with previous years? Serious ethics transgressions are more likely to occur at board level – how is detection being implemented?

Respond

Having completed the *monitor* step, the business needs to understand if it should be responding to the findings. Has the type of risks changed? Has the risk exposure increased in some way? Have departments created new working practices to circumnavigate legislation on the payment of foreign officials to secure new commissions, and do they need responding to? Is there sufficient visibility of payment patterns in overseas offices, and do they require modification? Have historical patterns of non-conformance to ethical practices within regional offices been addressed? Do new more bureaucratic procedures need to be implemented to mitigate the risk of non-compliance with anti-money-laundering requirements? The key question is what, if anything, needs to be changed. Are there areas of compliance that are weak, lack transparency, require a change in mindset or cultural shift or rectification to address clear deficiencies?

Evaluate

For a business ethics programme to be effective, its performance must be evaluated. Having a code of conduct, bespoke procedures and ethics training is not an end in itself. Consideration needs to be given to what should be evaluated. Typically evaluation is multifaceted. It must establish if the context of the business has changed and, if so, whether the ethics programme has been amended to suit. Have new regulations and standards emerged, and have they been reflected in processes and procedures? Have stakeholder needs been reassessed, and have reasonable demands from stakeholders for information been met? How many employees have been through the training programme? Has the ethics system been integrated into the everyday business practices, particularly where the business may be vulnerable to prosecution? Is there less misconduct? Have the actions that the programme said would be implemented been implemented? In addition, the evaluation process should also revisit the business's vision and goals and, in the light of the evolving context (in terms of stakeholder expectations, the regulatory regime, concerns over the environment, society's demands, competitor behaviour and the business's evolving culture), consider if they should be changed.

19.9 SUMMARY

This chapter has provided a lens on the world of business ethics. It is a vast subject and warrants close scrutiny. A breach of ethics, depending on its severity, can erode reputation and share price, but also lead to lost opportunities when potential business partners disengage and government departments place bidding restrictions on new contracts. Both at the time of the collapse of Enron and WorldCom, and after the global financial crisis, shareholders called for businesses to exhibit and practice business ethics in all their dealings. Examples of unethical behaviour were provided such as bribery, money laundering, inadequate reporting, environmentally irresponsible practices and insider trading. The media reports on what appears to be a never ending stream of unethical behaviour from collusion between airlines over price fixing, through false advertising claims, to the use of child labour. A risk management strategy for businesses is to design and implement a business ethics programme that meets emerging global standards. The programme proposed by the US Department of Commerce (2004) was described. Each of the seven steps of the programme was described in turn. However, rather than being yet another burden on corporations, some businesses see it as a way of achieving competitive advantage.

19.10 REFERENCES

Collins, J.C. and Porras, J.I. (1994) *Built to Last: Successful Habits of Visionary Companies*. Harper-Business, New York.

FSA (2002) *An Ethical Framework for Financial Services*. Discussion Paper 18, October. http://www.fsa.gov.uk/pubs/discussion/dp18.pdf

OECD (2011) *OECD Guidelines for Multinational Enterprises: Recommendations for Responsible Business Conduct in a Global Context*. http://www.oecd.org/dataoecd/43/29/48004323.pdf

US Department of Commerce (2004) *Business Ethics: A Manual for Managing a Responsible Business Enterprise in Emerging Market Economies*. US Department of Commerce, Washington, DC.

20 Health and Safety Management

Working without safety is a dead end job.
(Author unknown)

This chapter examines *health and safety*, the sixth of six classes of risk exposure that businesses face, relating to internal processes within the risk taxonomy proposed in Chapter 9. All six classes of risk are considered to be controllable by businesses to a large degree. Health and safety issues are part of the rich tapestry of enterprise risk management. While the UK has a strong safety culture and one of the lowest rates of fatalities at work, no enterprise can ignore health and safety. Even small businesses in the UK have supply chains that stretch across Europe and the Far East. According to the European Agency for Safety and Health at Work, every year in the European Union there are over 5000 fatal work-related accidents and millions of people are injured or have their health seriously harmed in the workplace. While health and safety management was once seen as a peripheral activity, it is now an integral part of the fabric of society. While history has demonstrated that the attitude of society to the moral obligation of employers has fluctuated over time, the pendulum has clearly swung towards an increasing awareness and expectation of moral behaviour. Society now puts pressure on employers to modify and constrain behaviour. The public outrage at rail, sea and air disasters clearly illustrates that if risk impinges on the general public, it is an intolerable face of society. Employers are now exposed to common and statute law that is criminally enforceable. Enterprises have to consider the consequences of failing to properly comply with health, safety and environmental legislation, approved codes of practice, accepted standards and other relevant legislation concerning, for example, fire prevention, pollution and product liability. Enterprises must recognise that losses may result from the requirement to implement preventive measures from enforcement notices, punitive damages through criminal sanctions and the compensatory consequences of the imposition of legal rulings. As with many of the subjects touched on in this book, which can be considered as disciplines in their own right, a brief perspective of health and safety is introduced here as it is part of the complex jigsaw of enterprise risk management. The structure of this chapter is illustrated in Figure 20.1.

20.1 DEFINITION OF HEALTH AND SAFETY RISK

What is health and safety risk? Health and safety risk for an organisation may be defined as the risk of injury or ill health to or the fatality of an employee (or employees). The primary aim of health and safety in the workplace therefore is the prevention of death, injury and ill health to those at work and those affected by work activities.

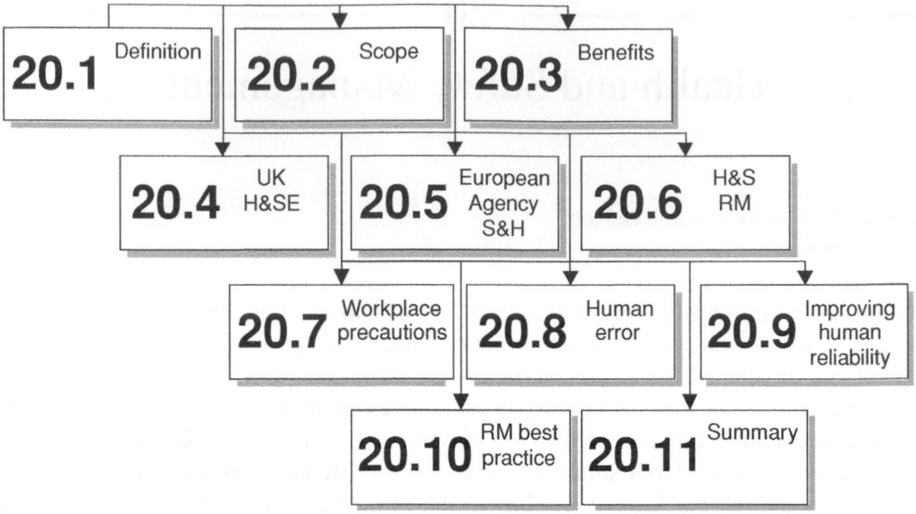

Figure 20.1 Structure of Chapter 20

20.2 SCOPE OF HEALTH AND SAFETY RISK

The sources of risk considered to be embraced within the term "health and safety risk" in a business context are as follows:

- more onerous government policies, legislation and/or standards;
- reputational damage;
- criminal prosecution and/or civil law suits;
- loss of market share, customers, suppliers, shareholders, funding, partners and/or employees;
- disruption to production, research or use of advanced technologies;
- negative media, press and/or internet coverage;
- increased insurance premiums and/or erosion of profits;
- increased staff turnover and associated recruitment costs.

20.3 BENEFITS OF HEALTH AND SAFETY RISK MANAGEMENT

There is a general consensus that a health and safety risk management system is good business management and improves bottom-line profitability wherever in the world the business operates. A health and safety risk management system helps to avoid:

- health and safety incidents or an increase in the number of incidents and/or their impact;
- non-compliance convictions, criminal prosecutions or enforcement notices;
- civil claims;
- adverse media attention and damage to reputation;
- increase in insurance premiums;
- compensation payments;
- the need to investigate the cause(s) of an accident;
- the need to prepare accident reports, attending hearings or inquest courts;

- the need to arrange for the injured employee's work to be continued by another employee;
- the need to make staff rehabilitation and return to work arrangements (recognising that returning staff may need to work at a reduced capacity, at least initially);
- loss of productivity, business, early completion bonuses or future orders;
- the need to engage solicitors and barristers to represent the organisation;
- loss of board, management and supervisor time in responding to incidents.

There are a number of additional benefits such as:

- increased productivity;
- greater production reliability and reduction in the chance of losing sales to a competitor;
- improvement in staff morale, together with staff retention and recruitment rates;
- reduced staff absenteeism;
- meeting increasingly stringent lending criteria;
- improved success rate in bidding for contracts;
- improved shareholder satisfaction from meeting increasingly higher health and safety standards.

20.3.1 Business Benefits

Does a health and safety risk management system improve business performance? A business survey funded by the Health and Safety Executive (HSE) attempts to answer that question (Cowling and Bevan 2007). The report's introduction states that its aim is "to provide robust empirical evidence concerning any linkages and impacts of health & safety strategy and expenditure on an array of hard and soft performance measures of intermediate and final business performance". Data was derived from a telephone survey of 3000 UK businesses, allowing for geographical representation according to government office regions. The telephone interviews were carried out in June and July 2004, and the sample was generated by random digit dialling. Interviews were carried out with chief executive officers, managing directors, chief finance officers or human resource directors in the UK. The study was conducted using computer-assisted telephone interviewing. In its conclusions the report highlights its starting hypothesis that the key to achieving high levels of business performance is to develop complementary strategies across all areas of the business because it is the overlapping and mutually reinforcing effects of multiple synergistic practices that have, potentially, the largest impact. The general approach of the survey, the authors state, was underpinned by the *a priori* belief that health and safety is a key area of strategic decision making that cannot be considered in isolation by businesses, and one which should be integrated into other areas of strategy to ensure not only consistency in terms of planning, but to achieve maximum impact on business performance outcomes. The report claims that evidence was obtained that suggested that health and safety risk is an important feature for large numbers of businesses across an array of sectors. In particular, it states that construction, retail/hotels/catering, agriculture and utilities are sectors that have to manage perceived risks surrounding health and safety. Interpreting the study findings, it would appear that, across all sectors, there was only a weak association between health and safety practices and business performance. Regarding health and safety expenditure specifically, the study noted that there was no evidence that spending more is associated with a lower level of business performance. The study considered that this could imply that higher spending is associated with a proportional increase in performance which balances this out. The study observed that in three cases a higher health and safety expenditure was associated with superior performance. It found that it was correlated with businesses having a greater capacity to attract quality employees from the industry pool, with higher employee commitment and faster sales growth. The study concluded that there was a suggestion that a strategic commitment to good health and safety practice did businesses no harm, and a spending commitment was strongly associated with tangible improvements in employee related aspects of the business.

20.3.2 The Enterprise Context: AstraZeneca

In its promotion of health and safety, the HSE provides a number of examples where businesses that have adopted health and safety practices and the "direct" benefits that have been obtained. It is not straightforward to isolate the application of health and safety practices as the only driver behind business improvement. An example of a company embracing health and safety management is AstraZeneca which (the HSE states) recognises that personal well-being is essential for employees to build the company's innovation and creativity, adding competitive advantage in an increasingly competitive global environment. AstraZeneca is a global pharmaceutical manufacturing company with 58 000 employees. Through the implementation of a health and safety system, the HSE advise it has realised the following benefits:

- UK health insurance spend is lower than benchmarked, saving £200 000 a year.
- UK absence levels are 31% lower than average levels for the UK quoted by the Confederation of British Industry.
- Employees note significant improvements in concentration and productivity at work,
- The company ranks in top 10% of Dow Jones sustainability performers worldwide, in the top 20% in Europe, and was recently listed in the FTSE4Good series.

20.4 THE UK HEALTH AND SAFETY EXECUTIVE

The HSE is a non-departmental public body with Crown status, sponsored by the Department for Work and Pensions (DWP) and accountable to its ministers. It is Great Britain's independent watchdog for work-related health, safety and illness. The HSE's primary function is to secure the health, safety and welfare of people at work and to protect others from risks to health and safety from work activity. It is responsible for regulating health and safety in Great Britain and works in partnership with local authorities. The HSE's strategy, *The health and safety of Great Britain: Be part of the solution*, defines the goals that the organisation and all stakeholders in the health and safety system strive to achieve.[1] The HSE's mission is the prevention of death, injury and ill health to those at work and those affected by work activities. The HSE seeks to influence people and organisations, duty holders and stakeholders, to embrace high standards of health and safety, demonstrate health and safety leadership and promote the benefits of employers and workers working together to manage health and safety sensibly. The HSE also investigates incidents and complaints about health and safety practices, and develops new or revised health and safety legislation and codes of practice.

The HSE regulates health and safety across a range of sectors and industries, including major hazard sites such as nuclear installations, offshore gas and oil installations and onshore chemical plants through to more conventional sites, quarries, farms, factories, waste management sites and hospitals. It accomplishes this through continued targeting of resources to

[1] http://www.hse.gov.uk/strategy

priority risks and sectors such as the agriculture and waste management industries and by applying an appropriate and proportionate mix of intervention techniques such as inspection, communication campaigns, advice and support and, where necessary, enforcement action. The HSE works together with local authorities, who are responsible for regulating half of the workplaces in Great Britain, in particular, commerce, retail, hospitality, entertainment and other services. Both regulators are responsible for many other aspects of the protection both of workers and the public in accordance with the Health and Safety at Work etc. Act 1974. Health and safety matters dealt with by the HSE have not been devolved to the administrations in Scotland and Wales. Effective working arrangements have been developed, however, between HSE and the devolved administrations to ensure that areas of "common and close interest" are managed appropriately.

20.4.1 The UK Perspective: Health and Safety Record

In 2010 the HSE released provisional data which showed that 151 workers had suffered fatal injuries between 1 April 2009 and 31 March 2010, a significant improvement on the 178 fatal injuries recorded in the previous year. As will be seen from the figures below the drop was impressive compared with the last five years:

- 2008/09 – 178
- 2007/08 – 233
- 2006/07 – 247
- 2005/06 – 217
- 2004/05 – 223

In addition to fatal injuries are deaths arising from occupational illnesses contracted by, for example, working in the mining industry (inhaling coal dust), the nuclear sector (being exposed to airborne radioactive waste products) and employment in the construction industry (inhaling asbestos dust). In other industries isolating employment activities as the sole cause of a fatal illness is not so straightforward.

The reporting of health and safety incidents at work is a statutory requirement, set out under the Reporting of Injuries, Diseases and Dangerous Occurrences Regulations 1995. A reportable incident includes: a death or major injury; any accident which does not result in major injury, but the injured person still has to take three or more days off their normal work to recover; a work related disease; a member of the public being injured as a result of work-related activity and taken to hospital for treatment; or a dangerous occurrence, which does not result in a serious injury, but could have done. While "dangerous occurrences" are quite specific, it is also important to consider near misses as their effective reporting has the potential to drastically reduce the likelihood of events materialising.

20.5 THE EUROPEAN AGENCY FOR SAFETY AND HEALTH AT WORK

The European Agency for Safety and Health at Work (EU-OSHA) aims to be the central provider of health and safety information and to ensure that it is relevant to every user, regardless of size of enterprise or sector of activity. Set up in 1996 by the European Union and located in Bilbao, Spain, EU-OSHA is the main EU reference point for safety and health at work.

The central role of EU-OSHA is to contribute to the improvement of working life in the European Union. Its website states:

- We work with governments, employers and workers to promote a risk prevention culture.
- We analyse new scientific research and statistics on workplace risks.
- We anticipate new and emerging risks through our European Risk Observatory.
- We identify and share information, good practice and advice with a wide range of audiences, such as social partners – employers' federations and trade unions.[2]

The European Commission has launched a new five-year strategy for Safety and Health at work. It aims to cut work-related accidents across the EU by a quarter. The new European Strategy covers the period 2007–2012 and is pursuing the positive trends of the previous Community Strategy 2002–2006 which has already borne fruit. Over the period 2002–2004, the rate of fatal accidents at work in the EU-15 fell by 17%, while the rate of workplace accidents leading to absences of more than three days fell by 20%.

20.5.1 Main Challenges Concerning Health and Safety at Work

In spite of the progress achieved, the latest results of the fourth European survey of working conditions show that many workers in Europe continue to perceive that their jobs pose a threat to their health or safety:

- almost 28% of workers in Europe say that they suffer from non-accidental health problems which are or may be caused or exacerbated by their current or previous job;
- 35% of workers on average feel that their job puts their health at risk.

Moreover, occupational hazards are not being reduced in a uniform way:

- some categories of workers are still overexposed to occupational risks (young workers, workers whose jobs are insecure, older workers and migrant workers);
- certain types of companies are more vulnerable (small and medium-sized enterprises, in particular, have fewer resources to put complex systems of worker protection in place, while some of them tend to be more affected by the negative impact of health and safety problems);
- certain sectors are still particularly dangerous (construction/civil engineering, agriculture, fishing, transport, health care and social services).

20.6 IMPLEMENTATION OF HEALTH AND SAFETY RISK MANAGEMENT

Health and safety systems are a formal and logical way of planning and controlling the management of risk in an organisation. The models can be based on the HSE model contained in its guidance referred to as HSG65 and called *Successful Health and Safety Management* (HSE 1997), or the environmental model BS EN ISO 14001, or OHSAS 18001, which models

[2] http://osha.europa.eu/en/about

Figure 20.2 Components of a health and safety management system

risk on the quality model ISO9001.[3] HSG65 may be considered to have three key components (Figure 20.2), as follows:

- Management arrangements – essentially the management of the system.
- Risk control systems (RCS) – to make sure workplace precautions are implemented and kept in place.
- Workplace precautions – to prevent harm at the point of risk.

Each of these three components is summarised below.

20.6.1 Management Arrangements

This typically consists of a five-step approach:

- Setting the policy
- Organising staff
- Defining and setting standards
- Measuring performance
- Audit and review.

20.6.2 Risk Controls

These are measures to ensure workplace precautions are implemented and in place:

- input controls such as purchasing and recruitment;
- process controls such as maintenance and change management;
- output controls such as product packaging/labels and disposal.

20.6.3 Workplace Precautions

These are very considerable and cover both multiple industries and multiple activities.

[3] HSG65 is considered advantageous if an organisation wishes to have a "stand alone" health and safety risk management system, whereas 18001 is more suited to an "integrated" approach due to its commonality with 9001 and 14001.

20.6.4 System Implementation

Positive indicators of the implementation of a health and safety risk management system would be:

- a publicly stated health and safety management system to monitor performance;
- display of the health and safety policy in a prominent location;
- line responsibility for health and safety clearly defined at board, committee and business unit level;
- the formal health and safety committee attended by representatives of the workforce;
- recognition of emerging risks and the creation of policies to respond to them, such as stress in the workplace;
- appropriate assessment of workplace safety and health exposure and the provision of the correct equipment for tasks which can save lives, reduce liabilities and increase staff morale;
- varying tasks in type and duration, particularly where staff carry out tasks which are highly repetitive, to improve concentration and avoid safety lapses;
- robust reporting and "no blame" culture including near misses;
- active health monitoring programmes (initial and ongoing).

20.7 WORKPLACE PRECAUTIONS

The types of health risk posing a threat to organisations and their staff include the following, as drawn from *Essentials of Health and Safety at Work*, 4th edition (HSE 2006), a publication that aims to enable organisations to prevent workplace accidents and ill health and comply with the law.

Fire	Lifting and handling
Work at height	Noise
Building work	Vibration
Machinery safety	Electricity
Plant and equipment maintenance	Radiations
Gas and oil filled equipment	Harmful substances
Pressurised plant and equipment	Flammable and explosive substances
Workplace transport	Managing health.

The last topic, managing health, covers such issues as workplace stress, drugs and alcohol and violence at work, but makes no reference to subjects such as HIV/AIDS or vision impairment.

20.8 CONTRIBUTION OF HUMAN ERROR TO MAJOR DISASTERS

The following accounts of accidents all highlight the importance of introducing risk controls, context-based training and conveying in-depth understanding. Many of the more serious health and safety incidents relate to the transport sector simply because of the number of passengers being moved at one time at speed, and frequently across hazardous terrain.

20.8.1 Tenerife, 27 March 1977

Incident. A KLM Boeing 747-206B, while attempting take-off, collided with a taxiing Pan Am Boeing 747-121 at Los Rodeos Airport on the island of Tenerife, Spain. Both aeroplanes were destroyed. All 234 passengers and 14 crew members in the KLM plane died, while 326 passengers and nine crew members aboard the Pan Am flight were also killed, predominately due to the fire and explosions resulting from the fuel spilled and ignited in the impact. The other 56 passengers and five crew members aboard the Pan Am aircraft survived, including the captain, first officer and flight engineer. The crash remains the deadliest accident in aviation history.

Details. Spain's Canary Islands (seven islands, of which Tenerife is the largest) are situated 250 nautical miles off the Moroccan coast of North Africa. On the afternoon of 27 March 1977 the passenger terminal at Las Palmas Airport, on the island of Grand Canaria (another of the seven islands) was thrown into chaos and panic after a small bomb planted by a terrorist exploded in a florist's shop on the terminal concourse. Eight people were injured, one seriously. The airport subsequently received a threat that there was a second bomb planted somewhere in the airport. As a result the airport was closed while a thorough search was made for the second device. All international incoming flights were then diverted to Tenerife's Los Rodeos Airport, less than an hour's flying time away. Among the flights to be diverted was a charter trip flown by KLM, a Boeing 747-206B piloted by Captain van Zanten, which had departed Amsterdam's Schiphol Airport carrying 234 passengers, and a Pan American flight from John F. Kennedy International Airport, a Boeing 747-121 piloted by Captain Grubbs, with 380 passengers on board. On landing, the controller directed both flights to vacate the runway via the last intersecting taxiway and to park their aircraft on the holding area. By the time the two aircraft were ready to depart the weather had deteriorated, with thick fog descending on to the airport. Following the tower's instructions, the KLM aircraft was cleared to back-taxi the full length of runway 30 and make a 180° turn to put the aircraft into a takeoff position. Shortly afterwards, Pan Am 1736 was instructed to also back-taxi, to follow the KLM aircraft down the same runway, to exit the runway by taking the "third exit" on its left and then using the parallel taxiway. Taking off in heavy fog on the airport's only runway, the KLM flight crashed into the top of the Pan Am aircraft back-taxiing in the opposite direction. The Pan Am had followed the back-taxiing of the KLM aircraft, under the direction of air traffic control. According to the cockpit voice recorder, Captain Grubbs spotted the KLM's landing lights just as the plane approached exit C-4, exclaiming, "[he] is coming straight at us!" with the co-pilot Robert Bragg yelling, "Get off! Get off! Get off!" The Pan Am crew applied full power and took a sharp left turn towards the exit to avoid a collision. The KLM pilot attempted to avoid a collision by climbing away, scraping the tail of the plane along the runway for 20 m. As the KLM left the ground its steep angle of attack allowed the nose gear to clear the Pan Am but the lower fuselage and aft landing gears struck the upper right side of the Pan Am's fuselage at approximately 160 mph, tearing apart the centre of the Pan Am jet almost directly above the wing, and its right engines crashed through Pan Am's upper-deck passenger cabin, immediately behind the cockpit.

Contribution of human error

- Knowledge base. For Captain van Zanten, the KLM pilot, the flight was one of his first after spending six months training new pilots on a flight simulator, where he had been in charge of everything (including simulated air traffic control), and having been away from the real world of flying for an extended period.

- Misperception. The Dutch authorities were reluctant to accept the Spanish report blaming the KLM captain for the accident. The Netherlands Department of Civil Aviation published a response that, whilst accepting that the KLM aircraft had taken off "prematurely", argued that he alone should not be blamed for the "mutual misunderstanding" that occurred between the controller and the KLM crew, and that limitations of using radio as a means of communication should have been given greater consideration.
- Routine violations. Sounds on the cockpit voice recorder suggested that during the incident the Spanish control tower crew had been listening to a football game on the radio and may have been distracted.
- Human error. Many contributing factors led up to the crash, but the probable cause, cited by the Air Line Pilots Association (ALPA), was the KLM pilot taking off without takeoff clearance from the tower. However, if the Pan Am aircraft had not taxied beyond the third exit, the collision would not have occurred. In addition, if Captain van Zanten had not decided to fully refuel at Los Rodeos instead of Las Palmas, the plane would not have been carrying the extra weight, which greatly inhibited the aircraft's liftoff ability, and a collision could have been avoided.
- Logical thinking. The KLM's flight crew had been aware of Pan Am back taxiing behind them on the same runway. Despite lack of visual confirmation due to the fog, the KLM captain thought that Pan Am had cleared the runway and, without waiting to verify, attempted to take off.

20.8.2 Chernobyl, 26 April 1986

Incident. The Chernobyl disaster was a nuclear accident that occurred at the chernobyl nuclear power plant in the Ukrainian SSR (now Ukraine). It is considered the worst nuclear power plant accident in history and it is the only one classified as a level 7 event on the International Nuclear Event Scale. Thirty-one deaths are directly attributed to the accident, all among the reactor staff and emergency workers. Estimates of the number of deaths potentially resulting from the accident vary enormously. The World Health Organisation (WHO) suggests it could reach 4000.

Details. As reported in the press at the time, the first warning of the disaster came from the detection of abnormally high levels of radiation by engineers at the Swedish Fosmark nuclear plant, 60 miles south of Stockholm. At first, suspecting difficulties in their own reactors the engineers desperately searched for a leak but found nothing. However, it was clear somewhere a mysterious source was releasing dangerous radiation into the atmosphere. Clearly something was terribly wrong. From an assessment of the prevailing winds it was clear it must be coming from the Soviet Union. For six hours, as officials throughout Scandinavia maintained that something was dangerously wrong, the Soviets steadfastly maintained that nothing untoward had happened. Finally an expressionless TV newscaster on Moscow television read a four-sentence statement from the Council of Ministers which left many questions unanswered. The terse, almost grudging announcement said in full: "An accident has taken place at the Chernobyl power station, and one of the reactors was damaged. Measures are being taken to eliminate the consequences of the accident. Those affected by it are being given assistance. A government commission has been set up." The Chernobyl nuclear power plant is located in Ukraine, 20 km south of the border with Belarus. At the time of the accident, the plant had four working reactors. The accident occurred when operators of the power plant ran a test on the electric control system of one of the reactors, it was the result of a combination of basic engineering deficiencies in the reactor and faulty actions of the operators. The safety systems had been switched off and the reactor was being operated under unsuitable and unstable conditions resulting in an uncontrollable power surge. This led to a sequence of events resulting in a series of explosions and fires that severely damaged the reactor building, completely destroyed the reactor, and released a plume of highly radioactive smoke into the atmosphere and over an extensive geographical area. It drifted over large parts of the western Soviet Union, as well as eastern, northern and western Europe.

Contribution of human error

- Knowledge base. The power station's operators wanted to test the efficiency of its electricity generators after the supply of steam coming off their nuclear reactor had been switched off. They knew this was dangerous, but pressed ahead even when a problem started to develop in the reactor's core.
- Misperception.The reactor was more resilient than it was under the prevailing operating conditions.
- Routine violations. Left to itself, the Chernobyl reactor should have shut itself down automatically. But the operators overrode the safety systems.
- Human error. The reactor was being operated under unsuitable and unstable conditions.
- Logical thinking. Had the reactor been better designed, the accident would have had less disastrous consequences. It had no containment and the resulting explosion scattered radioactive debris all over Europe. Such a design would not have been permitted in the West.

20.8.3 Kegworth, 8 January 1989

Incident. A British Midland Airways B737-400 aircraft crashed at Kegworth, Leicestershire, England, on 8 January 1989. As a result of the accident 39 passengers died and a further eight persons subsequently died from their injuries. Of the other 79 occupants, 74 suffered serious injury.

Details. As the aircraft was climbing through 28 300 feet the outer panel of one blade in the fan of the no. 1 (left) engine detached. This gave rise to a number of compressor stalls in the no. 1 engine, which resulted in airframe shuddering, ingress of smoke and fumes into the flight deck, and fluctuations of the no. 1 engine. Believing that the no. 2 engine had suffered damage, the crew throttled the engine back and subsequently shut it down. The shuddering caused by the surging no. 1 engine ceased as soon as the no. 2 engine was throttled back, which persuaded the crew that they had dealt correctly with the emergency. The no. 1 engine apparently operated normally after the initial period of severe vibration and during the subsequent descent. However, the engine suffered a major thrust loss due to secondary fan damage when power was increased during the final approach to land. The crew had initiated a diversion to East Midlands Airport and had received radar direction from air traffic control to position the aircraft for an instrument approach to runway 27. When the major thrust loss occurred the crew lost control of the aircraft. The aircraft initially struck a field adjacent to the eastern embankment of the M1 motorway and then suffered a severe impact on the sloping western embankment of the motorway.

Contribution of human error

- Knowledge base. The level of training the pilots received to fly the new plane was bought into question. It was said they were still using trial and error to find out what different instruments did.
- Misperception. Because it was known that this type of plane can fly on one engine, it was assumed the pilots had not gone through every procedure available to them. The loss of one engine would not be viewed as a really serious situation.
- Routine violations. In the emergency situation, the crew fell back on well-learnt behaviour which was not suitable for the design of the new plane.
- Human error. The flight crew made the incorrect response to the situation. The crew reacted to the initial problem prematurely and in a way that was contrary to their training. They did not assimilate the indicators on the engine instrument display before they throttled back the no. 2 engine.
- Logical thinking. The three cabin attendants did not inform the flight crew of the flames that had emanated from the no. 1 engine. In addition, when the captain informed the passengers (and hence the cabin attendants) that they had shut down the right hand engine (no. 2 engine) the passengers that had seen smoke coming out of the left-hand engine had assumed the pilots had a different way of seeing left and right.

20.8.4 *Herald of Free Enterprise*, 6 March 1987

Incident. Within half an hour of the ferry called the *Herald of Free Enterprise* sailing from Zeebrugge, Belgium for Dover, England, it capsized with 459 passengers on board. As a result, 188 lives were lost – 150 passengers and 38 crew members.

Details. The *Herald* left the harbour with both the inner and outer bow doors open. Water flooded into "G" deck as the ferry increased speed, thereby causing it to capsize and settle onto the sandbanks with its starboard side above the waterline.

Contribution of human error

- Knowledge base. There did not appear to be any knowledge base problems, but there was evidence of cultural problems. The inquiry found fault with the management system, stating in particular that "the Board of Directors did not appreciate their responsibility for the safe management of ships". It stated that "the entire body corporate was infected in sloppiness" and "there was a lack of thought about how the Dover—Zebrugge run should be organised".
- Misperception. Speed on leaving the harbour was given prominence in the company's communication. The priority was perceived as being speed rather than safety.
- Routine violations. Setting sail with the bow doors open was against the rules, but became routine. This behaviour appeared to be condoned by management.
- Human error. The assistant bosun had a duty to close the doors but had fallen asleep after relief from maintenance and cleaning duties. The bosun had noticed that the doors were

open but did not close them, as it was not his duty. The captain assumed that the bow doors were closed unless told to the contrary.

- Logical thinking. There was clear evidence of illogical thinking. No indicators were installed on the bridge to show the captain that the bow doors were closed. The ship routinely sailed with excessive numbers of passengers, sometimes well in excess of the life-saving capacity of the vessel. There was no monitoring equipment (e.g. CCTV) to observe problems in critical areas such as the car deck, engine room or superstructure doors.

20.8.5 *Piper Alpha*, 6 July 1988

Incident. A series of explosions ripped through the *Piper Alpha* oil production platform in the North Sea (standing 100 feet above the surface of the water), killing 167 people.

Details. A pump had been shut down using a permit to work during the day shift in order to remove a safety pressure valve for rectification. A blind flange assembly was put in its place. At the end of the working day the suspended permit was returned to the control room but not displayed. During the night the pump failed and the night shift started the pump that was fitted with the blind flange. Condensate entered the relief line and gas escaped from the flange assembly, which was not airtight. The gas exploded, cutting the main power supplies. Two other rigs feeding into the same oil export line did not shut down until one hour after the initial mayday, which meant that oil from the other rigs flowed back towards *Piper* and fuelled the fire. Gas pipelines ended in the area where the oil fire started, they were eventually ruptured in the heat and the explosion engulfed the rig in thousands of tonnes of burning gas.

Contribution of human error

- Knowledge base. Policies and procedures existed. However, decisions were taken based on an inadequate knowledge base, as personnel were not adequately trained.
- Misperception. The product was perceived to have priority over safety.
- Routine violations. Breaking the rules regarding the permit to work system had become common practice. Management ignored the rule breaking.
- Human error. This resulted predominantly from the organisation's negative health and safety culture.
- Logical thinking. The personnel who travelled up to the accommodation block following their meagre training did not think through that smoke and heat rise and perished. A helicopter could not reach them. Those personnel who went downwards and then into the sea survived.

20.8.6 Ladbroke Grove, 5 October 1999

Incident. Two passenger trains collided two miles outside London Paddington station at Ladbroke Grove Junction, leaving 31 people dead and hundreds injured, with some critically injured due to the rapid outbreak of fire in some of the carriages of the faster of the two trains.

Details. One train was a Great Western inter-city high-speed train (HST) running from Cheltenham Spa to Paddington, the other was a Thames Trains three-car diesel unit travelling from

Paddington to Bedwyn in Wiltshire. The crash happened when the driver of the latter train went through a red light outside Paddington during the rush hour and collided head on at 130 mph with the Great Western HST.

Contribution of human error

- Knowledge base. The driver did not have the required knowledge base or experience to make an adjustment to his behaviour when approaching the Ladbroke Grove junction signals. He had received no training from Thames Trains to provide him with an awareness of signals passed at danger and had not been warned that the signal outside Paddington was a black spot. The company had not tested his knowledge of the labyrinthine route that led into Paddington. He had not been given a map showing accident-prone signals. He had only been qualified for 13 days when the accident happened.
- Misperception. Eight drivers in the previous five years had failed to see the signal.
- Routine violations. An official report into the crash by Lord Cullen (2001) said that the Thames Trains safety culture in relation to training was "slack and less than adequate". It criticised "significant failures of communication within the organisation".
- Human error. The driver went through a red light outside Paddington during the rush hour. Lord Cullen concluded in the report that it was likely that the driver was unable to read the signal accurately because of sunshine glinting on the track.
- Logical thinking. If the driver had known the junction was a black spot he would have been more aware that there could be problems and to watch out for them. He may have slowed down in preparation for the possibility of a problem.

20.9 IMPROVING HUMAN RELIABILITY IN THE WORKPLACE

There are a number of commonly recognisable methods to improve human reliability in the workplace. These are briefly described below:

- *Reward schemes*. Employers need to ensure that health and safety is not sacrificed for productivity and that reward schemes (pay and conditions) are designed to ensure that safety performance is rewarded as well as productivity.
- *Job satisfaction*. Modern management theory places an increasing emphasis on employee consultation and highlighting employee concerns which collectively assist in uncovering the causes behind absenteeism and the incidence of accidents and work related illnesses.
- *Appraisal schemes*. Staff appraisal (including the setting of performance targets and the measurement of their achievement) is a tried and tested mechanism for setting goals, assessing performance against those goals and instigating corrective action where required.
- *Selection*. Recruitment and placement procedures should ensure that employees at all levels have the necessary mental and physical abilities to carry out their jobs. An assessment of individual physical fitness by medical examination may be required, together with an assessment of ability and aptitude. If any necessary skills are absent then an assessment must be made as to whether they can be compensated for by training.
- *Training*. Training is an essential element to ensure that staff have the skills, knowledge and attitude to make them competent in the health and safety aspects of their work. Training can never be completed just the once and assumed to be sufficient. The vast majority of organisations have some degree of turnover and so new recruits have to be trained.

In addition, training material needs to be refreshed to address new legislation, working practices, equipment, responsibilities, software and performance targets.

- *Human reliability analysis* (HRA). This can be used by management to devise management controls. HRA is being increasingly used in the development of expert and machine intelligent systems to improve the performance of humans and equipment. HRA can be used to develop training schemes for skills-based behaviour and associated physiological factors, the design of controls, workplace settings, buildings, environmental conditions and bad communication.

20.10 RISK MANAGEMENT BEST PRACTICE

Risk management best practice is implemented through the development of a risk management system, policy and procedures to provide safe systems of work, defining targets, measuring performance and revisiting procedures in the light of experience. The following are the main risk mitigation factors:

- Establishing a greater awareness of the legal liabilities of the organisation.
- Gaining an awareness of existing guidance such as BS 8800 which provides guidelines for an effective occupational health system, the International Labour Organisation's 2001 Guidelines on Occupational Safety and Health Management Systems (ILO-OSH), which were the result of extended international consultations held over 2000–2001, and the OHSAS 18000 Series of International Standards for Occupational Health and Safety Management Systems.
- Implementation of a health and safety management system – companies that have an occupational safety and health management system (OSH-MS) set up according to the ILO-OSH have both better safety and productivity records.
- Involvement of the workforce in both planning and running the organisation's OSH-MS creates improved ownership and participation.
- A functioning, recording, notification and indicator system provides a better picture of the problems and the follow-up that is necessary.
- Measurable targets for reducing occupational accidents and work-related diseases by targeting their causal factors.
- Workplace mapping techniques are an effective tool to identify health and safety problems in the workplace and define the measures necessary to resolve them.
- Development of a public relations response management plan and crisis management plan.

20.10.1 Crisis Management Plan

With the rapid pace of technological change, an organisation's ability to respond to an environmental health and safety crisis is more critical than ever. As a result of the widespread availability of mobile phones (particularly those that can take pictures), e-mail and other forms of immediate communication, combined with improved transportation, there is a growing belief that there is no excuse for a delayed response. In addition, photographic, computer and satellite technology has enabled media companies to put out news coverage and place pictures, video clips and text on their websites with startling speed. Organisations watch their problems being aired on the evening news the same day they occur with on-the-scene commentary of the chronology of events, pictures, interviews with specialists and a summary of the organisation's past performance. When a news reader says "the organisation was contacted but no one was available for comment", any disaster immediately appears worse. Any organisation's crisis management plan needs to be comprehensive and incorporate lessons learnt either from internal events or from events within the same industry. The plan must cover all aspects of an incident such as legal, financial, environmental, health and safety, shareholder liaison, community liaison, watchdog notification and, where appropriate, the police and emergency services.

20.11 SUMMARY

This chapter has provided a definition of health and safety risk, accompanied by a description of the scope of health and safety risk and the benefit of safety risk management. While a survey funded by the UK Health and Safety Executive in 2007 found only a weak association between health and safety practices and improved business performance, society considers businesses to have a moral obligation to implement a health and safety regime and legislation imposes a legally enforceable minimum standard of health and safety practice. A series of serious accidents were described involving a high number of fatalities which were the result of human error. These were tragic for those who lost their lives and an inconsolable loss for the family, relatives and friends of the diseased, and the lives of the employees who survived were changed for ever, scarred by the scenes they witnessed, the colleagues they lost and the media coverage and inquiries that followed. The businesses that endured the disasters suffered loss of staff, assets, reputation, customers, profits and future business and were the subject of public inquiries. Clearly health and safety management systems must focus on improving human reliability in the workplace. A number of approaches were described which can assist in addressing employee reliability. The key benefits of health and safety risk management can only be described in terms of not suffering a negative event. Hence, they cannot be seen, held or described in the same way as other business benefits. Yet health and safety practices are inseparable for contemporary business practice.

20.12 REFERENCES

Cowling, M. and Bevan, S. (2007) *Work and Enterprise Panel 2: Business Survey*. Research Report 589 prepared by the Institute for Employment Studies and the Work Foundation for the Health and Safety Executive. http://www.hse.gov.uk/research/rrpdf/rr589.pdf

Cullen, Lord (2001) *Ladbroke Grove Rail Inquiry, Part: 1 Report*. HSE Books, London.

HSE (1997) *Successful Health and Safety Management*, 2nd edition. HSE Books.

HSE (2006) *Essentials of Health and Safety at Work*, 4th edition. HSE Books.

Part IV

External Influences – Macro Factors

Businesses clearly do not operate in a vacuum but in an ever-changing scenario where changes in the operating environment are beyond the control of any individual business. This part of the book examines the way in which *external* "macro" influences impact businesses. These macro factors are distinct from micro factors inasmuch as they are events that occur at both national and international level. Micro factors influence individual businesses or consumers in the domestic market. An understanding of how these external macro influences impact on a business is important as it provides an appreciation of how a business is subject to constraints and exposed to opportunities. Macro factors include the state of the economy, the environment, the legal framework, political structure, market conditions and social factors. These subjects are discussed in the following chapters and describe the sources of risk included in the risk taxonomy included in Chapter 9 (see Table 9.1). The sequence of the chapters follows the sequence in which the subjects appear in the taxonomy, as shown in Figure P4.1. An extract of the taxonomy is given in Table P4.1 for ease of reference and assimilation.

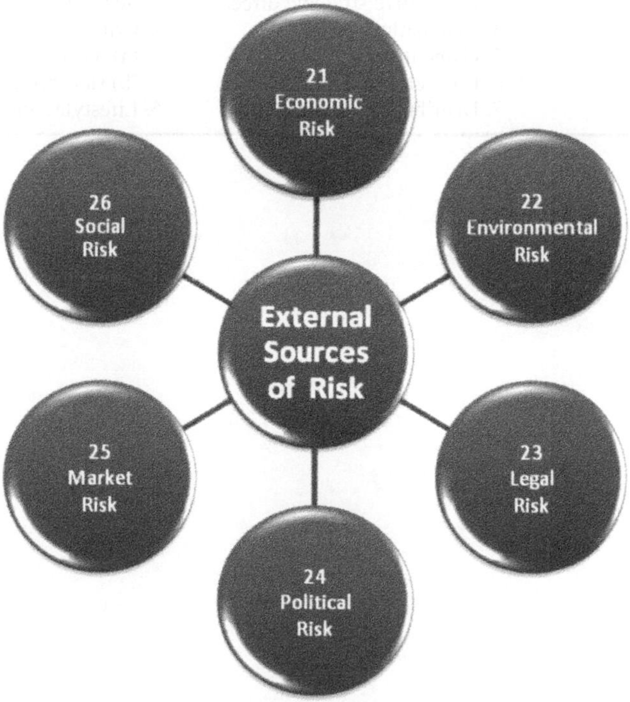

Figure P4.1 Structure of Part IV

Table P4.1 Business operating environment

Economic	Environmental	Legal
1. Macroeconomics	1. Energy sources	1. Companies
2. Microeconomics	2. Use of resources	2. Intellectual property
3. Government policy	3. Pollution	3. Employment law
4. Aggregate demand	4. Global warming	4. Contracts
5. Aggregate supply	5. Levies/emission controls	5. Criminal liability
6. Employment levels	6. Environmental sustainability	6. Computer misuse
7. Inflation		
8. Interest rate		
9. House prices		
10. International trade + protection		
11. Currency risk		

Political	Market	Social
1. Contracts	1. Market structure	1. Education
2. Transition economies	(a) number of firms	(a) general level
3. UK gov. fiscal policies	(b) barriers to entry	(b) language skills
4. Pressure groups	(c) new entrants	2. Population movements
5. Terrorism and blackmail	(d) homogeneous goods	(a) location
	(e) knowledge	(b) age mix
	(f) relationships	(c) pensions
	2. Product life cycle stage	(d)"grey market"
	3. Alternative strategic directions	3. Socio-economic patterns
	4. Acquisition	4. Crime
	5. Game theory	(a) business vulnerability
	6. Price elasticity	(b) staff relocation
	7. Distribution strength	5. Lifestyles and social attitudes

21 Economic Risk

Everything that can be counted does not necessarily count; everything that counts cannot necessarily be counted.
(Albert Einstein)

This chapter examines *economic risk*, the first of the six macro influences within the section of the risk taxonomy called "business operating environment" described in Chapter 11. Businesses do not operate in a vacuum. Hence, it would be difficult to comprehend an enterprise risk management process that did not address economic risk. The subject of economics relates to the allocation of resources. Economic business risk emanates from the performance of the national economy within which a business operates and the way its government elects to influence the economy and solve the basic economic problem of scarce resources and competing needs. Economics has such a large influence on business performance that managers must take steps to understand and predict economic phenomena and respond accordingly. While the subject is vast and it is only possible to examine some of the rudimentary elements of the body of economic theory here, any review of enterprise risk management that did not address those aspects of economics regularly encountered by businesses would clearly be incomplete. The structure of this chapter is illustrated in Figure 21.1. The next chapter examines the second of the business operating environment risk categories, environmental risk.

21.1 DEFINITION OF ECONOMIC RISK

What is economic risk? There does not appear to be any universally accepted definition. Economic risk is defined here simply as the influence of national macroeconomics on the performance of an individual business. Implicit within national macroeconomics is the modifying influence of government policy through the manipulation of aggregate demand and consumer spending. The essence of economic risk is that any one individual business has no control over national influences on aggregate demand.

21.2 SCOPE OF ECONOMIC RISK

The sources of risk considered to be embraced within the term "economic risk" are as follows:

- fall in demand (a shift in the aggregate demand curve);
- government policies (including interest rates and trade protectionism);
- movement in house prices;
- exchange rates;
- inflation.

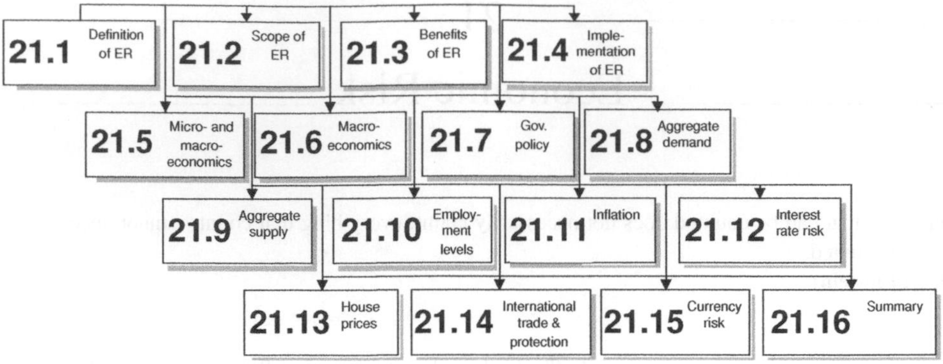

Figure 21.1 Structure of Chapter 21

21.3 BENEFITS OF ECONOMIC RISK MANAGEMENT

Economic risk management affords a business benefits as it:

- improves knowledge of where the government is planning public spending;
- provides an understanding of the impact of inflation and interest rates on demand;
- provides an understanding of how the short-term behaviour of the gross domestic product (GDP) impacts employment, prices and standards of living;
- promotes more rigorous market research when entering new markets in both domestic and international markets.

21.4 IMPLEMENTATION OF ECONOMIC RISK MANAGEMENT

The development of a sound system of economic risk management will depend on a number of issues such as:

- an understanding of the drivers and consequences of inflation;
- an understanding of the impact of changes in exchange rates on the demand curve;
- tracking planned government spending;
- an understanding of government fiscal and monetary policies;
- the taxation regime.

21.5 MICROECONOMICS AND MACROECONOMICS

To make sense of economic risk for businesses it is worth examining the difference between micro- and macroeconomics. Microeconomics is described by economists as being driven by households whose members have needs for goods and services. These members have resources – incomes, assets, time and energy – with which to satisfy their wants. However, the limitations of their resources forces them to make choices, which they do through markets where they are offered many ways to spend their money, energy and time. The trigger to which households respond is market prices. Given a set of prices, each household will make a given set of choices. In doing so, in aggregate they affect those prices. The prices signal to businesses what goods they may profitably provide. Given available technology and the known costs of production, businesses choose among (1) the products they might produce, (2) the ways of producing them and (3) the various quantities (and qualities) they can supply. In doing so, they affect prices. Businesses demand factors of production[1] in quantities that depend on their output decisions, which in turn depend on consumers' demands. These derived demands[2] for factors affect the prices of land, labour and capital goods. The owners of factors respond to factor prices by deciding how much of their services to offer and where to offer them. These choices determine factor supplies. Payments by businesses to factor owners provide factor incomes. The recipients of these incomes are members of households (who have needs and desires for goods and services),[3] and we have come full circle. This paragraph describes a circular flow of income. Money passes from households to businesses in return for goods and services produced by businesses and money passes from businesses to members of households in return for factor services provided by those members of households. Microeconomics studies individual parts of this flow in infinitesimal detail.

The basic problem of macroeconomics is the determination of total employment, output and price level. Macroeconomics studies the total amount of deployment of each of the major factors of production (with special attention to the total amount of labour employed); the total volume of output produced and income earned in the whole economy; the average level of prices in all product markets (called the price level); and the growth of the economy's total output, both actual and potential.

21.6 MACROECONOMICS

The three most important macroeconomic concepts are output, income and expenditure. They are the main indicators of a nation's economic performance. Firms produce the goods and services, which combined are the nation's output (O). Production requires factors of production whose owners are paid for services provided, and it thus generates income (Y). When the nation's output is sold, people spend money to purchase it, the value of expenditure (E) being the amount required to purchase the nation's output. They should all give the same total as they all measure the flow of income produced in the economy. Hence, $Y \equiv O \equiv E$. Figure 21.2 illustrates the circular flow of income in a simplified model of a national economy. The value of the output is equal to the incomes that it generates, that is, wages, rent, profit and interest. If it is assumed that all income is spent, expenditure will equal income and in turn, by definition, equal output.

21.6.1 Gross Domestic Product

The most important empirical measure of these variables (output, income and expenditure) is called the *gross domestic product*. This is the value of total output actually produced in the

[1] The factors of production are the inputs to the production process: land, which is all natural resources (including, for example, forestation and minerals from below the earth); labour, which is the workforce; capital goods, which are any man-made resources used in the production of goods and services (from the spade to the modern assembly plant); and entrepreneurs, individuals who seek out profitable opportunities for production (land, labour and capital goods) and take risks in attempting to exploit these.

[2] Businesses require land, labour, raw materials, machines and other inputs to produce the goods and services that they sell. The demand for any input therefore depends on the existence of a demand for the goods that it helps to make. Economists describe this demand as derived demand.

[3] Goods are tangible items such as cars and televisions. Services are intangible items such as education and haircuts. Goods and services combined are called commodities. The total output of all commodities in one country over some period, usually taken as a year, is called the gross national product.

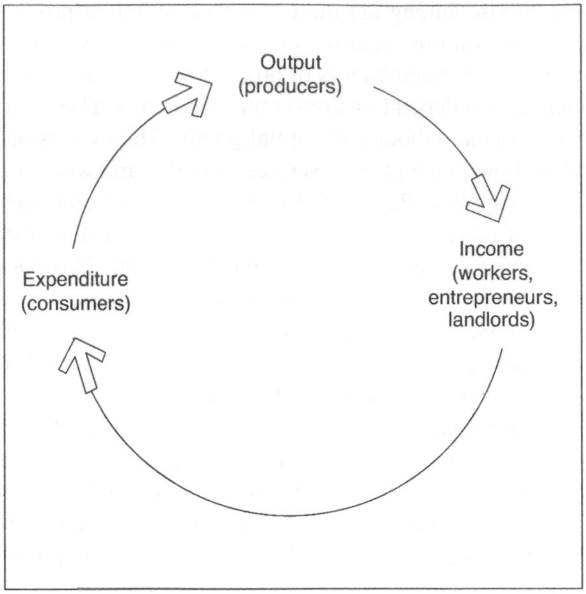

Figure 21.2 The circular flow of income in a national economy

whole economy over some period and commonly a year. A loose term for GDP is "national income". GDP is an important measure of the standard of living in a country such as the UK. Past short-term behaviour of GDP has been characterised by oscillations, which have given rise to some serious problems. Rapidly rising GDP often causes labour shortages, a balance of payments deficit and severe inflation. Declining GDP often causes bouts of heavy unemployment, static or falling living standards and isolated pockets of severe poverty (reported to exist in the UK as recently as the 1980s). Nominal (or money) GDP is GDP measured in terms of prices operating in the year in which the output is produced. It is sometimes referred to as GDP at current prices and is a measure that has not been adjusted for inflation. Nominal GDP may give a misleading impression of how a country is performing. This is because the value of nominal GDP may rise not because more goods and services are being produced but merely because prices have risen. For example, if 200 million goods are produced at an average price of £2, GDP will be £400 million. If in the next year the same output of 200 million goods is produced but the average price rises to £3, nominal GDP will rise to £600 million. So to get a truer picture of what is happening to output, economists convert nominal into real GDP. They do this by measuring GDP at constant prices by adopting the prices operating in a selected base year. For example, in 2003 a country's GDP is £900 billion and the price index is 100. In the following year the nominal GDP rises to £973 billion and the price index is 105. To calculate the real GDP the following expression is used:

$$\text{real GDP} = \text{nominal GDP} \times \frac{\text{price index in base year}}{\text{price index in current year}}.$$

So for 2004:

$$\text{real GDP} = 973\text{bn} \times \frac{100}{105} = 926.67\text{bn}.$$

21.7 GOVERNMENT POLICY

As described in the definition of economic risk, macroeconomics is influenced by government policy. The main government macroeconomic policy objectives are: low unemployment, low and stable inflation, a satisfactory balance of payments position, avoidance of excessive exchange rate fluctuations and steady economic growth. In striving to manage the economy, successive UK governments have been acutely aware of the problems that can arise when, say, the rate of inflation is high, the economy slips towards recession leading to an excessive rate of unemployment, or there is a surge of imported goods. Such problems have a consequence for the economy as a whole. In managing these problems, though, the dilemma governments face is that not all of their stated objectives can be achieved simultaneously.

21.7.1 Fiscal Policy

Fiscal policy is the name given to government policies which aim to influence government revenue (taxation) and/or government expenditure, in the pursuit of particular policy objectives. It is used by governments to influence the level of aggregate demand and supply in the economy. In simple terms it can be described as changes in the income or expenditure sides of government accounts. The UK government has been responsible for between 40% and 50% of national expenditure over the past 30 years. The main areas of public spending are the National Health Service, defence, education and transportation (primarily roads). In addition, the government is responsible for transferring large sums of money around the economy through its spending on social security and national insurance benefits. Income is mainly derived through HM Revenue and Customs (with income tax, corporation tax, value added tax and fuel duties being the largest contributors). The government has to make decisions about how much to spend, tax and borrow. It also has to decide on the composition of its spending and taxation. Should it spend more on defence than education and so on? Should vehicle excise duties be cut and fuel duties be increased? These decisions about taxation, borrowing and spending are called the *fiscal policy* of the government. Major changes in fiscal policy in the UK are normally announced at the time of the Budget, when the spending and taxation forecast for the following year is delivered. The government financial year commences on 6 April and ends on 5 April the following year. The Budget traditionally takes place before the end of the financial year.

21.7.2 Monetary Policy

Monetary policy is the attempt by government or the central bank of an economy to manipulate the money supply, the supply of credit, interest rates and other monetary variables to achieve the fulfilment of policy goals.[4] When the government wishes to stimulate the economy, it may choose to increase the money supply. When the government wishes to dampen down the economy it is likely to seek to reduce the money supply. Rates of interest tend to move upwards or downwards in line with the rate of interest set by the Bank of England. In 1997 the Bank of England was given independence by the government and is now responsible for setting the interest rate each month. This is done at a monthly meeting of the nine members of

[4] Narrow money is defined as the combination of currency in the form of notes and coin, together with balances that are available for use in normal transactions. Broad money is money not immediately accessible such as balances at banks and building societies that stand in the name of individuals.

the Monetary Policy Committee (MPC) at the Bank of England. The MPC takes into account the government's "inflation target" and the future inflation projections when deciding on the rate of interest. Any lowering of interest rates at a given level of prices will lead to a rise in aggregate demand. This is because a fall in interest rates leads to a rise in spending on consumer durables and an increase in investment.

The money supply, the rate of interest, the public sector borrowing requirement (see Section 21.14.4) and the exchange rate are all interlinked. If the government fixes a value for one, it cannot fix a value for the others. Hence a government may choose not to control the money supply in order to control other variables such as the rate of interest.

21.7.3 Competing Theories

There are two opposing schools of thought on economic policy: whether the economy is best managed through the application of *demand-side* or *supply-side* policies. The traditionalists believe that governments must intervene in the affairs of the economy to achieve their objectives, by the management of aggregate demand through the application of both fiscal and monetary policies (which are discussed below). This intervention they consider is crucial to achieving full employment, low inflation, balance of payments stability and economic growth. Other economists have taken a different view. Their belief is that macroeconomic objectives can be best met when the economy is allowed to function naturally, with little interference from government. In other words, the forces of *aggregate demand* and *aggregate supply* combined will allow the market economy to operate in the most effective way to allow governments to meet their objectives for the macroeconomy. Supply-side policies, which focus explicitly on managing the total amount of output in the economy, are combined with policies which manage aggregate demand, with the aim of providing the most appropriate ways of allowing governments to meet their macroeconomic objectives.

21.8 AGGREGATE DEMAND

Aggregate demand (AD) is the term used by economists to denote spending on goods and services produced in an economy. It consists of four elements: consumer spending (C), investment expenditure (I), government spending (G) and net expenditure on exports and imports ($X - M$) and is usually represented by means of the expression $AD = C + I + G + (X - M)$. Figure 21.3 shows a typical aggregate demand curve, which is downward-sloping. Real GDP is shown on the horizontal axis and the average price level is on the vertical axis. Any point on the AD curve therefore shows total real GDP required in an economy at a specific overall price level. As the curve relates to real GDP it takes inflation into account.

There are three main reasons why the AD curve is assumed to be downward-sloping, from left to right. Economists believe aggregate demand is increased when: (1) the domestic market switches from imports to domestic commodities as a result of prices appearing more attractive (an effect which can be negated by adverse exchange rates); (2) prices fall and domestic customers with bank or building society savings or stocks and shares use their savings to buy an increased quantity of goods and services; and (3) prices fall and customers make purchases in the belief that it is advantageous to buy at the current low prices, anticipating future increases will take place.

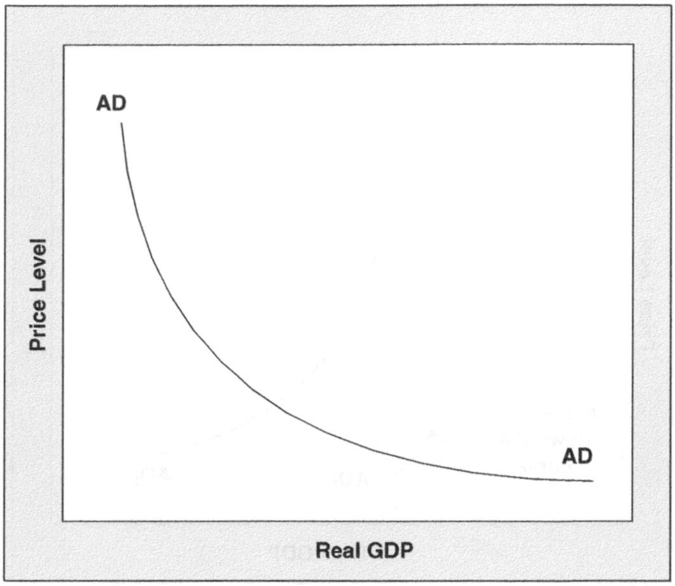

Figure 21.3 An aggregate demand curve

21.8.1 Using Aggregate Demand Curves

There are some circumstances that lead to movements along the AD curve as discussed above; however, if the basic circumstances underlying those causal factors alter then there will be shifts of the AD curve. For example, if the world economy moves into a boom there will be a boost to a country's exports at any given price and the AD curve may be said to shift to the right. However, in the instance where a tax on wealth holdings is imposed, so that stocks of monetary assets fall dramatically, then the AD curve may shift to the left. If, at any given price level, customer's price expectations alter, then the AD curve will shift (see Figure 21.4).

Businesses must be aware that more dramatic changes in the aggregate demand may arise from changes in the underlying constituents of aggregate demand, that is, consumer spending, investment, government spending, exports and imports.

21.8.2 Determinants of Consumer Spending

Consumer spending is the most important component of aggregate demand. It is the amount of money spent by individuals on goods such as food, clothing, housing, cars, personal computers, travel, entertainment and other consumables. A key determinant of consumer spending is the level of disposable income (income after payment of compulsory taxes and National Insurance contributions). For the economy as a whole, therefore, an increase in the level of income will lead to an increase in consumer spending. While in essence true, the statement is an oversimplification as additional spending will reflect the marginal propensity to consume (mpc), that is, the proportion of any increase in income that is spent on consumption. For low-income families the mpc is 100% as all additional income is spent on food and clothing and other essential items. At the other end of the spectrum the mpc falls to 50–60%, where income not spent is saved. However, the key issue is that when income increases so does

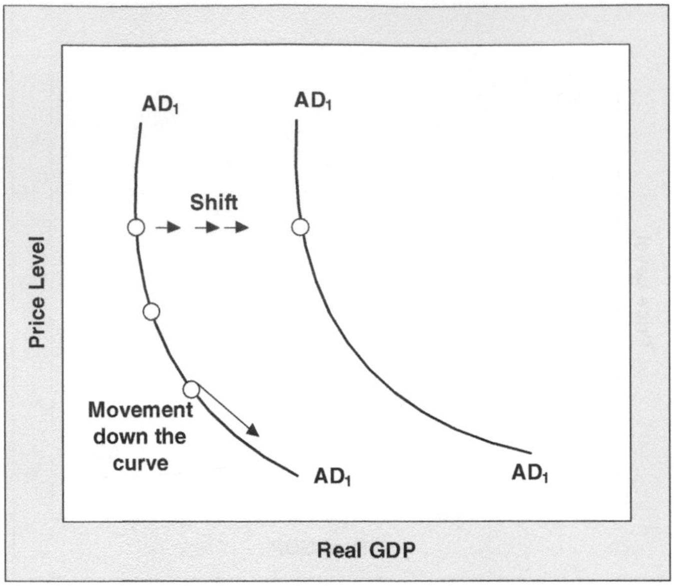

Figure 21.4 Shifts of and movements along the aggregate demand curve

consumer expenditure and the AD curve shifts to the right. If income tax is raised consumer spending declines and the AD curve shifts to the left. If the interest rate increases mortgage borrowers have to make higher monthly payments, consumer spending falls and again the AD curve shifts to the left.

21.8.3 Determinants of Investment Expenditure

Investment expenditure can be split into three main categories: public investment, private investment and household investment in houses, flats and other types of accommodation. Analysis of investment in the private sector is complex due to the number of variables which influence decision making; however, a key inducement is a fall in interest rates when projects become more profitable and the AD curve shifts outwards to the right. Another factor on which businesses base investment plans is future expectations, such as winning the bid to host the 2012 Olympics. These expectations can be very volatile and investment expenditure may fluctuate markedly. Public expenditure such as investment in transportation, defence and schools is rather less volatile but still dependent on the state of the economy. For household investment in housing, the rate of interest has a substantial effect on mortgage payments and in turn on disposable income and aggregate demand.

21.8.4 Determinants of Government Spending

This element of aggregate demand mainly includes current government spending on goods and services in the private sector and spending on employment in central government services. As government spending is relatively constant there is little change in the AD curve but if spending is curtailed and suddenly the "cork is taken out of the bottle" with a sudden resultant boost to expenditure, then there is a shift in the AD curve to the right.

21.8.5 Determinants of Net Expenditure on Exports and Imports

There is a multitude of issues which impact on net expenditure on exports and imports, and hence there is no simple way to summarise them in a single all-encompassing phrase. The strength of a country's exports is dependent on issues such as the availability of natural resources, the labour market and the transportation system. It is also dependent on whether: the country's industries are efficient and competitive; its companies produce good-quality products and deliver when promised; the companies provide good after-sales service; if research and development departments produce a stream of innovations; if marketing is properly taken into account; and if sales people can speak the language of the target countries. The level of the exchange rate is also crucial to the business exporter. Imports of goods and services do not contribute to aggregate demand directly, but they do so indirectly by diminishing demand overall. If imports penetrate the home market to a significant degree, then there is less scope for the domestic producers to capture the home market. Beyond these issues is the overall level of demand in the economy. When GDP rises, there is an automatic increase in imports (as a result of an increase in consumer demand). This is particularly the case for those imports that are needed for manufacturing. If a manufacturer of light aircraft needs steel, copper and rubber, all of which are imported, then an increase in the production of aircraft will require additional imports. Furthermore, an increase in living standards will also be accompanied by consumer demand for commodities.

21.9 AGGREGATE SUPPLY

Aggregate supply (AS) is the total output of the economy. In the UK economy this would be anything produced by the factors of production (land, labour and capital goods). Aggregate supply is simply the total output of the economy at a given price level at a given point in time. The AS curve shows the relationship between the total quantity supplied in an economy and the price level. The shape of the curve can be rationalised as follows. At low prices businesses collectively do not consider that it is worth producing goods and services as they do not expect to be able to sell them or break even. As the overall price level increases, more and more businesses consider that it is worthwhile to start producing or produce more and so supply expands. When the economy starts to run at capacity in terms of the available land, labour, capital goods and entrepreneurship, the AS curve starts to appear vertical (see Figure 21.5). In the short term, changes will be along the curve. If there is spare capacity in the economy that businesses respond to, an increase in demand will have a much bigger effect on output than it will have on prices. Looking at Figure 21.6, this is shown by the movement along the AS curve from A to B. Beyond B, businesses are reaching their full capacity and costs are rising which results in the AS curve being steeper. So, as illustrated by movement along the curve from C to D, small increases in output will have a larger impact on prices. From the long-term perspective, there are many factors which may shift the whole of the AS curve. Some examples are:

- an increase in the capital stock due to a reduction in interest rates;
- an improvement in the expectations of business executives;
- continuing technological change;
- increased investment in education and training;
- a reduction in unemployment benefits; and
- schemes to improve the geographical mobility of workers.

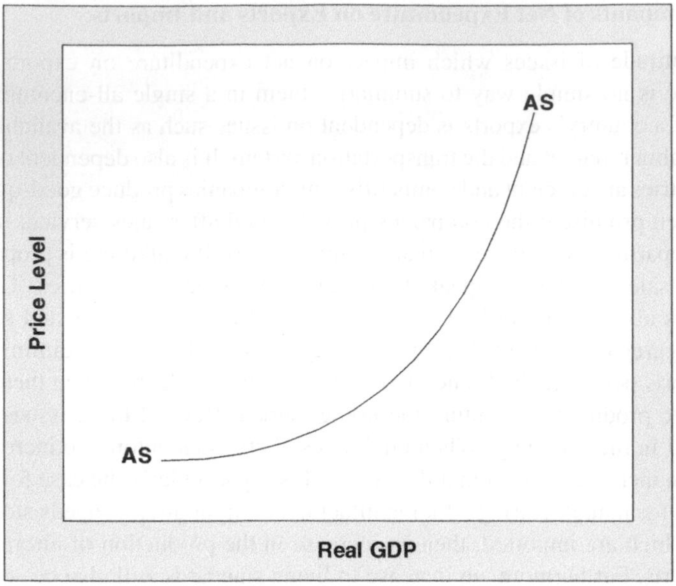

Figure 21.5 An aggregate supply curve

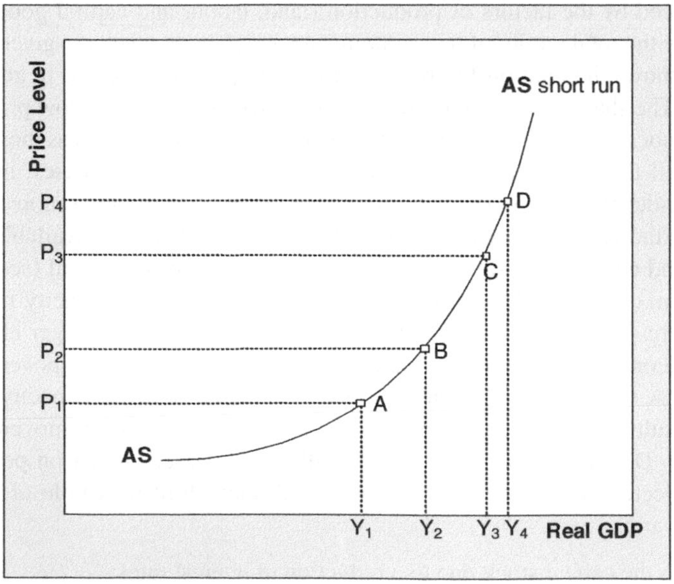

Figure 21.6 Short-run aggregate supply curve

21.10 EMPLOYMENT LEVELS

The social and political importance of the figure that expresses the unemployment[5] rate is enormous. It is widely reported in newspapers and on television. The government is blamed when it is high and takes credit when it is low. It is often a major issue in elections and few macroeconomic policies are formed without consideration of their effect on it. Unemployment, the number of people out of work, is measured at a point in time but is in a constant state of flux. Young people leave school, college or university seeking work. Former workers who have taken time out of the workforce, for instance to travel or bring up young children, seek to return to work. Workers, who have lost their jobs, either because they have resigned or because they have been made redundant, search for new jobs. Equally many hundreds of thousands of workers leave their jobs as a result of retirement, childbirth, disabilities or ill health. Long-term unemployment is generally considered to be a great social evil. Apart from the loss of income, studies suggest that the unemployed suffer from a wide range of social problems, including above average incidence of stress, marital breakdown, suicide, physical illness and mental instability and they have higher death rates. High unemployment has a negative impact on the domestic market for businesses. Those out of work rely on the welfare state, resulting in a structural deterioration in public finances. Income from taxation falls, and hence less government money is available for expenditure on areas such as roads, schools and defence. In addition, those in work become nervous and begin to spend less. Government action attempts to regulate unemployment, with mixed results. Between 1979 and 1986 it can be argued that the government kept aggregate demand too low through a combination of high interest rates (restrictive monetary policy) and tight control of government borrowing (restrictive fiscal policy). Policy makers are concerned about short-term fluctuations in *national income* (or GDP; see Section 21.6.1) because of their consequences for unemployment and lost output. Employment levels are important to businesses in terms of the availability, cost and ability (education and training levels) of potential employees. As explained above, employment levels are also important to businesses in terms of government spending.

Government actions, while designed to protect a country's interests, can sometimes have adverse side effects. In the United States the visa regime and strict immigration rules have generated concern among employers. Bill Gates stated that the decline in the influx of foreign computer science students was so severe that it was threatening to undermine America's position in the global software industry (Larsen and Gapper 2005). The US' status as "the IQ magnet of the world" was thought to be in jeopardy. At the World Economic Forum in Davos early in 2005, Mr Gates pointed to the sharp difference between emerging markets such as India and China, where about 40% of students take engineering degrees, and the US, where the proportion is about 4%.

21.11 INFLATION

For the purposes of this examination of the economy, *inflation* is defined as a sustained general rise in prices. The opposite of inflation – *deflation* – strictly speaking is a fall in the *price level*. However, it can also be used to describe a slowdown in the rate of growth of output of the economy. This slowdown or recession is often associated with a fall in the rate of inflation.

[5] The term "unemployment" used here means involuntary unemployment where a person is willing to accept a job at the going rate but no such job can be found.

A general rise in prices may be quite moderate. *Creeping inflation* would describe a situation where prices rose a few per cent on average each year. *Hyperinflation*, on the other hand, describes a situation where inflation levels are very high. The inflation rate is the change in average prices in an economy over a given period of time. The price level in an economy is measured in the form of an *index*. So if the price index were 100 today and 110 in one year's time, then the rate of inflation would be 10%. In the UK the most widely used measure of the price level is the Retail Price Index. It is calculated by recording the price of a representative range of goods and services. Individual prices are weighted before the final index can be calculated, as prices for some goods are more important than others.

Inflation is generally considered to be a problem. Some economists, mainly monetarists, have claimed that inflation creates unemployment and lowers growth. Inflation increases costs of production and creates uncertainty. This lowers the profitability of investment and makes businesses less likely to take risk associated with any investment project. Lower investment results in less long-term employment and long-term growth. Monetarists argue that inflation is caused by excessive increases in the money supply.

Some Keynesians[6] believe that excess demand in the economy is the principal cause of inflation – the *demand/pull* theory of inflation. Other Keynesians argue that inflation is primarily *cost/push* in nature. The demand/pull theory of inflation says that inflation will result if there is too much money chasing too few goods in the economy. The cost/push theory of inflation says that inflation is caused by increases in cost of production. There are four major sources of increased costs: wages and salaries, an increase in the cost of imported goods, increases in prices (to boost profits) and an increase in taxes which increases prices. Put another way, monetarists argue that inflation can only be controlled through monetary policy. Keynesians argue that monetary policy is ineffective. Inflation is best controlled through policies such as fiscal policy and incomes policy.

An effect of unanticipated inflation is to redistribute wealth from lenders to borrowers. To see how this happens, suppose that X lends to Y £100 at 5% interest for one year. If the price level rises by 10% over the year, X will actually earn a negative rate of interest on the loan. The £105 that X gets back from Y will buy fewer goods than the £100 X originally parted with. X's loss, however, is Y's gain. Y did not even have to use the £100 in any productive business enterprise to show a gain. All Y needed to do was to buy and hold the goods whose price rose merely by the average rate of inflation. At the end of the year Y can sell the goods for £100, pay back the £100 borrowed plus £5 interest, and show a gain of £5 for having done nothing more than hold goods instead of money. This type of inflationary redistribution occurs not just on borrowing and lending contracts, but on any form of contract that is stated in terms of monetary units.

21.12 INTEREST RATE RISK

While any single business has no control over the setting of interest rates, businesses can predict how consumers are going to react as trends are well established and can respond accordingly. In the UK this is now more important than in the past as the government is increasingly using interest rates to control the demand for money and expenditure. The risk for businesses whose

[6] Keynesian economics are based on the work of John Maynard Keynes, who is referred to as the creator of modern macroeconomics. His ideas, for which he is most famous, are recorded in his very popular book called *The General Theory of Employment, Interest and Money*.

income is quickly and significantly affected by changes in interest rates is not monitoring and responding to projected and actual changes in the rates. Changes in the interest rate affect business and consumer behaviour in a number of ways:

- *Exchange rate*. Increasing interest rates in the UK tend to make holding sterling deposits in the UK more attractive. An increase in demand for sterling is likely to raise the exchange rate for sterling. Raising the exchange rate will make exports more expensive abroad and imports cheaper in the UK. Lowering interest rates will have the opposite effect, reducing the exchange rate for sterling, thereby making exports cheaper abroad and dearer in the UK.
- *Discretionary expenditure*. For most homeowners, their mortgage is the most important item of expenditure. To avoid losing their home, most maintain their mortgage repayments. The majority of homeowners with mortgages are on variable rate mortgages so that if interest rates rise they must pay back more per month, leaving less disposable income. Likewise if interest rates fall, there will be more income left in the family budget for expenditure on other things.
- *Savings*. Higher interest rates encourage saving, since the reward for saving and therefore postponing consumption has increased. Lower interest rates discourage saving by making spending for current consumption relatively more attractive.
- *Borrowing*. Higher interest rates discourage borrowing as it becomes more expensive, while lower interest rates encourage borrowing as it becomes cheaper. Borrowing on credit has played an important role in the growth of consumer spending.

21.13 HOUSE PRICES

Expenditure on residential structures (as economists like to label them, but housing to you and me) is a significant part of gross private investment. House sales are often treated as an economic barometer. As these expenditures are both large and variable they exert a major impact on the economy. Influences on residential construction are both non-economic and economic. Non-economic influences include changes in demography, changes in the age profile of the population and increases in population. Economic influences include increases in family earnings, changes in tax incentives, stamp duty, the number of repossessions and changes in the interest rates. Interest payments are a large part of total mortgage payments. At an interest rate of 8%, about half of the money paid on a 20-year mortgage is interest. Only half is for repayment of the principal. As interest payments are such a large part of the total payments on a mortgage, small changes in interest rates cause a relatively large change in the annual payments. For instance, a rise in the rate of interest from 8% to 10% increases the monthly payments on a 20-year mortgage by nearly 15.4%. Changes in interest rates can therefore have a large effect on the demand for new housing. Hence, a fall in the rate of interest will lead to a temporary spurt of investment in new housing and a rise will reduce the demand for new housing.

21.14 INTERNATIONAL TRADE AND PROTECTION

21.14.1 Trade

Businesses need to have an understanding of the mechanisms of international trade and protectionism imposed by governments, to understand the risks and opportunities associated with the production of goods for export. There needs to be an awareness of the benefits and disadvantages of protectionism so that lobbying of government is done on an informed basis. Obviously production and distribution not only occur in the context of domestic markets but are also influenced by trading relations with foreign countries. While discussion on international trade frequently refers to trade between nations, what should not be overlooked is that in free-market economies most of the decisions determining the size, content and direction of foreign trade are taken by households and businesses. Businesses may see an opportunity to sell goods abroad and arrange to have these goods exported; other businesses may see an opportunity to sell foreign goods in the home market and arrange to have these goods imported. If households (either at home or abroad) find such goods attractive and purchase them, the ventures will be successful. If they do not, the goods will remain unsold and will no longer be imported or exported. Hence, in free-market economies foreign trade is determined just as domestic trade is – mainly by thousands of independent decisions taken by businesses and households and coordinated more or less effectively by the price system. Overlaid on this trade at household and business level is the frequent desire of governments to try to influence the process. They may provide subsidies for exports seeking to encourage foreign sales of domestically produced goods by making their prices more attractive; they may put tariffs on imports, seeking to discourage domestic sales of foreign produced goods, by making their prices less attractive.

21.14.2 Methods of Protectionism

There are basically three means by which a country can reduce its imports of some goods: (1) it may place a tax on imported commodities, called a tariff; (2) it may impose an import quota that limits the quantity of the commodity that may be shipped into the country in a given period; and (3) it may adopt domestic policies that reduce its demand for the imported commodity. For example, it may require potential importers to acquire a special licence, or it may restrict the ability of its citizens to use their funds to purchase the foreign exchange needed to pay for the commodity. Although each means of restricting trade has different effects, they all achieve the same reduction in the quantity of imports.

21.14.3 Trade Policy

A government may choose to impose or tighten currency controls. These are controls on the purchase of foreign currency by domestic citizens and firms. In the late 1960s, for instance, the UK government limited the amount of currency that could be taken abroad on holiday. Governments could equally restrict finance for investment abroad or even for imports. The government abolished exchange controls in 1979, and today the UK government is unable to impose currency controls because of its membership of the European Union.

21.14.4 Balance of Trade

The *balance of trade* is the difference between the value of exports and imports in a given year. An excess of exports over imports is referred to as a *surplus* and an excess of imports over exports as a *deficit*. A situation where exports are just equal in value to imports is a situation of *balanced trade*. Any balance of trade, surplus or deficit, requires financial flows between the countries involved. This is not to be confused with the balance of payments, which takes account of all transactions between one country and another in terms of receipts and payments. The three main components of the balance of payments are the current account (covering investment, income, exports and imports), the capital account and the balancing account (which records changes in revenue).

Since World War II governments have rarely balanced their budgets (i.e. they have rarely planned to match their expenditure with their receipts). In most years they have run *budget deficits*, spending more than they receive. As a result, in most years the UK government has had to borrow money. In the UK the borrowing of the public sector (central government, local government and other state bodies such as nationalised industries) over a period of time is called the *public sector borrowing requirement*. In two periods, 1969–1970 and 1988–1990, the UK government received more revenue than it spent. The normal budget deficit was turned into a budget surplus. The UK budget surplus is called the *public sector debt repayment* (PSDR). The debt incurred from overspending is called the national debt.

21.15 CURRENCY RISK

Understanding exactly what assets and liabilities of a company are exposed to foreign exchange risk is crucial to exposure management (Perry 1997). There is always a risk that the expected cash flows from overseas investments will be adversely affected by fluctuations in exchange rates. A fall in the value of the local currency may completely alter the feasibility of an investment project. It is important, therefore, to assess the level of exchange rate risk and to consider ways in which this risk can be minimised or overcome prior to commencing the project. Perry (1997) describes two types of foreign exchange risk:

- accounting or translation exposure, arising from the need to translate the balance sheet of a multinational's subsidiary into home currency terms;
- economic exposure, arising from a company's future cash flows rather than merely its "snapshot" accounting exposure.

Economic exposure (previously known as *transaction* exposure) arises from both the direct effect of having to pay or receive currency at a future date and currency fluctuations. Currency fluctuations may cause changes in the relative prices of goods across countries, which can affect a company's international competitiveness, altering its sales, profits and ultimately its value. Giddy and Dufey (1992) state that the concept of *accounting exposure* arises from the need to translate accounts that are denominated in foreign currencies into the home currency of the reporting entity. Most commonly, the problem arises when an enterprise has foreign affiliates keeping books in the respective local currency. For the purposes of consolidation these accounts must somehow be *translated* into the reporting currency of the parent company. Perry (1997) provides an example of *translation* exposure where, for instance, a UK company has a translation exposure in US dollars and it decides to sell dollars forward. If the dollar subsequently rises and it never actually receives the dollars from the subsidiary, it is forced to buy expensive dollars in the spot market without an offsetting transaction. This could lead to insolvency or at the very least a cash flow problem.

21.15.1 Risk Mitigation by Hedging

Where a business is engaged in overseas transactions involving large sums, an adverse movement in exchange rates can be catastrophic and so it will usually adopt some form of "hedging" to minimise the risk. There are various ways in which hedging can be carried out and the simpler of these are described below.

Netting

Netting is a simple technique which can be used where a business both imports and exports goods to the same country. If such a business has an amount owed and an amount owing to be settled at the same date, it is possible to offset one amount against the other. For example, a UK business which buys goods from and sells goods to Germany may find that it owes €5 million and is owed €6 million, with the settlement for both amounts occurring in one month's time. The amount owed can be used to pay part of the amount owing. Although in this case the major portion of the exchange rate risk has been dealt with, €1 million still remain. This will have to be dealt with/hedged by using some other technique. Netting is more readily undertaken by businesses that have operating divisions in different countries which trade with each other. Offsetting of the interdivisional debts will often be done through the headquarters.

Leading and Lagging

Another technique is leading and lagging. Assume a scenario where a UK business is owed $1 million by a US business which is due for settlement in three months' time. However, it is expected that the US dollar will fall in value against sterling during this period. In such a situation, the UK business may try to obtain earlier payment so as to avoid this adverse movement in the exchange rates. To achieve this some form of financial incentive will probably have to be offered to the US business as part of a mitigation action. This reduction of the settlement period is referred to as "leading".

Forward Market Hedge

A forward market hedge involves a business entering into a contract to buy or sell a fixed amount of currency at an agreed future date and at an agreed rate. The actual exchange or delivery takes place at a specified time in the future. While the amount of the transaction, the value date, the payments procedure and the exchange rate are all determined in advance, no exchange of money takes place until the actual settlement date. This commitment to exchange currencies at a previously agreed exchange rate is usually referred to as a *forward contract*. The contract is usually offered through the foreign exchange department of a bank. The pricing of a forward contract is a wholly arithmetical process, involving no assumptions or guesswork as to the likely exchange rate movements. The seller of the contract simply adjusts the current spot rate of exchange to allow for the difference in interest rates between the two currencies covered by the contract (Holliwell 1998). "Broken dates" are contracts where the expiry is fixed for other than complete months. The difference between the forward and the spot exchange rate is called the "differential". If the rate quoted for the forward contract is better for the client of the bank than the spot rate in the market, then it is at a "premium". If the forward rate is worse for the client of the bank than the spot rate then it is at a "discount".

To illustrate this technique, assume a scenario where a UK business is owed $1 million in two months' time for goods or services supplied to a US business. The UK business may hedge against the prospect of an adverse movement in exchange rates (i.e. a rise in the US dollar against sterling) by agreeing to sell $1 million in two months' time at an agreed rate of exchange. By doing this, the UK business is now certain about the amount receivable in sterling, as the dollars to be received in two months' time can be sold at the agreed exchange rate.

The bank offering the forward enters into a contract with the UK business to buy the $1 million in two months' time and pay the business in sterling. For the purpose of this illustration, the bank borrows $1 million which it immediately sells at spot rate in the market for sterling, placing the proceeds on deposit to earn interest in sterling for the next two months. At expiry of the forward contract in two months the bank receives from the business the $1 million it has agreed to buy under the contract and uses that money to repay the $1 million it has borrowed. The bank pays the business the sterling it had received from selling the $1 million at spot two months earlier, which has been held in the interest earning deposit account.

If the bank knows that it will earn a higher rate of interest on the sterling deposit than it will have to pay on the US dollar borrowing then the bank would give the client the benefit of the difference by buying the US$1 million at a premium to the spot rate, that is, the forward contract rate at which the bank will buy the US dollars is better for the business than the spot rate in the market on the day the forward contract is agreed. If the rate of interest the bank will earn on the sterling deposit is less than the rate of interest it will pay on the US dollar borrowing then the forward rate the bank quotes will be worse for the business than the spot rate in the market on the day the forward contract is agreed, that is, the forward rate will be at a discount to the spot rate.

In practice the bank is unlikely to actually borrow and sell currencies, but the above illustrates the principles on which the calculations for a forward contract are based. However, there is still a risk that the US business will not pay the amount owing on the due date. The problem with this technique of course is that a business will not be able to benefit from any beneficial movement in exchange rates, as the future exchange rate is fixed.

Fuel Market Hedge

A fuel market hedge involves a business entering into a contract to buy or sell a fixed amount of fuel at an agreed future date and at an agreed rate. Fuel prices are critical to commercial airlines. This is explained by an article in the London *Evening Standard* newspaper describing the potential impact of rising fuel prices on the budget airline Ryanair (Lea 2004):

> Ryanair will fail to make a profit this winter as its aircraft fuel bill soars by as much as 50% because it has not taken action against the spiralling oil price. With the price of oil at all-time highs above $50 a barrel, the Irish budget airline admitted today it has gone into the second half of its financial year – October to March – completely unhedged and in a position to do nothing other than pay through the nose for its fuel. Most airlines attempt to hedge their future bills by taking out forward contracts, sometimes over a number of years. But Ryanair boss Michael O'Leary said he decided not to buy fuel at a forward price of $40 a barrel 12 months ago because he thought it was too high. As a result, the Irish budget airline's fuel bill will shoot up by €55 million (£38.3 million), half as much again [as] it would normally have expected to pay.

A lack of fuel hedging has also left struggling American Airlines exposed. As reported by Daniel (2005), the airline had fuel hedges for just 4% of its fuel requirements for 2005.

Currency Futures

Outside of the interbank forward market, the best-developed market for hedging exchange rate risk is the *currency futures* market. In principle, currency futures are similar to foreign exchange forwards in that they are contracts for delivery of a certain amount of foreign currency at some future date and at a known price. In practice, they differ from forward contracts in important ways. A significant difference between forwards and futures is *standardisation*. Forwards are for any amount, as long as it is large enough to be of interest to the dealer, whereas futures are for standard amounts, each contract being far smaller than the average forward transaction. Futures are also standardised in terms of delivery date. The normal currency futures delivery dates are March, June, September and December, while forwards are private agreements that can specify any delivery date that the parties choose. Both of these features allow the futures contract to be tradable. Another difference is that forwards are traded by phone and are completely independent of location or time. Futures, on the other hand, are traded in organised exchanges such as the London International Financial Futures Exchange (LIFFE) in London, Singapore International Monetary Exchange (SIMEX) in Singapore and the International Monetary Market (IMM) in Chicago. The most important feature of futures contracts is the time pattern of the cash flows between the parties to the transaction. With futures cash changes hands every day during the life of the contract, or at least every day that has seen a change in the price of the contract. This daily cash compensation feature largely eliminates default risk.

Currency Hedging

Currency hedging eliminates the risk of foreign exchange movements adversely affecting profits. The principal financial instruments currently used by companies for hedging are currency forwards, swaps and options.

- A currency forward is simply an agreement to buy or sell foreign exchange at an agreed rate and at a certain time in the future.
- A currency swap can be seen as a series of forward contracts.
- Currency options, on the other hand, give a company the right, rather than the obligation, to buy or sell foreign exchange at a certain time in the future at an agreed price, thereby allowing the company to lock in a fixed amount of foreign currency to buy or sell, while allowing it to benefit from any favourable exchange rate movements.

The risks associated with hedging are that strategies are ineffective or at worst calamitous. Hedging is costly not only in terms of transaction costs, spreads and premium but also in terms of management time and effort. Highly trained personnel are required.

Money Market Risk

Money markets are the markets in which money is borrowed and lent in large quantities for relatively short periods (often very short). Effectively they are a market in deposits and advances. However, they are also a market in various forms of short-term security that are almost the equivalent of money. In the money markets banks need to adjust the interest rates they offer and charge until they have the right balance between those that are prepared to lend to them and those who want to borrow. Building societies need to adjust the rates they pay to savers and charge to borrowers so that they keep a balance between the money coming in and the amount going out as mortgage loans. However, the private sector commercial banks on occasion need to borrow from the government via the Bank of England or may be put in a position where they need to do so. In other words the Bank of England as the country's central bank is the lender of last resort to the banking system. The level at which it is prepared to lend and the terms can be used to influence the level of interest rates across the economy.

The Purchasing Power Parity Theory of Exchange Rates

There is an argument that disputes the need for hedging. *Purchasing power parity* (PPP) claims that exchange rate changes are offset by price level changes. If PPP exists, then a given amount of currency in one country, converted into another currency at the current market rate, will buy the same bundle of goods in both countries. For instance, if £1 = $2, and consumers only buy jeans, then purchasing power parity will exist if a £20 pair of jeans cost $40 in the USA. It will not exist if a pair of jeans priced at £20 in the UK is priced at $50 or $30 in the USA. If there are only two goods in the economy, food and clothing, then purchasing power parity will exist if an identical bundle of food and clothes costs £100 when it costs $200 in the USA, or £500 when it costs $1000 in the USA.

The PPP theory states that exchange rates in the long term change in line with different inflation rates between economies. To understand why exchange rates might change in line with inflation rates, assume that the balance of payments of the UK is in equilibrium, with exports equal to imports and capital outflows equal to capital inflows, but is suffering from a 5% inflation rate (i.e. the prices of goods are rising on average by 5% per year).

Assume also that there is no inflation in the rest of the world. At the end of one year, the average price of UK exports will be 5% higher than at the beginning. On the other hand, imports will be 5% cheaper than domestically produced goods. At the end of the second year, the gap will be even wider.

Starting from a PPP rate of $2 = £1, this change in relative prices between the UK and the rest of the world will affect the volume of exports and imports. UK exports will become steadily less price competitive on world markets. Hence sales of UK exports will fall. Imports into the UK, on the other hand, will become steadily more price competitive and their sales in the UK will rise. The balance of payments on the current account will move into the red.

A fall in the volume of UK exports is likely to lead to a fall in the value of exports (this assumes that exports are price elastic – price elasticity is discussed in Chapter 23) and therefore the demand for pounds will fall. A rise in the value of imports will result in a rise in the supply of pounds. A fall in demand and a rise in the supply of pounds will result in a fall in the pound's value.

So the PPP theory argues that in the long run exchange rates will change in line with changes in prices between countries. An example of a company's approach to foreign exchange is described in Vodaphone Group Plc's Annual Report 2004, p. 94:

> **Foreign exchange contracts, interest rate swaps and futures**
> The Group enters into foreign exchange contracts, interest rate swaps and futures in order to manage its foreign currency and interest rate exposure.
>
> **Hedges**
> The Group's policy is to use derivative instruments to hedge against exposure to movements in interest rates and exchange rates. Changes in the fair value of instruments used for hedging are not recognised in the financial statements until the hedged exposure is itself recognised.

Currency exposures

Taking into account the effect of forward contracts and other derivative instruments, the Group did not have any material financial exposure to foreign exchange gains or losses on monetary assets and monetary liabilities denominated in foreign currencies at 31 March 2004.

21.16 SUMMARY

This chapter has examined elemental macroeconomic theory together with UK government fiscal and monetary policies aimed at modifying aggregate demand, to achieve its objectives of full employment, low inflation, a stable balance of payments and economic growth. This examination provides the context to business economic risk. It provides a perspective of demand and supply and the modifying influences. Inflation clearly impacts business in terms of the amount paid for goods. Interest rate rises impact the pay rises that have to be made and are particularly felt in the housing industry, where significant rises deter the purchase of new homes, and house builders, the building trades and the retail sector, particularly the home improvement market, suffer. While there are now fewer trade restrictions for those trading overseas, currency risk can be a major risk and risk mitigation is undertaken through hedging practices.

21.17 REFERENCES

Daniel, C. (2005) In hard times saving dollars makes sense. Interview with Gerard Aspey, Chief Executive of American Airlines. *Financial Times*, 15 March.

Giddy, I.H. and Dufey, G. (1992) The management of foreign exchange risk. http://pages.stern.nyu.edu/~igiddy/fxrisk.htm

Holliwell, J. (1998) *The Financial Risk Manual: A Systematic Guide to Identifying and Managing Financial Risk*. Pearson Education, UK.

Larsen, P.T. and Gapper, J. (2005) Visa Rules Hurting US, Warns Gates. *Financial Times*, 31 January.

Lea, R. (2004) Ryanair profit grounded by spiralling fuel prices. *Evening Standard*, 2 November.

Perry, D. (1997) Living life on the hedge. In T. Dickson and G. Bickerstaffe (eds), *Financial Times Mastering Management*. Financial Times/Prentice Hall, London.

22
Environmental Risk

In a months-long investigation, the panel found that mistakes and "failures to appreciate risk" compromised safeguards "until the blow-out was inevitable and, at the very end, uncontrollable". (Presidential Commission examining the BP Gulf of Mexico oil spill, January 2011)

The previous chapter examined economic risk management, the first of six classes of risk exposure that businesses face relating to business operating environment within the risk taxonomy proposed in Chapter 9. This chapter examines the second of the macro influences of the business operating environment, *environmental risk*. It is often suggested that business behaviour that seeks to be overtly ethical or to give considerable weight to environmental concerns must do so at the expense of profit. However, many firms are now seeing ethical and environmentally responsible behaviour as being in their self-interest. Indeed, a business that proactively and positively addresses environmental issues may succeed in contributing to the national effort to protect the environment and be able to exploit upside risk (opportunities) to enhance business success. The introduction of genuine environmental policies within its marketing activities may enable a business to attract customers, investors and employees, and lower production costs through resource efficiencies, recycling and increased demand. The degree of success of the introduction of environmental policies will be dependent on the appeal to customers, the current or planned activity of competitors, the cost benefit ratio of implementation and comprehension of the prevailing legislation. In the UK, legislation governing environmental protection as advised by the Environment Agency[1] is very extensive and includes (but is not limited to) the following subject areas: air, chemicals, energy, land, noise and statutory nuisance, pollution, plant protection, radioactive substances, waste and water. The structure of this chapter is illustrated in Figure 22.1. The next chapter examines the third of the external processes, legal risk.

22.1 DEFINITION OF ENVIRONMENTAL RISK

What is environmental risk? Environmental risk for the majority of companies is the deterioration of bottom-line performance from:

- increased regulation of energy usage;
- eroded reputation, brand name and market share from an environmental incident;
- increased operating costs from the effects of global warming;
- higher fuel costs as natural resources are depleted; and
- loss of market share to more environmentally "savvy" competitors with marketing campaigns that portray social responsibility.

Environmental risk for businesses (commonly working in regulated industries) is reduced profit as a result of exposure to pricing restrictions, operating restrictions, investment

[1] http://www.environment-agency.gov.uk/netregs/legislation

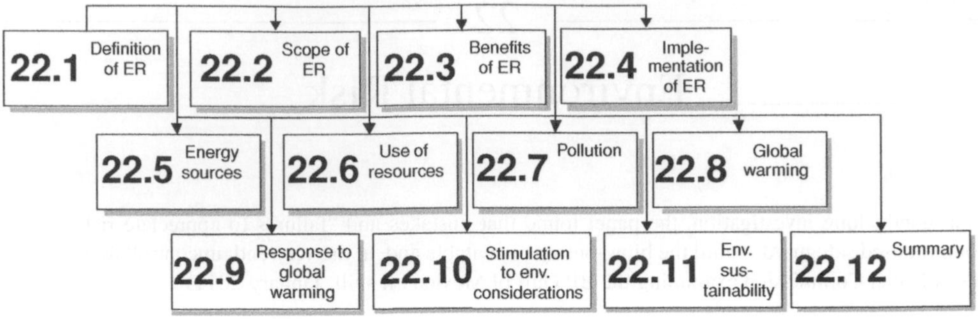

Figure 22.1 Structure of Chapter 22

obligations or fines resulting from prosecution due to regulation breach. In the market place, a business may suffer from:

- the inability to attract new customers;
- loss of existing customers from the publication of deficiencies in environmental performance;
- the imposition of increased operating costs arising from compliance with government legislation; or
- fines imposed by such organisations as the Environment Agency.[2]

For government regulators and environmental activists, environmental risk has a different meaning and relates purely to damage to the environment or the endangering of public health arising from a man-made environmental incident commonly in the form of pollution. The UK Environment Protection Act 1990 defines the "environment" as consisting of all, or any, of the following media: air, water and land (where the medium of air includes the air within buildings and other natural or man-made structures above or below ground).

- "Pollution of the environment" is defined as pollution due to the release (into any environmental medium), from any process, of substances which are capable of causing harm to people's health or any other living organisms supported by the environment.
- "Harm" for humans means harm to any of the senses or property, and harm to living organisms includes interference with the ecological systems of which they form part.
- "Process" means any activities carried out on premises or by means of mobile plant.

Reinhardt (2001) makes the interesting point that environmental risk is "asymmetric". He explains "asymmetric" by using the example of exchange rate risk, which has both an upside and a downside, where for instance the value of sterling can either rise or fall against other currencies. He expands on this theme by comparing environmental risks to security risks such as war and kidnapping, where there are no short-term upside risks.

[2] The Environment Agency was set up by the 1995 Environment Act. It is a non-departmental public body, sponsored largely by the Department for Environment, Food and Rural Affairs (DEFRA) and the National Assembly for Wales (NAW). The industry regulates industrial pollution, land quality, radioactive substances, waste management and water quality.

22.2 SCOPE OF ENVIRONMENTAL RISK

The sources of risk considered to be embraced within the term "environmental risk" are very considerable. Environmental risk for businesses may be considered to include, but not limited to:

- pollution of land, water or air (as defined, for instance, by the Environmental Protection Act 1990);
- increased regulation ("red tape") and higher operational costs;
- prosecution arising from the lack of observance of rules set by a regulatory body;
- reputational risk from adverse publicity as a result of pollution events, resulting in a reduced customer base;
- destruction of facilities or loss of manufacturing as a result of severe weather conditions;
- loss of oil production, resulting in higher energy costs.

With regard to the last of these, hurricanes Katrina and Rita in the United States in 2005 led to a reduction in oil production, which in turn resulted in higher fuel costs across Europe. As reported in *The Times*, London-listed food processing companies warned of higher input costs and stated that over and above the known increase in power bills are the less transparent pressures on packaging and transportation costs (Klinger 2005).

22.3 BENEFITS OF ENVIRONMENTAL RISK MANAGEMENT

Environmental risk management affords a business benefits as it:

- encourages examination of business continuity issues stemming from possible climate change events;
- prompts closer examination of the risks of adverse environmental incidents and response actions should they occur;
- stimulates marketing initiatives to promote products and brands in the context of the environment, sustainability, renewable energy and preservation of natural resources;
- affords management the opportunity to focus on revenue generating activities rather than fire-fighting adverse publicity;
- contributes to management actions which prevent attracting greater regulatory supervision and intrusion;
- increases competitor advantage where customer buying preferences favour businesses with a better environmental performance;
- reduces exposure to prosecution.

22.4 IMPLEMENTATION OF ENVIRONMENTAL RISK MANAGEMENT

The development of a sound system of environmental risk management will depend on a number of issues such as:

- the risk management system not overly constraining risk taking, slowing down decision-making processes or limiting the volume of business undertaken;
- the implementers of the risk management framework being distinct from the managers of the individual business units;

- risks being managed at an appropriate level in the organisation;
- the development of a culture which rewards the disclosure of risks when they exist, rather than encouraging managers to hide them.

22.5 ENERGY SOURCES

Businesses today face five known energy problems: the cost, quality, reliability and longevity of supplies and the control of emissions. Businesses tomorrow will face a myriad of energy risks which may or may not materialise, together with a number of evolving uncertainties which will most probably lead to change. BP's *Energy Outlook 2030* (BP 2011), which contains its projections for future energy sources and supplies, recognises the potential for change: "Forward-looking statements involve risks and uncertainties because they relate to events, and depend on circumstances, that will or may occur in the future". The document states that events may differ from projections depending on a variety of factors, including "product supply, demand and pricing; political stability; general economic conditions; legal and regulatory developments; availability of new technologies; natural disasters and adverse weather conditions; wars and acts of terrorism or sabotage; and other factors discussed elsewhere in this presentation".

Clearly traditional sources of supply are being depleted across the world. In its 2008 World Energy Outlook, the International Energy Agency (IEA 2008)[3] advises that, based on its World Economic Model,[4] the world's primary energy demand will increase by 45% between 2006 and 2030, an average annual growth rate of 1.6%. The Model describes world demand for coal advancing by 2% a year on average, its share in global energy climbing from 26% in 2006 to 29% in 2030. Most of the increase in demand for coal comes from the power-generation sector. China and India together contribute 85% to the increase in world coal demand over the projection period. Oil remains the dominant fuel in the primary energy mix, but its share drops to 30% in 2030, from 34% in 2006. Oil demand grows far more slowly than demand for other fossil fuels, mainly because of high final prices. Gas demand increases at 1.8% per annum over the projection period and its share in world primary energy moves up slightly, to 22% in 2030.The volumetric increase in China's energy demand in 2006–2030 dwarfs that of all other countries and regions, the result of its rapid economic and population growth. The almost 2000 million tons of oil equivalent increase in demand in 2006–2030 is nearly four times bigger than the combined increase in all of the countries in Latin America and Africa, and more than three times as large as the increase in the OECD.

Oil forms the foundation of our current way of life. China will be the world's biggest economy by 2030, and the question has often been raised as to whether energy growth can be sufficient to support high economic growth in China, Brazil and India or indeed in the

[3] The IEA is an autonomous body which was established in November 1974 within the framework of the OECD to implement an international energy programme. It carries out a comprehensive programme of energy cooperation among 28 of the OECD's 30 member countries. The basic aims of the IEA are: to maintain and improve systems for coping with oil supply disruptions, to promote rational energy policies in a global context through cooperative relations with non-member countries, industry and international organisations, to operate a permanent information system on the international oil market, to improve the world's energy supply and demand structure by developing alternative energy sources and increasing the efficiency of energy use, to promote international collaboration on energy technology and to assist in the integration of environmental and energy policies. The OECD is a unique forum where the governments of 30 democracies work together to address the economic, social and environmental challenges of globalisation.

[4] The World Energy Model is a large-scale mathematical model designed to replicate how energy markets function. It was first produced in 1993, is constantly reviewed and updated to ensure completeness and relevance, and contains huge quantities of historical input data (nearly 16 000 equations) on economic and energy variables.

wider group of industrialising countries. China is the largest source of oil consumption growth in BP's *Energy Outlook*, with consumption forecast to grow from 8 million barrels per day (Mb/d) to 17.5 Mb/d by 2030, overtaking the US to become the world's largest oil consumer. The pace of Iraqi capacity expansion, combined with production growth, is a key source of uncertainty in terms of both world supply and price. Iraq is expected to account for 20% of global supply growth from 2011 to 2030. A rapid increase in Iraqi output could have an impact on oil prices. OPEC is likely over time to seek to reintegrate Iraq into the quota system, which is an additional source of uncertainty.

The location of our energy is another problem – energy sources are often long distances from the point of consumption. Centuries ago virtually everyone would have depended on the fuel they could find within a short distance from home. Now, the energy for our fuel, heat and light travels vast distances to reach us, sometimes crossing not only continents, but diverse political and cultural societies on the way. These distances create a whole host of challenges from oil-related political instability to the environmental risks of long-distance pipelines.

The most obvious threat to the continued burning of fossil fuels is intensifying natural climate change and heating the Earth to dangerous levels. Can the greenhouse effect be ignored? There are still real costs that go with the search for and use of energy: air and water pollution, impaired health, acid rain, deforestation and the destruction of traditional ways of life. But the tank is running dry. It does not have to be like this. While our energy use is unsustainable, we already know what a less intrusive alternative would look like. Some scientists believe all we have to do is decide that we will get there, and how. It will make vastly more use of renewable energy, from inexhaustible natural sources like the sun and the seas.

22.5.1 Renewable Energy

Renewable sources of energy – wind power, solar power (thermal, photovoltaic and concentrated), hydro-electric power, tidal power, geothermal energy and biomass – are essential alternatives to fossil fuels. Their use reduces our greenhouse gas emissions, diversifies our energy supply and reduces our dependence on unreliable and volatile fossil fuel markets (in particular, oil and gas). The growth of renewable energy sources also stimulates employment, the creation of new technologies and improves our trade balance. The UK Department of Energy and Climate Change published its "UK Renewable Energy Roadmap" in July 2011. This document, the UK's first renewable energy roadmap, describes the approach to releasing the UK's renewable energy potential. It is seen that an increase in renewable energy across the UK can provide more security through a greater degree of independence helping to shield the UK from fossil fuel price fluctuations. Renewable energy will contribute to the UK's 2050 emissions target of an 80% fall in emissions. The roadmap sets out targeted practical actions to accelerate the adoption of renewable energy in the UK across the electricity, heat and transport sectors. While the roadmap considers that renewable deployment across all technologies is important, it focuses in particular on eight technologies that it believes have the greatest potential to help the UK meet the interim 2020 target in a cost-effective way, or offer considerable potential for the decades that follow. The eight technologies are onshore and offshore wind power, marine energy (such as wave and tidal devices), biomass electricity and heat, ground and air source heat pumps and renewable transport. One interesting development is the installation of a wind turbine located close to the M4 motorway in Reading's Green Park to provide a top-up charge on electric cars. Other charging points have been created at Michaelwood Services on the M5 in Gloucestershire, and at South Mimms Services on the

M25 in Hertfordshire. According to Ecotricity (the owner of the turbine), the Department for Transport said it expects "to see tens of thousands of plug-in vehicles on the roads in the UK" by 2015. The UK government has the laudable aspiration of being the location of choice for inward investment and a world-class centre of energy and expertise. However, it has just a little bit of competition.

In 2010 the *New York Times*[5] identified from its research that China's wind turbine production had surpassed that of Denmark, Spain and the United States; furthermore, China was (at the time of writing) the largest manufacturer of solar panels. The article made the interesting observation that with its efforts to dominate renewable energy technologies it raised the prospect that the West may someday trade its dependencies on oil from the Middle East for a reliance on solar panels, wind turbines and other equipment manufactured in China. Later in the same year the American Energy Innovation Council,[6] composed of executives such as Bill Gates (co-founder, Microsoft), Jeffrey Immelt (CEO, General Electric) and Ursula Barns (CEO, Xerox), urged the government to more than triple spending on energy research and development and to create a national energy board to guide investment. Gates said that the US would not come close to achieving an 80% reduction in greenhouse gas emissions by 2050 without an immense breakthrough. In addition, the executives said the lack of investment was crippling the US' international competitiveness. This view was perceptive. As reported in *CNN Money*[7] in 2011, China recently overtook the United States as the world's top market for attracting private capital for renewable energy investments. According to the *China Briefing* in June 2011,[8] mergers and acquisitions in the green energy sector between Chinese and foreign companies have been dramatic as the former has looked to capitalise on foreign technology and expertise and the latter on a bigger role in the Chinese market. Reportedly merger and acquisition contracts in the green energy sector in China amounted to US$2.126 billion in 2010. At the same time China is investing heavily. The Chinese government invested $120 billion in renewable energy in 2010 while the US invested just $20 billion.

Concern over global warming has renewed interest in nuclear energy as a means of generating commercial electricity, and some countries, such as France and Japan, are heavily dependent upon it. Nuclear energy is even being considered as an environmentally friendly source of renewable energy. However, nuclear energy is not renewable. It is reliant upon a finite source of fuel that can be exhausted. The uranium (and sometimes plutonium) used in nuclear power is a natural resource in the same way that oil, coal and natural gas are. The explosion at the Fukushima nuclear power station in north-east Japan that resulted from damage caused by an 8.9 magnitude offshore earthquake on 11 March 2011 (and the 7 metre tsunami that it triggered), set alarm bells ringing around the world over the safety of nuclear power. As a result of the disaster and amid escalating public concern, the European Union hastily convened emergency talks in Brussels between energy ministers, national nuclear safety officials and nuclear companies to address anxiety over nuclear safety.

The EU energy commissioner announced after the talks that the EU states had reached a ground-breaking accord to conduct "stress tests" on the 143 reactors across the Union. He said the tests would be conducted on a "voluntary" basis to assess whether the plants could resist earthquakes, tsunamis, floods, electricity cuts and terrorist attacks, taking into account both

[5] China leading global race to make clean energy, *New York Times*, 30 January 2010.
[6] A call to triple U.S. spending on energy research, *New York Times*, 9 June 2010.
[7] U.S. wins when China invests in green, *CNN Money*, 7 April 2011.
[8] An overview of China's renewable energy market, *China Briefing*, 16 June 2011.

the age of the plants and their location. Nuclear-free Austria, a constant critic of its nuclear neighbours, was insistent that stress tests were conducted across Europe, where 14 of 27 EU states have nuclear plants. It was felt a Japan-like disaster could happen again, anywhere. Japan's nuclear emergency prompted Germany and Switzerland to halt nuclear programmes. Germany's decision to phase out nuclear power by 2022 is attributed to Chancellor Angela Merkel, who is understood to have made her sudden U-turn based on advice received and the mood of the public, while fending off opposition from industrial and energy interests. Interestingly, Merkel intends to rapidly lead Germany into a new era of wind and solar power combined with enhanced efficiency. However, Barack Obama reaffirmed his commitment to nuclear energy, despite the disaster at Japan's Fukushima nuclear plant,[9] as nuclear power was part of the US' approach to reducing carbon emissions into the atmosphere. While the US could not take nuclear power off the table, he said the administration would incorporate lessons from Japan in designing and building new plants.

Some analysts suggest that nuclear power will be needed to bridge the gap between now and the renewable future. Many environmentalists (but not all) are deeply unhappy with the idea – fission technology has been in use for a generation, but concerns remain about radioactive waste disposal and the risk of accidents. Nuclear fusion – a new form of nuclear power, which combines atoms rather than splitting them – could be ready by around 2040, but that is too long to wait. The UK government estimates that 56% of energy used in UK homes could be cut using currently available technologies. We can install "power generators" on our roofs by covering our houses with solar tiles, or buying miniature wind turbines the size of a satellite dish. Practically, it is thought, the energy crisis is soluble. However, making significant widespread changes to domestic energy generation will require a dramatic change in thinking for politicians, suppliers and consumers alike.

22.6 USE OF RESOURCES

Businesses need to appreciate and understand the trends in energy consumption for a number of reasons. Energy costs are a significant component of manufacturing and transportation and a key element of the unit cost of production for many products. A reduction in energy costs can produce competitive advantage. An understanding of the trends in energy sources and energy demand provides opportunities to exploit.

In 2010, total UK overall primary energy consumption in primary energy terms (i.e. fuels obtained directly from natural sources) was 218.5 million tonnes of oil equivalent, 3% higher than in 2009. However, it should be noted that in 2009 UK primary energy consumption was at its lowest level in the last 20 years as a result of the downturn in the economy.

In 1970, fuel consumption was dominated by solid fuels use (47% of all energy consumption in the UK) and petroleum (44%), with gas contributing a further 5% and electricity 4%. By 1980 the fuel mix had evolved, with natural gas making up 20% of all energy consumption in the UK, solid fuels 36% and petroleum 37%. In 1990 the split between fuels was similar to that in 1980; however, by 2000 with changes in electricity generation, natural gas consumption had become the dominant fuel, responsible for 41% of all energy consumption in the UK, whilst solid fuels had fallen from 31% in 1990 to 17% in 2000. By 2010 more renewable fuels had entered the energy mix.

[9] Obama outlines energy plan, *Wall Street Journal*, 31 March 2011.

Additionally, in 1970, the industry sector was responsible for 43% of total UK consumption, followed by the domestic sector 25%, transport 19% and other final users (mainly agriculture, public administration and commerce). However, by 1990 industrial consumption had fallen to 26% of total final energy consumption in the UK, while transport consumption had risen to 33%. Domestic use had increased slightly to 28% and other final users to 13%. The decreasing trend in industrial consumption continued and in 2010 was 18% of total final energy consumption in the UK, with transport consumption responsible for 37% and domestic 32%.

22.7 POLLUTION

Businesses are at risk from prosecution for breaching environmental legislation and pollution. The UK Environment Agency regularly secures prosecutions for air, water and noise pollution. According the Agency's website, the following incidents occurred between 2006 and 2010.

> In 2010 DS Holdings Ltd, trading as Envirogreen, was prosecuted under the Water Resources Act 1991, Environmental Permitting (England & Wales) Regulations 2007 and the Hazardous Waste (England and Wastes) Regulations 2005. The offences included causing pollutants to enter the Chalvey Ditch near Cippenham, operating a regulated facility without an environmental permit, failing to keep a record of the hazardous waste transported and failing to complete a hazardous waste consignment note. The court heard that a tanker belonging to DS Holdings Ltd accidentally discharged approximately 4500 litres of hazardous chemicals into the Chalvey Ditch in Cippenham in September 2009. The discharge caused near total fish mortality in the watercourse as far as its confluence with the River Thames.

Examples of other incidents resulting in prosecution include pollution of a stretch of the River Thames with oil in 2008 by VK Transport affecting some 40 swans, and pollution of an 8-mile stretch of the River Coln with chlorine in 2006 by Biolab (UK) Limited as a result of a fire at their premises, killing over 2500 fish. In most cases the Agency issues businesses with multiple warnings prior to a prosecution.

A less reported problem is light pollution. Members of the Campaign to Protect Rural England have claimed that light pollution caused by commercial and domestic premises in East Yorkshire and North East Lincolnshire is now 30% greater than ten years ago.[10] They say that the Milky Way, northern lights and shooting stars are all but invisible to the naked eye because of the amount of ambient light. Satellite images of the area show whole swathes covered in an orange-yellow glow, which signifies intense light output from man-made structures. Ian McKechnie, the senior Environmental Health Officer at the East Riding of Yorkshire Council, says the light pollution has now got out of control: "Light pollution is now 28–30% worse than it was ten years ago and most alarmingly half of the areas in East Yorkshire that had dark skies ten years ago are now polluted with light." He said one of the main sources of light in rural areas is domestic security lights and the lighting in commercial glasshouses.

22.8 GLOBAL WARMING

Climate change is widely recognised as one of the key environmental challenges facing the world today. There is a growing scientific consensus over the potential impact on the climate

[10] BBC News (2003) Black mark over light pollution. 7 November. http://news.bbc.co.uk

from increasing concentrations of greenhouse gases in the atmosphere. The greenhouse effect is the natural process by which the atmosphere traps some of the Sun's energy, warming the Earth sufficiently to support life. From the press it would appear that most mainstream scientists believe a human-driven increase in "greenhouse gases" is increasing the effect artificially. These gases include carbon dioxide, emitted by fossil fuel burning and deforestation, and methane, released from rice paddies and landfill sites. According to the latest scientific research, Europe is already condemned to regular heat waves. In 2003 there was a heatwave across southern Europe with temperatures surpassing 40°C leading to forest fires notably in Portugal, Spain and Italy and the loss of over 20 000 lives. In 2007 southern Europe was again subjected to heatwave, with Greece, Hungary and Bulgaria severely affected. In addition, increasingly countries will be exposed to flash floods. An early indication was the flood which occurred in India in July 2005.[11] However, in 2010 there were significant floods in China, India, North Korea, Guatemala, Brazil and Poland, the worst being in Pakistan when over 1400 people lost their lives. In 2011 the pattern continued with floods in Brazil, Australia, South Africa, Sri Lanka and, most prominently, in China where over 550 000 people had to be evacuated. Increasingly unstable weather patterns are predicted to make forest fires, crop losses and water shortages commonplace, as global warming takes hold later in the century.[12]

An increase in the area of land covered by desert is threatening to make refugees of millions of people and send dust storms across the oceans. Recent research has found that degradation of dry lands resulting from climate change and human activities (such as farming methods) is increasing the amount of area covered by desert, causing a rising number of large and intense dust storms which without proper management will lead to unproductive desert. The impact may arise in the form of striking contrasts across the world, with the simultaneous occurrence of rising sea levels and flooding on some continents and water shortage and famine on others. The United Nations reported in 2011 that the horn of Africa had again been hit by drought, leaving 12 million people at risk from starvation.

The climate research division of the Scripps Institute of Oceanography in San Diego, California, believes it has conclusive proof that global warming is due to human activity.[13] Its study used several scenarios to try to reproduce the observed rise in ocean temperatures over the last 40 years, such as natural climate variability, solar radiation and volcanic emissions, but none fit the bill. The lead author of the study, Dr Tim Bartlett, said colourfully that "what absolutely nailed it was greenhouse warming". The study, announced in February 2005, highlighted that regional water supplies will be dramatically affected with the South American Andes and western China severely affected. Already, in England, four of the five warmest years in the last 340 years have occurred in the last decade. But government agencies advise we could see a very much greater rise over the course of the next century unless action is taken to significantly reduce greenhouse gas emissions. Commenting on BBC Radio, the then Environment Secretary Margaret Beckett said "we could be coming to the tipping point where change could be irreversible".

[11] On that occasion more than 65 centimetres fell in Mumbai, which, according to officials, was the heaviest rainfall ever recorded in India in a single day. As a result of these rains in Mumbai and other parts of the state of Maharashtra, 430 people lost their lives and over 150 000 were stranded. BBC News (2005) Indian monsoon death toll soars. 28 July. http://news.bbc.co.uk

[12] Walsh, C. (2004) Emissions impossible for cbi to stomach, *The Observer*, 21 March.

[13] BBC News (2005) Greenhouse gases "do warm oceans". 17 February. http://news.bbc.co.uk

22.9 RESPONSE TO GLOBAL WARMING

22.9.1 Earth Summit

In response to increasing concerns about climate change, the United Nations Framework Convention on Climate Change was agreed at the Earth Summit at Rio de Janeiro in 1992, attended by 170 heads of state. At the Earth Summit all developed countries agreed a voluntary target to return their emissions of greenhouse gases to 1990 levels by 2000.

22.9.2 The Kyoto Protocol

With the passing of time it quickly became apparent that the UN Convention commitments could only be a first step in the international response to climate change. Successive climate prediction models illustrated that deeper cuts in emissions were required to prevent man's serious interference with the climate. The Kyoto Protocol was formulated to address this issue. The Protocol was signed on 11 December 1997 in Kyoto, Japan, based on principles set out in a framework United Nations Convention signed at the Earth Summit in Rio de Janeiro. As the then Secretary-General of the UN, Kofi Annan, said in his video message on the activation of the Protocol:

> Today we celebrate the entry into force of the Kyoto Protocol. This is a great stride forward in our struggle to confront one of the biggest challenges we face in the twenty-first century: climate change. Scientists continue to tell us that the first signs of climate change are already visible. If this change is not addressed, sustainable development will be out of reach. Climate change is a global problem. It requires a concerted global response. The Kyoto Protocol provides a truly global framework.

This emphasis on climate change being a global problem requiring a global response influenced the content of the Protocol. According to Article 25 of the Protocol, it could only come into force after it had been ratified by at least 55 parties to the Convention agreed at the Rio Summit (where these parties, referred to in Annex 1, accounted in total for at least 55% of the total carbon dioxide emissions for 1990) by depositing their instruments of ratification, acceptance, approval and accession. Following the decision of the US and Australia not to ratify, Russia's position became crucial for the fulfilment of the 55% emissions condition. Russia had deliberated over ratification it is thought due to political and economic considerations. It finally did ratify on 18 November 2004, and the Kyoto Protocol came into force 90 days later – on 16 February 2005.

22.9.3 Pollution Control Targets

The Kyoto Protocol is significant as it is the first treaty of its kind to impose legally binding commitments on countries to improve pollution control. In this instance, to reduce greenhouse gas emissions by 5.2% below 1990 levels over the period 2008–2012. The six targeted gases (listed in Annex A of the Protocol) are carbon dioxide (CO_2), methane (CH_4), nitrous oxide (N_2O), hydrofluorocarbons (HFCs), perfluorocarbons (PFCs) and sulphur hexafluoride

(SF_6).[14] These gases are considered to at least contribute to global warming – the rise in global temperature which may have severe consequences for life on Earth. However, these reductions in harmful gases are considered modest, and at its inception the Protocol did not have the support of the US. This has led to criticisms that the agreement is toothless, as well as being virtually obsolete without US backing. The Bush administration dubbed the treaty "fatally flawed", partly because it does not require developing countries to commit to emissions reductions (which includes China and India). The lack of US backing, it has been implied, is a result of business pressure on the President. The 2004 US presidential elections, with an estimated cost of $4 billion, were classified as the most expensive ever. It was reported that Larry Noble, of the Washington-based Center for Responsive Politics (CRP), had said: "Companies make political contributions and support because they want to have someone in office that is sensitive to their needs. Most look at it as investment, and they expect a return."[15] The pharmaceutical industry heaved a huge sigh of relief that Democrat plans to limit refundable drug costs on state health programmes would not be enacted for at least another four years. Pfizer and similarly GlaxoSmithKline employees must have felt that their donation to the Republican cause was well spent. Car makers (who donated nearly $1 million to the Republicans) were also relieved that Kerry's emissions proposals would not see the light of day.

22.9.4 Sufficiency of Emission Cuts

The head of the UN Environment Programme, Klaus Toepfer, said at the time of the Protocol coming into force that Kyoto was only a first step and much hard work needed to be done to fight global warming.[16] Developed countries have only committed to cut their combined emissions to 5.0% below 1990 levels by 2008–2012, whereas there is a consensus among many climate scientists that in order to avoid the worst consequences of global warming, emission cuts in the order of 60% across the board are required.[17] Pearce (2004) advised that the clock is ticking and that every year the world is releasing almost 7 billion tonnes of carbon into the atmosphere. He commented that before the industrial age the CO_2 level was steady at around 280 parts per million (ppm), whereas in 1997, when the Kyoto Protocol was drawn up, the CO_2 level had reached 368 ppm and by 2004 it had reached 379 ppm. Pearce reported that most predictions of soaring temperatures, floods, droughts, storms and rising sea levels are based on a concentration of 550 ppm and that, on current trends, this figure is likely to be reached by the second half of this century. Furthermore, as there are time lags in natural systems such as ice caps and ocean circulation, changes would continue to take place for a considerable time even after CO_2 levels had stabilised.

22.9.5 US Climate Pact

On 28 July 2005 the US entered into a joint agreement with Australia and others to cut greenhouse gas emissions in a pact which rejected the Kyoto Protocol on climate change. The alternative to the Kyoto Protocol includes China, South Korea, India and Japan as well as Australia. The signatories argue that it complements rather than weakens the 1997 Kyoto

[14] Kyoto Protocol to the United Nations Framework Convention on Climate Change. http://unfccc.int/resource/docs/convkp/kpeng.html

[15] Morgan, O. (2004) The big donors. *The Observer*, 7 November.

[16] BBC News (2005) Kyoto Protocol comes into force. 16 February. http://news.bbc.co.uk

[17] BBC News (2005) Q+A: The Kyoto Protocol. 16 February. http://news.bbc.co.uk

agreement. Reticence on the part of US President Bush to support Kyoto is based on his belief that it would damage America's industry and economy despite the stance of many Western governments who argue it is a critical step to address global warming (Reid 2005). Both Australia and the US believe that the best way to cut government emissions is through the development of new technology rather than setting emission caps. However, the problem is that this new technology is not just round the corner and pollution is not confined to its source, it affects us all. China's lack of attention to emissions control is already having an impact on Hong Kong. While Hong Kong has some of the most environmentally friendly policies in the world, every autumn its sky is turned brown and smoggy as dust and poisonous gases are blown over from mainland China by the prevailing winds. Hong Kong is now a service economy dependent on sectors such as finance and tourism, and neither expatriate bankers nor tourists are keen on smog (Mallet 2005). Additionally, pollution and smog are exacerbating problems for those with respiratory ailments.

22.9.6 The Copenhagen Accord

In December 2009, the 15th session of the Conference of Parties to the United Nations Framework Convention on Climate Change (COP-15) was held in Copenhagen, following preparatory talks in Bonn, Bangkok and Barcelona. Although COP-15 did not produce a legally binding agreement to cut emissions as some participants wanted, delegates of 193 countries represented approved a motion to "take note of" the Copenhagen Accord. The Accord was drafted by the US, China, India, Brazil and South Africa. As described by the respected United States Energy Information Administration, the emissions mitigation pledges submitted by countries as part of the Copenhagen Accord fall into two general categories: absolute reductions and intensity reductions. Absolute reductions reduce greenhouse gas emissions independent of economic or material output. Japan, Russia, the European Union, the US and Brazil have announced absolute reduction goals, which are expressed as percentage reductions below historical base-year amounts. (For example, Japan announced its goal was to reduce carbon dioxide emissions to a level 25% below its 1990 levels by the year 2020.) China and India announced intensity reduction goals, which typically are expressed as reductions in emissions per unit of output as measured by GDP. (For example, China announced its intention was to reduce its carbon emissions intensity by 2020 to a level 40–45% below its emissions intensity recorded in 2005). According to the UN website, the Accord recognises that:

- climate change is one of the greatest challenges of our time;
- deep cuts in global emissions are required according to science to hold the increase in global temperature below 2°C;
- the adaption to the adverse effects of climate change and responding to them is a challenge faced by all countries;
- international cooperation is required to implement adaptation actions aimed at reducing vulnerability and building resilience in developing countries;
- there is a crucial role to be adopted in reducing emission from deforestation and forest degradation;
- there was a need to pursue various approaches to enhance the cost effectiveness of mitigation actions, including the use of the markets;
- there needs to be provision of funding to developing countries.

In addition, the Accord recognised the need to create both a green climate fund to support projects and programmes in developing countries and a technology mechanism to accelerate technology development and transfer for mitigation.

22.9.7 European Union

International negotiations are under way to draw up a UN agreement to govern global action on climate change after 2012, when the first commitment period of the Kyoto Protocol expires. The EU has taken a leading role in these negotiations and wants them to result as soon as possible in a comprehensive, ambitious, fair and science-based global agreement that is legally binding. The Copenhagen Accord is considered to be a step towards such an agreement. The EU sees the Accord as a basis for further progress and is working to make it operational at the earliest opportunity. Scientific evidence shows that for the world to have a 50% chance of keeping within the 2°C ceiling, global emissions of greenhouse gases need to peak by 2020 at the latest, be cut by at least 50% of their 1990 levels by 2050, and continue to decline thereafter. Industrialised nations must take the lead by making deep emission cuts of 25–40% below 1990 levels by 2020 and of 80–95% by 2050. In this context the EU has made a unilateral commitment to cut its emissions in 2020 to at least 20% below 1990 levels, and is offering to scale up this reduction to 30% provided other major emitters in the developed and developing worlds take on their fair share of the mitigation effort under a global agreement.

22.9.8 Cancún Agreements

The UN Climate Change Conference in Cancún, Mexico, ended on 11 December 2010 with the adoption of a package of decisions that set all governments more firmly on the path towards a low-emissions future and support enhanced action on climate change in the developing world. The package dubbed the "Cancún Agreements" was welcomed by the final plenary.

The United Nations Framework Convention on Climate Change Executive Secretary, Christiana Figueres, said:

> Cancún has done its job. The beacon of hope has been reignited and faith in the multilateral climate change process to deliver results has been restored. . . . Nations have shown they can work together under a common roof, to reach consensus on a common cause. They have shown that consensus in a transparent and inclusive process can create opportunity for all. Governments have given a clear signal that they are headed towards a low-emissions future together, they have agreed to be accountable to each other for the actions they take to get there, and they have set it out in a way which encourages countries to be more ambitious over time.

A set of initiatives was launched to protect the poor and the vulnerable from climate change and to deploy the money and technology that developing countries need to plan and build their own sustainable futures. In addition, they agreed to launch concrete action to preserve forests in developing nations. Critically they also agreed that countries need to work to stay below a 2°C temperature rise and they set a clear timetable for a review to ensure that global action is adequate to meet the emerging reality of climate change. Ms Figueres emphasised: "This is not the end, but it is a new beginning. It is not what is ultimately required but it is the essential foundation on which to build greater, collective ambition."

The next conference of the parties is scheduled to take place in South Africa, from 28 November to 9 December 2011.

As advised by the United Nations, elements of the "Cancún Agreements" include the following:

- Industrialised country targets are officially recognised under the multilateral process, and these countries are to draw up low-carbon development plans and strategies and assess how best to meet them, including through market mechanisms, and to report their inventories annually.
- Developing country actions to reduce emissions are officially recognised under the multilateral process. A registry is to be set up to record and match developing country mitigation actions to finance and technology support from industrialised countries. Developing countries are to publish progress reports every two years.
- Parties meeting under the Kyoto Protocol agree to continue negotiations with the aim of completing their work and ensuring there is no gap between the first and second commitment periods of the treaty.
- The Kyoto Protocol's clean development mechanisms have been strengthened to drive more major investments and technology into environmentally sound and sustainable emission reduction projects in the developing world.
- Parties launched a set of initiatives and institutions to protect the vulnerable from climate change and to deploy the money and technology that developing countries need to plan and build their own sustainable futures.
- A total of $30 billion in fast start finance from industrialised countries to support climate action in the developing world up to 2012 and the intention to raise $100 billion in long-term funds by 2020 are included in the decisions.
- In the field of climate finance, a process to design a green climate fund, with a board with equal representation from developed and developing countries, is established.
- A new "Cancún Adaptation Framework" is established to allow better planning and implementation of adaptation projects in developing countries through increased financial and technical support, including a clear process for continuing work on loss and damage.
- Governments agree to boost action to curb emissions from deforestation and forest degradation in developing countries with technological and financial support.
- Parties have established a technology mechanism with a Technology Executive Committee and Climate Technology Centre and Network to increase technology cooperation to support action on adaptation and mitigation.

22.9.9 Domestic Government Response to Climate Change

According to the Department for the Environment, Food and Rural Affairs (DEFRA), the Climate Change Act 2008 made the UK the first country in the world to have a legally binding long-term framework to cut carbon emissions. The Act introduced a binding reduction target requiring the UK to reduce its emissions by at least 80% by 2050 against 1990 levels. It also introduced a long-term framework for managing emissions through a system of national carbon budgets: caps on the total quantity of greenhouse gases permitted in the UK over a specified time. Each carbon budget covers a five-year period, with the first three carbon budgets covering 2008–2012, 2013–2017 and 2018–2022. During these periods, emissions must be reduced (from 1990 levels) by 22%, 28% and 34% respectively. The Act also created a framework for building the UK's ability to adapt to climate change, including:

- a UK-wide climate change risk assessment that must take place every five years;
- a national adaptation programme which must be put in place and reviewed every five years to address the most pressing climate change risks to England;
- a mandate giving the government the power to require "bodies with functions of a public nature" and "statutory undertakers" (e.g. water and energy utilities) to report on what they are doing to address the risks posed by climate change to their work.

The Act also introduced an Adaptation Sub-Committee (ASC) of the independent Committee on Climate Change. The role of the ASC is to provide advice, analysis, information and other assistance in relation to:

- requests from the national authorities (of England, Wales, Scotland and Northern Ireland) for advice on adaptation;
- the preparation of the UK Climate Change Risk Assessment including methodology and conclusions;
- the implementation of the government's Adaptation Programme (for England and reserved matters).

Under the Act the government is required to conduct a Climate Change Risk Assessment (CCRA) in order to understand the level of risk posed by climate change to the UK. This must be presented to Parliament by 26 January 2012. Work is ongoing to produce reports on the key climate risks for each of the 11 sectors identified for priority action – agriculture, water, flood and coastal management, marine and fisheries, forestry, biodiversity, business, built environment, transport, energy and health. A synthesis report is also being produced, pulling together the analysis from each sector and drawing out the interdependencies.

The Adaptation Sub-Committee of the Committee on Climate Change will undertake their statutory review of the draft findings and will provide their final advice on the CCRA in their 2nd Report on the UK's level of preparedness for climate change.

22.9.10 Levy

HM Revenue and Customs, which now administers it, describes the UK Climate Change Levy as being part of a range of measures designed to help the UK meet its legally binding commitment to reduce greenhouse gas emissions. The government initially published a draft UK climate change programme (available from DEFRA), which shows the policies intended to deliver the UK's objective of reducing greenhouse gas emissions. The government required that all sectors make a contribution in reducing greenhouse gas emissions, and drew up a programme to this effect, with a key element being the climate change levy announced in the 1999 Budget. The Climate Change Levy is an environmental tax which came into force on 1 April 2001. The primary law on Climate Change Levy is contained in the Finance Act 2000 Part II, clause 30 and in Schedules 6 and 7. The Act also provides for secondary legislation that deals with the implementation aspects of the tax, such as registration and accounting procedures.

The levy is chargeable on the industrial and commercial supply of taxable commodities for lighting, heating and power by consumers in the following sectors of business: industry, commerce, agriculture, public administration and other (non-specific) services. The levy does not apply to taxable commodities used by domestic consumers, or by charities for non-business use. All revenue raised through the levy is recycled back to business through a 0.3% cut in employers' national insurance contributions, introduced at the same time as the levy, and support for energy efficiency and low-carbon technologies. The levy is charged on taxable supplies. Taxable supplies are certain supplies of the following taxable commodities: electricity, natural gas as supplied by a gas utility, petroleum and hydrocarbon gas in a liquid state, coal and lignite, coke, and semi-coke of coal or lignite and petroleum coke. The following are not taxable commodities for levy purposes: oil, road fuel gas, heat, steam and waste as defined in statute.

22.9.11 Emissions Trading

Emissions trading works by allowing countries to buy and sell their agreed allowances of greenhouse gas emissions. Highly polluting countries can buy unused "credits" from those which are allowed to emit more than they actually do. Countries are also able to gain credits for activities which boost the environment's capacity to absorb carbon. These include tree planting and soil conservation, and can be carried out in the country itself, or by that country working in a developing country. More than 2000 UK installations – from factories to power plants – will be able to trade in the new "pollution permits". Different sectors face different restrictions: a 13% cut for electricity generators, for example, and more than 30% for offshore oil and gas operators.

22.9.12 Impact on Business

Global warming (and its influence on existing and planned legislation), is likely to impact production in both anticipated and unanticipated ways. The beleaguered coal industry looks especially vulnerable, with several operators reportedly considering closing their coal-fired plants altogether (Walsh 2004). The move by government to increase energy supplies from greenhouse gas-free sources may provide business with new exploitable opportunities, although wind power, for instance, has proved more expensive to generate than hoped. Multinationals supplying global markets may well move production plants to countries where emission levies and energy costs are less onerous, if they consider the respective government's policies will remain unchanged for a number of years. The motor industry, for instance, has claimed that car makers may be encouraged to relocate to less stringent regimes in mainland Europe, undermining the scheme's environmental benefits and robbing the UK of jobs in the process. In response to the business community's disquiet, ministers argue that an enlarged renewable energy sector will create jobs. They emphasise that the changes will be phased in gradually and point out emissions cuts are nothing new. Tom Delay, CEO of the Carbon Trust, predicts that: "Those that act now by cutting carbon emissions will be tomorrow's winners, whereas those that wait are risking their future business success". Those companies that are willing to demonstrate their green credentials are likely to increase market share. The Prius hybrid car produced by the Japanese car maker Toyota was predicted back in 2005 to be "a sales smash hit" (Mackintosh and Milne 2005). The Prius saves petrol by using a battery and electric motor as well as an engine. The hybrid has been helped by incentives in London where hybrid drivers are exempt from the £8 a day congestion charge. While the vehicles cost more to make, manufacturers have to balance whether to pass the whole of the increase onto the consumer or accept lower profits. Toyota has sold more than 315 000 Toyota Prius cars in 2010 in Japan (where petrol costs are high), thus making this the best-selling model in the domestic market over the past 20 years. More and more manufactures are examining alternative energy sources. BMW have joined the ranks of environmentally friendly producers through the recent launch of their ActiveE, the all-electric four-seater family sedan.

22.10 STIMULATION TO ENVIRONMENTAL CONSIDERATIONS

22.10.1 FTSE4Good Index

FTSE is an independent company jointly owned by The Financial Times and the London Stock Exchange. It does not give financial advice to clients, with the aim of having the freedom to provide objective market information. FTSE indices are used extensively by a range of investors such as consultants, asset owners, fund managers, investment banks, stock exchanges and brokers. The indices are used for purposes of: investment analysis, performance measurement, asset allocation and portfolio hedging. As they declare on their website, FTSE is a leader in the creation and management of over 120 000 equity, bond and alternative asset class indices. It has offices in London, Frankfurt, Hong Kong, Beijing, Shanghai, Madrid, Milan, Mumbai, Paris, New York, San Francisco, Sydney and Tokyo, and works with partners and clients in 77 countries worldwide. The FTSE4Good selection criteria were designed to reflect a broad consensus on what constitutes good corporate responsibility practice globally. The criteria originate from common themes of ten sets of declared principles. Using a widespread market consultation process, the criteria are regularly revised to ensure that they continue to reflect standards of responsible business practice, and developments in socially and environmentally responsible investment, as they evolve. Since the launch of the index series the environmental and human rights criteria have both been strengthened. FTSE's in-house Responsible Investment Unit undertakes an extensive global engagement programme with all FTSE4Good companies affected by new criteria requirements. Companies are made aware of the deadlines and what steps they have to take. The FTSE4Good inclusion criteria are intended to be challenging but achievable, in order to encourage companies to try to meet them. The data used in assessing companies' eligibility for inclusion in the FTSE4Good indices is independently researched by EIRIS and is available for scrutiny by companies and their advisers.

22.10.2 Carbon Trust

The Carbon Trust is a not-for-profit company whose mission is to accelerate the move to a low-carbon economy. It provides specialist support to help business and the public sector cut carbon emissions, save energy and commercialise low-carbon technologies. By stimulating low-carbon action, it strives to contribute to key UK goals of lower carbon emissions, the development of low-carbon businesses, increased energy security and associated jobs.

It is working with industry and academia to accelerate the development and deployment of low-carbon technologies. The Carbon Trust declares that it targets support where it can make the biggest difference. As a consequence it runs customised projects for particular low-carbon technologies. The Trust's technology accelerators aim at opening markets for low-carbon technologies, while its research challenges are about commercialising promising technologies which have not yet entered the market.

The Trust's technology accelerators aim to accelerate deployment of technology sectors that are on a path towards full commercialisation but where there are significant barriers to uptake.

It identifies and addresses market and regulatory barriers to deployment, and this often involves large-scale monitored demonstrations and widespread dissemination of findings to industry and government. An example of a technology accelerator is the Offshore Wind Accelerator, declared to be a ground-breaking research and development initiative whose aim is to reduce the cost of energy by 10%. The initiative has its roots in the UK's desire to achieve the 15% renewable energy target set for the UK by the EU. The Trust calculates that 40% of electricity must come from renewables by 2020. Currently, the figure is just 5%, so an eightfold increase is required. To achieve this, mass-deployment of offshore wind will be required. The Trust believes that it has the potential to supply 25% of the UK's electricity by 2020.

In the 2006 report *CTC610 - Policy frameworks for renewables*, the Carbon Trust concluded that offshore wind power has the greatest potential to deliver renewable electricity power by 2020 in the UK. Now with the step change implied by the EU target, this study builds on the Carbon Trust's knowledge and experience in offshore wind to assess:

- how much offshore wind power capacity could reasonably be required to reach the 2020 renewable energy target;
- what would be required to deliver this, cost effectively and to the maximum benefit of the UK;
- what the UK government, industry and other stakeholders should do to achieve the above.

The extent of industry transformation and the long timescales demand a strategic perspective. The Carbon Trust worked together with the strategy consultancy the Boston Consulting Group (BCG) and commissioned new analyses from technical consultancies. The study draws these together with interviews with leading industry and government stakeholders into a cohesive set of insights and recommendations. The study demonstrates that the UK will need to build 29 GW of offshore wind capacity by 2020. Whilst this represents a challenge similar in scale to developing North Sea oil and gas, it is considered technically feasible. Given the amount of investment and public support required, the government has a major role making it possible, minimising costs to the consumer and maximising the economic benefit to the UK. This study has been developed with strong collaboration from both government and industry.

22.10.3 Public Pressure

Within its report on ethical consumerism entitled "Ten Years of Ethical Consumerism: 1999-2008", the Co-operative Bank describes the growth in expenditure on ethical goods and services. As the title of the report suggests, it has been produced for ten consecutive years and acts as a barometer of ethical spending in the UK. In this report, ethical consumerism is defined as personal allocation of funds, including consumption and investment, where choice has been informed by a particular issue – be it human rights, social justice, the environment or animal welfare. The key message of the report is that expenditure on ethical goods and services has grown almost threefold in the period examined. Overall the ethical market in the UK was worth £36 billion in 2008, compared to £13.5 billion in 1999. Whilst most sectors have outstripped the market, which has seen overall consumer spending increase by 58% in the ten-year period, Fairtrade[18] in particular enjoyed significant success with sales up 30-fold. The data shows that sales of energy-efficient electrical appliances and boilers, which had grown 12 and 9 times

[18] Fairtrade is a market-based social movement and approach that aims to improve the trading conditions for producers in developing countries and promote long-term prosperity and sustainability. The movement advocates the payment of a higher price to producers

respectively, had also seen exceptional growth, while the mature financial services market had seen ethical banking and investments triple over the course of the decade. The average spend per household on ethical products and services reached £735 in 2008. Of this, spending on products and services to tackle climate change reached £251 per household, a tenfold increase over 1999. Total expenditure on environmentally friendly products and services such as energy-efficient appliances, green energy and carbon offsetting reached £6417 million. However, critically this still represents less than 1% of total household expenditure. Although this report shows that the idea of ethical purchasing is now well established among many consumers, there is still a long way to go in collectively tackling climate change. The growth in energy efficient products such as boilers, white goods and, more recently, lightbulbs has been underpinned by government intervention. The Co-op advises in its report that in order for the UK to reduce its carbon emissions by 30% by 2020 there will need to be a step-change in take-up of low-carbon technologies, and this will need a new contract between business, government and the consumer.

22.11 ENVIRONMENTAL SUSTAINABILITY

"Sustainable" and "sustainability" are now key trigger words in the world of advertising for positive, emotive images associated with words such as "green", "wholesome", "justice", "goodness" and "environment". They are used with sophistication to sell holidays, cars and even lifestyles. Businesses are capitalising to affect a progressive groundswell towards sustainability. Marketing is indirectly reflecting national and international initiatives to preserve natural resources and the environment we collectively enjoy. The catalyst for these initiatives was the United Nations Earth Summit, where the term "sustainability" came into popular use. The term is used to recognise that social, economic and environmental issues are linked and must be addressed together, rather than in the fragmented way they are often dealt with. It was perhaps the first time countries around the world admitted that our way of life was not sustainable in the long term. With growing concern over environmental pollution and long-term sustainable use of environmental resources, the global community had come together to develop objectives to progress sustainable development. This resulted in Agenda 21, the Rio Declaration on Environment and Development and the Statement of Principles for the Sustainable Management of Forests, being developed and adopted by more than 178 governments at the United Nations Conference on Environment and Development (UNCED) held in Rio de Janeiro in 1992. Agenda 21 is a plan of action addressing the human impact on the environment to be implemented globally, nationally and locally by organisations and governments associated with the United Nations. Agenda 21 was followed by the Commission on Sustainable Development (CSD) created in December 1992 to monitor and report on implementation of the agreements at the local, national, regional and international levels. Progress was reviewed in 1997 at the United Nations General Assembly meeting and a full implementation of Agenda 21, the Programme for Further Implementation of Agenda 21 and the Commitments of the Rio principles were reaffirmed at the World Summit on Sustainable Development (WSSD) held in Johannesburg, South Africa, from 26 August to 4 September 2002.

---

as well as the creation of higher social and environmental standards. It focuses in particular on exports from developing countries to developed countries.

An important issue in today's business environment is that a firm must be seen to be "green" among the local community, customers, potential customers and stakeholders in the business. A lack of attention to environmental and sustainability issues will pose a risk to potential growth. Unfortunately, the majority of people believe that ecological (green) marketing refers solely to the promotion or advertising of products with environmental characteristics. Terms such as "recyclable", "refillable", "reusable", "environmentally friendly" and "ozone friendly" arc some of the expressions that consumers most often associate with green marketing. While these terms are green marketing claims, in general green marketing is a much broader concept, one that can be applied to consumer goods, industrial goods and even services. Thus, green marketing incorporates a broad range of activities, including product modification, changes of the production process, packaging changes, as well as modifying advertising. Ottman (2000) suggests the following green marketing strategies in order to get the message across:

1. Adopt a thorough approach to corporate greening. This includes all functions of the business. Everything from being energy efficient to introducing environmentally friendly fuel.
2. Appoint a highly visible chief executive officer (CEO) with environmental leanings and make him/her the centrepiece of your corporate social image (e.g. Anita Roddick of the Body Shop).
3. Be transparent. Allow stakeholders access to information so that they know exactly what the level of potential health risk associated with various projects is.
4. Work cooperatively with third parties, such as government agencies and environmental pressure groups.
5. Vigorously communicate your company's commitment to accountability and continuous improvement. This can include "cause-related marketing". For example, the UK supermarket giant Tesco works with a different charity every year.
6. Act now. Do not wait to get the green message across.

22.12 SUMMARY

This chapter has examined energy sources, renewable energy and current energy consumption. The world is running short of traditional sources of energy and the development of renewable energy particularly in emerging economies such as China and India is too slow. The chapter described global initiatives to address greenhouse emissions and global warming. It looked at the stimulation to control adverse impacts on the environment in the UK. The creation of the FTSE4Good Index was described, which ranks corporate performance in terms of attention to the environment, human rights and social issues. Additionally, it looked at the creation of the Carbon Trust, a government-funded company, to move the UK to a low-carbon economy. Words like "sustainability" are now key trigger words in the world of advertising for positive, emotive images associated with words such as "green" and "environment" and are used with sophistication to sell holidays, cars and consumables. Businesses are capitalising to effect on a progressive groundswell towards sustainability. There are a number of diverse sources of both risk and opportunity from the environment. Energy supplies and costs are a threat, whereas the development of renewable energy sources is an opportunity. Levies on pollution control are a threat, whereas the proactive management of production and product composition can lead to brand development and competitive advantage.

22.13 REFERENCES

BP (2011) *BP Energy Outlook 2030*, January. BP, London.

IEA (2008) *World Energy Outlook 2008*. International Energy Agency, Paris. http://www.iea.org/weo/2008.asp

Klinger, P. (2005) Energy costs and EU outlook sour Tate & Lyle's outlook. *The Times*, Friday, 30 September 2005.

Mackintosh, J. and Milne, R. (2005) Hybrid makers yearn to fuse green credentials with profit. *Financial Times*, Tuesday, 20 September 2005.

Mallet, V. (2005) Bad smells and smog from over the border. *Financial Times*, Tuesday, 20 September 2005.

Ottman, J. (2000) *In Business*, 22, 6.

Pearce, F. (2004) Kyoto Protocol is just the beginning. *New Scientist* print edition, 10 October, www.newscientist.cm/article.

Reid, T. (2005) US in plan to bypass Kyoto climate control. *The Times*, Thursday, 28 July 2005.

Reinhardt, F. (2001) Tensions in the environment. In J. Pickford (ed.) *Mastering Risk Volume 1: Concepts*. Financial Times, Harlow.

Walsh, C. (2004) Emissions impossible for the CBI to stomach. *Observer*, Sunday, 21 March 2004.

23
Legal Risk

It seems to be a law of nature, inflexible and inexorable, that those who will not risk cannot win.
(John Paul Jones)

The previous chapter examined environmental risk management, the second of six classes of risk exposure that businesses face relating to business operating environment within the risk taxonomy proposed in Chapter 9. This chapter examines the third of the macro influences of the business operating environment, *legal risk*, and the legal context of a business. Businesses do not operate in a vacuum and by their nature must engage with other businesses to exploit opportunities. The activities of business organisations are subject to a wide range of legal liabilities and obligations. Legal liability describes a situation where a person is legally responsible for a *breach of an obligation* imposed by the law. Such obligations may arise from the operation of either civil or criminal law. The activities of business are subject to a wide range of potential liability. *Contractual liability* arises when two or more persons enter into a legally enforceable agreement with each other, *tortuous liability* consists of the breach of a duty imposed by the law and *criminal liability* arises from committing a crime as defined by criminal law. The legal system can be both a source of risk exposure (e.g. breach of contractual obligations) and risk mitigation (e.g. by the use of patents). The purpose of this chapter is to draw out the categories of legal risk that a business is exposed to for the purpose of developing a risk taxonomy and understanding the sources of risk. It is not intended to be a comprehensive guide to business law, and readers seeking specific guidance on such matters as statutes, case law, legal services and contract law should refer to the standard reference texts and, where appropriate, seek legal opinion. The structure of this chapter is shown in Figure 23.1. The next chapter examines the fourth of the business operating environment risk categories, political risk.

23.1 DEFINITION OF LEGAL RISK

What is legal risk? Legal risk for a business may be defined as the risk of failing to: operate within the law, be aware of its legal obligations, honour contractual commitments, agree remedies for compensation with a supplier in the event of default, show evidence that it has operated within the law, or recognise and effectively manage legal threats.

23.2 SCOPE OF LEGAL RISK

The sources of risk considered to be embraced within the term "legal risk" are very considerable. Legal risk for businesses may be considered to include, but is not limited to:

- breach of environmental legislation (see Chapter 22);
- inaccurate listing information in terms of misstatements, material omissions or misleading opinions;
- breach of copyright;
- loss of business as a result of senior management time being "lost" through a protracted legal dispute;

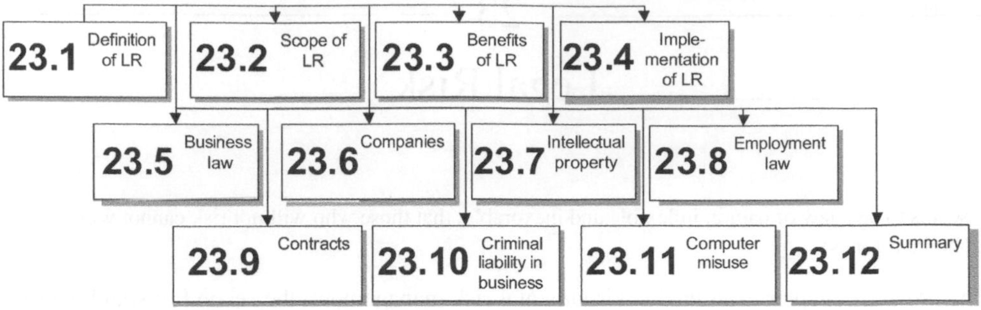

Figure 23.1 Structure of Chapter 23

- prosecution for breach of the law;
- legal disputes with overseas trading partners arising from, say, a lack of appreciation of the difference between the local laws and English law;
- loss of reputation as a result of a prosecution or a dispute with a customer, partner or supplier;
- lost legal disputes through poor record keeping.

23.3 BENEFITS OF LEGAL RISK MANAGEMENT

Legal risk management affords a business benefits as it:

- reduces the amount of management and external support time involved in legal disputes;
- provides for greater contractual, regulatory and statutory compliance;
- reduces the risk of reputational damage;
- promotes a more thorough review of contracts engaged in at home and overseas (where appropriate).

23.4 IMPLEMENTATION OF LEGAL RISK MANAGEMENT

The development of a sound system of risk management will depend on a number of issues such as:

- understanding the legal framework within which companies operate;
- having legal representatives review major contracts before completion;
- maintaining legal representation;
- ensuring annual reports and accounts are accurate;
- ensuring compliance with copyright, trademark and patent law;
- ensuring compliance with client confidentiality requirements;
- reviewing current product law prior to the release of new products into existing and new markets;
- maintaining systems and processes which adhere to employment law;
- ensuring employees are aware of the laws that they have to adhere to in the fulfilment of their role and duties, and providing training and monitoring where required;
- providing effective legal defence against challenges;
- understanding the requirements of legal discovery and impact of changes in technology.

23.5 BUSINESS LAW

An understanding of the sources of legal risk arising from business activities requires examination of the basic features of the English legal system. It should be noted that English law refers to the law as it applies to England and Wales. Scotland and Northern Ireland have their own legal systems. This chapter is exclusively concerned with English law.

It is common practice to divide English law into categories. These categories are used because they are convenient divisions of the law. The categories are not black and white. Categories may overlap and there are sometimes differences of opinion as to the category into which some areas of the law fall. The primary categories of law may be considered to be public and private law. Their sub-elements are shown in Figure 23.2.

Public law is concerned with the relationship between the state and its citizens. This comprises three key areas:

- *Constitutional* law is concerned with the workings of the British constitution. It concerns such matters as the position of the Crown, the functioning of central and local government, the composition and procedures of Parliament, citizenship and the civil liberties of individual citizens.
- *Administrative* law is concerned with the resolution of complaints made by individuals against the decisions of administering agencies, such as government agencies dealing with child benefit, income support and state pensions.
- *Criminal* law relates to kinds of wrongdoing which pose such a serious threat to the good order of society that they are considered crimes against the whole community. The criminal law makes such anti-social behaviour an offence against the state and offenders are liable to punishment. The state accepts responsibility for the detection, prosecution and punishment of offenders.

Private law is primarily concerned with the rights and duties of individuals towards each other. The legal process is commenced by the aggrieved citizen and is often contrasted with criminal law.

Another major distinction can be drawn between civil and criminal law. *Civil* actions are brought by an individual (known as the plaintiff), who is seeking compensation for the loss he/she has suffered as a result of the actions of the defendant, and the case is heard in a civil court. The plaintiff will be successful if he/she can prove his/her case on the balance of

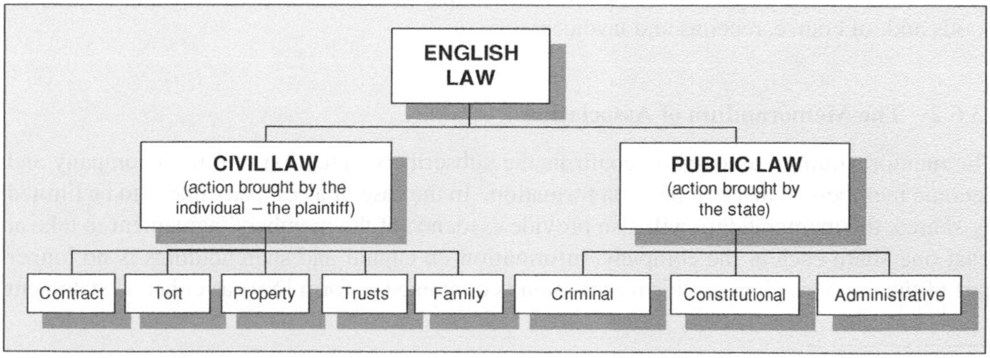

Figure 23.2 Division between public and private law

probabilities – that is, the evidence weighs more in favour of the plaintiff than of the defendant. If the plaintiff wins his/her action the defendant is said to be liable. If damages are awarded as a result of a successful civil action, they are payable to the plaintiff and are generally assessed on the basis that they should compensate and not punish. A plaintiff is not required to commence a civil action and he/she can discontinue it at any time before judgement. Many of the laws affecting businesses and business people are part of the civil law, especially property, contract and tort law.

In contrast, *criminal* cases, which are called prosecutions, are normally initiated by the state but may be brought by a private citizen; however, this is rare. If a prosecution is successful the accused or defendant is liable to punishment. This affords no direct benefit to the victim of the crime since he/she does not receive fines paid or the fruits of a criminal's labours in prison. Although punishment does not compensate victims, it is now possible for the criminal courts to make reparation payable directly to his/her victim under the Powers of Criminal Courts Act 1973.

23.6 COMPANIES

23.6.1 The Company Name

The Companies Act 2006[1] includes a system for controlling the names and business names of companies (from 1 October 2009 this Act replaced the Business Names Act 1985). A private company, whether limited by shares or guarantee, must end its name with the word "limited". A public company must end its name with the words "public limited company" or "plc". Private companies limited by guarantee may apply for exemption to omit the word "limited" from the name. Automatic exemption can be granted if the relevant conditions are satisfied. A name will not be accepted by the Registrar if it is the same as one that already exists on the index of names. Similar names can be registered. A name will not be registered if it is, in the opinion of the Secretary of State, offensive, its publication would be a criminal offence, it suggests a connection with Her Majesty's government or it includes a sensitive word or expression which may mislead the public. A name which includes "bank" in its title has to have the approval of the Financial Services Authority (until it is replaced by another body). For the use of "charity" or "charitable" the approval of the Charity Commission Direct is required. The Act requires that the company's full name must be shown in an obvious place and in readable form outside the registered office, on all places of business and on all business letters, business e-mails, websites, notices, official publications, cheques, orders for money or goods and, of course, receipts and invoices.

23.6.2 The Memorandum of Association

The memorandum of association confirms the subscribers' intention to form a company and become members of that company on formation. In the case of a company that is to be limited by shares, the memorandum will also provide evidence of the members' agreement to take at least one share each in the company. Information on capital and shareholdings is no longer part of the memorandum, as this information is contained in form IN01 as either a "statement

[1] Reference should also be made to the Company and Business Names (Miscellaneous) Regulations 2009 and the Company and Business Names (Sensitive Words and Expressions) Regulations 2009.

of capital and shareholdings" or, for those companies limited by guarantee, a "statement of guarantee". Once the company has been incorporated, the memorandum will no longer affect the ongoing operation of the company and it cannot be amended. It will become, to a large extent, a historical document.

23.6.3 Articles of Association

A company's articles of association are its internal rulebook. Every company formed under the Companies Act 2006 will have articles of association – commonly referred to simply as "articles". The articles are chosen by the members and form a contract between the company and its members with the aim of ensuring the company's business runs as smoothly and efficiently as possible. They will set out how the company will make decisions and include various matters connected with the shares. Every company is required to have articles by law and the articles are legally binding on the company and all of its members. A company may draft its own articles or adopt model articles (which it may amend or leave unchanged). Articles specifically prepared must be printed, divided into paragraphs numbered consecutively and signed by each subscriber to the memorandum in the presence of at least one witness. The articles together with the memorandum when registered are a contract which binds the company and the members as if signed and sealed by each member. The members are bound to the company by the provisions of the articles. The company may alter or add to its articles by a special (or written) resolution subject to certain restrictions.

23.6.4 Financing the Company

The share capital of a company may be divided into preference and ordinary shares. In addition, both of these classes of share may be issued as redeemable by the company at a future date. Holders of preference shares have the right to payment of a fixed dividend (e.g. 10% of the nominal value) before any dividend is paid on the other shares. However, there is no right to such dividend unless the company has sufficient distributable profits to pay it. This is why preference shares differ from loan capital. Interest on loan capital must be paid whether the company has distributable profits or not. Ordinary shares rank for dividend behind preference shares, and also sometimes the terms of issue provide that the preference shares shall have the right to claim repayment on capital before the ordinary shares if the company is wound up. Ordinary shares therefore carry most risk. Generally they have most of the voting rights in general meetings and therefore control the company. Trading companies have an implied power to borrow and "charge" their assets as security for a loan, that is, to give the lender a right to appoint, for example, a receiver to sell the company's assets in order to repay the loan if the company does not otherwise pay it. The memorandum usually gives an express power to borrow and details of the extent to which the company can charge its assets as security. A debenture is a written document (in the form of a deed), which forms the evidence that a lender has made a loan to a company. It is a type of fixed-interest security, issued by companies (as borrowers) in return for medium and long-term investment of funds. A debenture is evidence of the borrower's debt to the lender. Those who subscribe for the debenture stock receive a stock certificate rather like a share certificate. The word debenture derives from the Latin *debeo*, meaning "I owe". Debentures are issued to the general public through a prospectus and are secured by a trust deed which spells out the terms and conditions of the fundraising and the rights of the debenture holders. Debenture stock is found where the loan is to come from the public.

23.6.5 The Issue of Shares and Debentures

In accordance with the Companies Act the directors of public and private companies cannot issue shares without the express authority of the members. This is usually given by the members by ordinary resolution at a general meeting of the company. The authority may be given for a particular allotment of shares or it may be a general power (limited for a maximum period of five years).

23.6.6 The Official Listing of Securities

The Financial and Services Act 2000 gives express liability to those responsible for the listing particulars for misstatements, material omissions and misleading opinions. The Act makes statutory only "financial condition" information. This is information which investors and their professional advisers would reasonably require in order to make an informed assessment of the company's financial position, its assets and liabilities and prospects. The remedy given is for persons suffering loss to sue for financial compensation. As regards civil claims, the Act sets out who can be responsible for all or some of the listing particulars. These include the issuing company and its directors and anyone who expressly takes responsibility for part or parts of the particulars, such as an expert who authorised the contents of the particulars or part of them. However, this does not preclude a plaintiff from suing for fraud or misrepresentation under the Misinterpretation Act 1967.

23.6.7 The Remedy of Rescission

The main remedy for loss resulting from a misstatement in a prospectus is damages based either on breach of a statutory duty under the Financial Services Act 1986 (or the Public Offers of Securities Regulations 1995) or the Misinterpretation Act 1967, or at common law citing case law which lays down the principles of liability for negligent misstatements. The remedy of rescission involves taking the name of the shareholder off the register of members and returning money paid to the company by him/her.

23.6.8 Protection of Minority Interests

To fully explain the protection of minority interests the case law that needs to be referred to is beyond the scope of this book. However, it is appropriate to state here that a minority with a small shareholding has redress through the courts where they believe their interests are not being protected. One such instance is fraud on the minority, "fraud" in this context meaning some sort of improper behaviour by the majority which amounts to an abuse of their voting control. There are two primary examples. The first is where the minority itself may be defrauded, for example where directors propose to use their powers (or those of the company) for improper use; then a minority shareholder can bring a representative claim, asking that the transaction in question be set aside or stopped. The second is where the company is defrauded. Actions here are of an entirely different nature. The member sues to put right a wrong done to the company and not (except indirectly) a wrong to its members.

23.6.9 Duties of Directors

The relationship between a company and its directors is that of principal and agent, and as agents the directors stand in a fiduciary relationship to their principal, the company. In addition, directors owe a duty of care at common law not to act negligently in managing the company's affairs. Examples of fiduciary duties include the requirement for directors to use their powers for the proper purpose, that is, for the benefit of the company. Directors must not take secret profits and benefit from the company. There is nothing in the Companies Act which sets out the standard of skill and care which a director must bring to his work, and hence where remedies are sought reference is made to case law. Non-executive directors without business qualifications or experience are only required to "do their best". Non-executive directors with relevant qualifications and/or experience in business must exercise such reasonable skill and care as may be expected from a person of his/her professional standing and/or experience. Executive directors such as finance directors are normally employed for their expertise in company matters as stated under their contracts of service. There is commonly an implied term in the contract of service that the director will exercise the reasonable skill and care which a person in his or her position ought to have.

23.7 INTELLECTUAL PROPERTY

Most advanced industrial economies are progressively becoming "knowledge based" so that questions of intellectual property rights are becoming ever more important. "Intellectual property" is the term used to refer to a product or process that is marketable and profitable because it is unique. The value of intellectual property can quickly be destroyed unless businesses enforce their rights in this area. A clear risk to any business is that it does not use the law to protect its intellectual property, which can be an extremely valuable business asset. This uniqueness is protected by *patent law*, which gives protection to technological inventions. The law of *copyright* protects, for instance, rights in literacy, musical and artistic works. The law of *trademarks* and *service marks* protects the use of a particular mark if it is used in trade. The law relating to *registered designs* protects articles that are mass-produced but distinguished from others by a registered design which appears upon them. The law also protects those in business from competitors who maliciously belittle their products or who pass off their products as those of another business. There is also some protection for businesses with regard to, for instance, the use by employees of confidential information. The primary legislation addressing these issues is to be found in the Copyright, Designs and Patents Act 1988 (as amended) and the Trade Marks Act 1994.

23.7.1 Patents

Patents are a form of risk response planning. Patents are a powerful way of protecting market share (for a limited time) and have a direct impact on bottom-line performance. While in existence they are a form of monopoly, but when they expire market share can be severely curtailed. On the announcement in 2002 that a US court had overturned patents on one of its top-selling drugs, the share price of UK drugs giant GlaxoSmithKline (GSK) slumped to a two-year low. The US court had ruled that patents on GSK's antibiotic Augmentin were invalid, clearing the way for competitors to produce generic versions of the drug. At the time Augmentin attracted global sales of $2 billion (in 2001) and accounted for about 7% of GSK's annual sales.[2] In October 2004 the New York-based company Pfizer, one of the world's largest drugs companies (with a significant presence in England), warned the market that its profits could be affected by the expiry in 2005 of patents on four of its most popular medicines and the introduction of competitor drugs: the anti-fungal drug Diflucan, epilepsy treatment Neurontin, antibiotic Zithromax and hypertension treatment Accupril. The four drugs at the time had a combined sales value of more than $5 billion.[3] Patents are granted to inventors by government and give inventors the right for a limited period to stop others from making, using or selling their inventions without permission. When a patent is granted, the invention becomes the property of the inventor, which – like any other form of property or business asset – can be bought, sold, rented or hired. It also brings the right to take legal action against others who might be infringing the invention and to claim damages. UK patents last for up to 20 years and are available from the UK Patent Office. A patent empowers the owner (proprietor) of an invention to take legal action against others to prevent the unlicensed manufacture, use, importation or sale of the patented invention. This right can be used to give the proprietor breathing space to develop a business based on the invention, or another person or company may be allowed to exploit the invention and pay royalties under a licensing agreement. Patents are generally intended to cover products or processes that possess or contain new functional or technical aspects; patents are therefore concerned with, for example, how things work, what they do, how they do it, what they are made of or how they are made.[4] The Patent Office has found that the vast majority of patents are for incremental improvements in known technology, with innovation being "evolution" rather than "revolution". The only way to get protection for an invention is to apply for a patent. Patents are generally regarded as legal weaponry rather than publications of information.

Application

An application for a patent can be made by or on behalf of the inventor of a new process or device and the grant of a patent will be made to the inventor or to any person who is entitled to it, for instance in the situation where the inventor has sold the idea before patenting it. Before an application is made for a patent there are some fundamental rules that have to be followed in almost all countries which if broken, even in ignorance of the consequences, may land the applicant in an irretrievable situation. So, most importantly an applicant must not tell anyone about the idea before the patent application is filed. This may appear impractical as the idea does have to be assessed as to whether or not it is commercially sound, whether or not it can be manufactured and, if so, whether it can be sold for an acceptable price. However, somehow, the applicant has to find out these things without actually telling the person they are speaking to what it is the applicant wants to produce. If it is felt necessary to share the idea with someone, that someone must not disclose it to anyone else without the express permission of the applicant. Ideally the person with whom the idea is being shared should sign a confidentiality agreement. This is particularly important if the applicant is talking about a commercial contact to a potential business colleague. The applicant can talk to a chartered patent agent in confidence because all chartered patent agents work under strict rules of confidentiality. The reason for confidentiality is that, to be patentable, an invention must be

[2] BBC News (2002) Drugs giant hit by patent woes. 24 May. http://news.bbc.co.uk
[3] BBC News (2004) Patents set to hit Pfizer profits. 20 October. http://news.bbc.co.uk
[4] http://www.patent.gov.uk

novel. An invention is not novel if it has been made available to the public (any individual) before the date of filing the patent application. If the applicant shows their idea to a potential manufacturer without a confidentiality agreement in place, the novelty of the idea is destroyed and the applicant is no longer entitled to a patent for it. This allows the manufacturer to make the product itself without any acknowledgement of the applicant's contribution, financial or otherwise.

The next issue an applicant has to address for the idea to be patentable is an inventive step. This means that the invention must not be obvious to someone who is "skilled in the art" to which the invention relates. Patent agents advise whether something is inventive over what is already known by experience. The UK Patent Office will not give advice of this type as it does not provide this sort of function. To establish and decide whether an idea is inventive, it is necessary to find out what has been done before in the relevant field. Professional searchers can be instructed to look through earlier patents and patent applications, but this will inevitably be expensive and will probably generate documents in several different languages, which then have to be deciphered. Other sources of information are the internet and the British Library.

What Can be Patented?

It should be assumed that every bright idea can be the subject of a patent. When application is made for a patent four essential criteria must be met: (1) it must be shown that the applicant has an invention, (2) the invention must not be excluded (see below), (3) it must be something new (not published, made public in the world anywhere previously, or be something that would be obvious to lots of people) and (4) it must be capable of industrial application. The term "industry" is meant here to be interpreted in its broadest sense as anything distinct from purely intellectual or aesthetic activity. It does not have to imply the use of a machine or the manufacture of an article.

Exclusions

Under the Patents Act 1977 (and as amended by the Patents Act 2004), certain items cannot be protected by patent. Among these are discoveries – something you found out about but did not invent. This might include a scientific theory or mathematical method, a mental process, artistic or aesthetic creation, or a method of presentation of information or doing business.

Registration

An application for a patent can be made by or on behalf of the inventor of a new process or device and the granting of the patent will be made to the inventor or to any person who is entitled to it, in the instance where the inventor has sold the idea before patenting it. In the UK application is made to the Patent Office in London or Newport, Gwent. The Patent Office is part of the Department of Trade and Industry and it deals with the granting of patents, registered trademarks and registered designs. James Dyson remarks that applicants need either to put in a lot of hard work and research themselves, or need to pay a professional patent agent to do some of the work for them. Additionally, he says, using a patent agent does increase the cost of getting a patent quite dramatically, but you can then be sure of getting good advice and a well-drafted patent at the end of the day. On the robustness of patents Dyson says: "if you are going to use your patent as a legal weapon, surely you'd like it to be as sharp as possible?".[5] The Dual Cyclone™ was nearly never made due to patent and legal costs. Unlike a songwriter who owns the song he writes, an inventor has to pay substantial fees to renew his patents each year. During the development years when James Dyson had no income, this nearly bankrupted him. He risked everything, and fortunately the risk paid off.

Infringement

A UK patent will in general be infringed by making, using or selling something which is subject to the patent without the owner's consent. Patents are territorial rights; a UK patent will only give the holder rights within the UK and rights to stop others from importing the patented products into the UK. Hence, a French competitor could legally make the invention in France unless there was a French patent. Additionally, the goods produced in France could also legally be exported to any other country where there was no patent, though not of course to the UK. Infringement of a patent is the subject of civil rather than the criminal law and actions for an injunction, damages or an account of profits are brought in the Patents Court which is part of the Chancery Division of the High Court. Once a patent is issued, the patentee must enforce the patent without the aid of the Department of Trade and Industry. An example of infringement was the dispute between Dyson and its rival Hoover. In 1999, Hoover tried to imitate a Dyson and James Dyson was forced to take the matter to court to protect his invention. Dyson's patent relates to the basic principle of putting two cyclones of increasing efficiency into a vacuum cleaner to increase the overall separation efficiency of the cleaner. Any cleaner that incorporates the features of the patent will fall within the scope of the patent.

Inventions can relate to methods (like methods of manufacture or industrial processes) as well as to products or parts of products. James Dyson, who invented the bagless vacuum cleaner technology, told the press at the time of settlement: "when we discovered that Hoover had stolen one of our patents, we tried to settle the case to avoid a long and expensive court battle, but Hoover refused. So we were forced to defend our patent against Hoover's infringement." Furfthermore, he said: "I hope it encourages inventors who have their ideas stolen by multinational companies to fight for their patent rights."[6] Dyson settled the dispute by accepting a £4 million ($6.3 million) damages offer from Hoover. It followed a High Court ruling two years previously that Hoover's Triple Vortex cleaner infringed Dyson's patent for its Dual Cyclone™ vacuum cleaner. Dyson said the settlement was bigger than any previous court award in British patent case history. The company said it had offered Hoover the opportunity to settle the claim for just over £1 million before the case went to the High Court. Hoover took the case to the Court of Appeal but lost, and it was refused the right to appeal to the House of Lords.

Patents in the US

Patent legislation in the US is similar to that in the UK. The authority in the US for the granting of a property right to the inventor and issuing a patent for an invention is the United States Patent and Trademark Office (USPTO). Generally, in the US the term of a new patent is 20 years from the date on which the application for the patent was filed or, in special cases,

[5] http://www.dyson.co.uk/invent/

[6] BBC, 3 October 2002.

from the date an earlier related application was filed, subject to the payment of maintenance fees. United States patent grants are effective only within the US, US territories and US possessions. The right conferred by the patent grant is stated by the grant as being "the right to exclude others from making, using, offering for sale, or selling" the invention in the US or "importing" the invention into the US. What is granted is not the right to make, use, offer for sale, sell or import, but the right to exclude others from making, using, offering for sale, selling or importing the invention. The patent law specifies the subject matter for which a patent may be obtained and the conditions for patentability. The law establishes the USPTO to administer the law relating to the granting of patents and contains various other provisions relating to patents. The USTPO website states that the US Constitution gives Congress the power to enact laws relating to patents, in Article I, section 8, which reads: "Congress shall have power ... to promote the progress of science and useful arts, by securing for limited times to authors and inventors the exclusive right to their respective writings and discoveries." Under this power Congress has from time to time enacted various laws relating to patents.

The first patent law was enacted in 1790. The patent laws underwent a general revision which was enacted on 19 July 1952, and which came into effect on 1 January 1953. It is codified in Title 35, United States Code. Additionally, on 29 November 1999, Congress enacted the American Inventors Protection Act 1999, which further revised the patent laws.[7] Infringement disputes in the US can involve considerable sums of money. In April 2004 the software company Microsoft agreed to pay $440 million (£240 million) to California-based InterTrust Technologies (a maker of software to protect online delivery and payment for films and music) to settle a legal row over its use of anti-piracy patents.[8] InterTrust, which is part-owned by Sony and Philips, sued Microsoft in 2001 after licensing talks collapsed. The company accused Microsoft's new Windows XP program of infringing its patents and demanded damages that, at the time, were expected to run into billions of dollars. The agreement came one week after Microsoft paid $1.6 billion to a rival, Sun Microsystems, to end another patents battle.

23.7.2 Copyright

Copyright is a form of protection provided to the authors of "original works of authorship" including original literary works (e.g. novels, instruction manuals, computer programs, lyrics for songs, articles in newspapers, some types of databases), original dramatic, musical and artistic works (e.g. paintings, engravings, photographs, sculptures, collages, works of architecture, technical drawings, diagrams, maps, logos); published editions of works, sound recordings (which may be recordings on any medium), films, including videos; and broadcasts. The Copyright, Designs and Patents Act 1988 generally gives the owner of copyright the exclusive right to reproduce the copyrighted work, to prepare derivative works, to distribute copies of the copyrighted work, to perform the copyrighted work publicly, or to display the copyrighted work publicly. The copyright protects the form of expression rather than the subject matter of the writing. Copyright does not protect ideas. For example, a description of a machine could be copyrighted, but this would only prevent others from copying the description; it would not prevent others from writing a description of their own or from making and using the machine. The Act does not require the owner of a copyright to register it or to follow any formalities with particular regard to it.

[7] http://www.uspto.gov/

[8] BBC News (2004) Microsoft settles patents case. 12 April. http://news.bbc.co.uk

Ownership and Duration

The author of the work is the owner of the copyright. However, there are circumstances where, for instance, copyright features as an issue in the contract between an employer and a management consultant. Copyright of work carried out by the employee on behalf of the employer normally resides with the employer. The relevant rules on duration are now contained in statutory instrument SI 1995/3297, Duration of Copyright and Rights in Performance Regulations 1995, passed in order to harmonise UK law with that of the European Union. The Regulations increase the basic term of copyright in literacy, dramatic, musical and artistic works. This raises the former provision of the present life of the author plus 50 years after his death to life plus 70 years.

Infringement

The person infringing the copyright will usually have copied from the work and an action can be brought for an injunction and/or damages or for a share of the profits made from the wrongful use of copyrighted work. There are no statutory defences to copyright infringement. However, there are common law defences. The basis of common law defence and case law is beyond the scope of this book. Under the Copyright, Designs and Patents Act 1988, authors are given certain moral rights which exist quite independently of copyright. It provides protection alongside a copyright and is useful for an author who has sold the copyright to someone else. The right of paternity includes the right to be identified as the author and must be claimed by the author. The right of integrity enables the author to object to changes to his or her work by way of additions, deletions, alterations or adaptation which amount to what is termed mutilation or distortion of the work. The right of attribution gives a person (such as an author) the right to prevent a work which he has not produced being attributed to him/her. An author whose moral rights have been infringed is entitled to an injunction and damages.

23.7.3 Designs

A design right works similarly to a patent but involves the look, colouring, shape, texture and/or materials associated with a product. Hence, a design refers only to the features of shape or pattern applied to an article by an industrial process which is judged solely by looking at the article, such as the shape of a Coca-Cola bottle. It is now possible to register the Coca-Cola bottle as a trade mark under the Trade Marks Act 1994.

Registration

Designs may be registered at the Patent Office (Designs Registry) under the Registered Designs Act 1949 (as amended by the 1988 Act). Registration gives the owner of the design protection for five years, and this can be extended for four further periods of five years on payment of four further fees every five years, making 25 years in all.

Infringement

The registered design owner's rights for infringement are to sue the person responsible for damages and/or an injunction or an account of profits made from the wrongful use of the design or an order for the surrender of the infringing copies.

23.8 EMPLOYMENT LAW

Businesses must comply with employment law in the employment of staff. Failure to do so can lead to prosecution. The ordinary principles of the law of contract apply. So in a contract of employment there must be an offer and an acceptance, which is in effect the agreement. There must also be the intention to create legal relations, payment terms, together with proper consent by the parties. That is, there is no mistake, misinterpretation, duress or undue influence. In addition, the contract must not be illegal. A contract does not require any written formalities and can be made orally. However, certain written particulars of it are required to be given to the employee by the Employment Rights Act 1996. These particulars must be given to all employees within two months of starting work unless the employee has entered into a written contract with the employer containing all the relevant terms. The particulars are required to contain minimum information, with which we are all familiar. Additionally, an employer must prepare and revise when necessary a statement of his or her policy in regard to the health and safety of his or her employees and arrangements for carrying out the policy in accordance with the Health and Safety at Work Act 1974. An employer has other duties in terms of remuneration, holiday pay, sick pay, time for antenatal care, maternity leave (under certain conditions), dismissal procedures and other obligations under the prevailing legislation. Employment law is now very extensive and made all the more complex by the European Union. Businesses are at risk if employment law is not understood and adhered to.

23.9 CONTRACTS

This section is concerned with the legal framework governing the supply of goods and services.

23.9.1 Essentials of a Valid Contract

The essential ingredients of a contract are as follows:

- *Legality*. The purpose of the agreement must not be illegal or contrary to public policy.
- *Agreement*. An agreement is formed when one party accepts the offer of another.
- *Consideration*. The parties must show that their agreement is part of a bargain; each side must promise to give or do something for the other.
- *Intention*. The parties must have intended their agreement to have legal consequences.
- *Capacity*. The parties must be legally capable of entering into a contract.
- *Genuineness of consent*. The agreement must have been freely entered into.
- *Formalities*. In some cases, certain formalities must be observed.

A contract which possesses all of these requirements is said to be valid. If one of the parties fails to fulfil his/her promises, he/she may be sued for breach of contract. The absence of an essential element will render the contract void, voidable or unenforceable.

23.9.2 Types of Contract

Contracts may be divided into two broad categories. Speciality contracts and simple contracts. Specialty contracts require that the signature of the person making the deed must be witnessed and attested. It must be clear on the face of the document that it is intended to be a deed.

Simple contracts are not deeds or informal contracts and may be made in any way – orally, in writing or they may be implied from conduct.

23.10 CRIMINAL LIABILITY IN BUSINESS

Suppliers are subject to extensive criminal controls over their activities. Criminal law affects the supplier of goods and services, with regard to:

- misdescriptions of goods and services;
- misleading price indications about goods and services;
- safety of consumer goods;
- safety and quality of food.

23.10.1 Misdescriptions of Goods and Services

The Trade Descriptions Act 1968 prohibits the use of certain false trade descriptions by a person acting in the course of a trade or business. The main offences created by the Act are:

- applying a false trade description to any goods or supplying goods to which a false trade description is applied;
- knowingly or recklessly making a false statement in respect of the provision of services, accommodation or facilities.

False Trade Description of Goods

- *Strict liability*. Prosecution must normally establish two essential requirements, first, a prohibited act has been committed (*actus reus*), and second, the act was intentional and hence the person was of guilty mind (*mens rea*). However, some offences do not require *mens rea* for a prosecution. These are known as crimes of strict liability.
- *In the course of trade or business*. An offence can only be committed by a person acting in the course of a trade or business.
- *False trade description*. The Act defines a trade description as "an indication, direct or indirect, and by whatever means given" which includes such matters as quantity, size, gauge, method of manufacture and composition.
- *Applying a false trade description*. The ways in which a description may be applied to any goods is set out in the Act and includes labels, packaging, oral statements and advertisements.
- *Supplying and offering to supply goods to which a false trade description is applied*. To avoid the problem that the display of goods in a shop window or on a supermarket shelf is not an offer to sell in the contractual sense, the Act provides that "a person exposing goods for supply or having goods in possession for supply shall be deemed to offer to supply them".
- *Disclaimers*. Case law has provided that "for a disclaimer to be effective it must be as bold, precise and compelling as the trade description itself and must effectively be brought to the attention of any person to whom the goods may be supplied".

False Trade Description of Services

The Act states that it is an offence for any person in the course of any trade or business to make a statement which he knows to be false or recklessly to make a statement which is false with regard to the following issues: the provision of any services, accommodation or facilities; the nature of any services; the time at which, manner in which or persons by whom such services etc. are provided; the examination, approval or evaluation by any person of any such services etc.; and the location or amenities of any accommodation so provided.

The following points should be noted about the offence of false trade description of services:

- *Requirement of mens rea*. This subject is covered by two sections of the Act. First, the prosecution must show that the defendant made the statement knowing it to be false. Second, the prosecution must show that the defendant made the statement recklessly.
- *In the course of any trade or business*. As set out in the Act, the offence cannot be committed by a private individual.
- *Statements*. The Act regards as significant statements about existing facts, but not statements which are effectively promises about the future.
- *Services, accommodation and facilities*. The Act does not define these terms and hence interpretation can only be accomplished by studying case law. Services may be dry cleaning, accommodation might be a hotel, and facilities might be a car park.

23.10.2 Misleading Price Indications

Controls over false and misleading statements regarding prices are now contained in the Consumer Protection Act 1987 (as opposed to the Trade Descriptions Act 1968). Section 20(1) of the Act states that a person will be regarded as guilty of an offence, if in the course of any business of his or her, he or she gives (by any means whatsoever) to any consumers an indication which is misleading as to the price at which any goods, services, accommodation or facilities are available.

The following points should be noted about the offence of misleading price indicators:

- *Consumers*. The Act defines "consumers" as anyone who might want the goods, services, accommodation or facilities other than for business purposes.
- *Price*. The definition of "price" contained in the Act covers the total payable as well as any method of determining the total amount.
- *Misleading*. The Act sets out a list of circumstances in which the prices or the methods of determining the price will be considered misleading.
- *Services or facilities*. A list of items included in the definition provided by the Act is as follows: credit, banking or insurance services; the purchase or sale of foreign currency; the supply of electricity; off-street car parking; and holiday caravan parks. Services and facilities specifically excluded are those provided by an employee to his/her employer, by an authorised person or appointed representative under the Financial Services Act, and facilities for a residential caravan park.
- *Accommodation*. Part III of the CPA applies to short-term accommodation such as hotels and holiday flats and new freehold homes for sale and homes on lease for more than 21 years. Homes for rent are not covered. Fees charged by estate agents are also covered.

- *In the course of any business of his*. An offence can only be committed by a person acting in the course of a business of his or hers. Hence, this wording limits the scope of the offence to the owners of a business. Employees cannot be made liable for misleading price indications, even when acting in the course of their employment.

23.10.3 Product Safety

The legal framework for dealing with the problem of unsafe products is contained in the General Product Safety Regulations 1994 and Part II of the Consumer Protection Act 1987. The General Product Safety Regulations 1994 implement the provisions of the EC Directive on General Product Safety. They impose requirements concerning the safety of products intended for consumers or likely to be used by consumers where such products are placed on the market by producers or supplied by distributors. The following points should be noted.

- *Scope of the Regulations*. The Regulations apply to products intended or likely to be used for consumer use which have been supplied in the course of a commercial activity. A consumer is a person who is not acting in the course of a commercial activity. A commercial activity is defined as any business or trade. The Regulations apply whether the products are new, used or reconditioned. Products used exclusively in the context of a commercial activity, even if for or by a consumer, are not the subject of the Regulations. The Regulations do not apply to the following types of products: second-hand products which are antiques; products supplied for repair or reconditioning before use; products that are subject to specific provisions of EC law covering all aspects of their safety; and products that are subject to specific provisions of EC law which cover an aspect of safety.
- *General safety requirement*. The Regulations provide that a producer may not place a product on the market unless it is a safe product. It is an offence to fail to comply with the general safety requirement.
- *Safe product*. The Regulations set out what is meant by a "safe product". A product will be safe if under normal or reasonably foreseeable conditions of use (including duration) there is no risk or the risk has been reduced to a minimum. Any risk must be compatible with the product's use, considered acceptable and consistent with a high level of health and safety protection.
- *Producer*. A "producer" is defined by the Regulations as a manufacturer established in the EC; where the manufacturer is not established in the EC, a representative or the importer of the product; or other professionals in the supply chain but only to the extent that their activities might affect the safety of the product.
- *Information requirements*. A producer is required by the Regulations to provide consumers with information so that they can assess inherent risks and take precautions. The duty only arises where the risks are not immediately apparent without adequate warnings.
- *Duty of distributors*. A distributor must act with due care to help producers comply with the general safety requirement. In particular, a distributor will commit an offence if he or she supplies dangerous products.
- *Defence of due diligence*. It is a defence for a person accused of an offence under the Regulations to show that he or she took all reasonable steps and exercised all due diligence to avoid committing the offence.

- *By-pass provision*. The Regulations provide a by-pass provision to enable the prosecution of the person, in the course of a commercial activity of his or hers, whose act or default causes another to commit an offence.
- *Enforcement of penalties*. The Regulations are enforced by the weights and measures authorities in Great Britain except in relation to food, in which case enforcement is the responsibility of food authorities.

23.11 COMPUTER MISUSE

Businesses are protected against computer misuse to a degree, in that legislation such as the Computer Misuse Act 1990, which came into force on 29 August 1990, acts as a deterrent, but to bring a prosecution is sometimes like closing the stable door after the horse has bolted. The damage to customer loyalty, reputation, relations with business partners or income is already done. Prosecution provides little solace. Additionally, computer misuse is now a global problem with problems of "hacking" or virus infections being initiated beyond our shores and thus beyond our laws.

23.11.1 Unauthorised Access to Computer Material

It is an offence to knowingly cause a computer to perform a function with the intent to secure unauthorised access to programs or data held in a computer. This basic offence is designed to criminalise the activities of both external "hackers" who obtain access to computers using the public telecommunications system and internal business employees who knowingly exceed the limits of their authority to use a computer (such as a disgruntled employee wishing to harm his/her employer or an employee wishing to obtain information which would be useful when joining a new employer). The offence is triable summarily and is punishable by a maximum of six months' imprisonment or a fine of £5000. A fine of this level is trivial when considering the level of harm that can be inflicted on a business.

23.11.2 Unauthorised Access with Intent to Commit or Facilitate Further Offences

It is an offence triable either by magistrates or in the Crown Court to commit an unauthorised access offence with the intent to commit or facilitate the commission of any serious offence, the sentence for which is fixed by law or where the maximum sentence could be five years or more. These serious crimes would include theft and blackmail. This offence would cover the situation where a "hacker" obtains unauthorised access to a computer system with the intent of hijacking funds in the course of an electronic funds transfer. The maximum penalty for this offence if convicted on indictment is five years' imprisonment or an unlimited fine.

23.11.3 Unauthorised Modification of Computer Material

It is an offence to internationally cause the unauthorised modification of the contents of any computer with the intent to impair a computer's operation or to prevent or hinder access to any program or data held in a computer or impair the operation of such a program or the reliability of data. This offence is designed to cover interference with computer programs and data such as the deletion or alteration of material or the introduction of computer viruses. The offence is triable either by magistrates or in the Crown Court where the maximum penalty is five years' imprisonment and an unlimited fine.

23.12 SUMMARY

This chapter has examined some of the sources of legal risk that a business may be exposed to. The treatment has not attempted to be exhaustive, but to look at those issues that are common to most businesses. It has examined the division between public and private law and the headings under which any action might be bought. It has looked at aspects of the Companies Act, together with administrative issues that can expose businesses to litigation. With such a focus on corporate governance in recent years there has been renewed attention to the accuracy of listing particulars and the information contained in annual reviews. The chapter has examined the aspects of intellectual property in terms of patents, copyright and designs. While patents and copyright in particular can be useful protections to secure competitive advantage, if sometimes for only a limited period, they can also be a source of risk if infringed. Additionally, businesses must adequately address employment law, contracts and criminal liability. Now more relevant than ever, due to their ubiquity, is the management of computers, the information they hold and compliance with the Computer Misuse Act 1990.

24
Political Risk

The message is that there are known knowns; there are things we know that we know. There are known unknowns; that is to say there are some things that we now know that we don't know. But there are also unknown unknowns – the ones we don't know we don't know.
(Donald Rumsfeld)

The previous chapter examined legal risk management, the third of six classes of risk exposure that businesses face relating to the business operating environment within the risk taxonomy proposed in Chapter 9. This chapter examines the fourth of the macro influences of the business operating environment, *political risk*. Nearly all businesses that venture overseas face political risk in one form or another. The political environment of overseas markets will always play a key role in shaping the threats and opportunities of businesses seeking geographical expansion. Typically, political risk for a business has involved a single government. However, that is changing. While businesses may seek to expand into a single nation state (e.g. France) this may also entail them operating within a supra-national body, which comprises collections of nation states (e.g. the European Union). Decisions within both types of political entity have a major impact on the prospects of businesses achieving the performance targets they have set themselves. In addition, political factors that may impact on the business operating environment may well alter according to the "colour" of the incumbent government. While it may be a sweeping generalisation, it may be argued that right-wing parties (e.g. the Conservatives in the UK, Republicans in the US) traditionally favour free enterprise and market forces, while left-wing parties (Labour in the UK, Democrats in the US) traditionally favour tighter control over businesses and some government intervention to provide essential services for those that cannot afford them. However, the political risk profile of an individual state or states may vary considerably over time. A dramatic example is the events that unfolded in North Africa and the Middle East in 2010 and 2011. Triggered in North Africa in December 2010, there was a revolutionary wave of demonstrations and protests which ultimately were of historical consequence in Tunisia, Egypt and Libya. There were also major protests in Algeria, Bahrain, Iran, Iraq, Jordan, Oman, Yemen and Syria, with minor incidents in countries such as Saudi Arabia and Sudan. The protests shared techniques of civil resistance in sustained campaigns involving strikes, demonstrations, rallies and marches, as well as the use of social media such as Facebook and Twitter to organise, communicate and raise awareness. At the time of writing, two heads of state had been overthrown, Tunisia's on 14 January 2011 and Egypt's on 11 February 2011. Tunisia's president Zine El Abidine Ben Ali fled to Saudi Arabia and Egypt's President Mubarak resigned after 30 years of rule. Numerous factors led to the protests which some commentators have described as the refusal of youth to silently accept the previous status quo where dictatorship had been accompanied by human rights violations, government corruption, unemployment and poverty. This chapter examines the political context of business, reviews the various types of political risk and explores some of the techniques that might be used to mitigate them. The structure of the chapter is shown in Figure 24.1. The next chapter examines the fifth of the business operating environment risk categories, market risk.

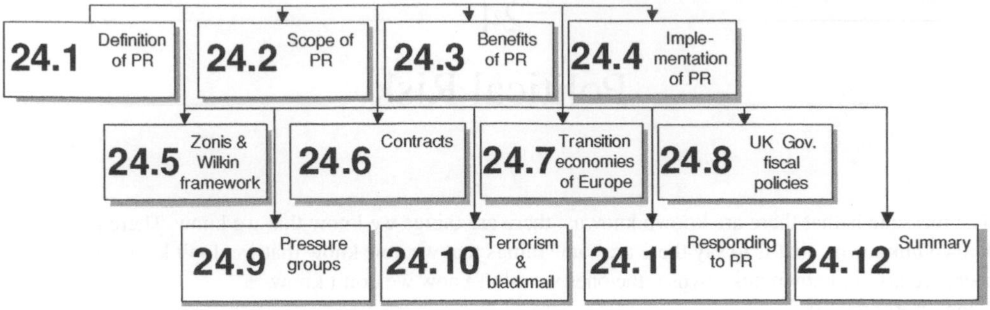

Figure 24.1 Structure of Chapter 24

24.1 DEFINITION OF POLITICAL RISK

What is political risk? The definition adopted here is that proposed by Zonis and Wilkin (2001): political risk is the uncertainty that stems, in whole or in part, from the exercise of power by governmental actors and the actions of non-governmental groups. This definition applies to both domestic and international markets, although it is more commonly associated with overseas exposure and, in particular, developing countries. Political risk can also be incurred through government inaction or direct action (of both national and local government). Inaction could be failure to issue permits as required, or government failure to enforce local legal provisions. Examples of direct action include the following: contract frustration, currency inconvertibility, tax laws, tariffs, expropriation of assets, or restriction in the repatriation of profits. The definition embraces sources of political risk such as political instability, politicised government policy and political violence. Political risk may also stem from increased credit risk if the government changes policies to make it difficult for the company to pay creditors.

24.2 SCOPE OF POLITICAL RISK

Two broad categories of political risk are often identified, *macropolitical* and *micropolitical* (Griffiths and Wall 2005). They might also be called country-specific and firm-specific political risk, as described below.

24.2.1 Macropolitical Risks

These potentially affect all businesses in a country. Threats may arise from dramatic actions such as terrorism, civil war, a *coup d'état* or military insurgence. Such risks may result in governments seizing the assets of the firm without compensation. However, the more common macropolitical risk specific to a country is the potential threat of adverse economic circumstances, leaving a business unsure of the security of a planned future investment, or, if a project has commenced, concerned over the outturn bottom-line performance. An example of an adverse threat is economic recession with less aggregate demand for a broad range of products. Similarly, higher general levels of inflation or taxation might adversely affect all businesses, as might escalating crime, labour disputes or the onset of a national recession.

24.2.2 Micropolitical Risks

These only affect specific firms, industries or types of venture. Such risks may take the form of new regulations or, say, taxation imposed on specific types of business in the country. However, this picture is changing. As Kobrin (1997) points out, "the 'new' political risk is likely to involve multiple companies and multiple governments where business partners are caught between three, four or five governments with different policy objectives and economic philosophies". In addition, he describes the emergence of an asymmetry between international business and international politics. The existence of this asymmetry arises from the emerging mode of business organisation and the nature of politics. While politics is still organised geographically in terms of territory and borders, business organisation is not confined by national boundaries. Companies such as IBM, Siemens and Toshiba have formed an alliance to pool knowledge to survive in a market place where competition is fierce and technologies are developing rapidly. As a consequence, organisation centres, hierarchies and borders lose relevance. Without clarity over the hierarchy of offices (headquarter or subsidiary) or the nationality of the business (is it American, German or Japanese?), what government can impose what controls? How is taxation resolved? If such alliances become commonplace, in the future political risk will be more complex.

Box 24.1 gives examples of both macro-and micropolitical risk. It is not an exhaustive list but does provide an indication of some of the common sources of risk. Each geographical region will have its own unique risk profile.

24.3 BENEFITS OF POLITICAL RISK MANAGEMENT

Political risk management affords a business benefits as it:

- provides a proactive systematic and methodical approach to the evaluation of alternative investment opportunities based on analysis of different geographical markets set in different political contexts;
- provides another tool with which to examine return on investment;
- supports more rational decision taking between competing choices;
- produces concrete mitigation actions to reduce investment exposure;
- contributes to a more holistic approach to risk management.

24.4 IMPLEMENTATION OF POLITICAL RISK MANAGEMENT

The development of a sound system of political risk management will depend on a number of issues such as:

- developing intelligence on the market the business wishes to penetrate;
- having a clear understanding of the historical and social environment of the country in which investment is planned;
- understanding the support that can be obtained from UK government agencies;
- developing an understanding of the sources of political risk;
- building political risk management capabilities.

Box 24.1 Types of macro- and micropolitical risk and their impact on businesses

	Government	Impact on businesses
Macropolitical	• Recession • Inflation or hyperinflation • Military insurgence, *coup d'état*, civil wars or other politically motivated violence	• Loss of sales • Higher operating costs • Lost sales • Disruption to production • Increased security costs • Lower productivity • Difficult staff retention
	• Campaigns against foreign goods	• Loss of sales • Reduction in market share • Increased public relations costs to recapture the market
	• Product contamination	• Loss of sales • Erosion of reputation
	• Bureaucratic incompetence • Change in government resulting in new fiscal policies	• Increased operating costs • Increased taxation
	• Corruption • Government policies on credit payments • Increased taxation • New or revised/more stringent legislation	• Increased operating costs • Difficulty in payment of creditors • Lower after-tax profits • Increased costs of production • Protracted approvals • Delays in getting product to market
	• Expropriation • Confiscation • Nationalisation of industry or project • Terrorism or kidnappings • Currency devaluations/ depreciation	• Loss of sales • Loss of sales and future profits • Economic loss • Disrupted production • Increased managerial costs • Increased security costs • Lower productivity
	• Currency revaluation/ appreciation • Restriction on repatriation of profits • Breach of contract/contract frustration	• Reduced valuation of repatriated earnings • Less competitive in overseas markets and in competing against imports in home market • Loss of profit • Lower productivity • Loss of sales
Micropolitical	• Industry-specific taxation • Tariffs and quotas • Politically motivated violence (e.g. against petrochemical or pharmaceutical industry)	• Lower after-tax profits • Volumes • Disruption to production • Disruption to research and development • Higher security costs • Staff retention difficulties

Source: Based on Griffths and Wall (2005).

24.5 ZONIS AND WILKIN POLITICAL RISK FRAMEWORK

This section describes the political risk framework proposed by Zonis and Wilkin (2001). The authors state that the key to developing a proactive, broadly focused political risk management strategy is to adopt a comprehensive and systematic view of the *factors driving political risk*. The authors believe this approach will allow a business to identify the exact position of problems, assess where improvements can most readily be made and lay out an action plan. They go on to say that the factors that drive political risk can be broken down into three basic areas: (1) external drivers such as political instability and poor public policy; (2) interaction drivers based on the relationships between the company and external actors; and (3) internal drivers such as the quality of the company's political risk management process. The drivers of political risk identified by the authors are included in Box 24.2. Zonis and Wilkin advise that a company can use a framework such as the one that they propose to identify the vital drivers behind the political risks it faces. The authors say that the deployment of such a framework can be made the responsibility of the either the company's political risk or government relations department or a management committee specifically convened by a top executive. More importantly, Zonis and Wilkin quite rightly express the view that having identified political risks (an action which is obviously an important part of the risk management jigsaw), managers should turn their attention to building risk management capabilities. These capabilities it is thought should include risk management policies, business processes, organisation, human capital, methodologies, reports, systems and data. The authors clearly consider that political risk management can affect business performance: "A company with a superb capability in political risk management can defuse risks before they escalate. A company with little capability can turn a low risk investment into a high risk one (particularly when the company provokes regulators to intervene)."

Box 24.2 Zonis and Wilkin political risk framework

External drivers

External drivers of political risk can be subdivided into several categories. The classic drivers are incidents of political instability (such as riots and coups) and poor public policy (such as hyperinflation and currency crises). These types of political risk attract the headlines (as in Indonesia, Russia and Brazil) and often dominate the attention of the political risk managers. However, an external driver of political risk that is frequently overlooked is a weak institutional framework. For direct investors, drivers in this area (such as corrupt regulatory agencies and ineffective legal systems) can be more critical than headline generating political and economic developments. Weak political institutions pose the threat of a "death of a thousand cuts" such as that suffered by the power plant in India. (In the early 1990s several US energy companies entered India. The country suffered no major economic crisis or political instability, but by 2000, a majority of the power generation projects in the country had been abandoned. One US company was forced to obtain 150 bureaucratic clearances for its project and the state government renegotiated its power-purchase agreement four times. In addition, politicians took the company to court on charges of corruption. The company spent some $27 million in legal and administrative fees before ultimately pulling out, after seven years of delay.) Weak political institutions include such hazards as failing legal systems, biased regulatory systems and the inability of the government to provide expected services (such as infrastructure and services). In most cases the company cannot influence these drivers of risk (it cannot make the host country more stable, or alter its basic macroeconomic policy decisions). Hence, the company needs to focus on assessing these risks accurately and managing their impacts (e.g. by buying political risk insurance).

Interaction drivers

Interaction drivers can be split into categories based on the relationship involved. Companies typically have many relationships that influence political risk levels. The most common include relationships with home-country and host-country governments, with local governments in the host country and with regulators. If these relationships turn sour, political risk levels will increase.

Other important relationships include those with local communities and the labour force. Community demands can lead to government intervention. (A Canada-based energy company decided to undertake a major expansion of a power project to produce new energy supplies for Canada and for export to the state of New York in the US. However, the project soon fell foul of both Indian tribes and environmentalist non-governmental organisations (NGOs). Both these groups waged a sophisticated media campaign against the project. The company attempted to respond with a public relations campaign of its own, but was unsuccessful. Eventually the New York Power Authority pulled out, depriving the project of 30% of its expected revenues, leaving the Canadian company with substantial sunk costs on its books.)

Unions. One of the most common reasons for unwanted host government intervention is labour issues. Several academic studies have found that companies with well-organised large labour forces experience higher levels of political risk. Labour unions can be extremely effective in seeking government action.

NGOs. As a result of the internet, information about a company's far-flung investment activities can be disseminated rapidly and easily, and multiple NGOs can effectively coordinate their action. When NGOs are involved a company's value at risk in an investment project exceeds the project itself, since NGOs can affect a company's global reputation as well as consumer behaviour.

Shareholders. When shareholders obtain information about foreign political events from the media, they are likely to respond to negative events by hammering the company's share price. It is therefore critical for the company to present its political risk management strategy to shareholders in a compelling fashion, before political risk crises hit, and to follow this with status updates when problems do occur.

Interaction-based drivers of risk differ from external drivers in that the quality of the company's relationship with external drivers strongly affects risk levels. Hence, the company can influence both the probability and the impacts of these political risks.

Internal drivers

Organisation. Problems in this area can have profound effects on the company's ability to deal successfully with political risks. For instance, there is often difficulty aligning management incentives with the political risk management goals of the company as a whole.

Information. Information problems have been difficult to overcome and one typical problem is that internally produced risk assessments were too lengthy or abstract to be useful. They were not company-specific and, even when correct, managers did not believe them and failed to act on their recommendations.

Policy. Policy-related risks include problems with managers failing to obey risk management policies or problems with the absence or incompleteness of these policies.

Techniques. Risk relating to the use of specific risk management techniques arises when companies select the wrong type of risk management strategy (as when they purchase political risk insurance coverage that does not completely cover their exposure).

Human capital risks. These include problems that arise when staff are not qualified to deal with political problems (such as when project managers attempt to avoid dealing with the host government, "stick to business" and fail to cultivate a positive relationship).

24.6 CONTRACTS

Companies that contract with either foreign governments or private entities located abroad, for either the purchase or supply of goods or services, often face a number of political risks that threaten the profitability of the transaction. Events to forestall the fulfilment of a contract can occur before or after shipment or delivery of the contracted goods, and before, during or after completion of the contracted services. Included below are the more common types of prevailing contract risk events:

- contract frustration prior to the shipment or delivery of goods;
- outbreak of a new war or civil war preventing the contract from being completed;
- unilateral contract cancellation by the other party to the agreement, where the party is a government entity;
- cancellation of legally granted import or export licences, or implementation of laws preventing import or export of goods;
- contract frustration after shipment or delivery of goods;
- failure of the government foreign exchange authority to transfer the transaction amount in the contract currency to the exporter, although the private local buyer has deposited the agreed-upon payment in equivalent local currency with the foreign exchange authority;
- payment default by a buyer that is a government entity;
- where contract disputes have arisen, a government buyer or bank does not honour an arbitration judgement in favour of the exporter in accordance with the arbitration procedure outlined in the transaction contract.

24.7 TRANSITION ECONOMIES OF EUROPE

While the appeal to invest in transition economies may be strong due to the emerging markets, such investments are prone to considerable political risk. This exposure exists despite the fact that transition economies seek to make foreign direct investment (FDI) from businesses from advanced market economies attractive. Their motivation stems from the desire to secure the transfer of both technology and management techniques, as well as capital. Japanese

FDI into Britain, for instance, brought with it management techniques such as just-in-time, which spread widely through UK businesses, improving efficiency. To appreciate the political risk it is necessary to understand both the starting point and the evolutionary path that the transition economies are travelling. Since 1989 some 28 independent states of central and eastern Europe and the Commonwealth of Independent States have embarked on a transition from "Marx to the market". While these states differ markedly in many economic respects (such as GDP per capita, population size and structure of the economy) they all share a legacy of a planned economic system. For Russia, for instance, the planning process determined what was produced, how production was organised and the distribution of production. The results of the transition to the market for the different states have been mixed. This has been due in no small measure to the enormity and complexity of the task of turning a centrally planned economy into a market economy. The key characteristics of centrally planned economies were self-sufficiency (in order to be independent of the capitalist economies); collective public ownership, with economic decisions being made centrally by the state; production by state-owned enterprises (SOEs); coordination of the activities of SOEs through input/output analysis (raw material sources and output); limited international trade (which was secondary to domestic production); and the banking system.

While communism guaranteed full employment and a reasonable standard of living for its citizens, there were a number of problems with the centrally planned economy. There were no rewards for managers in SOEs to produce output efficiencies. The planners gathered resource requirement information from the managers of SOEs. The SOEs would put in bids for more inputs than they required and bargain for production targets which were easier to achieve than proposed by the planners. Incentives and penalties were weakly linked to performance and tended to encourage just satisfying output targets. Planned targets were expressed in volume or weight, with little reference to quality. Workers were not individually incentivised; they were often given bonuses relating to total production. This led to workers obtaining a "free ride". They obtained their bonus without even achieving the basic output.

In a market economy high prices indicate a shortage, which is addressed by competition. In a planned economy scarce resources are indicated by queues. In summary, planned economies suffered from numerous problems and inefficiencies such as chronic shortages of consumer goods, repressed inflation due to underpricing of scarce goods and services, hidden unemployment due to overstaffing and the guarantee of full employment, and soft budget constraints caused by stringent financial control of SOEs. The transition process led to a rapidly declining domestic market, a significant increase in imports, inflation, unemployment and declining government expenditure due to a dramatic fall in tax revenues. It is against this backdrop that Western businesses have invested in business development in emerging markets. The key issues are uncertainty over fiscal policies, worker mentality, productivity rates, instability in the price of inputs, inflation, poor infrastructure and low domestic demand.

24.8 UK GOVERNMENT FISCAL POLICY

The UK government influences the economy to accomplish its four main aims: full employment, favourable balance of payments, stable prices and economic growth. The government strives to accomplish these aims by affecting total demand in the economy through the implementation of policies. Fiscal policy relates to changes in taxation and government spending. Taxes can be direct (taken directly from income, e.g. income tax or corporation tax), indirect (placed on goods and services such as VAT) and progressive (such as taking a higher proportion of income as income rises). Government spending includes social security, health, education, defence, public order and safety, housing and community amenities, transportation and communication.

24.9 PRESSURE GROUPS

Business development overseas may be exposed to pressure groups. These are organised groups of people with similar interests who attempt to influence others, notably governments and large businesses. They range in size from international organisations such as Friends of the Earth, Greenpeace and Amnesty International to small community groups concerned only with local matters, which exist just for the life of one particular issue. A pressure group's success largely depends on its level of financial, public and political support, as well as on its organisational ability. Depending on the planned business investment, in terms of whether it is unique or following a well-worn path, reaction from the more well-known pressure groups can be predicted to a degree. However, it is the groups that form just to oppose or restrict a single development that can be harder or impossible to predict.

24.10 TERRORISM AND BLACKMAIL

Food and drink, cosmetics, pharmaceuticals and tobacco products are susceptible to product contamination, both accidental and by deliberate interference through malice or for financial or political gain. Such incidents are increasingly common and when they occur they attract media attention that can have a disastrous impact on the product or brand name. Companies can incur significant costs: from recalling and replacing the damaged product; through lost sales; from advertising costs to regain the public's confidence; and from rehabilitating the brand name. Mobile and permanent investments located in emerging markets are increasingly vulnerable to terrorist attacks, war or other forms of politically motivated violence, resulting in physical damage to these assets. Outbreaks of military conflict are occurring on a more frequent basis, not only in less developed countries but also in industrialised nations, such as the unrest in former Yugoslavia and Libya. Along with more prevalent incidents of war and civil war, acts of terrorism are also on the rise, with targets more often on transportation systems or in the corporate sector than in military or diplomatic locations.

The militant group known by the name Al Qaeda, recognised by most countries in the world to be a terrorist organisation, warrants special mention. It is widely believed that this group has attacked civilian and military targets in various countries, the best known being the attacks on New York City and Washington, DC, on 11 September 2001. According to media channels it seeks to rid Muslim countries of Western influence, get rid of what it views as "corrupt" regimes and set up a pan-Islamic caliphate. However, what exactly Al Qaeda is, or was, remains in dispute. There are opposing views. Is it now a mythical entity with only disconnected regional movements in places like Yemen and Somalia with no umbrella organisation? Or is it a strongly led integrated network based in Pakistan with a clearly defined direction? As reported in the press at the time, when asked about the possibility of Al Qaeda's connection to the London bombings on 7 July 2005, Metropolitan Police Commissioner Sir Ian Blair said: "Al Qaeda is not an organisation. Al Qaeda is a way of working . . . but this has the hallmark of that approach. Al Qaeda clearly has the ability to provide training, to provide expertise and I think that is what has occurred here." Al Qaeda inspired activity resurfaced again in December 2007 when the former Pakistani Prime Minister, Benazir Bhutto, was killed in a suicide attack at an election rally in Rawalpindi along with more than 150 others. The Pakistan government later said it had intelligence that suspected Al Qaeda leader Baitullah Mehsud was responsible. Al Qaeda may have evolved, but the ability of inspired groups to plan and direct attacks remains a fundamental unanswered question. Whether such attacks are planned directly by Al Qaeda's leadership or not, time has shown that the danger from terrorism remains global, with serious attacks across a wide geographical spread from London to Algeria and Pakistan. The coming years are unlikely to be any different.

24.11 RESPONDING TO POLITICAL RISK

A common criticism of political risk analysis is that it usually takes place too late when projects are already under way (Griffiths and Wall 2005). This is reinforced by Zonis and Wilkin (2001), who cite a common problem of businesses taking a reactive approach to political risk. They go on to say that such businesses usually end up with a political risk management strategy that emphasises damage control. Such strategies are unnecessarily expensive. Political risks are generally far easier to handle before they evolve into full-blown crises. Companies that rely on damage control strategies after some event are driven to employ experienced management consultants, solicitors and ex-diplomats to limit losses and protect shareholder value. However, such expenditure may be fruitless; once political decisions are made they are hard to reverse, as credibility and public confidence are at stake. In addition, there are no standard techniques such as control charts that can be used to minimise political risk (Eppen 2001). Management time and effort must be directed towards political risk at the initiation stage of projects, before taking the step to invest overseas. Approaches that businesses may use to ensure that this risk is minimised and appropriately assigned include the following:

- Undertaking proper planning and due diligence. Although there are a number of ways to protect a business against political risks, too many businesses begin operations in an unfamiliar country without having taken the time and devoting the resources necessary to ensure a better-than-average chance of success. Developing solid relations with relevant governing authorities is the preferred approach, but this may not always be possible.
- Investing in projects or entering into contracts where the host government has adopted policies that encourage private sector involvement, particularly where this involvement will provide risk mitigation and promote risk transfer. Long-standing policies provide comfort in that, once introduced, such measures are less likely to be repealed.
- Investing in projects where the host government has clear and unambiguous statements of government support for the type of investment being made, particularly if linked to some form of guarantee. Deep (2001) cites the example where for the $616 million Liabin B coal-fired plant, the first Chinese infrastructure project to be financed entirely with foreign capital, the concession agreement provided sweeping government guarantees. This included clauses that entitled the project to compensatory payments in the event of any significant changes in law, including tax and environmental regulations that might prevent the company from fulfilling its obligations to lenders. Any statements of government support will, ideally, contain within it indications of the underlying benefits that the government wishes to accrue from supporting these types of investment. This will help potential investors ascertain the degree to which individual schemes are compatible with national programmes or aspirations.

- Obtaining insurance against political risk. National export credit agencies provide this to their exporters within limits. In the UK this is the Export Credits Guarantee Department.[1] Political risk insurance (PRI) can be obtained on the private market, from such multinational companies as Marsh & McLennan, the Aon Corporation and Willis. There are now more PRI providers, with greater capabilities than ever before. Whether a business wants to take out general coverage (against expropriation, currency inconvertibility or political violence) or create coverage tailored to specific circumstances, it is highly probable that one or more of the private-sector PRI providers can meet a business's needs. Businesses must remember to pursue coverage before a problem occurs. After it happens, coverage will be difficult to obtain. More extensive cover can be obtained from multilateral development agencies such as the World Bank and their Multilateral Investment Guarantee Agency. PRI has become particularly important to lenders in the wake of the Asian financial crisis and sovereign defaults by Russia, Indonesia and Pakistan.
- Entering into "hedging contracts" to protect themselves against fluctuations in interest rates and currency exchange rates. These are financial devices used to reduce losses as a result of future price movements.
- Creating a risk-friendly investment environment by establishing a good relationship with the workforce. Too often, foreign businesses are perceived as having uncaring managers who do not appreciate their workers. This can have dire consequences. One of the best ways to protect a business's assets is to engender a loyal workforce. Management can be replaced much more easily than can a workforce.
- Incorporating strong arbitration language into contracts to address labour disputes.
- Enhancing on-site security to protect against terrorist attacks.
- Being attuned to what is happening in your host country. This may sound obvious, but it is easy to lose sight of the bigger political picture once immersed in operational issues. After an operating environment has changed, it is often too late to do anything about it. Remain engaged with your local embassy and chambers of commerce. A collective voice is more powerful than that of an individual firm, even if the firm has a solid relationship with governing authorities.

24.11.1 Assessing Political Risk Factors

As stated earlier in this section, there are no standard techniques such as control charts that can be used to minimise political risk. However, progressive companies tend to use tools such as decision analysis based on decision trees and scenario analysis to help them appreciate the risks associated with a given business opportunity. Additionally, with fast computing capabilities available, businesses can use mathematical models to help the decision-making process and reduce the impact of uncertainty. For investors, the nature of the political risk they are exposed to varies significantly depending on the type of investment they are engaged in and hence

[1] The Export Credits Guarantee Department is a government department based in London and Cardiff. Its mission is to benefit the UK economy by helping exporters of UK goods and services to win business, and UK firms to invest overseas. It provides guarantees, insurance and reinsurance against loss, taking account of the government's international policies. It helps exporting UK companies compete in overseas markets by arranging medium- to long-term finance facilities and credit insurance cover for contracts ranging from around £20 000 up to hundreds of millions of pounds. The ECGD provides cover for over 120 countries. When supporting finance, it does not act as a lender but provides repayment guarantees to UK lending banks. ECGD insurance cover provides protection to UK exporters against a number of the commercial and political risks that arise during the manufacturing and credit periods. Similarly, ECGD investment insurance cover protects UK investors against certain political risks involved in investing overseas. These insurance covers are particularly useful for exports to, and investments in, emerging markets.

whether they fall into the direct or portfolio category. In general, portfolio investors are more likely to be affected by macropolitical risks, such as a sudden increase in interest rates or unanticipated currency devaluation, while direct investors (investing in individual businesses) tend to be affected more by business-specific risks. These investors therefore need to focus on those political dynamics that affect the overall business environment in a host country. When assessing political stability, the focus should be on the legitimacy of state authority, the ability of that authority to impose and enforce decrees, the level of corruption that pervades the system of authority, and the degree of political fractionalisation that is present (Wagner 2000). For investors, effective political risk management requires distinguishing between developments that pose true risks (i.e. well-defined threats to business performance from political events) and media headline-grabbing events.

24.11.2 Prioritising Political Risk Factors

Once identified and assessed in terms of their probability and impact, political risks can be prioritised so that management energy is used in the most efficient way. A way of communicating the varying likely impact of political risks to a management team or board is through the use of a risk map (as discussed in Appendix 1) or a risk register, where the risks have been recorded in descending order. Visual representations, which permit information to be readily and quickly assimilated, are commonly preferred by boards to reduce board meeting preparation time.

24.11.3 Improving Relative Bargaining Power

In an attempt to overcome political risk, some businesses seek to develop a stronger bargaining position in the country within which they are operating. For instance, a business might attempt to create a situation in which the government of the country loses more than it gains by taking action against the interests of the business. This could be the case when the business has technical knowledge that will be lost to the country if the business moves to another country to avoid, say, new regulations. The extent of bargaining power may be improved if the business is as fully integrated as possible with the local economy so that it becomes part of the country's infrastructure. A good example is the car industry, which typically uses local labour, suppliers and subcomponent manufacturers. If the business chooses to relocate, the local economy is hit hard by the loss of employment, loss of employee spending and the reduction in sales for manufacturers and suppliers. Risk management techniques here may include: developing good relations with the host government and other local political groups; producing as much of the product locally as possible; creating joint ventures and hiring local people to manage and run the operation; carrying out extensive local research and development; and developing good employee relations with the local labour force (Griffiths and Wall 2005). These techniques draw to the attention of the host country the disbenefits to the economy of overactive (and unwelcome) interference in the business's activities.

24.12 SUMMARY

In a fully developed enterprise risk management process, political risk will be addressed as a primary source of risk and opportunity. The degree to which a business trades and, in particular, plans to expand its business activities overseas will dictate the importance that it will attach to political risk exposure, reviewing market expansion prospects and preparing responses to them.

There is always somewhere in the world that is going through significant change which impacts on existing and potential markets. Recent examples of this include the ongoing political crisis in Thailand which commenced in 2008 and emanated from a conflict between the political parties, the wave of pro-democracy uprisings which commenced in the Middle East and North Africa in December 2010 and became known as the Arab Spring, and austerity measures which provoked social unrest in Greece, Spain and Italy in 2011.

This chapter has reflected on two broad categories of political risk, namely macro- and micropolitical risk. The former affects all businesses across a country, while the latter impacts a particular industry or sector. Conducting business with foreign governments or private entities is exposed to a series of events leading to contract frustration. The three elements of the Zonis and Wilkin political risk framework were examined: external drivers, interaction drivers and internal drivers. The related experiences of US energy companies in India were very poignant. While there may be broad appeal to invest in transition economies, the legacy of worker ethics, poor infrastructure and the instability of the price of raw materials, all overlaid by uncertain fiscal policies, takes the shine off potential returns. It was acknowledged that any business venture may be resisted by pressure groups and, depending on the scale and nature of the venture, may attract the attention of organisations such as Greenpeace or Friends of the Earth. Most important of all, any response to political risk needs to be initiated early enough to afford the opportunity to be able to initiate one or a combination of responses so that they can make a difference. This entails ensuring that the risk is identified and assessed early and this all depends on good intelligence.

24.13 REFERENCES

Deep, A. (2001) A firm foundation for project finance. In J. Pickford (ed.), *Mastering Risk Volume 1: Concepts*. Financial Times, Harlow.

Eppen, G.D. (2001) Charting a course through the perils of production. In J. Pickford (ed.), *Mastering Risk Volume 1: Concepts*. Financial Times, Harlow.

Griffiths, A. and Wall, S. (2005) *Economics for Business and Management: A Student Text*. Pearson Education, Harlow.

Hunt, B. (2001) Issue of the moment: The rise and rise of risk management. In J. Pickford (ed.), *Mastering Risk Volume 1: Concepts*. Financial Times, Harlow.

Kobrin, S.J. (1997) Globalization and multinationals. In T. Dickson and G. Bickerstaffe (eds), *Financial Times Mastering Management*. Financial Times/Prentice Hall, London.

Wagner, D. (2000) Defining "political risk". International Risk Management Institute, http://www.irmi.com/expert/articles/2000/wagner10.aspx

Zonis, M. and Wilkin, S. (2001) Driving defensively through a minefield of political risk. In J. Pickford (ed.), *Mastering Risk Volume 1: Concepts*. Financial Times, Harlow.

25
Market Risk

The essence of competitiveness is liberated when we make people believe that what they think and do is important – and then get out of their way while they do it.
(Jack Welch)

The previous chapter examined political risk management, the fourth of six classes of risk exposure that businesses face relating to business operating environment within the risk taxonomy proposed in Chapter 9. This chapter examines *market risk*, the fifth of the macro influences of the business operating environment. If market risk is to be fully understood, it must be examined through a wide-angle lens to appreciate the overlap with all other classes of risk, as part of a holistic approach to enterprise risk management. There is an overlap, for instance, with technology risk discussed in Chapter 17, which examined subjects like mechatronics, which can lead to new products, product development and/or increases in market share. Economic risk, described in Chapter 21, overlaps with market risk in that demand is clearly influenced by such issues as government policy, employment levels and inflation. Additionally there is an overlap with social risk, discussed in Chapter 26, where changes in demographics, lifestyle, standard of living and disposable income all create changes in demand. An appreciation of market risk entails gaining an insight into the market structure within which a business operates to understand obvious issues such as the size of the market and the number of competitors, but also any barriers to entry, company-specific competencies required, bargaining power of suppliers, product diversification and so on. The rate of change is now so rapid in some markets such as media and electronics that proactive risk management is vital to stay in business, let alone retain market share. Market risk policies should take into account the nature and complexity of the business's activities, objectives, competitiveness, the regulatory environment, together with its staff and technological capabilities. Market risk is treated differently here than the approach adopted by the financial sector, as discussed in Section 25.1. The structure of this chapter is described in Figure 25.1.

25.1 DEFINITION OF MARKET RISK

What is market risk? The simplistic definition adopted here is as follows: market risk is the exposure to a potential loss arising from diminishing sales or margins due to changes in market conditions, outside of the control of the business. All businesses are exposed to some form of market risk. The level and source of market risk differs from industry to industry, and from company to company within the same industry. Market risk is multifaceted and has to do with market structure, the strategic direction adopted for market growth, price variation, price elasticity and the behaviour of suppliers and buyers. Barclays Bank PLC, in its 2009 annual report, defines market risk as "the risk that the Group's earnings or capital, or its ability to meet business objectives, will be adversely affected by changes in the level or volatility of market rates or prices such as interest rates, credit spreads, commodity prices, equity prices and foreign exchange rates". While different industries face specific forms of market risk,

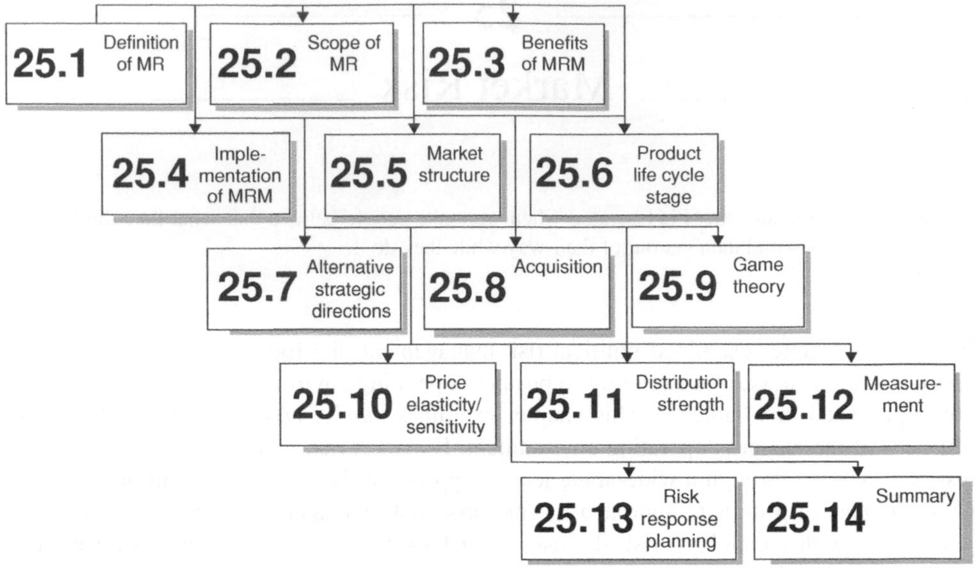

Figure 25.1 Structure of Chapter 25

there are some market risks that are faced by all companies – for example, erosion of market share, an increase in number of competitors, downturn in market size and substitute products. The definition of market risk offered here differs from other texts, which refer to market risk encompassing subjects such as interest rate, foreign exchange, equity and commodity risk. These risk types are considered here to be financial risks and are dealt with in Chapter 15. The financial sector in particular, it could be said, has a far more narrow and specific focus than, say, the software, telecommunications and electrical goods industries. Within the financial sector, market risk is defined as the risk of adverse changes of market prices of the trading portfolio during the period required to liquidate the transactions.[1] The definition adopted here leans towards the broad economist's definition, where goods and services are traded in a free market.[2]

25.2 SCOPE OF MARKET RISK

It is important for any business to recognise the extent of market risk in the environment and how it can be addressed. This naturally requires an understanding of the environment and its various components (see Figure 25.2). Each component has a direct impact on the welfare of a business. Demographic trends affect the size of the market, its location and, to a degree, the kind of goods and services required. The legal and political characteristics of the

[1] A transaction for a commercial bank is the collection of a deposit or the provision of a loan. All related transactions make up the "banking portfolio". For investment banks, which are in the business of market transactions within capital markets, transactions relate to derivates, foreign exchange and equity. These transactions make up the "market portfolio".

[2] A free market is one over which the central authorities exert no direct control. Buyers and sellers are free to arrive at any agreements on quantities to be traded and on prices at which trade will occur. A controlled market is one over which the central authorities exert some substantial, direct control by, for example, licensing buyers and sellers in the market, setting legal minimum or maximum prices at which trade can take place, or setting quotas controlling the amounts that individual buyers and sellers may trade in the market.

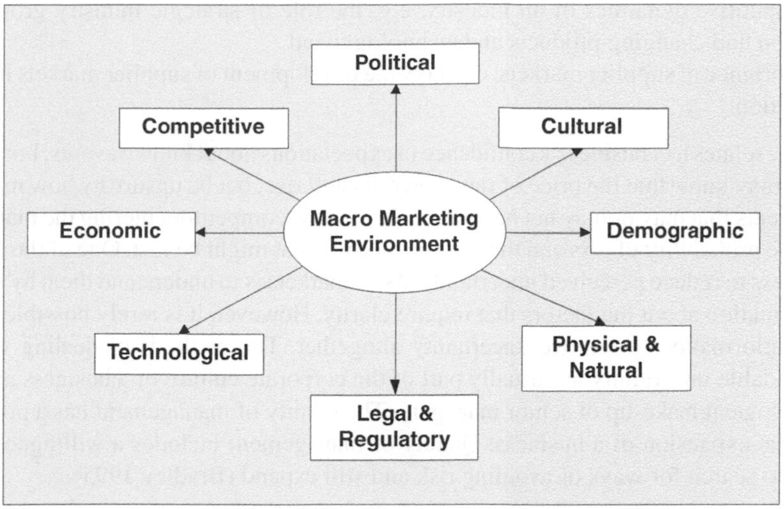

Figure 25.2 Sources of market risk and opportunity (Bradley 1995)

environment affect a business particularly with regard to the ability to participate in foreign markets and the ease with which foreign competitors are able to enter the domestic market. Changes in the economic environment affect marketing by way of the pattern of economic growth and movements in interest and exchange rates. Innovation and technological advances introduce standards for competition and opportunities for the marketing of new products and services. Competition can both limit and erode market share. Legislation introduced to protect the environment can raise the unit cost of production while at the same time creating new opportunities. It is necessary for any business to monitor the macro environment to ensure that an appropriate response is adopted at the micro level. Contextual analysis allows the business to respond to changes and cope with marketing uncertainty.

25.2.1 Levels of Uncertainty in the Marketing Environment

The marketing capability of a business will depend on its ability to cope with the elements of the external environment. The external environment includes the macro environment, the industry environment and the task environment. The business's macro environment consists of the general economic, social, legal, technological and cultural factors and their influence on business and industry. The industry environment refers to demand and competition in an industry, the system of distributors and suppliers and the network of relationships involving companies in several industries (Porter 1980). The task environment relates to the environment specific to the business such as the local labour market, transport infrastructure and supplier network. Monitoring, analysing and understanding the macro and industry environment is a key part of a business marketing capability and, according to Möller and Antilla (1987), includes the ability to understand:

- the complex influences affecting the industry;
- how demand arises in an industry, the role of primary customers, derived demand and the value of marketing channels;

- the competitive dynamics of an industry, e.g. the role of strategic industry groups, price formation and changing products and technology; and
- the importance of supplier markets, e.g. how the development of supplier markets influences competition.

Uncertainty relates to a business's confidence in expectations about known events. For instance, a business may know that the price of raw materials will rise, but be unsure by how much. Risk refers to events that may or may not happen, such as a new competitor entering the market. Risk includes the probability of loss and the significance of what might be lost. One of the best ways for a business to reduce perceived uncertainty about markets is to understand them by obtaining more information about the factors that require clarity. However, it is rarely possible to obtain sufficient information to remove uncertainty altogether. The methods of dealing with risks and unavoidable uncertainty are usually part of the corporate culture of a business and reflect the psychological make-up of senior managers. The quality of management has a pronounced effect on the expansion of a business. Quality of management includes a willingness to take risks, and to search for ways of avoiding risk and still expand (Bradley 1995).

25.3 BENEFITS OF MARKET RISK MANAGEMENT

Market risk management affords a business benefits such as, but not limited to:

- improving the ability to achieve its business objectives;
- encouraging a systematic and methodical review of the macro and micro market environment;
- supporting a proactive approach to seeking out opportunities from changes in market conditions;
- ensuring that market risks are identified, measured, monitored, controlled and regularly reported to senior management or the board of directors;
- providing analysis of risks and opportunities associated with barriers to market entry.

25.4 IMPLEMENTATION OF MARKET RISK MANAGEMENT

The development of a sound system will depend on a number of issues such as:

- the risk management system not overly constraining risk taking, slowing down decision-making processes or limiting the volume of business undertaken;
- whether those responsible for establishing and implementing the risk management framework are distinct from the managers of the individual business units;
- the development of a culture that rewards the disclosure of risks when they exist, rather than encouraging managers to hide them.

25.5 MARKET STRUCTURE

Market structures are the characteristics of a market, which determine firms' behaviour. Economists single out a small number of characteristics:

- the number of firms in the market and their relative size;
- the ease or difficulty with which new entrants might come in;

- the extent to which the goods are similar;
- the extent to which all firms in the market share the same knowledge; and
- the extent to which the actions of one firm will affect another firm.

25.5.1 The Number of Firms in an Industry

The number of firms in an industry may vary from one to many.

- A monopoly is said to exist where there is only one supplier in the market.
- A market structure is oligopolistic where it is dominated by a few large producers. In an oligopolistic market there may be a large number of firms, but the key characteristic is that most are small and relatively unimportant, while a small number of large firms produce most of the output of the industry.
- In perfect competition or in a monopolistic competition there are a large number of small suppliers, none of which is large enough to dominate the market.

25.5.2 Barriers to Entry

Market structures are not only affected by the number of firms in an industry and their relative output (to total demand) but also by the potential number of new entrants to the market. Businesses already in a market are not only exposed to the risk of competition from other businesses already in the market, but also from new entrants. JetBlue, the low-cost airline and a relatively new entrant to the US airline market, achieved a valuation on its fifth birthday in 2005 of $2 billion, $500 million more than American Airlines, the world's largest airline. (Its high valuation was based on the speed with which it reached major carrier status.) On the other hand, businesses considering entering a market for the first time must make themselves aware of the obstacles or threats to success. There are a number of barriers to entry, which reduce the potential risk of further competitors entering an industry. These include capital costs, scale economies, natural cost advantages, legal barriers, marketing barriers and restrictive practices. A key challenge for JetBlue, for instance, was securing slots at New York's John F. Kennedy airport. JetBlue's recognition of JFK's potential for a domestic operator was a key aspect of its success. The timing of new entrants can also be critical. As reported in the press during February 2005, JetBlue benefited from the fact that network carriers were more intent on saving themselves than killing rivals, and being one of the few companies to purchase aircraft during the downturn enabled them to buy them on enviable terms from Airbus.

Buying a baker's shop is relatively cheap, and the entry cost for most single-outlet retailers is small. Buying an aluminium smelter or a car plant, on the other hand, is extremely expensive. Entry costs to these industries are very high and only large companies on the whole can pay them. Capital costs therefore can represent a very important barrier to entry to a market. In some industries, economies of scale are very large. A few firms operating at lowest average cost (the optimum level of production) can satisfy all the demand of buyers. This will act as a barrier to entry because any new firm entering the market is likely to produce less and therefore have much higher average costs than the few established producers. Existing businesses in a market may also have natural cost advantages such as close proximity to natural resources, a motorway network or an energy source. As a result, they will either be able to produce at lower cost or be able to generate higher revenues than their potential competitors.

The law may give firms particular privileges. Patent laws can prevent competitor firms from making a product for a given number of years after its invention. The government may give a firm exclusive rights to production. It may, for instance, give broadcast licences to commercial television companies or it may make nationalised industries into monopolies by legally forbidding firms to set up in the industry, as used to be the case with the Post Office. Existing firms in an industry may be able to erect very high barriers through high spending on advertising and marketing. The purpose of this is to make consumers associate a particular type of good with the firm's product creating a powerful brand image. One example, which succeeded for some 50 years, was Hoover with its vacuum cleaner. Until recently a personal stereo was often called a "Walkman", the brand name of Sony which first put it on the market.

Marketing barriers can make an industry almost impossible to enter. Firms may deliberately hinder competition through restrictive practices. For instance, a manufacturer may refuse to sell goods to a retailer which stocks the products of a competitor firm. Firms may be prepared to lower prices for long enough to drive out a new entrant to the business. The ferry companies operating across the English Channel lowered their prices just before the opening of the Channel Tunnel to compete with Eurotunnel. Car insurance companies advertise aggressively to both maintain and increase market share and deter new entrants. The extent to which there is freedom of entry to a market varies enormously. Manufacturing industries with high capital costs and with extensive marketing power tend to have higher barriers than service industries. But many service industries have high barriers, too. Banking, for instance, has a high capital cost of entry, legal permission is required and marketing barriers are high.

New Entrants/Competition

Company-specific competencies determine the basis of competition and so determine the success or failure of a company. For example, retailers compete on distribution skills, consumer products depend on successful advertising, and consumer electronics depend on engineering design (Bradley 1995). These critical capabilities change over time, however, requiring businesses to adapt. Failure occurs if the business does not recognise the shift in the critical capabilities required, or if it cannot obtain competence in the new critical capability introduced as an innovation. Currently innovation (other than in the conventional sense) is also seen as attention to customer service. At a marketing meeting hosted by Proctor & Gamble Europe at its Schwalbach Technical Centre in Frankfurt in June 2005, innovation was described as simply improving products, services and business processes for customers. This is counter to the argument that as all products and services have become commoditised, the only way to break free of competition is to innovate your way to some form of uniqueness, either by achieving a breakthrough in product or service attributes or by adopting a branding strategy that associates the product or service with a unique set of ideas or emotions. Patrick Barwise (Professor of Management and Marketing at the London Business School) argues that it is the little things that count, that customers are much less interested in breakthrough technologies than in quality products, good service, on-time delivery and other benefits commonly provided by the competent company (Barwick and Meehan 2004).

Successful firms attempt to monitor their closest competitors very carefully to avoid being outmanoeuvred by technological developments. New technologies have the ability to blur the distinction between competitors, suppliers and customers. Changing technology may allow a firm that was once a customer to become a competitor. For these reasons, a business must be alert to the actions of other firms in the industry. New entrants to an industry add capacity.

However, if the capacity added is greater than growth in demand, this will reduce profitability (on the assumption that competition is purely through price). The threat of new entrants is low in cases where:

- industries are capital-intensive;
- economies of scale are a key factor;
- access to resources is a key factor (such as through government licences);
- access to distribution is difficult;
- investment by buyers is high and hence the appetite or ability to switch supplier is low.

New entrants may seek not to replicate the value chain of existing firms but to focus on certain activities where barriers to entry are lower. For example, a firm may enter the market for a product but subcontract the manufacturing to a low-cost producer and concentrate on research and development, marketing, sales and distribution (Friend and Zehle 2004). A business that has achieved economies of scale in one industry may be able to apply these economies in another one.

25.5.3 Product Homogeneity, Product Diversity and Branding

In some industries products are essentially identical whichever firm produces them. Coal, aluminium and gas are examples. This is not to say there are not different grades of coal or types of aluminium, but no producer has a monopoly on the production of any such grade or type. Goods that are identical are called homogeneous goods. Businesses find it much easier to control their markets if they can produce goods which are non-homogeneous. Rarely do businesses rely on one product or service. Generally a business is concerned with managing a number of products or services. Hence the problem for marketing managers is one of managing a portfolio. The portfolio concept is valuable as it classifies products or services according to where they are on the life cycle referred to above, relative to competing products.

When examining the product portfolio of most businesses, it is usually possible to find those products in which the business should invest for the future, those in which it should maintain existing investment levels, and those in which investment should be limited or even withdrawn. By classifying the portfolio of products in this way, it is easier to identify the opportunities – identify the gaps in the business's product line so that new ones might be developed and introduced. Differentiating their product from their competitors and creating brands allows them to build up brand loyalty. This in turn leads to a reduction in the elasticity of demand for their product. A branded good may be physically no different from its competitors, or it may be very slightly different. However, branding has value for the firm because consumers think that the product is very different, so different that rival products are a very poor substitute for it. This perception is built up through advertising and marketing and enables firms to charge higher prices without losing very much custom (i.e. demand is relatively inelastic, price increases lead to a small decrease in demand).

25.5.4 Knowledge

Buyers and sellers are said to have perfect knowledge or information if they are fully informed of prices and output in the industry. So if one firm were to put up its prices, it would lose all its customers because they would buy from elsewhere in the industry. So there can only be one price in the industry. Perfect knowledge also implies that a firm has access to all information that is available to other firms in its industry. In UK agriculture, for instance, knowledge is widely available. Farmers can obtain information about different strains of seeds, the most effective combinations of fertilisers and pesticides, and when it is best to plant and reap crops. Perfect knowledge does not imply that all firms in an industry will posses all information. Businesses that do not take the trouble to obtain the relevant readily available information will survive in the short term but will eventually be driven out of business by more efficient competitors. Businesses cannot predict a bear market, a recession, new market entrants, natural disasters or breakthroughs in technology. Perfect information only means that all businesses have the same access to information. Firms have imperfect knowledge where, for instance, there are industrial secrets. Individual firms may not know the market share of their competitors or they may be unaware of new technology or new products to be launched by rival companies. Information could then act as a barrier to entry, preventing or discouraging new firms from entering the industry.

25.5.5 Interrelationships within Markets

There are two possible relationships between businesses in an industry. Businesses may be independent of each other. This means that the actions of any one business will have no significant impact on any other single business in the industry. In agriculture, for instance, the decision of one farmer to grow more wheat one season will have no direct impact on any other farmer. This is one reason why perfect knowledge exists to some degree in agriculture. There is no point in keeping secrets if your actions will not benefit you at the expense of your competitors. If businesses are interdependent then the actions of one business will have an impact on other businesses. An advertising campaign for one brand of washing detergent, for instance, is designed to attract customers away from other brands. Businesses are more likely to be interdependent if there are few businesses in the industry.

Bargaining Power of Suppliers

The balance of power between suppliers and the supplied industry is a function of the degree of fragmentation of that industry. For instance, in an industry with many small suppliers and few large buyers, the bargaining power of the suppliers will be weak. However, when there are few large suppliers their bargaining power will be strong. In industries where inputs are standardised and there is ample availability of substitutes, provided the costs of switching are reasonable, the ability of suppliers to raise prices is limited. Certain supply chain strategies, such as just-in-time manufacturing or just holding low stocks, increase the dependency on suppliers. To reduce the bargaining power of suppliers, strategies are to maintain a diverse base of suppliers or to make a few suppliers dependent on your business, as has been the practice of Marks and Spencer. However, on a cautionary note, the squeezing of suppliers must not be too severe as there is a direct dependency between purchaser and buyer.

Bargaining Power of Buyers

The prices that business can obtain will have the largest impact on profitability. In most instances buyers shop around for the best prices and hence exert a downward pressure on prices. There are a number of factors that increase the power of buyers:

1. Buyers are large compared with the supplying industry. For examples, farmers selling to a few large supermarket chains suffer from this. Tesco, Asda and Sainsbury hold the majority of the market share between them.
2. In business-to-business markets where buyers produce the product in-house, extending their value chain backwards. This is not only a serious threat, it also increases a buyer's knowledge of the suppliers' costs. Knowledge of the suppliers' costs considerably increases the bargaining position of buyers.
3. Buyers can switch suppliers with minimal cost impact. Therefore the extent to which products can be differentiated will have a direct impact on prices.

25.6 PRODUCT LIFE CYCLE STAGE

It has been observed that markets for products grow in an S-shaped manner and eventually decline to be replaced by new products, as illustrated in Figure 25.3. For the purposes of market forecasting, the product life cycle is commonly analysed in five stages:

- *Introduction*. Sales volumes are low and increase in a linear fashion. There are few competitors, the product may suffer from quality problems and there is little variety between different versions of the product. Unit costs and prices are high.
- *Accelerating growth*. Buyer groups widen and sales increase rapidly. More suppliers enter the market and prices start to fall. A greater variety of product forms start to appear.
- *Decelerating growth*. Penetration is still increasing, but at a declining rate. Prices are falling more quickly and become a significant issue. Variety increases further and there is an increased focus on product quality. Late adopters buy the product.
- *Maturity*. Penetration is no longer increasing. There may be consolidation. Prices are declining further, but at a slower rate.
- *Decline*. Prices are low, but no longer declining. Some competitors may exit the market as the returns are too small or other emerging markets appear more rewarding.

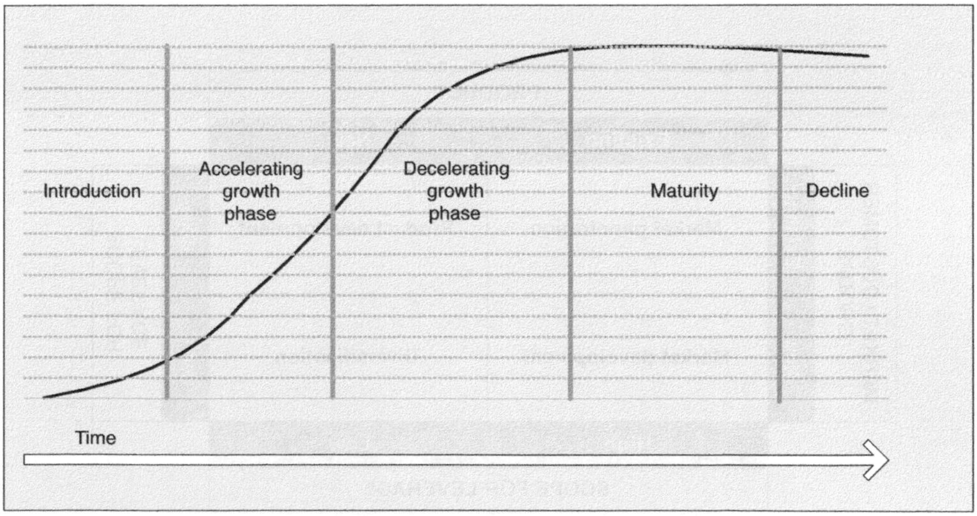

Figure 25.3 Product life cycle stages

This framework of product sales provides a tool to forecast sales volumes, prices and market share: the essential elements of the demand forecast. The framework is a way of viewing product opportunity or risk.

25.6.1 Sales Growth

A part of risk management is understanding demand risk, which in turn requires an understanding of the likelihood of a major downturn. How does a business prepare for a such a risk? Is a business able to withstand a fall in sales? An attractive scenario would be one where it was possible to take advantage of a downturn and gain market share at the expense of one or more competitors. An example of such a situation is the downturn experienced in the computer industry. Cossin (2005) cites the example where, as a result of a thorough cost focus, computer company Dell created a strong margin advantage during the boom years of the tech bubble (11.2% margin in 1998 versus Compaq's 4.5%). When the downturn hit, Cossin states, Dell was able to take over competitors by decreasing its prices by about the level of its competitors' margins. While others saw sales fall by 30% on average, Dell sales were flat in 2001, hence giving it a market leader position (with a remarkable return on operating assets of 38.8%) while others were seeing a fall in sales, income and profit.

25.7 ALTERNATIVE STRATEGIC DIRECTIONS

The alternative strategic directions for a business are: to grow the business, do nothing or withdraw. Generally business plans are developed to expand a business.

A business can be developed in four possible directions (see Figure 25.4):

- market penetration – sell more of the same to the same market;
- product development – sell new products to existing customers;
- market development – seek out new markets for existing products;
- diversification – sell new products to new groups of customers.

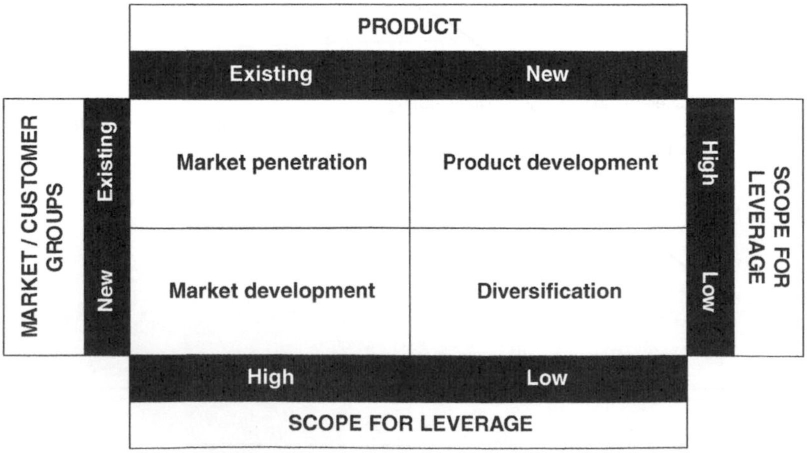

Figure 25.4 Alternative strategic directions for business development (Friend and Zehle 2004)

With all development strategies their success or otherwise will depend on the degree of success in leveraging core competencies and resources. The scope for leverage is highest for a market penetration and lowest for a minor diversification strategy.

25.7.1 Market Penetration

Selling more of an existing product to the same customers or market is generally regarded as the easiest development strategy. Products and markets are well known and this strategy potentially provides the best scope for leveraging existing skills and assets. In growing markets all competitors, to a greater or lesser extent, pursue penetration strategies to increase market share.

25.7.2 Product Development

The basis of a product development strategy is to sell new products in addition to existing products into the same market or to the same customer groups. The term "new products" is used here to describe genuine new products that satisfy different needs and generate incremental sales, not enhancements or new versions of existing products. As the new products are sold into existing markets or to existing customers, some aspects of the value chain can be leveraged, particularly distribution and customer knowledge.

A common example of product development is the move by car insurance businesses into home and travel insurance. There are risks attached to this strategy in that while the business will have extensive data upon which to set premium levels for car insurance, this data will not exist for the other business streams. Additionally, if the service provided by the new businesses attracts adverse publicity, it may reduce the customer base for the core business.

A brand may be extended to new products, but there is a risk that the brand is overstretched and its value is diluted. In this case the strategy becomes solely a diversification strategy as opposed to product development, which may lead to lower sales of existing products because the brand value has diminished.

Given that, on the one hand, many businesses suffer substantial and sometimes crippling losses from the failure of *new products* introduced by them, and, on the other hand, businesses such as Sony, Apple, DuPont, BMW, Hewlett-Packard and Toshiba (see Box 25.1) owe their success to the planned introduction of new products. Hence market development is a critical area of risk and opportunity management.

It is difficult to pin down what constitutes success or failure to facilitate risk management due to the absence of an agreed definition of what constitutes a new product and what amounts to failure. However, of assistance is a seminal study, on which most succeeding analyses have been based, undertaken by the National Industrial Conference Board (NICB) in the early 1960s and first reported in 1964 (National Industrial Conference Board 1964). In rank order of importance, the eight major causes of failure were cited as:

- Inadequate market analysis
- Product defects
- Higher costs than anticipated
- Poor timing
- Competitive reaction
- Insufficient marketing effort
- Inadequate sales force
- Inadequate distribution.

Robert Cooper of McGill University in Canada undertook a study called Project New-Product, which deserves particular attention. His findings were derived from a random sample of 177 firms located in Ontario and Quebec, representing a broad spectrum of industries and markets. Of the 177 firms, 103 agreed to participate, yielding 102 successes and 93 failures. From a literature review he derived eight research propositions which he hypothesised would positively relate to product success, as shown in Box 25.2.

To test these hypotheses, 77 variables were characterised and respondents were asked to rate each success and failure along these 77 dimensions, using scales of 0–10 indicating whether the factor was relevant or not. Preliminary analysis showed a high degree of intercorrelation among the 77 variables, and these were reduced to 18 summary variables using the factor analysis technique. Subsequently 11 of these factors were found to discriminate between success and failure, which are listed in order of importance below:

1. Introducing a unique but superior product.
2. Having market knowledge and marketing proficiency.
3. Having a technical and production synergy and proficiency.
4. Avoiding dynamic markets with many new product introductions.
5. Being in a large, high-need growth market.
6. Avoiding introducing a high-priced product with no economic advantage.
7. Having a good "product/company fit" with respect to managerial and marketing resources.
8. Avoiding a competitive market with satisfied customers.
9. Avoiding products "new to the firm".
10. Having a strong marketing, communications and launch effort.
11. Having a market-derived idea with considerable investment involved.

Box 25.1 Market for energy storage devices

A clear opportunity exists for new product development of energy storage devices. The ubiquitous laptop computers and mobile telephone, now considered essential for business life, are both dependent on energy storage devices – batteries. Product development resulting from technological innovation in battery design will lead to technological advances in the industries that depend on their products, computing and telecommunications. Kevin Fitzgerald, chief executive of Ener1 (a Florida-based battery company) has said: "because battery capacity is a bottleneck in product innovation, engineers work around the constraints when developing products ... [however] by developing storage cells that are light, powerful and long-lasting, companies such as ours can eliminate the energy bottleneck and enhance product development in a variety of industries". The pressure to produce more efficient energy storage devices is taking place on three scales:

- Computer, telephone and consumer electronics manufacturers are desperate for better power sources, as they cram more battery-draining features on to mobile devices.
- Vehicle manufacturers are under strong pressure to replace diesel and internal combustion engines with electrically powered alternatives for environmental reasons.
- The electricity supply industry needs to be able to store power generated from renewable sources that are only intermittently available, such as wind and sunshine.

Lithium battery development appears to be making significant strides. Previously they were criticised for slow energy release (for some high-powered applications), slow recharging and short life. However, two breakthroughs in recharging speed were announced in March 2005. Altair Nanotechnologies, a specialist Nevada-based company, said lithium batteries made with its new electrode "nanomaterials" had significantly reduced recharge times. Toshiba promptly followed suit by making similar claims for its new lithium battery which can recharge 80% of its capacity within a minute and 100% in "only a few more minutes", stating its breakthrough is also based on using nanoparticles. Batteries now represent a $48 billion a year world market, growing 6.5% annually according to Freedonia, the US-based business research group.

Source: Cookson (2005).

Box 25.2 Robert Cooper's product success factors

1. Products which are superior, have a differential or economic advantage or are unique relative to competing products	
2. Products where other elements of the commercial entity – selling, distribution, production, etc. – are proficient	Commercial entity
3. Projects where considerable technical and market knowledge is acquired	Information required
4. Projects where the technical, marketing and evaluative (process) activities are proficiently undertaken	Proficiency of process activities
5. Products entering mass, large, growing, dynamic and uncompetitive markets, with a high unsatisfied need for such products	Nature of the market place
6. Projects where a high degree of resource compatibility exists between the needs of the project and the resource base of the firm	Resource base of the firm
7. Projects familiar to the firm (do not involve new technologies, new markets, etc.)	Nature of the project
8. Market-derived projects (product idea came from the market place)	

25.7.3 Market Development

The common strategy adopted is to seek to sell existing products to new customer groups or markets. A classic market development strategy is to extend the geographical reach, either within the home country or by means of exports. This generally requires modifications to the marketing mix (product, price, promotion or place) such as adjustments to the products so that they appeal to new market segments, printing manuals in different languages and ensuring compliance with local standards. An example of marketing development is the diversification of PepsiCo into snack foods, making it a food and beverage company (see Box 25.3). However, extending geographical reach must be accomplished within an environment that commonly is predominantly uncontrollable. The uncontrollable elements are: the extent and rate of economic development in the market (e.g. sale of electrical goods in developing countries may be difficult); social and cultural influences which prohibit or restrict product sales; political and legal restrictions; business practices; and the competition policy. When a firm begins to do business in foreign markets, an additional set of uncontrollables come into play. Bradley (1995) suggests the differences arise under the following headings:

- Political and legal systems
- Economic trends and levels of development
- Physical features, geography and topography
- Distribution systems
- Government sponsorship.

A classic example of extending geographical reach is the operating of new routes by airlines. The emerging economies of Brazil, Russia, China and India are seeing domestic traffic increase year on year by over 10%. China is seen as a vast domestic air travel market. International travel to and from Asia and the Middle East continues to show strong growth. The aircraft maker Boeing, in its annual market outlook released in June 2011 raised its long-term forecast of demand for passenger and freight aircraft based on continued expansion of the Asian aviation market. The company projected that the number of commercial aircraft would more than double by 2030 and that passenger traffic would nearly triple over the same period.

Box 25.3 PepsiCo diversification

Nearly 100 years into what is arguably the world's most famous corporate rivalry, Coca-Cola was firmly in the ascendancy over the struggling PepsiCo. It was October 1996, and the cover story on the *Fortune* magazine explained "How Coke is Kicking Pepsi's Can". The article quoted Robert Goizueta, Coke's then chairman, dismissing Pepsi as a spent force. "As they become less relevant", he said "I don't need to look at them very much anymore". Eight years later, everything has changed. PepsiCo has become the investor's favourite, its shares more than doubled since their best of that month, while those of troubled Coke languish below where they stood then. Half of the story behind this role reversal is well known. A year after the *Fortune* article Mr Goizueta died, precipitating a period of management upheaval and sluggish growth from which Coke has yet to recover. Much less attention has been given to how PepsiCo emerged from its rival's shadow to become one of the best performing companies in the food and drink industry. In those eight years PepsiCo has increased its sales 45 per cent and net profits more than fourfold. While Coke derives all its revenues from beverages, PepsiCo is more diverse. The biggest part of the company is not cola but the Frito-Lay snack food business, which merged with Pepsi in 1965. Only two thirds of PepsiCo's beverage volume is carbonated soft drinks, compared with more than 80 per cent of Coke's. ... As Cola-Cola's share of the US beverage market began to shrink in the late 1990s PepsiCo's diversification started to look the smarter choice. PepsiCo's earnings growth was a third greater than Coke's last year at 18 per cent and its sales growth four times higher at 8 per cent. ... PepsiCo is [now] the world's fourth largest food and beverage company ranking behind Nestlé, Kraft and Unilever. [However, much] of the biggest challenge facing PepsiCo is consumer concern in North America and western Europe about obesity. A third of adults and one in six children in the US are clinically obese, according to the US Centre for Disease Control and Prevention. [But] Mr Reinemund insists the obesity crisis can be an opportunity [upside risk] rather than a threat [downside risk] for companies that embrace consumer demand for healthier food and drink.

Source: Ward (2005a).

25.7.4 Diversification

The strategy of diversification aims at selling new products into new markets. It can be subdivided into related and unrelated diversification.

Related diversification means that a business stays broadly within the industry but needs to acquire new competencies and resources:

- A business can pursue a strategy of related diversification by means of vertical integration: extending the value chain backwards and forwards. For example, an airline may choose to undertake the maintenance of its fleet of aircraft rather than outsource it, or a manufacturing business may choose to make components rather than buying them from a supplier.
- A successful restaurant chain may decide to leverage its brand name to get into the frozen ready meal market. The skills required to run a restaurant and manufacture convenience foods for distribution in supermarkets are entirely different, but the brand and possibly some recipes may provide the link between the two.

Unrelated diversification takes a business into a completely new field, a different industry. As it is unrelated, it is sometimes difficult to establish a strategic logic for such a move particularly where there is only a limited degree of synergy. Such a move, however, would be more comprehensible if the existing market is in terminal decline or where new rewarding markets are emerging where consumer demand remains unsatisfied. Conglomerates are the archetypical diversified business, where a holding company manages a diverse collection of companies, acting almost like an investment fund. An example is Virgin Group, which has over 30 companies including well-known brands such as Virgin Trains, Virgin Mobile, Virgin Radio and Virgin Atlantic.

- The financial management and planning skills are core competencies of the head office, and this expertise can be applied to make portfolio companies in different industries more successful.
- Diversified companies are less affected by a downturn in one industry.

An example of diversification is the strategy adopted by some luxury car makers of entering into the bottom end of the car market by producing the smaller car for volume sales. This strategy is not without risk. BMW's launch of the Mini was very successful, with the planned opening of another plant to cope with demand. Other ventures have not fared so well (see Box 25.4).

Box 25.4 The risk of diversification: Mercedes

"Smart is a disaster", Eckhard Cordes, Head of Mercedes, said this week. It is an unusually frank admission of the scale of the problem that has beset Daimler Chrysler's trendy small car brand ever since its troubled birth over a decade ago. [Daimler-Benz bought Chrysler, America's third largest car maker in 1998.] The marque has been owned by the German carmaker since it bought out Swatch, the watchmaker, from a joint venture in 1998 shortly before the first car launch. It has always been a disappointment. Smart has consistently missed its sales targets in spite of rapid model expansion and it quickly achieved iconic design status. It has also remained resolutely loss making, missing last year's break-even target and now further delaying the goal of returning to the black in 2006 until at least 2007. ... Mr Cordes forecasts Smart will break even in 2007 but for many that has a repetitive ring about it. Jürgen Schrempp, Daimler chief executive, set a goal in 2000 that the marque would stop losing money by last year. Some analysts estimate that instead Smart lost about €500 m (£655 m, £343 m) in 2004 – or €3300 for every vehicle sold. Many investors and analysts question the wisdom of continuing with Smart, warning that Mercedes runs the risk of becoming too thinly spread. "Why would you build a small car as a luxury carmaker?" asks one leading shareholder. "It's a segment with low profitability and ferocious competition." Senior Daimler executives will meet in April to discuss a strategic review of Smart's future.

Source: Mackintosh and Milne (2005).

Another more recent diversification failure, this time outside the motor industry, was Cisco's reported disastrous foray into alien markets. Cisco is perhaps best known for the design of technologies and services for Internet communication. CEO John Chambers put Cisco on a course of diversification that commentators have described as "ruinous" and which shareholders have criticised due to the resultant "miserable stock performance". Chambers considered that Cisco's core router business was not enough to attain the level of planned growth. His answer was to adopt a highly risky strategy and diversify from Cisco's core strengths into a number of unfamiliar intensely competitive markets that typically produced lower profits than the original business model. A puzzle for commentators was why Cisco ever diversified beyond its enterprise hardware and software businesses. Cisco has become a supplier of a collection of largely unrelated products which range from home video conferencing and living room set-top boxes to complex data centre virtualisation software. By implication Cisco must have several disparate sales forces, R&D groups and management divisions. It was thought Chambers needed to "see the light" and move in a new direction, re-focusing on the company's core strengths, cutting costs, reducing its bloated upper management structure and ejecting low-margin businesses. In November 2010 when Cisco announced disappointing first quarter results and Chambers said Cisco faced "unusual uncertainty", investors were rattled. Cisco's share price fell by 15.8% in a day of heavy trading. In 2011 Cisco announced in the press that it planned to cut operating expenses by $1 billion in 2012, to help future profit margins. In a memo to staff sent on April 04, 2011 Chambers acknowledged that Cisco had disappointed its investors, confused its employees, and tough decisions would have to be made going forward. He said that the company would now focus on its core switching and routing business.

25.8 ACQUISITION

Johnson and Scholes (1989) have identified alternative methods to implement a strategy once it has been selected, one of which includes acquisition. They describe the trade-off between cost, speed and risk exposure. Acquisitions can be a quick route for product development. Established companies often acquire smaller businesses to gain control of new technology and hence new products. Acquisition provides the opportunity to increase market share. Overseas acquisitions are a well-established way of building a foreign presence. An example is Coutts, which acquired Bank von Ernst in Switzerland in 2003, adding 1000 staff and increasing Coutts' assets by 20%. Coutts, founded in 1692, has enjoyed a reputation as the UK's most exclusive bank, promising unrivalled personal service. Another example is BAE Systems' acquisition of American company United Defence. However, as in this instance, acquisitions abroad can involve government agencies and as a result may be protracted affairs. The US regulatory agency, the Committee on Foreign Investment in the US (CFIUS), called in for approval the proposed BAE takeover as part of its role of investigating large foreign acquisitions of American companies for potential national security threats. The CFIUS clearance marked a watershed for the Pentagon, which is now likely to have a foreign company as a prime contractor for several large weapons systems.

Other notable acquisitions include Kraft Foods' purchase of Cadbury in 2010 for £11.9 billion (creating the world's biggest confectioner) and Hewlett-Packard's surprise £7 billion takeover of Autonomy Corporation in 2011, the largest European software company (a significant expansion for HP outside of their traditional personal computer business).

25.9 COMPETITION

Oligopoly is characterised by price stability. One explanation of this is that changing price is a very risky strategy for one business because it will provoke a reaction from other firms. Non-price competition is common in oligopolistic markets as it is a less risky strategy than price competition. Successful branding enables producers to charge a premium price and earn abnormal profit on a product. The large number of different market strategies available to oligopolistic businesses may result in permanent disequilibrium in the market.

Central to the understanding of oligopoly is interdependence. The actions of one large firm in the industry will directly affect all other firms in the industry. It is essential therefore in any examination of oligopoly to understand the nature and consequences of those reactions. Game theory is a powerful tool for the analysis of oligopolistic behaviour. It explores the reactions of one player to changes in strategy by another player and has a wide variety of applications. In a game the players are interdependent. An aspect of game theory is that by limiting your own options, for example, you can make it clear to opponents how you will respond to their actions, whatever they do, thereby increasing the chances the other side will back down. The best move for a player depends upon how the other players will react. Thomas Schelling and Robert Aumann, the pioneers of game theory, were awarded the Nobel Prize for economics in October 2005.

25.9.1 Price Stability

One commonly observed feature of an oligopoly is price stability. Businesses maintain stable prices over a pricing season which may last from six months to several years. Price stability may be a rational strategy for oligopolists. If an oligopolistic firm raises prices, it risks losing market share if its competitors do not follow suit. Lower market share could lead to lower profits and if investment and research and development budgets are cut, the ability to compete in the long run will be reduced. If it lowers its price it risks starting a price war. It could be that the size of the market will expand as consumers buy more of the industry's products. But the benefits in the form of larger sales could well be more than offset by losses of revenue due to lower prices. All businesses in the industry could see sharp falls in profits as they battle it out. Eventually prices will have to rise again to restore profitability and the firm that started the price war could have lost market share. So changing prices is a risky strategy. When prices change as a result of an increase of raw materials or government intervention, all firms tend to change their prices by the same percentage. A rise in petrol prices by one company is usually matched by other petrol suppliers.

Table 25.1 Change in profits from different marketing strategies

	Business B	
	Lower price	Leave price unchanged
Business A Lower price	– *£10* (A) **–£20 (B)**	+*£5m* (A) **–£27m (B)**
Leave price unchanged	–*£12m* (A) **+£5m (B)**	0 (A) **0 (B)**

Game theory can be used to explain business behaviour and the potential source of risk. Consider Table 25.1. There are just two firms in the industry (it is a duopoly). Each firm has two strategies. It can either lower the price of its product or leave it unchanged. The figures in the boxes represent the change in profits of firms A and B. The change in profits of business A for a particular combination of strategies is shown in italics, while the change in profits for business B is shown in bold. For instance, if both businesses cut prices, business A will lose £10 million in profits while firm B will lose £20 million.

It is clear from the table that it is in both businesses' interests to leave prices unchanged. If one business decides to lower its price in order to gain market share (and the other firm does not), it will increase its profits while the other firm would suffer a drop in profits. For instance, if business B lowers its price it will gain £5 million, while firm A will lose £12 million. The same is true for business A. If it dropped its price and firm B did not react, business A would increase its profits. However, if business B did react and lower its prices too, the resulting price war would have disastrous consequences for both sides. Their respective sales would remain the same or similar, while the profit per unit of production would fall.

25.9.2 Non-Price Competition

A characteristic of oligopology is the lack of price competition. Price wars can be very damaging for firms in an oligopolistic industry. So firms choose to compete in other ways than price. Firms decide on a marketing mix, a mixture of elements which form a coherent strategy designed to sell their products to the market. The marketing mix is often summarised as the "4 Ps" (Anderton 1995). Figure 25.5 describes a potential marketing mix composed of the commonly recognised elements of product, price, promotion and place.

Figure 25.5 The marketing mix composed of the four Ps (Friend and Zehle 2004)

Firms produce a product which appeals to their customers. The product may or may not be differentiated from rivals' products. A price needs to be set, but this could be above or below the price of competing products depending on the pricing strategy to be used. Promotion (advertising and sales promotion) is essential to inform buyers in the market that the good is on sale and to change their perceptions of a product in a favourable manner. A good distribution system is essential to get the product to the right place at the right time for the customer. An advertising campaign for instance by firm A is likely to be limited in cost and may increase market share. Other firms in the industry may react by launching their own advertising campaign, but there is a reasonable chance that the advertising campaign of the competitors may not be as good, plus advertising may expand the market as a whole. The reward for business A will be a small increase in market share and presumably profits as well. However, if the advertising campaign is unsuccessful and a competitor's advertising campaign is a runaway success, the resultant loss of market share and corresponding loss of profits may be significant. So before any initiative is taken to increase market share, consideration has to be given to the likely response from competitors.

25.9.3 Branding

Interdependence limits the ability of oligopolistic firms to exploit markets to their own benefit. Ideally oligopolistic firms would like to turn themselves into monopolists with full control of their markets. One way of doing this is by the creation of strong brands.

- A strong brand has few good substitutes so far as the buyer is concerned. The firm is therefore able to charge a premium price (a relatively high price for a good) and earn monopoly profit on the good without seeing too great a fall in demand for it.
- It is very difficult for competitors to challenge the supremacy of the brand. For instance, Mars bars, BMW cars and Omega watches all have stable demands at premium prices in the short run.

The Orange mobile phone brand is now hailed as one of Europe's most successful technology brands. The original branding strategy was launched in 1994 using the catch line "the future's bright, the future's Orange". The strategy revolved around an unusual brand name, distinctive colour coding and advertising that ran counter to the prevailing emphasis on technology nuts and bolts by emphasising emotions and not even featuring mobile phones. The launch stood for reassurance with its pledge to lead customers into a brave new world of technology. The strategy worked and by the end of 1995 Orange's UK customer base had more than doubled from 379 000 in 1994 to 785 000 in 2005 (Carter 2005). In 2005 France Telecom decided to revitalise the brand to overcome a branding problem, as over time Orange had lost ground to its competitors and in particular O_2, the mobile phone business spun off from the BT Group in 2001. The branding of Orange proved difficult to maintain. It had lost its differentiation; it had become a commodity buy.

As strong brands are so difficult to create in the first instance, many businesses prefer to take over other companies and their brands at very high prices, rather than attempt to establish new brands. To establish a new brand, a company usually has to produce an innovative product and then market it effectively. The cost of advertising to launch a new brand can be very considerable. Once a brand is created it needs to be protected. In 1985 the then chief executive of Coca-Cola, Roberto Goizueta, decided to relaunch its trade mark drink and replace the 99-year-old formula with a sweeter recipe. Seventy-eight days and over 400 000 protest letters and phone calls later, we are advised by Ward (2005b), "a chastened Mr Goizueta announced the return of the original formula, renamed Coke Classic". According to Ward, "it remains arguably the greatest marketing disaster in US corporate history". Mr Goizueta's actions had been a reaction to arch-rival Pepsi which had been close to deposing Coke as America's favourite cola. Pepsi had been running television advertisements that showed consumers choosing its sweeter product over Coke in blind taste tests. However, reverting, under protest, to the old formula allowed the company to turn disaster into triumph. Sales of Coke Classic increased dramatically above previous levels and customer loyalty was reinforced. No sympathy is required for Mr Goizueta. He was not fired. He received a $5 million bonus at the end of 1985 as a reward for increasing the company's stock price by more than a third over the year. Until recently Coca Cola was the world's most valuable brand; however, it is no longer even ranked in the top ten, having been pushed out by rising stars such as Google, Bank of America, Apple and Wells Fargo.

25.9.4 Market Strategies

The microchip industry might be argued to be an oligopoly. The industry is interesting from the perspectives of rapid advancements in technology and the behaviour of the competitors in the market to increase market share. Game theory[3] is a useful tool to understand the success or otherwise of the different market strategies adopted by the main players. Casual spectators might assume that the industry is a duopoly due to the headline grabbing activities of Intel and AMD, but this is not the case as there are a number of "players" in the market. In the last ten years competitors have used advertising, price-cutting, technology improvements, production and product reliability to improve market share at times with both spectacular and disastrous results. Chip production has been dependent on developments in software, computers, mobile phones and the internet, together with growth of the economy. Without understanding the opportunities to be seized and the risks associated with alternative marketing strategies, Intel and AMD would not have fared so well; however, they have not always got it right, as Box 25.5 explains.

Box 25.5 Marketing strategies

The microchip market

Intel made the first microprocessor in 1971, which was used by Bowmar to make calculators. Intel is the dominant world supplier of microprocessors, which back in 1995 had an estimated 80% of the $10 billion market. Its nearest rival, Advanced Micro Devices (AMD), had estimated sales of $700 million in the same period. Intel and AMD remain sparring partners today. One of the problems that Intel faced back then was that computer manufacturers, such as Compaq and IBM, bought its chips wholesale. These computer companies were sophisticated enough to shop around for the cheapest prices. Other chip manufacturers could, within a few years of a chip being put onto the market by Intel, produce a chip which would perform the same functions. Clone chips tended to be cheaper. In 1993–1994 Intel attempted to get around this problem by advertising its latest mass-market chip, the Pentium, directly to the final customer with the slogan "Intel inside". The Intel Pentium

[3] The branch of mathematics which became known as game theory was developed by John Von Neumann. Game theory provides a way to analyse conflicts of interest mathematically to establish the best cause of action in any situation.

brand, mentioned in comics and on television talk shows, became a household name soon after introduction. The computer manufacturers were furious. Sales of non-Pentium computers slowed down and manufacturers were left with more stocks of these products than they had anticipated. It also limited their ability to shop around for the cheapest chips. From a game theory perspective, Intel realised that pursuing a strategy which increased customer awareness of its product would lead to larger revenues and profits than a strategy which left customers thinking that the only branded product they were buying was the computer itself. Intel's gain was a loss for other chip manufacturers. In the longer term, Intel has realised that it needs to pursue an aggressive strategy of product development. In game theory terms, a failure to launch new products will allow other manufacturers to gain a technological lead. This would reduce revenue, market share and profit for Intel, while allowing other chip manufacturers to gain these. A race developed between Intel and AMD, which currently has no end in sight. They continuously attempt to increase market share at the expense of each other, through improvements in processor speeds. The two companies take it in turns to announce product developments.

In August 1999, for instance, AMD launched what was claimed at the time to be the world's fastest and highest-performing microprocessor for personal computers. Just the previous week Intel had unveiled a 600 MHz version of its flagship chip, the Pentium III. AMD said both IBM and Compaq, two of the world's leading PC manufacturers, had already committed themselves to using its new Athlon chip in some of their machines. AMD said the top version of the Athlon would have a clock speed of 650 MHz. Laurianne McLaughlin, senior associate editor, news, at *PC World* magazine stated at the time that "it clearly poses real competition to Intel at the high end. It will be very attractive to power desktop users." Many industry analysts remained cautious, however, as a result of AMD's previously problematical product launches. Unofficial reports of AMD's manufacturing appeared to be favourable.

Intel and AMD also strove to secure market share through price. In 1999 California-based AMD found a "price war" with Intel painful. In the first and second quarters of 1999 the company had reported losses and as a consequence had embarked upon a big marketing push around the Athlon in an attempt to turn around its results. It was helped by an independent assessment of the 600 MHz Athlon, which reported that this chip was, on average, 9% faster than the 600 MHz Pentium III from Intel. The name of the new chip line was a clear attempt to break away from the association with the K6 family of slower, cheaper processors.

The technology race was unrelenting. Only three months later AMD unveiled a new Athlon microprocessor that ran at a faster speed than Intel's latest Pentium III chip. Intel had led the race for the speediest computer chip with its latest Pentium IIIs running at speeds of 733 MHz. However, the new Athlon chip reached speeds of 750 MHz and was the first to be made using AMD's 0.18 micron manufacturing process, which creates transistors with line widths just 1/500th the width of a human hair. Intel also made chips using its 0.18 micron technology, considered important in a strategy to lower manufacturing costs for the higher profit margin chips. While processor speed often grabs the headlines, more important to the business success of chip manufacturers was the pricing of the chips and the ability to supply them. Intel had difficulty keeping up with demand for its higher-speed Pentium IIIs, particularly its Coppermine processor, a 733 MHz Pentium III chip. AMD said its new Athlon processor would be used by Compaq, IBM and other PC makers, largely in high-end consumer models.

In June 2000, escalating the war with its competitors, Intel unveiled details of its new Pentium 4 processor, which ran at speeds of up to 1.5 GHz. Intel claimed its newest chip beat the processing power of the recently announced 1 GHz Athlon processor of AMD, and its own 1 GHz Pentium III. The problem for Intel was that most existing software required much less processing power than this newest generation of chips. Presenting the new chip at the time, Intel's Chief Executive, Craig Barrett, argued that faster computers were essential in a world where new generations of internet commerce, gaming and entertainment applications required ever greater processing power. Industry analysts, though, cautioned consumers that there was little reason for them to rush into the shops and upgrade their computers. They said that users should wait for software developers to catch up with chip makers. It was mainly professional video editing systems and some financial applications that could potentially make the greatest use of ultra-fast processors. At the time, Intel supplied processor chips for about 90% of the world's computers, cementing its dominant position through its close alliance with Microsoft, the maker of the Windows operating system.

In August 2001, AMD issued a profits warning blamed on the slowing economy. The problem is twofold, Robert R. Herb, Executive Vice President and Chief Marketing Officer, told a technology conference in California. On the one hand, sales of flash memory chips – which are at the heart of mobile phones and other consumer electronic devices like digital cameras – looked likely to fall 30% or as much as $100 million that quarter. On the other, sales of central processing chips for PCs were flat from the previous quarter. AMD had recently gained market share from its arch-rival Intel, but the downturn in the economy had hit technology spending particularly hard, triggering massive price cuts from both companies. From games theory we know that this would hurt both companies. AMD alone cut prices for some product lines by as much as 50%. Intel, meanwhile, continued to claim that it was seeing a seasonal rise in demand in the second half of the year. The company was betting that its newest and fastest chips, coupled with the launch of Microsoft's new Windows XP operating system, would galvanise millions of users with (by modern standards) elderly, slow machines to ditch them and upgrade.

In October 2001 Intel stuck to its large-scale spending plans, despite the slump in chip sales that had hit both its share price and its bottom line. Intel predicted that its markets would soon recover and remained committed to capital expenditure of $7.5 billion in 2001, with a further $3.9 billion being poured into research and development. "Our strategy here is new products, new technology, building blocks for the internet", said Chief Executive Craig Barrett. "The internet is still growing. We see it as being the vehicle for our growth over the next five to 10 years," he said. Intel's plans to keep up spending were seen as crucial to boost its market share which was under threat from firms such as Broadcom and 3Com as well as AMD. "We are absolutely deadly focused on growing our market segment share", said Intel Architecture Group's Paul Otellini. The companies were fighting an expensive battle for their stakes in markets that had been shrinking for months. Intel's profits fell 77% during the third quarter of 2001 and its share price lost 69% in just over a year as the company's traditional markets virtually collapsed. The company predicted that demand would pick up again, in part due to mobile internet developments. This, it said, would bring about growth of 15–20% per year.

In August 2004 Intel announced that it had created a more powerful chip without increasing its size, wrong-footing analysts' warnings that the firm had hit its technical limits. The breakthrough came after Intel managed to shrink the size of its transistors by 30%, enabling more of them to fit on a single chip. That increased the processing power of each standard-sized chip, paving the way for ever more powerful computers. The new device also allowed Intel, still the world's biggest maker of memory chips, to steal a march on AMD. This latest development was hailed as further confirmation of Moore's law, a guiding principle of the technology sector for the last 40 years. Intel founder Gordon Moore predicted in the late 1960s that the number of transistors on a chip – and therefore its processing power – would double every two years. (This prediction soon proved too conservative; a few years later Moore had to revise his projection to every 18 months.) "This is evidence that Moore's law continues", said Mark Bohr, Intel's Director of Process Architecture and Integration. Intel and its rivals have thrived on their ability to pack more transistors onto a chip, or semiconductor, but sceptics had questioned whether they could maintain the pace of progress. The new chip had been made possible by a process which limits power consumption by parts of the chip when not in use, reducing heat emissions, Intel said.

In 2011 Intel launched its Oak Trail chip in Beijing, designed for tablet computers, in an attempt to take on the UK-based ARM, whose chip architecture is now incorporated in the majority of smartphones and tablets (such as Apple's iPhone and iPad) – an important and fast-growing market. The chip was the first product to be released by its newly formed "netbooks and tablets" group. Intel predicted that more than 35 devices would soon be using the Oak Trail chip, including products by Fujitsu and Lenovo. Intel had unsuccessfully tried to enter the smartphone market in 2010 with a chip codenamed Moorestown, but not a single smartphone company chose to use it as it was considered too large and power hungry.

25.10 PRICE ELASTICITY/SENSITIVITY

25.10.1 Elasticity

Elasticity of demand measures the responsiveness of demand to a change in a variable such as price, income or advertising. It measures how much demand changes in percentage terms, compared to the variable. If demand changes more in percentage terms than the variable, it is sensitive or elastic. For instance, if demand increases 30% following a 10% price cut, the price elasticity of demand is said to be 3. It is elastic because demand has changed in percentage terms three times more than price has changed. If demand changes less than the variable, it is insensitive or inelastic. For instance, if demand changes 5% following a 10% price cut, the price elasticity is 0.5. It is inelastic because demand changes half as much as the change in price.

25.10.2 Price Elasticity

Price elasticity is the sensitivity of demand to changes in price, and is measured by dividing the percentage change in demand by the percentage change in price:

$$\text{price elasticity} = \frac{\%\text{ change in demand}}{\%\text{ change in price}}.$$

If demand is not sensitive to price (price inelastic) the business is more likely to increase price to increase revenue, because the increase in price leads to a smaller decrease in quantity demanded (in percentage terms). If demand is sensitive to price the business will lower price to increase revenue because a lower price will lead to a larger increase in quantity demanded (in percentage terms). Whether an increase in revenue will also increase profit will depend on what happens to unit costs as output changes. If unit costs increase (perhaps because the firm has had to expand capacity or because material costs increased) the profit margin and indeed overall profits may fall. Elasticity is a useful concept as it helps with planning. A business can estimate the impact of changes in price and income on demand. This will allow a business to plan for sales, staffing, cash flow, production and stock levels. However, there is a risk that estimates are incorrect and the buyers do not behave as anticipated. It is not possible to predict all the variables such as changes in income levels, competitors' advertising and advances in technology. Risk mitigation may be accomplished by test marketing, for instance changing the price and measuring the effect. It is only risk mitigation rather than risk removal as it may not be safe to generalise about the results and assume buying behaviour will be the same nationally or internationally.

25.11 DISTRIBUTION STRENGTH

The type of distribution policy adopted by a business depends on the product itself and on other marketing policies. The nature of the product determines to a large extent whether users rely on providers or distributors, as primary sources of technical information and supply availability. If the product is technically complex, users desire a direct relationship with the source of product technology. For industrial products, the source is usually the original equipment manufacturer. However, it could be other suppliers if they have accumulated the necessary technical skills. Caution is exercised over uncontrolled distribution as this is likely to bring with it some serious long-term sales problems. In developing a distribution policy for consumer products, for instance, it is necessary to examine the relationship between product type, the likely form of distribution and customer buying preferences. Detailed knowledge of buyer behaviour is therefore essential in forming distribution policy, and an absence of this knowledge is a risk to the business.

25.12 MARKET RISK MEASUREMENT: VALUE AT RISK

25.12.1 Definition of Value at Risk

Value at risk (VaR) calculates the worst loss that might be expected at a given confidence level over a given time period under normal market conditions. VaR is, essentially, an expansion and application of modern portfolio analysis as developed over the last half century by Harry Markowitz and many others. There is now a wealth of information on VaR and a recommended reading list is given in Appendix 15.

25.12.2 Value at Risk

VaR is one of the most common measurements of market risk in the financial sector. It gives a fixed probability (or confidence level) that any losses suffered by the portfolio over the holding period will be less than the limit established by VaR. Hence it should also be said that there is also a fixed probability that the losses might be worse. Critically, the VaR limit does not give an indication of how severe the losses could be or specify the worst possible loss. VaR simply states how likely (or unlikely) it is that the VaR measure will be exceeded. VaR is now recognised as a standard measure of market risk, expressed in terms of the money that might actually be lost. For instance, a bank might report that the daily VaR of its trading portfolio is £50 million at the 95% confidence level. Described another way, there is only a 5% chance that a daily loss greater than £50 million will occur, under normal market conditions. The appeal of VaR relates to its ability to provide a consistent and comparable measure of risk across all products and business streams. Cossin (2005) makes the valid observation that as risk models become more complex, board members sometimes unfortunately treat them as black boxes. The purpose of the modelling is to inform decision makers, but if the model builder is unable to communicate how the model arrives at its answers, the board members cannot grasp the statistical basis of the model, or if both description and comprehension are poor, the value of modelling is either diluted or lost altogether, and board members revert to instinctive beliefs and judgement.

25.12.3 VaR Model Assumptions

Before the financial crisis VaR models were based on seven assumptions.

1. Short observation periods (of, say, 12 months) provide a sufficient time span to be able to make robust defendable predictions about likely future events.
2. Robust inferences can be drawn from past asset-price volatility to guide thinking about the probability of future events.
3. The distribution of likely values will most likely be normally distributed (shaped like a bell curve), as values will be grouped equally above and below the mean and then tail off symmetrically from the mean.
4. Risk can be reliably projected to enable businesses to make informed investment decisions and implement controls according to an investor's or firm's appetite for risk.
5. The actions of one firm are independent of the actions of other firms in the same market, and a single player is incapable of affecting market equilibriums by inducing similar and simultaneous behaviour.
6. Top management and boards understand how VaR models are constructed and the reliance that can be placed on the results, and are able to assess and appropriately exercise judgement over the risks being taken.
7. The increasing sophistication of VaR models matches and makes safe participation in the increasingly complex securitised credit market through the ability to both measure and manage risk.

Unfortunately, in times of crisis none of the assumptions are appropriate. The distribution of asset-price volatility has much fatter tails – that is, the likelihood of extreme events in asset-price swings is much higher than the normal distribution models, including VaR, would predict. In addition, as has became painfully apparent, volatility today can exceed anything that has occurred in the past. The limitations of banks' risk models were painfully exposed in August 2007. As recorded in the *Financial Times* at the time, the Chief Financial Officer of Goldman Sachs, David Viniar, stated that the bank witnessed a vast shift in risk exposure repeatedly, one day after another. In other words, the unthinkable was happening on a daily basis. Alan Greenspan stated, when testifying to the Congressional Committee for Oversight and Government Reform in October 2008, that while the modern risk management paradigm had held sway for some time, "the whole intellectual edifice … collapsed in the summer [of 2007] because the data inputted into the risk management models generally covered only the past two decades, a period of euphoria". Nocera (2009) wrote that while VaR was very popular and heavily relied upon by risk managers and banks, it exacerbated the crisis by giving a false sense of security to bank executives and regulators. He portrayed VaR as both easy to misunderstand and dangerous when misunderstood. The FSA identified four of the problem areas of the application of VaR in the *Turner Review* (FSA 2009), described in Box 25.6. These failed assumptions mean that reliance on such models can lead to calamitous consequences.

Box 25.6 Contribution of VaR to the global financial crisis.

The financial crisis has revealed, however, severe problems with these [sophisticated mathematical] techniques. They suggest at very least the need for significant changes in the way that VAR-based methodologies have been applied: some, however, pose more fundamental questions about our ability in principle to infer future risk from past observed patterns. Four categories of problem can be distinguished:

- *Short observation periods.* Measures of [VaR] were often estimated using relatively short periods of observation e.g. 12 months. As a result they introduced significant procyclicality, with periods of low observed risk driving down measures of future prospective risk, and thus influencing capital commitment decisions which were for a time self-fulfilling. At very least much longer time periods of observations need to be used.
- *Non-normal distributions.* However, even if much longer time periods (e.g. ten years) had been used, it is likely that estimates would have failed to identify the scale of risks being taken. Price movements during the crisis have often been of a size whose probability was calculated by models (even using longer term inputs) to be almost infinitesimally small. This suggests that the models systematically underestimated the chances of small probability high impact events. Models frequently assume that the full distribution of possible events, from which the observed price movements are assumed to be a random sample, is normal in shape. But there is no clear robust justification for this assumption and it is possible that financial market movements are inherently characterized by fat-tail distributions. This implies that any use of VAR models needs to be buttressed by the application of stress test techniques which consider the impact of extreme movements beyond those which the model suggests are at all probable. Deciding just how stressed the stress test should be, is, however, inherently difficult, and not clearly susceptible to any mathematical determination.
- *Systemic versus idiosyncratic risk.* One explanation of fat-tail distributions may lie in the importance of systemic versus idiosyncratic risk i.e. the presence of "network externalities". The models used implicitly assume that the actions of the individual firm, reacting to market price movements, are both sufficiently small in scale as not themselves to affect the market equilibriums, and independent of the actions of other firms. But this is a deeply misleading assumption if it is possible that developments in markets will induce similar and simultaneous behaviour by numerous players. If this is the case, which it certainly was in the financial crisis, [VaR] measures of risk may not only fail adequately to warn of rising risk, but may convey the message that risk is low

and falling at the precise time when systemic risk is high and rising. According to [VaR] measures, risk was low in spring 2007: in fact the system was fraught with huge systemic risk. This suggests that stress tests may need (i) to be defined as much by regulators in the light of macro-prudential concerns, as by firms in the light of idiosyncratic concerns; and (ii) to consider the impact of second order effects i.e. the impact on one bank of another bank's likely reaction to the common systemic stress.

- *Non-independence of future events; distinguishing risk and uncertainty.* More fundamentally, however, it is important to realise that the assumption that past distribution patterns carry robust inferences for the probability of future patterns is methodologically insecure. It involves applying to the world of social and economic relationships a technique drawn from the world of physics, in which a random sample of a definitively existing universe of possible events is used to determine the probability characteristics which govern future random samples. But it is unclear whether this analogy is valid when applied to economic and social relationships, or whether instead, we need to recognise that we are dealing not with mathematically modellable risk, but with inherent "Knightian" uncertainty. This would further reinforce the need for a macro-prudential approach to regulation. But it would also suggest that no system of regulation could ever guard against all risks/uncertainties, and that there may be extreme circumstances in which the backup of risk socialisation (e.g. of the sort of government intervention now being put in place) is the optimal and the only defence against system failure.

Source: FSA (2009, Section 1.4(iii)).

25.12.4 Use of VaR to Limit Risk

Even after the heavy criticism that VaR received after the financial crisis it continues to be used today. The approach adopted depends on the lessons learned and boards receptiveness to risk management

Financial institutions continue to manage market risk in different ways, depending on how risk management has evolved over time and the working practices of risk management specialists employed to support its implementation. At the time of writing, market risk management at JPMorgan Chase is an independent risk management function, aligned primarily with each of the firm's business segments. Market Risk works in partnership with the business segments to identify and monitor market risks throughout the firm as well as to define market risk policies and procedures. The risk management function is headed by the firm's chief risk officer. Market Risk seeks to facilitate efficient risk/return decisions, reduce volatility in operating performance and make the firm's market risk profile transparent to senior management, the board of directors and regulators. Market Risk is responsible for the following functions:

- establishing a comprehensive market risk policy framework;
- independent measurement, monitoring and control of business segment market risk;
- definition, approval and monitoring of limits;
- performance of stress testing and qualitative risk assessments.

Its application of VaR is described in Box 25.7.

Box 25.7 Application of VaR: an example

JPMorgan Chase's primary statistical risk measure, VaR, estimates the potential loss from adverse market moves in a normal market environment and provides a consistent cross-business measure of risk profiles and levels of diversification. VaR is used for comparing risks across businesses, monitoring limits, and as an input to economic capital calculations. Each business day, as part of its risk management activities, the Firm undertakes a comprehensive VaR calculations that includes the majority of its market risks. These VaR results are reported to senior management.

To calculate VaR, the Firm uses historical simulation, based on a one-day time horizon and an expected tail-loss methodology, which measures risk across instruments and portfolios in a consistent and comparable way. The simulation is based on data for the previous 12 months. This approach assumes that historical changes in market values are representative of future changes; this assumption may not always be accurate, particularly when there is volatility in the market environment. For certain products, such as lending facilities and some mortgage-related securities for which price-based time series are not readily available, market-based data are used in conjunction with sensitivity factors to estimate the risk. It is likely that using an actual price-based time series for these products, if available, would impact the VaR results presented. In addition, certain risk parameters, such as correlation risk among certain instruments, are not fully captured in VaR. In the third quarter of 2008, the Firm revised its reported IB Trading and credit portfolio VaR measure to include additional risk positions previously excluded from VaR, thus creating a more comprehensive view of the Firm's market risks. In addition, the Firm moved to calculating VaR using a 95% confidence level to provide a more stable measure of the VaR for day-to-day risk management. The Firm intends to present VaR solely at the 95% confidence level commencing in the first quarter of 2010, as information for two complete year-to-date periods will then be available.

Source: JPMorgan Chase & Co, Annual Report (2009).

25.12.5 Calculating Value at Risk

There are three common methods of calculating VaR, and each has its own benefits and drawbacks. They are known as the historical method, variance–covariance (or analytical) method and Monte Carlo simulation.

Historical Simulations Method

Historical simulation is intuitive, easy to understand and is the simplest and most transparent method of calculation. The fundamental assumption of the historical simulations method is that the past performance of a portfolio is a good indicator of the near-future. In other words, the recent past will reproduce itself in the near-future. This involves running the current portfolio across a set of historical price changes to yield a distribution of changes in portfolio value, and computing a percentile (the VaR). There is also no need to estimate the volatilities (and to some degree the correlations) between the various assets as they are implicitly captured by the actual daily realisations of the assets. The fat tails of the distribution and other extreme events (it is argued by proponents) are captured (as long as they are contained in the dataset and the dataset covers a sufficiently long enough period). The main disadvantages of this method are that it relies completely on a particular historical dataset and its idiosyncrasies and it cannot handle sensitivity analyses easily. In addition, if a simulation is run in a bull market the VaR may be underestimated; and conversely, if a simulation is run after a crash, the falling returns which the portfolio has experienced may distort VaR. Also it may not always be computationally efficient when the portfolio contains complex securities or a very large number of instruments.

Variance–Covariance or Analytical Method

This method assumes a normal distribution of portfolio returns, which requires estimating the expected return and standard deviation of returns for each asset. A distribution is described as normal if there is a high probability that any observation from the population sample will have a value that is close to the mean, and a low probability of having a value that is far from the mean. This method assumes that asset returns follow a normal pattern. The VaR model uses the normal curve to estimate the losses that an institution may suffer over a given time period. The main benefits of the variance–covariance method are that it requires very few parameters, it is easy to implement and it is quick to run computations (with an appropriate mapping of the risk factors). However, as the number of securities in a portfolio increases, these calculations can become unwieldy. As a result, a simplifying assumption of zero expected return is sometimes made. This assumption has little effect on the outcome for short-term (daily) VaR calculations but is inappropriate for longer-term measures of VaR. The advantage of this method is its simplicity. The disadvantage is that the significant assumption that price changes in the financial markets follow a normal distribution, which can be unrealistic.

Monte Carlo Method

Monte Carlo simulation involves developing a spreadsheet-based model for future stock price returns and running multiple hypothetical trials or simulations based on the model. As with historical simulation, Monte Carlo simulation allows the risk manager to use actual historical distributions for risk factor returns rather than having to assume normal returns. It is a flexible tool that can be quickly updated. Each simulation is one potential outcome. A number of trials are run and it is the statistical analysis of this group of aggregated trials that enables predictions to be made about price volatility. Due to the speed of current computers, more complex models populated with extensive data can be run moderately quickly. Each time the simulation is run, the result is different, as it will be the summation of a new set of random numbers drawn from the distribution of each variable. This method is more realistic than the previous two models and therefore is more likely to estimate VaR more accurately. However, for many uninformed users, Monte Carlo simulation appears to have an inherently opaque or "black box" nature and hence they are sceptical of its merits. Those who are conversant with the method, depending on the software used, are able to view the inputs and the statistical analysis undertaken and understand how the results were derived.

25.13 RISK RESPONSE PLANNING

There are no hard and fast rules for establishing risk management practices for a business. The golden rule is that the practices adopted should be tailored to suit the business and the market within which it operates. The approach followed should facilitate management discussion of market risk and opportunity and clearly set out how market risk management will be performed throughout the business. There should be a clear designation of roles in and responsibilities for each aspect of market risk management within the company. Authority levels should be determined for implementing risk management actions. These will commonly entail considerable sums of money and be part of a broad strategy such as advertising, research and development, product development and diversification. An integral part of the practical risk management steps will involve risk identification, measurement and reporting. For broad comprehension and acceptance of the approach adopted, the metrics, methodologies and assumptions within any analysis must be transparent. An insurance policy must be established which defines the type of risks that are to be insured, the target risk levels and the products and strategies that can be used.

25.14 SUMMARY

As with the other classes of risk examined, in a fully developed enterprise risk management process, market risk must be addressed as a primary source of risk and opportunity. Clearly market risk has a strong overlap with several of the other classes of risk such as technological, economic and social risk. Of primary importance to any understanding of market risk for a business is a grasp of the market structure and with it both the opportunities and the threats from existing and potential competition. Hence, the market forecasting process is crucial to addressing sources of risk. It must be acknowledged that (1) demand is not homogeneous but may be disaggregated into a series of distinct submarkets and segments, and (2) a competitive strategy increases the importance of product development, which leads to a more rapid product obsolescence and a shortening of product life cycles.

It could be argued that there is a fundamental inevitability about the accelerating rate of change. An understanding and acceptance of change is crucial to long-run survival and success. The concept of the product life cycle was examined, and while this conventional representation must not be taken too literally (as the length of the various phases may vary considerably), it provides a good forecasting and risk management tool. As competition is a major source of risk, monitoring competitive activity has assumed increasing importance. This is particularly so in recent years, with the growing popularity of what is termed "the fast second". This relates to firms depending more on their ability to copy or improve on a new product and cash in on the growth phase than on being the first to market with a new product. Game theory was explored in terms of the risks associated with pricing strategy particularly in an oligopoly and the development of brands to create competitive advantage. Similarly, price elasticity is examined to explore the use of price to increase demand and hence income; however, there is risk attached to estimating how the variables such as competitor behaviour will change. Value at risk was examined as a way of measuring market risk exposure.

25.15 REFERENCES

Anderton, A. (1995) *Economics*, 2nd edition. Causeway Press, Ornmskirk.

Barwick, P. and Meehan, S. (2004) *Simply Better: Winning and Keeping Customers by Delivering What Matters Most*. Harvard Business School Press, Boston.

Bradley, F. (1995) *Marketing Management – Providing, Communicating and Delivering Value*. Prentice Hall, London.

Carter, M. (2005) Orange rekindles the emotional ties. *Financial Times*, 20 September.

Cookson, C. (2005) Industry shakes off lethargic image. *Financial Times*, 18 April.

Cossin, D. (2005) A route through the hazards of business. FT Mastering Corporate Governance, 10 June.

Friend, G. and Zehle, S. (2004) *Guide to Business Planning*. Profile Books, London.

FSA (2009) *The Turner Review: A Regulatory Response to the Global Banking Crisis*. FSA, London. http://www.fsa.gov.uk/pubs/other/turner_review.pdf

Johnson, G. and Scholes, K. (1989) *Exploring Corporate Strategy*. Prentice Hall.

Mackintosh, J. and Milne, R. (2005) The little car that is causing big problems. *Financial Times*, 4 March.

Marcelo, R. (2005) Budget airlines set to take over Indian skies. *Financial Times*, 27 April.

Möller, K.E. and Antilla, M. (1987) Marketing capability in small manufacturing firms: a key success factor? Paper presented to the 11th Annual Conference of the European Marketing Academy, York University, Toronto.

National Industrial Conference Board (1964) Why products fail. Conference Board Record, New York, October.

Nocera, J. (2009) Risk mismanagement. *New York Times*, 4 January.

Porter, M.E. (1980) *Competitive Strategy*. Free Press, New York.

Ward, A. (2005a) A better model? Diversified Pepsi steals some of Coke's sparkle. *Financial Times*, 28 February.

Ward, A. (2005b) Classic error that continues to make bubbles 20 years on. *Financial Times*, 25 April.

26
Social Risk

There are risks and costs to a program of action. But they are far less than the long-range risks and costs of comfortable inaction.
(John F. Kennedy)

The previous chapter examined market share risk, the fifth of six classes of risk exposure that businesses face relating to business operating environment within the risk taxonomy proposed in Chapter 9. This chapter examines *social risk*, the sixth of the macro influences of the business operating environment: those aspects of society that impact on business performance over which businesses have no ability to control and minimal opportunity to influence. Existing and emerging trends in lifestyle choices and social attitudes are explored in terms of the risks and opportunities emanating from the evolving characteristics of the workforce. Against a backdrop of an inexorable increase in living standards, changes are occurring in drinking, smoking, eating and exercise habits. Various methods of classifying the UK workforce are considered, with particular attention being paid to socio-economic groups. Of course the demographic and socio-cultural dimensions make an important contribution to the PEST assessments discussed in Chapter 8. The structure of this chapter is shown in Figure 26.1.

26.1 DEFINITION OF SOCIAL RISK

What is social risk? From a business perspective, social risk emanates from changes in society, which create changes in demand, open new market opportunities or alter a business's responsiveness to demand, as a consequence of the characteristics of its workforce. The availability, education, health and outlook of the workforce all influence a business's performance capabilities. Workforces are assumed to take on the habits, behaviours and culture of the society within which they work and live. Social risk is defined here as a society's impact on business, and not vice versa. Hence, social risk in this context should not be confused with (1) corporate social responsibility,[1] (2) supporting vulnerable individuals or households who are susceptible to damaging welfare losses, (3) government social policy, social security and welfare state policy or (4) social risk management as defined by the World Bank Group.[2]

[1] The government views corporate social responsibility (CSR) as the voluntary actions that business can take (over and above compliance with minimum legal requirements) to address both its own competitive interests and the interests of wider society. The government is looking to businesses to contribute to the government's sustainable development goals. However, there is still some debate as to what CSR means. As declared by the Department of Trade and Industry (2004), "we remain a long way from consensus on what it means and its value. Some suggest that it is just about glossy reports and public relations. Some see it as a source of business opportunity and improved competitiveness. Some see it as no more than sound business practice. Others see it as a distraction or threat. Is it a framework for across the board regulation of all of the relationships between business and the rest of society, nationally and globally?" The government anticipates that the debate will continue.

[2] The World Bank Group defines social risk management (SRM) as "a framework that can be used to analyse the sources of vulnerability, how society manages risks and the relative costs and benefits of various public interventions on household welfare", where vulnerability describes the exposure to uninsured risk leading to a socially unacceptable level of well-being in the future. As such, SRM addresses how vulnerable individuals and households can be helped to better manage risks and become less susceptible to damaging welfare losses (http://www.worldbank.org/srm).

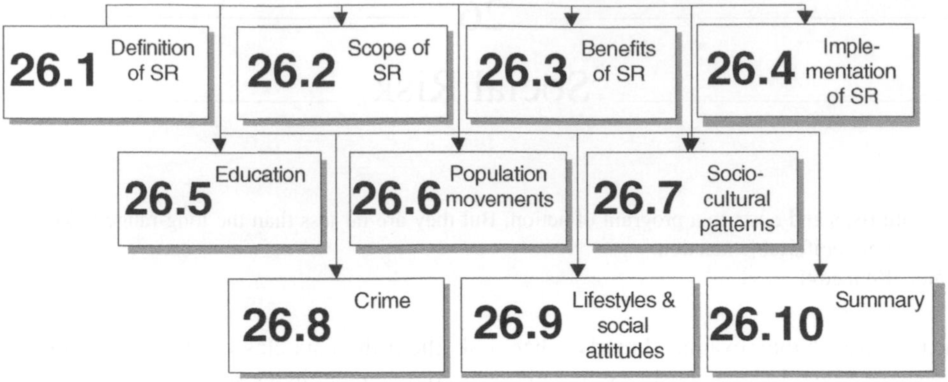

Figure 26.1 Structure of Chapter 26

26.2 SCOPE OF SOCIAL RISK

The sources of risk that are considered to be embraced within the term "social risk" are as follows:

- poor standard of education of new recruits, particularly with regard to language skills;
- linguistic barriers to international trade;
- the shrinking percentage of the working population that is of working age;
- misaligned marketing strategy through a lack of appreciation of the new socio-economic groups, their geographical dispersion, income range or the percentage of the overall population that they represent;
- the value of the home improvement market and the way it responds to interest rates;
- growing obesity (particularly among teenagers) and its potential impact on the future workforce;
- loss of market share through lack of attention to the "grey market".

26.3 BENEFITS OF SOCIAL RISK MANAGEMENT

Social risk management affords a business benefits such as, but not limited to:

- understanding the risks to its business premises and personnel from crime including theft, arson, vandalism and violence;
- identifying the risks associated with employment and, in particular, the level of education of new recruits and the promotion of a proactive approach to understanding and addressing any shortcomings, particularly with regard to foreign language capabilities;
- comprehending the evolving socio-economic groups and the changes this makes to market sectors and hence product markets;
- identifying trends such as public response to increases in the interest rate and corresponding decrease in spending and diminishing retail sales, such as the home improvement market;
- understanding the risk emanating from employees' use of alcohol and drugs – abusers threaten their own physical and mental health, increase the risk of accidents, lower productivity, raise insurance costs and reduce profitability; they are characterised by erratic work patterns, higher absenteeism and lower productivity.

26.4 IMPLEMENTATION OF SOCIAL RISK MANAGEMENT

The development of a sound system will depend on attention to a number of trends, such as:

- Growing environmentalism
- Growing interest in health and fitness
- Concern over obesity
- Growing concern about the ethics of organisations
- More leisure time
- Earlier retirement
- More women in the workforce
- Increasing improvement in living standards
- Slow population growth
- Ageing population
- Drug and alcohol abuse.

26.5 EDUCATION

Education is seen as a contributor to economic growth. Government education policies have attempted to deal with unemployment and our lack of competitiveness in the market place. In a speech in 1976 the Labour Prime Minister, James Callaghan, suggested that the education system was not giving enough concern to the needs of industry and it should be made more accountable to popular wishes and desires.

National education in the UK is important to businesses as the quality of education of school and college leavers and graduates from university directly impacts on their ability to compete in the market place. Government attention to education fluctuates but events in the market place and the status of the economy bring the subject to the fore. Economic developments, in particular the persistence of high unemployment and the contraction of manufacturing industry, have repeatedly forced questions on the content and purpose of education onto the agenda of public debate. The report published by CILT[3] entitled *Talking World Class: The Impact of Language Skills on the UK Economy* brought into sharp focus the detrimental impact of the UK's complacency over our inability to speak foreign languages (CILT 2005). In his introduction to the CILT report, Sir Trevor McDonald states:

> The growing internationalisation of business life in the global economy is well documented. Britain's economy has always benefited from foreign trade, but now, with new patterns of business ownership and the growing creation of strategic alliances across borders, overseas connections are becoming part of the very fabric of economic life. What effect are these trends having on our competitiveness, and where will they take us in the future? With the widespread use of English in international business, it is easy to overlook linguistic and cultural factors. Yet these have a crucial effect – surveys of exporting businesses show that nearly half have experienced linguistic or cultural barriers and one in five have lost business as a result.

Kathryn Board, CEO for CILT, wrote in her article for the *Independent* newspaper in October 2009 that Professor James Foreman-Peck, (an economist at Cardiff Business School who has studied the role of language in determining patterns of world trade), has shown the UK is

[3] CILT is the government's recognised centre of expertise on languages. The organisation's mission is to promote a greater capacity among all sectors of the UK population. It is a registered charity and it derives its finance from central government and other secondary sources.

underperforming in its trade with non English-speaking countries. He considered not sharing a language a barrier to trade for countries the world over, but for the UK the effect is nearly double the world average. Foreman-Peck calculated by improving our language skills we could add £21 million a year to GDP, and concluded we are significantly under-investing in them.

CILT's "Language Trends Report" of 2010, (the latest in a series of reports which explore aspects of the teaching of languages in secondary education in England) again makes alarming reading. This report is based on responses to a questionnaire sent to a representative sample of 2000 secondary schools in England (1500 maintained schools and 500 independent schools). CILT's reports have become a barometer for the health of language teaching in secondary schools in England. The key findings of the report stated that:

- An increasing number of maintained schools (one in five compared to 14% last year) have shortened KS3 language studies (for 11 to 14 year olds) to two years. This represents a significant reduction in the amount of time spent on languages for those pupils who do not take the subject to GCSE.
- The number of students studying languages at GCSE level (15 to 16 year olds) continues to decline in state secondary schools, with the trend remaining downward.
- Recruitment to language courses post 16 appears to be very fragile, with more schools reporting a decline than an increase in numbers in both maintained and independent sectors.
- The recent government decision to step back from making languages statutory in the primary curriculum may have added a stimulus to secondary schools to play down any initiatives that they had with regard to including languages in their forward planning.
- Employers rank foreign language skills as second only to IT when it comes to finding the right candidate.
- Where the use of languages is important for a role, employers will favour young multi-lingual professionals from outside the UK for jobs over mono-lingual Brits.

26.6 POPULATION MOVEMENTS: DEMOGRAPHIC CHANGES

Populations can be perceived in two broad ways: first, as a market; and second, as a source of labour. Thus changing demographics have both market and resource implications for businesses. Demographics is the study of how populations change over time. Demographers analyse changes in births, deaths and the age composition of the population as well as patterns of disease and other health-related issues. In the UK, the Government Actuary's Department (GAD)[4] collects population statistics and strives to produce projections of how the population will change over time, particularly over short-term horizons. Population changes are important to businesses as they affect demand for goods and services, especially in the medium and long term. Careful analysis of these patterns and trends is therefore important to marketing managers in business and planners in government. Populations change naturally over time. For any given period, the change in population is given by:

$$\text{population change} = \text{births} - \text{deaths} + \text{net migration}.$$

Net migration is the difference between the number of people entering the country to live (immigration) and the number leaving permanently (emigration). Since the 1980s net migration

[4] Further information on the Government Actuary's Department can be found at http://www.gad.gov.uk

in the UK has been positive and is expected to remain so for at least the immediate future. However, what concerns many Western countries is not so much the size of their population as the problems associated with their ageing population.[5] The percentage of those over 65 years of age is gradually increasing. The population is ageing as more people are living longer and there is a fall in the birth rate. Life longevity is considered to be the result of better health care, improved standards of living, and improved nutrition and housing. The rise in the life expectancy of the old has been accelerating over the past two decades. For example, in 1980 male expectancy at 60 was a further 16 years. In 2004 it was a further 20 years. The United Nations has observed that the median age is increasing across the whole world with, not surprisingly, a greater increase in the more industrialised economies. This has major implications for international business in terms of productive location (available supply of labour of working age) as well as the range of products likely to be in demand. It also has implications for the government as state pensions are funded by contributions of current workers.

The proportion of workers to pensioners is known as the support ratio, sometimes expressed simply as taxpayers per pensioner. Therefore if there are more and more pensioners and fewer and fewer workers to support them, it will be difficult to maintain the real value of pensions. Currently in the UK there are four workers per pensioner, giving a support ratio of 4:1. In 50 years' time this is expected to fall to somewhere around 2.5:1, a situation described as a "demographic time bomb" (Griffiths and Wall 2005). The pattern would appear to be the same outside the UK. A study into the impact of an ageing workforce on labour markets by the Organisation for Economic Co-operation and Development warns of rising wage inflation, increased pressure on public finances and declining growth unless this demographic situation is addressed. The report considers that in Europe the ratio of active to inactive workers is expected to fall to 1:1 by 2050. It estimates that GDP growth in OECD countries will fall by 30% compared to 1970–2000 unless more employees work longer to overcome skill shortages (Taylor 2005).

26.6.1 The Changing Market

Although most Western populations are expected eventually to decline, this will not happen for many years to come. At one time marketing departments largely ignored the older generation, stereotyping them as frail inactive individuals with little spending power. However, market analysis is waking up to the "grey market", a term now in widespread use to describe the over-50s. Some 44% of Britain's adults are over 50, and this proportion is rising rapidly. It was estimated that by the end of 2005, this group would have increased by more than 13% in just 10 years. Estimates vary but it is commonly considered that this population is set to increase to one third of the populace by 2020. This demographic group is not only large it is also rich. It is currently 20 million strong, growing fast and holds 80% of the nation's wealth. It has the largest proportion of outright homeowners and more savings than the adult population at large. House price increases and maturing life insurance policies have added to this wealth. One of the key ways this group is spending its time and money is on holidays, with the majority taking two holidays a year. The older generation now accounts for a substantial proportion of the UK tourism market, approximately 22% of domestic holidays and 26% of foreign holidays.

[5] India is an interesting exception; the second most populous country in the world, it has an astonishing demographic dividend, with 65% of the population below the age of 35.

26.7 SOCIO-CULTURAL PATTERNS AND TRENDS

Most societies have some form of hierarchy, which ranks people according to their social standing. Traditionally, UK society was seen to exist of three "classes" – working class, middle class and upper class – with most of the wealth and influence concentrated in the land-owning aristocratic upper classes. It is estimated that in 1911 the richest 1% held around 70% of the UK's wealth. Reference to these three classes is now rarely made. Times have changed, and inherited wealth is not the dominant factor it once was. The UK Office of National Statistics (ONS) Social Trends report shows that the richest 1% hold 23% of the UK's wealth.

26.8 CRIME

National crime statistics influence overseas investment, mergers and acquisitions, tourism, property speculation and choice of university. Regional crime statistics enable companies to make specific decisions such as determining the site of new or relocated premises, and relocation of staff (and their families). Crime influences a manufacturer's choice of location for new production plants, businesses selection of office rental space and families' choice of location for a new home. There is a clear variation between towns, cities and districts across the UK in terms of their desirability as places of residence or locations for businesses. The most favoured locations fluctuate; however they tend to be where crime is low, employment is high, education levels are high and salaries are above average.

26.8.1 Key Facts

The Home Office figures published in January 2011 show that:

- There were 619 deaths currently recorded as homicide in England and Wales in 2009/10 (this is the lowest number of deaths since 1997/98, when 606 were recorded).
- Taken overall, the risk of being a victim of homicide was 11 offences per million population.
- In 2009/10, as in previous years, more than two-thirds of homicide victims (68%) were male.
- The most common method of killing continues to be by sharp instrument.
- There were 41 shooting homicide victims in 2009/10 compared with 38 in 2008/09.
- Female victims were more likely to be killed by someone they knew. Around three quarters (76%) of female victims knew the main suspect, compared with half (50%) of male victims.
- Based on 2004 Eurostat figures, England and Wales are in the middle of the European table of homicide rates at 13.5 deaths per million population. Finland topped the table at 23.4, whereas Austria had the lowest murder rate in Europe at 6.1 per million (Home Office 2011a).

There is currently a decline in most crimes:

- In 2010 robbery, burglary, vehicle theft, drug offences and firearm offences were all reported to be on the decline compared with the previous year, whereas there was an increase in sexual offences (Home Office 2011b).

Additional information, such as on the prison population and firearm use, is available through regional data:

- The prison population has increased year-on-year between 2003 and 2008, reaching over 80 000 in 2008.
- Statistics show firearm offences decreased steadily between 2002 and 2008.
- Police operations in which firearms were authorised are on a steady increase (Home Office).

26.9 LIFESTYLES AND SOCIAL ATTITUDES

26.9.1 More Home Improvements

Businesses in the retail sector supplying the home improvement market are susceptible to swings in demand which can have a dramatic effect on bottom-line performance in a market where annual consumer spending in the UK exceeds £7 billion. There are a number of drivers for home improvements. Research conducted by the retailer Homebase found that the main reason why people embarked on home improvements was to make their homes more comfortable to live in, increase their value prior to putting them on the market to sell, carry out necessary repairs and create more space. However there are a number of pressures on consumer spending, such as job security, the cost of living and the state of the economy. The years between 2004 and 2010 have proved to be difficult for the UK home improvement market, more recently due to the world financial crisis and the subsequent turmoil in the European Union. Different research agencies are predicting the market is likely to shrink from the aggregate sales figure of approximately £9.0 billion in 2008 to £7.8 billion in 2013. Indicators of the struggle emerge, for instance, when the media announces job losses such as occurred in 2005 when B&Q shed 400 jobs. The overall UK home improvement market shrank by around 1% in 2006, recovered marginally in 2007 and then shrank by some 3.6% in 2008. In 2010 Chancellor George Osborne increased VAT from 17.5% to 20%, which exacerbated the fall in sales. The leading players in the market are B&Q (a subsidiary of Kingfisher). Homebase, Focus and Wickes. They all have to consider new entrants to the market, buyer power, supplier power, the threat of substitutes and rival behaviour.

26.9.2 Motherhood, Marriage and Family Formation

The UK Office for National Statistics publication "Social Trends No 40, 2010 edition" provides a view of family formation and motherhood, and specifically evolving changes in society. This report, while published in 2010, refers to statistics for 2007 and 2009. There are very few surprises. Previous trends appear to be continuing down the same path.

- Since the 1970s, there has been a fall in the proportion of babies born to women aged under 25 in England and Wales.
- There is an increasing trend towards smaller household sizes. The average household size in Great Britain has fallen from 3.1 people per household in 1961 to 2.4 people in 2Q 2009.
- Apart from a small number of exceptions, the number of marriages per year has continued to fall since the 1970s.

- The number of divorces has increased year on year since the 1970s, other than the period 2003 to 2007 where there has been a small percentage decline.
- As a result of the choice women are making in terms of living independently, continuing their education and participating more fully in the labour market, the average age of women marrying for the first time reached 29.8 years in 2007. Between 1971 and 2011 there has been an upward trend in the proportion of women in employment and a downward trend in the employment rates for men (ONS 2011b).
- Along with changing attitudes to motherhood, the average age of mothers having their first child within marriage in 2007 was 30.3 years.

26.9.3 Health

The government report "Health, Social Trends 41" published by the Office for National Statistics in 2011 provides a summary of the key aspects of health matters in the UK as follows:

- In the UK, in 2008, health expenditure per head and life expectancy were ranked 16th and 17th respectively by the 32 Organisation for Economic Co-operation and Development (OECD) countries at $3281 and 79.9 years.
- In 2008, within the 32 OECD countries, the highest health expenditure per head was in the United States where life expectancy was ranked 24th. The highest life expectancy was in Japan where health expenditure was ranked 20th.
- In 2008 life expectancy at birth in the UK was at the highest level recorded for both males and females at 77.6 years and 81.7 years respectively: both had increased by about 20 years since 1930.
- Not only are life expectancies increasing, but both men and women are staying healthy and free of disability for more of their lives.
- The proportion of adults reporting that a long-standing illness or disability limited their daily activities changed very little between 1981 and 2009.
- In 2008/09 more than half of males and females in Great Britain were classed as overweight or obese (60% and 52% respectively).
- In the UK, between 2006 and 2008, the highest incidence of cancer in males was of the prostate (100.0 per 100 000 population) and in females was of the breast (123.0 per 100 000).

While UK companies need to take cognizance of local social trends, those that have interests overseas will have to reflect within their international operations on the local prevailing social issues. Of prominence is the HIV/AIDs pandemic (see Box 26.1).

Box 26.1 The HIV and AIDS pandemic

HIV was first identified in 1983. HIV stands for Human Immuno-deficiency Virus (HIV), and is the virus which causes Acquired Immunodeficiency Syndrome (AIDS). HIV attacks, and slowly destroys, the immune system by entering and destroying the cells that control and support the immune response system. After a long period of infection, usually 3–7 years, enough of the immune system cells have been destroyed so as to lead to immune deficiency. As a result the virus can be present in the body for several years before symptoms appear. When a person is immuno-deficient the body has difficulty defending itself against many infections and certain cancers, known as "opportunistic infections". Around half of all people who become infected with HIV do so before they reach the age of 25 and most will die of AIDS or related illnesses before they are 35. This means that HIV/AIDS is affecting some of our most creative and economically active people.

Companies are recognising the impact that the virus is having, in terms of the human, financial and social costs, to its operations and host communities. While the countries most affected are on the African continent (such as Botswana, Ethiopia, Ghana, Kenya, Namibia, South Africa, Uganda, Zambia and Zimbabwe), global mobility and high levels of immigration in the UK has meant that UK firms have not escaped workforce problems arising from the pandemic. The rapid spread of the disease is directly linked with the development of the global economy and the significant growth in international trade and travel. Without such extensive and interlinked transportation systems, the disease would never have moved as quickly as it has. HIV and AIDS present companies with enormous and varied challenges. The degree to which HIV and AIDS will affect a company will depend on the degree of adoption of globalisation of production, and the search for new workforces and markets around the world. In particular it will relate to the number of employees infected, their respective roles in the company, the ability of the company to cope with absenteeism, mortality rate, and the benefits provided by the company such as health insurance and pensions. A survey completed by Stellenbosch University, South Africa in 2004 found that more than a third of the companies surveyed indicated that HIV and AIDS had reduced labour productivity or increased absenteeism, and raised the cost of employee benefits. The problems encountered by firms with subsidiaries or manufacturing plants in the worst affected countries have experienced the following problems:

- Higher absenteeism and staff turnover.
- Reduction in the productivity of the workforce.
- Increased production costs and an increased production cycle.
- Overall labour costs increase.
- Recruitment costs increase.
- Training costs increase.
- Insurance premium costs increase.
- Manager's time and resources eroded by HIV-related issues.
- Payouts from pension fund cause employer and/or employee contributions to increase.
- Returns to training investments are reduced.
- Morale, discipline, and concentration of other employees are disrupted by the frequent death of colleagues.
- Additional recruiting of staff required.
- Fall in revenue and profitability.

26.9.4 Less Healthy Diets

Health experts are alarmed at the unhealthy lifestyles of children in the UK. Doctors attending the National Obesity Forum conference in London in 2004 called for smaller portions and for play areas for children. Professor Thomas Wadden, of Pennsylvania University, said: "People, particularly children, are being 'swept along' with the environment [and] obesity is always treated as if people have a lack of will-power. But we live in a society where there is a high-fat, high-sugar diet and low physical activity." He said, around 22% of adults in the UK were obese and the same proportion again were overweight. "It's an epidemic where everyone is being affected." He considered society had to make it easy for people to make the right choices.

More recently a British Medical Association (BMA) report issued in 2005 called for a junk food advertising ban and rules for the nutritional balance of school meals and pre-prepared food. The BMA warned that without strong action, children would increasingly develop adult diseases such as type-2 diabetes, cancer and bone problems. In 2005 it was estimated that in the UK, about 1 million children under 16 are now obese. In addition, the BMA's Board of Science has warned that if current trends continue, it is estimated at least a fifth of boys and a third of girls will be obese by 2020. The BMA report called for the government to intervene in a range of areas, including schools, the food industry and advertising. Unhealthy children will grow up to be unhealthy adults. Hence, there is a concern that businesses will experience sickness levels and lost man-days that will far exceed current levels.

26.9.5 Smoking and Drinking

Government figures show that in 2009[6] 21% of the adult population of Great Britain were cigarette smokers. The overall prevalence of smoking has been at this level since 2007. The prevalence of cigarette smoking fell substantially in the 1970s and the early 1980s, from 45% in 1974 to 35% in 1982. The rate of decline then slowed, with prevalence falling by only about one percentage point every two years until 1994, after which it levelled out at about 27% before resuming a slow decline in the 2000s. Smoking prevalence was higher among men than women from 1974 until 2008, but in 2008 the difference between men and women was not statistically significant. In 2009, smoking prevalence was slightly higher among men (22%) than among women (20%). There were striking differences between the various socio-economic groupings. Smoking prevalence is lower among those in higher professional households (10%) and highest, at 32%, among those in a routine occupation. As smoking is the leading cause of preventable illness and premature death in Great Britain, reducing its prevalence is a key government target in improving public health. Legislation came into force in February 2003 banning cigarette advertising on billboards, in the press and in magazines in the UK, and further restrictions on advertising at the point of sale were introduced in December 2004. A ban on smoking in enclosed public places came into force in Scotland during the spring of 2006 with similar bans in England and Wales being introduced in 2007. On 1 October 2007 it became illegal in Great Britain to sell tobacco products to anyone under the age of 18. Since smoking is estimated to be the cause of about a third of all cancers, reducing smoking is one of three key commitments at the heart of the government's *NHS Cancer Plan*, which was published in 2000. In particular, the *Cancer Plan* focused on the need to reduce the comparatively high rates of smoking among those in manual socio-economic groups, which result in much higher death rates from cancer among unskilled workers than among professionals.

[6] The General LiFestyle Survey (GLF) is a multi-purpose continuous survey carried out by the Office for National Statistics (ONS). It collects information on a range of topics from people living in private households in Great Britain. The survey started as the General Household Survey (GHS) in 1971 and has been carried out continuously since then, except for breaks to review it in 1997/1998 and to redevelop it in 1999/2000. The survey presents a picture of households, families and people living in Great Britain. The GLF and GHS have been monitoring smoking prevalence for over 35 years.

The Department of Health estimates that the harmful use of alcohol costs the NHS around £2.7 billion a year and 7% of all hospital admissions are alcohol related. Drinking can lead to over 40 medical conditions, including cancer, stroke, hypertension, liver disease and heart disease. The government's statistics illustrate that during the 1990s there was a slight increase in overall weekly alcohol consumption among men and a much more marked one among women. Following an increase between 1998 and 2000, there has been a decline since 2002 in the proportion of men drinking more than, on average, 21 units a week and in the proportion of women drinking more than 14 units. The proportion of men drinking more than 21 units a week on average fell from 29% in 2000 to 23% in 2006. There was also a fall in the proportion of women drinking more than 14 units a week (from 17% in 2000 to 12% in 2006). The risk for businesses is that drinking habits lead to absenteeism, poor performance, accidents and behavioural problems. Sustained problems of excessive drinking may lead to long-term illness. On the reverse side of the coin, an increase in the popularity of drinking increases demand for suppliers in the brewing industry.

26.9.6 Long Working Hours

Studies have found that working long hours can greatly increase employees' exposure to the risk of injury or illness. Research by the University of Massachusetts[7] found that workers who do overtime are 61% more likely to become injured or ill, once factors such as age and gender were taken into account. Plus working more than 12 hours a day raised the risk by more than a third. The study of US records from 110 236 employment periods found that a 60-hour week carried a 23% greater risk. The study looked at data from 1987 to 2000. Report co-author Allard Dembe said that risk was not necessarily associated with how hazardous the job was. "Our findings are consistent with the hypothesis that long working hours indirectly precipitate workplace accidents through a causal process, for instance, by inducing fatigue or stress in affected workers." Also, he said the findings, published in the *Occupational and Environmental Health* journal, supported initiatives such as the 48-hour European Working Time Directive to cut the number of working hours. From the records, researchers found 5139 work-related injuries and illnesses, ranging from stress to cuts, burns and muscle injuries. More than half of these injuries and illnesses occurred in jobs with extended working hours or overtime. The researchers concluded, not surprisingly, that the more hours worked, the greater the risk of injury.

26.9.7 Stress Levels

Stress is a common adverse reaction to excessive pressure. It is the duty of employers to make sure that employees are not made ill by their work. Stress can make employees ill. The costs of stress to a business may show up as high staff turnover, an increase in sickness absence, reduced work performance, poor time keeping and more customer complaints. Stress in one person can lead to stress in other staff who have to cover for their colleague. Employers who do not take stress seriously may leave themselves open to compensation claims from employees who have suffered ill health from work-related stress. Outward signs of stress in individuals are changes in mood or behaviour such as irritability, deteriorating relationships with colleagues, indecisiveness, absenteeism or reduced output. Those suffering from stress may also smoke

[7] BBC News (2005) Long working hours "health risk". 17 August. http://news.bbc.co.uk

or drink alcohol more than usual or even turn to drugs. They may, not surprisingly, complain about their health, which may suffer in a number of ways.

Stress is considered a major contributor to the initiation and continuation of addiction to alcohol as well as to subsequent relapses. Addiction is considered a complex problem determined by multiple issues, including psychological and physiological factors. While many studies have demonstrated an association between alcohol addiction and stress, they have been unable to establish a direct causal relationship between the two. However, stress, and the body's response to it, are thought to play a role in the vulnerability to initial alcoholic misuse and relapses after treatment.

26.9.8 Recreation and Tourism

Living standards, disposable income and the amount of time available for leisure pursuits for the working population have a direct impact on the amount of time spent on recreation, sportswear, sport and tourism. This includes fitness centres, sports centres and clubs, attendance at sports fixtures and holidays and the whole of the tourist industry. The tourist industry includes tourist operators, hotels, airlines and airports.

26.10 SUMMARY

This chapter has described the social threats and opportunities that impact businesses as a result of a combination of risks emerging from the characteristics of the workforce and opportunities arising from demand from products and services to meet social needs. For the majority, the standard of living is constantly improving and as a result spending patterns change to reflect disposable income. The lifestyle habits of employees and their priorities reflect on the attributes of the workforce. The health of the workforce, in particular, is considered to directly impact on business performance. When the smoking, drinking and/or eating habits of any employee are excessive, it can have a debilitating effect on their ability to perform through loss of concentration, energy and commitment. In extreme cases it can lead to long-term health problems and frequent absences from work due to sickness. While employers are sympathetic, it can be very disruptive, presenting repetitive resource management difficulties.

While the English have a propensity to work long hours (unlike some of our European cousins), it is not always productive as tiredness can lead to accidents and injuries. Employers also need to ensure that their staff are not under excessive pressure, which can lead to stress. The visible signs of stress can surface as changes in behaviour such as irritability and absenteeism and, where pronounced, long-term sickness. Employees deal with stress in different ways, with some drinking or smoking more and in some cases taking drugs. It is considered that British people spend more money on alcohol to help them cope with stress than any of their European counterparts.

As a nation our reluctance to learn foreign languages translates into poor linguistic skills and significantly reduces our ability to compete internationally for overseas trade. Regional crime figures may influence where workers move to and where businesses establish themselves. Regional variations in crime figures directly translate into distinguishable regional insurance costs for home, car and business premises insurance.

The home improvement market was singled out for examination as a visible sign of lifestyle trends due to its size (annual expenditure) and its tendency to act as an economic barometer, as it is impacted by interest rates, the housing market and consumer confidence. The home improvement market is always buoyant when the house market is very active as homeowners are buying materials to improve their home to move up the housing ladder, and those that have recently bought a new home commonly redecorate to tailor their property to their own tastes.

26.11 REFERENCES

CILT (2005) *Talking World Class: The Impact of Language Skills on the UK Economy*. CILT, London.

Chamber of Commerce/LSC (2004) *The Impact of Foreign Languages on British Business*. London.

Department of Trade and Industry (2004) *Corporate Social Responsibility: A Government Update*. http://www.bis.gov.uk/files/file48771.pdf

Dorling, D., Mitchell, R., Shaw, M., Orford, S. and Smith, G.D. (2000) The ghost of Christmas past: Health effects of poverty in London in 1896 and 1991. *British Medical Journal*, 321, 1547. http://www.bmj.com/content/321/7276/1547.full

Europa (2005) Eurobarometer. http://www.europa.eu.int/comm./education/policies/lang/languages/lang/europeanlanguages_en.html

Frean, A. (2005) Abortions soar as careers come first. *The Times*, 28 July.

Frith, M. (2005) North–South divide emerges over the UK's best places to live. *The Independent*, 9 August.

Grant Thornton (2004) *International Business Owners Survey*. London.

Griffiths, A. and Wall, S. (2005) *Economics for Business and Management: A Student Text*. Pearson Education, Harlow.

Hagen, S. (2005) *Language and Culture in British Business*. CILT, London.

Home Office (2011a) *Homicides, Firearm Offences and Intimate Violence 2009/2010*. Home Office, London, 20 January.

Home Office (2011b) *Crime in England and Wales: Quarterly Update to December 2010*. Home Office, London, 20 April.

Home Office *Home Office Crime Statistics, Regional Trends*. http://rds.homeoffice.gov.uk/rds/pdfs09/hosb1309.pdf

ONS. (2011). *Smoking and drinking among adults 2009. A report on the 2009 General Lifestyle Survey*. UK Office for National Statistics, London.

ONS (2011b) *Labour Market, Social Trends 41*. UK Office for National Statistics.

The NHS Cancer Plan, Department of Health (2000) Available at www.dh.gov.uk/assetRoot/04/01/45/13/04014513.pdf

Taylor, A. (2005) Global growth to fall unless people work longer. *Financial Times*, 11 October.

Part V
The Appointment

This part of the book presupposes that not all businesses will have the knowledge, experience or in-house resources to implement enterprise risk management, and may wish to engage external support in the form of an independent impartial party to assess the maturity of their existing practices, address particular stakeholders' concerns, introduce best practice or carry out specific risk management activities. This part of the book is devoted to the consultant appointment process. It is subdivided into four chapters, which cover an introduction to consultancy services, interview with the client, appointment and implementation. The purpose of these chapters is to help consultants and in-house departments understand the communication process that has to be engaged in with sponsors of enterprise risk management to establish what is required, what should be offered and how implementation should be carried out. It will also enable procurers and sponsors of risk management services to structure their thinking about what they require, interfaces with other sectors of the business, benefits, the process to be followed and, to some degree, consultants' expectations.

27
Introduction

From time to time businesses are faced with having to engage in activities which are non-routine, expose the business to more risk than day-to-day business operations and in which the business has limited experience. Such activities might include an acquisition, the construction of a new building, product diversification, the purchase and installation of a new IT system, organisational change or a combination of these activities. In such a situation, the business has to decide if it will undertake the risk management of these activities in-house, using its own resources, or engage an external specialist consultant. In addition, it will need to decide what scale or extent of risk management activity is necessary. This will normally be decided by the board, a member of the board or the risk committee (if one exists), and any pressure brought to bear by the non-executive directors, to follow a particular solution. The option selected is commonly a function of whether in-house expertise exists, whether this expertise has the time available to undertake this additional activity, or the importance of the assignment. It may also be based on the value attributed to the risk management activity, whether major stakeholders have expressed a preference that external help be sought, whether staff would be more responsive to a consultant, the commercial sensitivity of the activity, or the cost if it was outsourced. However, regardless of which route is adopted, any risk management activity will have to follow a common pattern. This chapter briefly describes the common steps within a generic process.

27.1 CHANGE PROCESS FROM THE CLIENT PERSPECTIVE

27.1.1 Planning

All businesses at some stage engage in some form of organisational change in order to realign themselves with the market, facilitate expansion, increase market share or create new markets. This may include IT investment, product development, premises rationalisation, embarking on e-commerce or an overseas investment. Any proposed major change that a business intends to carry out must be planned as a project in its own right. There must be a clear set of objectives, timeframe, budget, sponsor, project manager, brief, designated participants and desired outcomes. There must be an understanding of the complexity, degree of novelty and how the activity will impact existing and potential customers. There must be clear recognition of the significance of the programme to the success of the business and the speed with which the activities are to be completed. All change programmes entail risk. Hence, there must also be an understanding of the risks and opportunities associated with the change. What are the ramifications if the change is implemented late? What are the implications if the change does not accomplish all of its objectives? What would the effect be if the cost of the change significantly exceeded expectations?

27.1.2 Timely Information

A prime objective of risk management is to enable both improved and timely decisions to be made. While risk management informs initial investment decisions, it must be an ongoing process. Finding out at the end of a change programme that the change did not accomplish its objectives robs a board of its decision-making powers. It is denied the opportunity to abandon, delay or change the programme of activity. There must be regular reporting so that the progress of the implementation of the change is understood and conveyed to the board to facilitate choices. Deloitte Research, a part of Deloitte Services LLP, undertook a survey to analyse the causes of major shareholder value losses between 1994 and 2003 (Kambil *et al.* 2005). Kambil *et al.* found from their study that the absence of, or inadequate, risk management was also compounded by the lack of timely information. While early risk management can steer investment decisions in the right direction, risk analysis must be carried out during implementation, to inform boards of the ramifications of late delivery, partial delivery or programme failure. The study found that senior executives and boards of directors had been deprived of information on the causes, financial impact and possible resolution of the problems that had arisen. The authors found from their research that the lack of timely information reflected adversely on the senior executive team and their control of the organisation, and often led to their departure. The shock to investors, who suddenly learnt of previously undisclosed problems at a "late hour", drove share prices down. Hence, while risk assessments can assist boards with programme choice from among a series of alternatives, interim results on an ongoing change programme must be released on a regular cycle to permit both prompt decision making and reporting to shareholders. This need for ongoing risk management support places demands on risk management resources.

27.1.3 Risk Management Resources

Businesses embarking on a change programme must assess their risk management capabilities. If the expertise or the resource is not available in-house, and the choice is made to engage a risk management consultant, it is common for a selection process to be undertaken. This entails engaging in some or all of the following activities:

- understanding which department or cost centre will fund the consultant's costs and who will be the designated sponsor;
- describing the objectives of the activity;
- preparing a description of the activity and how it interfaces with the day-to-day activities of the business;
- deciding who will manage the appointment process and over what timeframe it will be concluded;
- deciding on the method of inviting proposals;
- drawing up a long list of consultants;
- sending out letters to the long list, inviting expressions of interest;
- preparing pre-selection criteria to assess the long list, to derive a short list;
- short-listing interview;
- evaluating returns and interviews and preparing a short list to invite tenders from;
- agreeing on the number of consultants to be invited to submit proposals;
- preparing an exclusion notification – a letter advising those who will not be invited to tender;

- agreeing on the information to be submitted as part of the technical and financial proposals, the quality/price ratio and the scoring matrix;
- pre-selection interview (discussed in Chapter 28);
- inviting technical and financial proposals from short-listed firms;
- agreeing on the interview process in terms of timing, purpose, structure, attendees and overall format;
- agreeing on the individual or individuals who will make the selection;
- appreciating who from the business will have to be involved in the assignment and assessing their availability;
- determining who (within the business) the primary point of contact would be for the consultant if appointed;
- deciding if a lump sum or day rates are required, depending on the degree of certainty of the duration of the assignment.

The actual sequence of activities followed will be dictated by the size, complexity and value of the commission and the time available for a formal tender process.

27.2 SELECTION OF CONSULTANTS

27.2.1 Objectives

The objectives of a change programme must be clear and unambiguous. They must not be too wordy. They should be small in number so that they can be readily committed to memory. Being readily recalled and articulated radically improves communication and commitment. They must be aligned with both the goals of the business and the goals of other change programmes already commenced. They must be recorded and agreed upon by the sponsor and the implementers.

27.2.2 The Brief

A primary activity in the selection and appointment of a consultant is the preparation of a description of the change activity, sometimes called the *brief*. If the change is organisational restructuring, the brief must clearly set out the high-level drivers for the change (the objectives), any subsidiary objectives and the goals to be accomplished (the success criteria). The brief will also contain, among other things, the economic, market and historical context of the change, the required timeframe, cost, implementers, roles, responsibilities, current business activities and activities post-implementation, the section of the organisation affected, the geographical spread of the business premises affected and the internal business interfaces.

27.2.3 Describing Activity Interfaces

The most compelling argument for introducing risk management practices is to reduce the risk of a reduction in shareholder value. Deloitte Research undertook a survey to analyse the causes of major shareholder value losses between 1994 and 2003 (Kambil *et al.* 2005). While the authors of the survey recognised that the past does not automatically predict the future, the study sought to better understand the factors underlying corporate value losses and to suggest better ways to reduce vulnerability and disarm the "value killers". The authors say that to their surprise they found that major value losses were not always driven by a failure to respond adequately to a single category of risk failure – but often due to the failure to respond to many different types of interdependent risks over a short period of time. That many major value losses had been the result of several types of risk interacting. They cite the following case as an example of risk interdependencies:

> After a fourth profit warning in five quarters a major manufacturer saw its shares plunge by more than 25%. In total the firm lost more than half its market value over the course of one year. Traditionally a market leader, the manufacturer was initially slow to respond to the strategic risk posed by competitors aggressively introducing products with new features. But its effort to reduce costs through massive reorganisation left it vulnerable to further losses from inadequately managing operational risk. The firm consolidated more than 30 administrative centres into just three, which slowed order fulfilment and billing, and increased customer administration costs and accounts receivable, leading to further losses.

Any briefing to consultants should as far as possible describe how the activity (for which the risk assessment is being undertaken) interfaces with other business activities. This may include subsidiaries, other premises, suppliers, partners and so on. The briefing should also refer, when appropriate, to how the activity relates to any forthcoming deliveries to major customers.

27.2.4 Appointment Process Management

A manager needs to be allocated the responsibility of managing the appointment process. The individual selected needs to have sufficient time to dedicate to the process to be sufficiently familiar with the objectives of the assignment and the skills required to implement the assignment. This individual will be responsible for preparing a programme for the selection and appointment process. The discipline of preparing a programme forces consideration of the activities to be undertaken, their likely duration, the sequence in which they will be implemented, who will be involved in each activity and the overall duration. It will inform the implementers of the change programme how quickly a consultant can be brought on board.

27.2.5 The Long-Listing Process

The production of the long list is the first step towards establishing a viable tender list, which will ensure genuine competition among capable suppliers. The long list can be compiled from a list of consultants who have supplied services to the business before and recommendations from non-executive and executive directors together with recommendations from professional bodies. It may be useful at this stage to ask consultancies that are not known to the business to provide brochures and possibly references, as well as to examine their websites. When the long list is compiled, a preliminary enquiry should be sent to each firm on the list to ask them to confirm without obligation that they are interested, that they have the resources and that they will submit a tender if invited to do so. It is normal to request that replies are made by e-mail or in writing by a given date. The preliminary enquiry usually covers the following points:

- Title of the commission
- Intended date of the invitation to tender
- Intended date of the contract award
- Commencement date of the commission
- Duration of the commission

- A request for written confirmation of interest in being included in the tender list
- A request for information to support the enquiry recipient's inclusion in the tender list, or expression of interest, against which consultants will be assessed
- Advice that neither the enquiry nor their positive response to it would in any way guarantee that they would be included on a tender list or that the commission would proceed at all.

27.2.6 Short-List Selection Criteria

The supporting information that consultants are requested to provide with their expression of interest must be adequate to differentiate them and give sufficient information against which to judge their competency. Hence, in order to form a final short list, it is necessary to establish the criteria and associated weighting, to be able to evaluate responses. The greater the extent to which the brief has been defined, the more focused and specific the pre-selection criteria can be.

The key criteria for pre-selection are as follows:

- Experience of similar commissions in terms of
  - change programme type contemplated;
  - business context;
  - approach (method of working and management);
  - technical capability;
  - completion within proposed timeframe;
  - professional resources and support facilities available
- Qualifications and experience of staff
- Financial standing
- Adoption of a quality management system
- Location of consultant's offices
- Corporate membership of professional organisations
- IT policy
- Knowledge of risk management software
- Outline suggestions for the approach to the commission in terms of identification, assessment and management
- Risk management process.

27.2.7 Request for a Short-Listing Interview

The short-listing interview can be held at the consultant's offices or the client's offices. If the consultant and client have worked together recently, then an interview may be unnecessary. If the two parties have worked together before but not recently, the interview could be held in the client's office. Where an interview is to be held, then a letter can be issued to those who have submitted favourable responses stating the details of the interview. This would include the date, time, location, duration, format, number of attendees, topics for discussion and possibly samples of completed commissions.

27.2.8 Compilation of Short List

When all pre-selection information has been gathered and evaluated, the short list can be produced. The number of tenderers selected for the short list will depend on the size and complexity of the commission. Those who are not short-listed should be advised in writing. The short-listing should be carried out by those identified at the outset of the selection process within the timeframe established in the selection process programme. The number of consultancies selected for the short list will depend on the nature of the commission, but it is usually recommended that three or four firms are invited to tender. It is thought that three is the absolute minimum, just in case a firm declines or drops out part way through the tendering process. Other than in very unusual circumstances, such as where tendering costs may be very high, consultants are not directly reimbursed for the costs they incur in preparing tenders. This is an overhead for each consultancy. Hence, they have to build their tendering costs into their tenders. If clients regularly have tender lists of (say) ten, then consultants would have to increase the fees they charge to cover a larger number of unsuccessful tenders. It is therefore in clients' interests to keep tender lists low (below ten) but sufficiently high (say, four) to obtain competition and value for money.

27.2.9 Prepare an Exclusion Notification

Consultancies who have not been short-listed should be advised in writing. A standard letter can be prepared which can be tailored to suit specific circumstances. It may contain information such as the following:

- Title of the commission
- The commission reference, if one exists
- The date of their letter
- The date of an interview, if attended
- Notification that, after careful consideration, the consultancy will not be invited to tender
- Notification that the decision to exclude them from the tender list does not affect their prospects of being invited to tender for suitable future work (provided the submission was not so poor that any future involvement would be out of the question)
- An offer to advise them by phone why they had been unsuccessful.

27.2.10 Prepare Tender Documents

The tender invitation will commonly require a technical submission and a price. Instructions may be given for them to be submitted together or separately. The technical submission typically calls for a methodology – how the assignment will be undertaken. The invitation will provide instructions on how the price/tender is to be broken down, together with advice on the quality/price mechanism. This mechanism advises tenderers of the weighting that will be applied to the quality of the submission and the price when assessing the tenders. Indicative quality/price ratios are 60/40 and 70/30. These ratios assume that experience, technical competence, management skill and methodologies will take on far more significance than price. Advising tenderers of the quality/price mechanism in advance provides a sponsor with tenders that are more aligned to his/her priorities. The documents will describe the business activity, the objectives of the assignment, the consultant duties, the timetable for the tender process, including the duration of the assignment and any other pertinent information. At the time of preparing the tender documents, the evaluation criteria should be prepared. It is common for a label (or labels) to be prepared and issued to tenderers for the tender returns, to enable tenders to be opened simultaneously. In addition, these documents will be accompanied by an agreement which will describe the terms of the appointment. The content of a typical agreement is described below.

27.2.11 Agreement to be Issued with the Tender Invitation

The content of any agreement needs to be adjusted to suit the circumstances of the commission and requires careful deliberation. The agreement terms must reflect the common events that may occur during the life of a commission. Examples of the subjects to be addressed are as follows:

- Definition of terms
- Duty of care
- Consultants' duties
- Changes to consultants' duties
- Payment
- Subcontracting
- Personnel
- Copyright and publication
- Assignment
- Suspension of services
- Insurance
- Health and safety
- Termination
- Intellectual property
- Additional services
- Confidentiality
- Publicity
- Conflict of interest.

27.2.12 Tender Process

Ensure sufficient time is allowed both for the tenderers to prepare their returns and for the tender review process. Make clear the specific date, time and location for the return of the tenders. Prepare a tender return summary sheet as part of the audit trail and for future reference, if the need arises again to recruit a risk consultant. Issue the tender invitations on the programmed date. Deal with tenderers' queries promptly during the tender period, and advise all other tenderers of both the questions and the answers. Record the tenders received, not received, opening time and the prices received. Ensure the tender reviewers are notified of the return date, the assessment criteria and their responsibilities. Select the preferred tenderer.

27.2.13 Award

Resolve any outstanding agreement issues with the preferred tenderer. Notify the successful tenderer of the award of the commission, reconfirming the start date, the point of contact and the time and place of the "kick-off" meeting. Notify the internal assignment participants and the accounts department and also, where appropriate, the IT and security departments.

27.2.14 Notification to Unsuccessful Tenderers

The unsuccessful tenderers should be notified promptly after the successful tenderer has been informed. This may be accomplished by a simple single-page letter containing information such as the following:

- Title of the commission
- The commission reference, if one exists
- The date of the tenderer's submission
- Notification that, after careful consideration, the consultancy has been unsuccessful
- An expression of gratitude for tendering
- An offer to advise why they have been unsuccessful.

27.3 SUMMARY

This chapter has described in summary a generic process for the appointment of consultants for implementing a risk management assignment. It has described the selection process from long-listing to short-listing, creation of assessment criteria, determining the contents of the agreement and the award process. The assignment's results are only ever likely to be as good as the original briefing and objectives-setting the consultant was provided with at the outset of the assignment.

27.4 REFERENCE

Kambil, A., Layton, M. and Funston, R. (2005) Disarming the value killers. *Strategic Risk*, 27.

28 Interview with the Client

The previous chapter provided an introduction to the process of the appointment of consultants and the common steps involved. This chapter examines the interview between the consultant and the client prior to the preparation of a proposal, which is described in the next chapter. This initial meeting may take place at the behest of the sponsor as a result of his/her need to address a specific issue, as part of a tender process (similar to that described in the previous chapter) or at the request of a consultant seeking to sell a particular service. It will be instrumental in the development of the relationship between the client and the consultant and, where the client has a need, establishing exactly what that need is. The purpose of this chapter is (for the client procuring services) to inform information sharing and (for the consultant) to suggest the questions that should be raised and the information that should be collected. It is a communication process where both parties must fully engage if a satisfactory outcome is to be achieved from both perspectives.

28.1 FIRST IMPRESSIONS/CONTACT

The first meeting between a client and a consultant will be critical to whether any future contact is made. During the first meeting between the two parties, a series of important issues are decided, such as:

- the interpersonal nature of the relationship and whether the client could work constructively with the consultant;
- whether the consultant organisation is considered to have appropriate experience and expertise;
- whether the consultant(s) who will be assigned to the task have sufficient experience;
- whether the consultant organisation has a good reputation for delivery;
- whether there is mutual respect for the professionalism of both parties;
- whether the consultant wants to take on the assignment;
- whether the client wants to employ the consultant;
- whether the proposed project will add value to the business; and
- what levels of risk both parties are exposed to by entering into a contract.

It is imperative for the consultant to apply sufficient time and energy to understanding the client representative as well as the problem. It is important to establish a solid relationship. The client–consultant relationship is the foundation stone that will support any future assignment. Successful consulting is driven by the extent to which the consultant can get close to the client to arrive at a position where he/she becomes more of a confidant or trusted adviser than purely a supplier of services. How far a consultant can build a sense of rapport with a client before the consulting process begins will affect the degree of success of the outcome of the assignment. Rapport is a multi-headed beast that needs to be tamed. The key components of rapport are:

- *Rectitude*: providing a supportive behaviour underpinned by company core values.
- *Approach*: a receptiveness to new paradigms of working.
- *Problem*: from the very beginning, seeking to view the problem as the client sees it, not as you, the consultant, see it.
- *People*: professional, experienced, personable consultants who achieve success through goodwill and a collective will to succeed.
- *Outcome*: establishing from the outset, by the use of questions, the client's wished-for outcomes.
- *Resolve*: a determination to focus on the dominant issues and accomplish the assignment goals.
- *Trust*: the development of trust which is the cement that builds and sustains any client relationship.

28.2 CLIENT FOCUS

In any dialogue between the consultant and the client it will be important for the consultant to establish what is of vital importance to the client:

- Statutory compliance (employment law, health and safety legislation, etc.)
- Regulatory compliance (Sarbanes-Oxley Act, etc.)
- Maintaining shareholder value
- Corporate governance/internal controls (Combined Code 2003)
- Financial stability
- Market share
- Information security
- Business continuity
- Project risk management (project risk is related to corporate risk, since the profile of the latter will change if management invests in risky projects)
- Satisfying reporting requirements.

28.3 UNIQUE SELLING POINT

A "unique selling point" is a sales proposition that the competition cannot or does not currently offer. It is one of the basics of effective marketing for consultancies that has stood the test of time. The consultant will be looking to differentiate him/herself from the competitors in what is commonly a crowded market place. The client will be wishing to understand a consultant's general experience and competency, his/her standing (financial, insurance, reputation), whether he/she can adequately complete the assignment planned and whether he/she can offer something above the norm. Table 28.1 provides an example of a table constructed by a consultancy to describe its unique selling point. The consultants have gone through the common steps of identifying the features of their firm (step 1), converting the features into benefits that customers will value (step 2), ranking the benefits in order of importance on a scale of 1–10, where 10 is very important and 1 is insignificant (step 3) and identifying the benefits as being "standard", "uncommon" or "different", where "standard" represents

Table 28.1 Analysis of unique selling point(s)

Feature	Benefit	Ranking (1–10)	Standard, uncommon (few) or different
Consultancy firm			
General knowledge of the industry or sector (such as government departments) – commissions already completed for businesses in the same market	The consultancy has specific experience of the customers' industry	9	F
Reputation in the market place	The consultancy is well respected	9	D
Previous professional indemnity claims	Customers will take comfort that no PI claims have been made by other clients	8	S
Unresolved disputes	Customers can be confident that, due to an absence of disputes, consultancy staff will not be distracted by participation in dispute resolution	8	S
Year of establishment	The consultancy has been in existence longer than 5 years, the period thought to be a minimum for stability	7	S
Financial standing	The consultancy shows consistent growth	7	S
Number of staff	Customers can be confident that the assignment will be adequately resourced	7	S
Staff retention	Customers can be confident that there will be minimal changes in personnel	7	S
Number of years of corporate experience	The consultancy has been in existence longer than 25 years, during which time it has accumulated a wealth of experience	6	S
Current workload	Does not prohibit taking on more assignments	6	S
Location of offices	The consultancy has offices across the UK	6	S
Price	Customers will not see a significant difference in price	5	S
Consultant			
Experience: • Size (and/or complexity) of commissions completed • Breadth of experience across different industries • Number of years	Customers can see that the individual's experience is highly beneficial to the outcome of assignments	10	D

Table 28.1 Analysis of unique selling point(s) (*Continued*)

Feature	Benefit	Ranking (1–10)	Standard, uncommon (few) or different
General knowledge of the industry or sector (such as government departments)	Staff have specific experience of the customer's industry	9	F
Knowledge of the drivers of business success and failure in the industry or sector within which the organisation operates	The staff have knowledge of the reasons behind failure and success	9	F
Knowledge of the regulatory framework within which the organisation operates	The consultancy has knowledge of the regulatory framework	9	S
Approach to assignments	Customers can see that the approach is highly beneficial to the outcome of assignments	9	F
Research into best practice	The consultancy is continually looking at new publications, software and news reports	8	F
Qualifications	Staff are very highly qualified	7	F
Ability to apply appropriate industry norm modelling tools	Staff are experienced in the use of the common modelling tools	7	S
Extent of published works	Staff have published a number of papers and articles	5	F

features commonly offered by competitors in the market place, "uncommon" means just that (few competitors offer this feature) and "different" means unique. While many of the features have been marked as standard, their absence would preclude the consultancy from competing. Also more than one feature can be awarded a 10, in terms of being very important.

28.4 PAST EXPERIENCES

To help establish what would be an acceptable approach to the assignment and to avoid repeating the actions of previously appointed consultants which were not well received, during the interview, consultants need to include in the conversation the following type of questions, without making it an all-out inquisition.

- What is the client's previous experience of using consultants?
- What good results came from the experience?
- Were there any problems?
- What, if anything, would they seek to do differently this time?
- Do they have any fears or concerns about having consultants in the company now?
- Were there any communication difficulties?

28.5 CLIENT INTERVIEW

The meeting between the client and the consultant will afford the consultant the opportunity to ask a series of questions and gather important information. The seven Ss listed below (based on Cope's seven Cs of consulting (Cope 2003)) are a series of questions to be posed by the consultant to enable him/her to learn more about the client organisation, the assignment and the client representative.

28.5.1 Scene/Overview

- What is the reason for seeking external support?
- Why is the assignment thought to be necessary?
- What are the implications of doing nothing?
- Who is the ultimate client (financier) and what is their level of "buy-in" to the proposed assignment?
- Which client representatives would be involved and where are they located?
- How will things be different or better once the assignment is complete?
- How will success be measured?

28.5.2 Situation/Context

- Is there an existing risk management process?
- What is the organisational structure and the relationship between the audit committee, internal audit, the board and the risk management committee?
- What are the roles and duties of the departments/committee involved in risk management?
- What are the principal gaps between these functions?
- Are the business objectives clear?
- What is (are) the group subsidiary relationship(s)?
- What changes are currently taking place in the business environment?
- What, if any, changes are currently in progress in the organisation?
- What major changes have taken place in the business over the last 12 months?
- Are there any concerns about factors that might impact the assignment?
- What concerns would any participants in the assignment have?
- Are there any side effects that could arise from undertaking the assignment?
- Who can stop the assignment from being successful?
- What are the unspoken or shadow issues that might cause the assignment to fail?

28.5.3 Scheme/Plan of Action

- What constraints are there on any proposed methods of working?
- What are the criteria for a successful solution?
- What is the budget and timescale?
- What has the client commenced already?
- What will be the availability of non-executive directors?
- What will be the availability of the functional heads?

28.5.4 Solution Implementation

- Who will be involved in the assignment?
- What will be the response of the intended participants?
- What methods will be acceptable to implement the solution?
- Is there a standard engagement/deployment process that will have to be followed?
- Are there any aspects of the assignment that will be managed by a third party?
- Where is the power to effect change held?
- Have those people who will not resist and those who will be the key influencers been identified?

28.5.5 Success, Measurement of

- How important is it for measurement to take place?
- Is the client prepared to pay for the measurement to take place?
- Are qualitative or quantitative measures required?
- How will the business's buy-in to the change be measured?
- Who will understand the measurement?
- What measures has the organisation used in the past?
- How will the organisation measure the assignment's performance?
- For how long will measurement continue?

28.5.6 Secure/Continue

- How long does your organisation want the change to last?
- Has the organisation tried this before, did it last and if not why?
- What can we do to help ensure that the change will last?
- Is your organisation prepared to invest in things that will make it last?
- Does your organisation have the resources in place to support any change?
- Are responsibilities defined to maintain the change once it is complete?
- Is there anyone who will try to eradicate the change once it is complete?

28.5.7 Stop/Close

- What does "good" look like?
- Once the change is complete, what differential value will we have added?
- What can be learnt from the assignment?
- How can this learning be used elsewhere?
- What can we do to ensure that your organisation is not dependent on us once the change is complete?
- What would we have to do for your organisation to recommend us to a colleague?
- What else might we be able to help you with?

28.6 ASSIGNMENT METHODOLOGY

The consultant will be required to spell out their methodology for the assignment. In simple terms, the methodology is the collective term for the activities that will be undertaken, the sequence in which they will be carried out and the interdependencies. This might reflect the following path, depending on the requirements of the assignment: desktop study, review of methodology and background to the assignment with the sponsor, interviews and data collection, report writing, presentation of findings and finalisation of report.

28.7 CHANGE MANAGEMENT

To survive and prosper, most businesses cannot stand still but have to morph to reflect the evolving changes in the market place. As Stewart (1982) correctly summarises, the degree of change that businesses have to endure can vary enormously. For instance, for a top manager in a whisky distilling company, the only major change for several years may be the redesign of the label, whereas for a top manager in a technically innovating industry like electronics, change may be frequent and extensive. How managers react to change will have an important impact on a business's future. The tempo of change has speeded up and hence the demands on managers to plan for and adjust to change are greater. The challenge is recognising the changes and understanding if they are a threat or an opportunity and how they should be responded to. Risk management is a tool for coping with change. But it cannot be a once-only activity.

28.8 SUSTAINABLE CHANGE

To be effective, the changes that are implemented have to endure. Where organisations need external support to identify and implement the necessary changes (to maintain or improve business performance), effective consultancy support must deliver "value through sustainable change" (Cope 2003). The three factors that Cope describes as having to be present and managed to accomplish this goal are as follows:

- *Change*. A change must always take place. If the client or consumers do not think, feel or behave any differently at the end of the engagement, then what is the value? It is imperative that the consultant and client are both clear as to the change that is required.
- *Value*. There must be explicit value realisation. Only by understanding and taking responsibility for the change and the value derived from the change can the consultant and the client develop their capacity to repeat the activity and so enhance performance further at a later date.
- *Sustainability*. There is little point in making a change that has value, if it does not stick. This is at the root of problems with so many change programmes. The consultant and client may enjoy their shared working experience, make amazing leaps in performance, celebrate their success and then move on, only to find that little value remains three to six months later.

Cope goes onto say that "by looking at the change engagement as a battle of reinforcing and representative forces it becomes easier to map and measure what factors will enable the change to live beyond a short-term fix and deliver value through sustainable change". The primary determinant that drives any successful outcome will be the balance between the repressive forces that cause the client and business to revert to the old way of operating and the positive forces that help them hold onto the gains. The key parties within these opposing forces are likely to be members of the sponsor group, employees or members of the end-user group, as illustrated in Figure 28.1. While every assignment will have its own unique characteristics, it

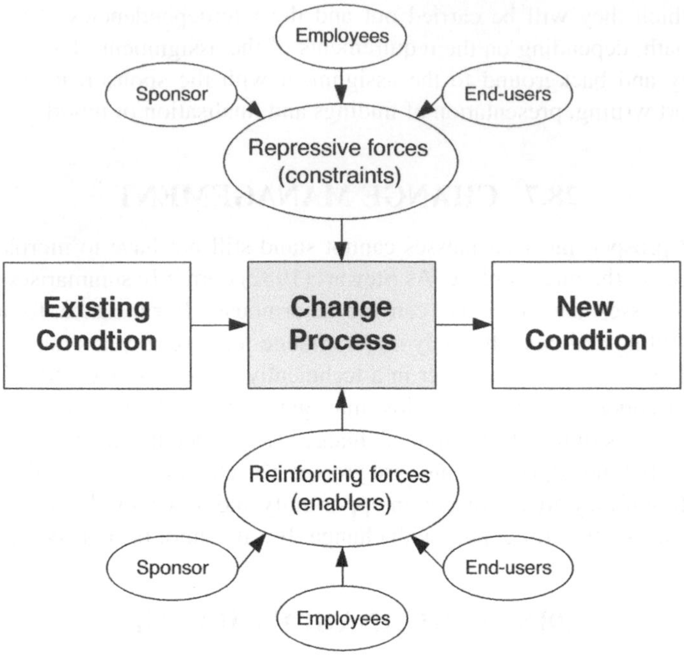

Figure 28.1 Influences on a change process

is recognised that there are a number of repressive forces that occur time and time again which either threaten or cause the derailment of an assignment to deliver sustainable value:

- The client does not fully appreciate the current position and is unwilling to make the changes identified as necessary.
- There is a failure to understand the root causes behind the current position.
- Those affected by the change are not helped to work through the pain of letting go of the former methods of working.
- There is no accurate measurement and confirmation that the change has delivered the desired results.
- When the assignment is over, and the pressure of work returns, staff revert to the methods of working that they remember, understand and are comfortable with.

28.9 SUMMARY

The focus of this chapter has been on the first meeting between the client and the consultant who may be engaged to provide risk management support. It has examined the importance of this first meeting during which those very first impressions are formed and both parties decide whether they wish to develop the working relationship. It is important for any consultant to establish what a client's specific focus is, what is important to them and how much it means to their business. The consultant has to understand what their unique selling point is and the client needs to understand a consultant's core competencies and what they can offer over and above their competitors. As part of the interview, the consultant needs to blend into the conversation, without making it too intrusive, questions about the client's past experiences of the use of consultants, to avoid repeating those actions which would not be well received. The client interview, it is suggested, should follow a pre-planned sequence covering the subjects addressed under the headings collectively called the seven Ss. A lot of reference has been made to change, as risk management involves undertaking new processes and activities and approaching problems in a more systematic way. At this stage of dialogue with the client, the consultant would need to consider the methodology that would be likely to be adopted, and there are a series of steps that are commonly selected. Any assignment on its own (or when it forms part of a programme) forms a change process. For the assignment to contribute value to the organisation on whose behalf it is being conducted, it must contribute sustainable change.

28.10 REFERENCES

Cope, M. (2003) *The Seven Cs of Consulting: The Definitive Guide to the Consulting Process*. Pearson Education, Harlow.

Stewart, R. (1982) *The Reality of Management*, revised edition. Pan Books, London.

29
Proposal

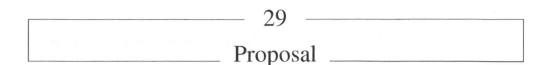

The previous chapter provided an introduction to the process of the appointment of consultants and, in particular, the initial meeting between the client and the consultant. This chapter examines the preparation of a proposal for enterprise risk management services and is for the benefit of consultants working within the discipline of risk management. It is largely based on Chapman (2003). The next chapter looks at the steps involved in implementation of assignments.

29.1 INTRODUCTION

Commissions are the lifeblood of any consultancy. Proposals are a vital ingredient in the securing of new commissions. The degree of success of a consultancy will therefore hinge in part on the quality of its proposals. (Winning commissions is obviously essential for survival but, equally, they have to be well executed. Poorly executed commissions will not secure repeat business or, worse still, will damage any hard earned reputation.) A proposal, then, will need to satisfactorily address all those issues a client has specifically requested responses to, build a persuasive argument, be readily understood and clearly articulate the benefits that will be derived from the assignment.

29.2 PROPOSAL PREPARATION

29.2.1 Planning

Prior to embarking on the written proposal it is prudent to plan the preparation of the proposal as if it were a project in its own right. This entails obtaining answers to some basic questions:

- Who is going to write the proposal – will it be a single individual, will marketing be involved in providing information on previous experience, is support required for the preparation of diagrams/charts/organograms, is background research required, are CVs required to be updated?
- What will the subject matter be, or was the information gathered during the interview with the client so complete and precise that there is no need for follow-up questions?
- When does the proposal have to be submitted – how many copies are required, is it to be bound or is it to be sent electronically by e-mail; if by e-mail, should the proposal exclude photographs to cut down the file size?
- How is the proposal to be prepared? What software is required and how will it be accessed? And if the proposal is to be restricted to a limited number of pages, what subjects will receive a more comprehensive treatment?

29.2.2 Preliminary Review

One of the main sections of the proposal will be how the consultant proposes to carry out the assignment. This entails working back from the deliverables and understanding the incremental steps that will have to be undertaken. It is often helpful to prepare a task list recording the assignment tasks in the order in which they would be completed. Keep the list to one page if possible. The task list is essential to determine the resources required, the time required to complete the assignment, any software that will have to be purchased (either by the consultant or by the client), travel and accommodation requirements and the need for a partner, if all of the work cannot be undertaken in-house.

Proposals often fall into the trap of presenting their expertise in the form of a "recipe" which tells the client in considerable detail how to turn the assignment into a do-it-yourself job. Or worse, where the client is completely without principle, the detailed and comprehensive proposal finds its way into the hands of a competitor who is either cheaper or has a long-standing relationship with the client.

29.3 PROPOSAL WRITING

29.3.1 Task Management

This technique treats the proposal as if it were a project in its own right to guide the tasks to be carried out, the resources required to complete the activities, the sequence in which the activities will be carried out and the outputs of each task.

29.3.2 Copying Text

Consultancies have experience of writing proposals and commonly have a store of previously completed proposal documents. As writing proposals can be very time-consuming, combined with the fact that proposals normally have several similar elements, it is common practice for some material to be copied and pasted from previous proposals. However, there is a danger in this practice, as the reused text may have been evaluated as poor by previous recipients, be time-elapsed or contain erroneous information. If the text is poorly edited, reference to other clients, projects or locations may be left in, giving the impression the proposal was produced in haste, was not considered important or the author's organisation is incapable of producing carefully crafted documents.

29.3.3 Master Copy

For large proposals, and particularly where there are a number of contributors, it may be appropriate to maintain a hard copy in a hardback folder subdivided with numbered dividers. This way contributors can see how their element fits within the whole document to avoid repetition and the use of inconsistent language or terms. Browsing can highlight terminology, readability, presentation and sequencing issues. Regular reviews will provide a good indication of progress. Independent assessment will identify errors and omissions.

29.3.4 Peer Review

Ensure the proposal is reviewed by peers. The proposal should be read by colleagues experienced in preparing proposals, with a view to their offering constructive criticism as to how it could be improved. If you have concerns about any aspect of the proposal, draw these to the attention of the reviewer to ensure that they focus on and address your concerns. Adequate time should be allowed for the review. The reviewer should be primed in advance that the document is coming, its likely length and complexity and the timeframe within which they will have to operate. Ensure that the final proposal is proof-read and a check is carried out that the language is clear and lucid.

29.4 APPROACH

State the activities that will be undertaken:

- Describe what the output will be.
- Indicate the order in which things will be done.
- Develop a benefit statement for each activity.
- Describe how each main activity represents a benefit to the client.
- Structure the list of benefits logically so that it is clear how each activity contributes to the achievement of the desired outcome.
- Avoid indicating through unnecessary justification of a benefit, how it will be achieved.

29.5 PROPOSAL

The successful implementation of enterprise risk management requires the preparation of a proposal, sometimes referred to as the *terms of reference* (ToR). A ToR is the key document used to capture the problem definition. It is a critical communication tool and vital to the success of any risk management exercise (Figure 29.1). Regrettably, for a host of reasons, they are either not prepared, lack clarity, are incomplete, not understood, not read in full by the recipients or not disseminated.

29.5.1 Identify the Parties – the Who

The first section of the ToR should identify the *who*: who within the client organisation has initiated and requires the risk management process to be undertaken, who is funding the study, who within the client organisation will provide the necessary support to the process, who will participate in the study and who will facilitate it.

- It should be made explicit who has initiated the study and their contact details should be recorded.
- It should be made clear who is funding the study and their contact details should be recorded.
- The responsibilities of the client should be stated together with the individual(s) who will undertake the tasks listed. The Office of Government Commerce (2007) provides very good guidance on client responsibilities. It is essential that the client advises the participants in advance of the study of its purpose, the timing, attendance required, the data that will be collected, the location of meeting rooms booked, the funding of any travel and subsistence, the facilitator(s) and the study outputs. The participants should not be surprised to be contacted by the facilitator. Participation should not be discretionary.
- The key stakeholders should be identified – commonly the contract parties and those who may influence the project. For consortia projects the stakeholder list may be extensive. Key

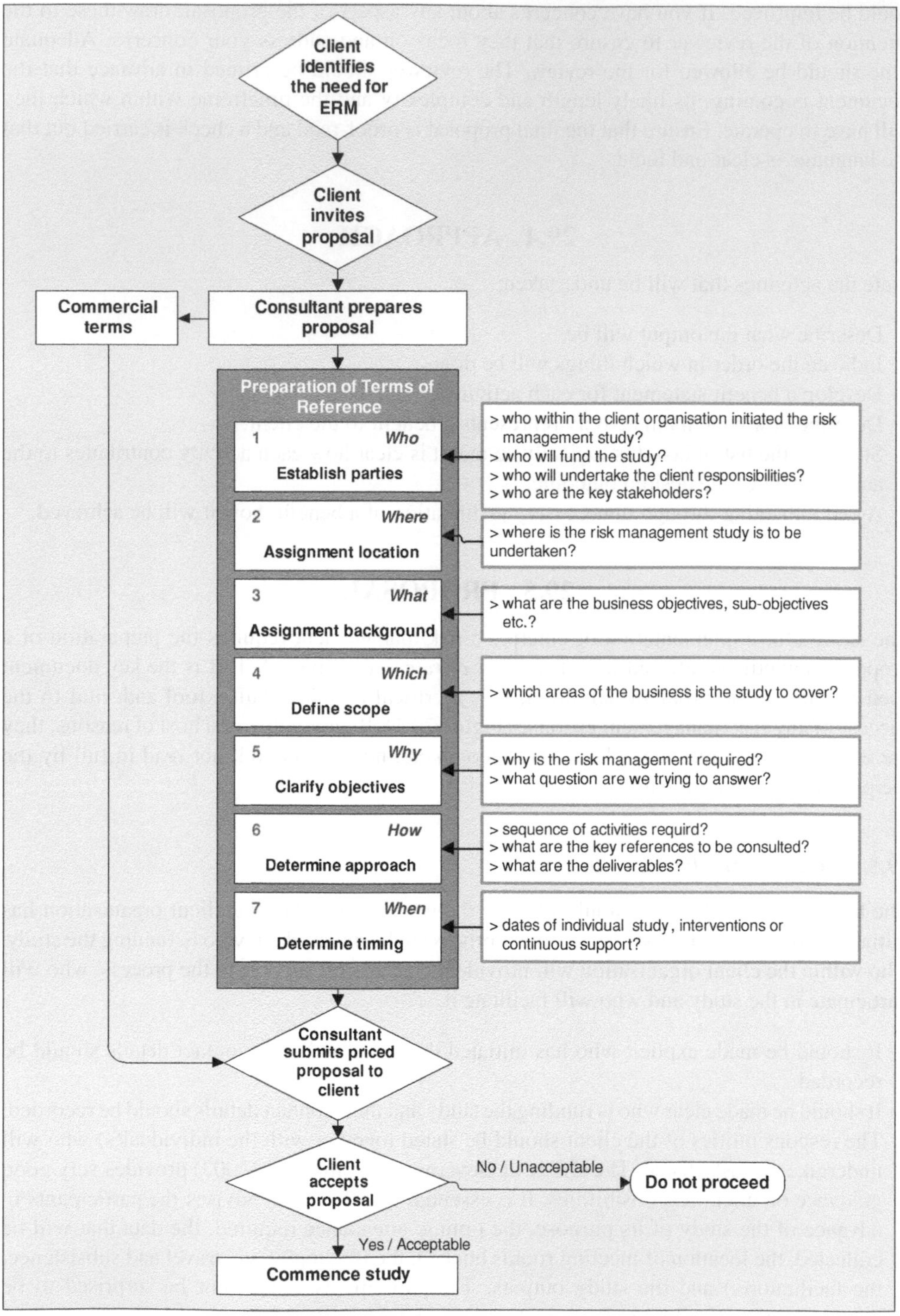

Figure 29.1 Preparation of a proposal

stakeholders' aims should have been identified, discussed, aligned and incorporated into the project objectives prior to the commencement of the risk management study.
- The interviewees should be listed – commonly the project team members together with representatives of the key stakeholders.
- The risk management study facilitator(s) should be noted. This is the individual(s) who will structure the process, facilitate data gathering, provide modelling, provide guidance on implementation and provide data for decision making.

29.5.2 Identify the Location – the Where

The second section of the ToR should identify the *where*: where is the risk management process to be undertaken? This can be significant on a number of fronts. If the analyst has to be co-located with the project team which is overseas, then travel, accommodation and subsistence costs have to be factored into the cost of the risk study. If the interviewees are geographically dispersed and face-to-face interviews are considered essential, again similar expenditures will have to be factored into the risk management study costs. If a consortium is undertaking the project, will two project offices have to be visited? What reliance can be placed on conference calls, and is video conferencing available? In addition, ethical considerations have to be taken into account such as data protection and storage, confidentiality and safety of the participants.

29.5.3 Understand the Project Background – the What

The third section of the ToR relates to the *what*: what are the project details? Understanding the project background means establishing as a minimum the project objectives, project catalysts, project stages and any key milestones, the key stakeholders and the organisational relationships – the contract parties. Unless the project objectives are established, how would it be possible to establish the threats and opportunities?

29.5.4 Define the Scope – the Which

The fourth section of the ToR should identify the *which*: which part of the project is the risk management process to address? The risk management scope must be established, for instance whether it relates to just the business case, just the feasibility options or all of the decision gates. The scope will affect the cost of the study, the duration, the resources required and the extent of the information to be reviewed. Prior to finalising the scope, the benefit to the client must be assessed and a view taken as to what external support is required and what elements can be undertaken in-house.

29.5.5 Clarify the Objectives – the Why

The fifth section of the ToR should identify the *why*: why should the risk management study be undertaken? What question are we trying to answer? What is the problem? Problem definition is the most important step of the ToR. Problem definition should clarify the problem being examined for all those involved (the stakeholders or actors) and be worded in such a way as to give insight into the problem. Two important concepts used in problem definition are the gap and problem owner. A problem can be seen as a gap between what we have and what we want (or where we are and where we want to be). The problem owner may be the project sponsor or the owner of a specific risk.

29.5.6 Determine the Approach – the How

The sixth section of the ToR should describe the approach or the *how*: how should the risk management assignment be undertaken? A corollary to understanding why an assignment should be undertaken is that there is no single best way to undertake all risk studies – the need to vary the approach will be driven by why it needs to be undertaken. A basic axiom of those who undertake successful model construction and risk management for decision-making purposes is that the approach (the how) adopted must suit the specific needs of the business decision or project under examination. This means responding to the question to be answered at the same time as understanding the constraints of the assignment. Constraints may include the financial constraints of the client, the timing of the board meeting to which the analysis has to be supplied or the availability of participants.

Planning for the risk management process begins with understanding the data to be analysed and selecting an appropriate model. A "model" in this context is the conscious simplification of a real situation we use to carry out analysis. Modelling involves a transformational process where outcomes are explained by a range of inputs and their associated assumptions. Most models of interest have a mathematical form, and those commonly of interest to risk management have some element of uncertainty and require an understanding of probability. Of particular interest is their graphical representation, which will be a key tool in communicating the results. The outputs must be clear and precise, as these will be required for the decision, which was the catalyst for the risk management assignment. Risk analysis does not need to be and should not be carried out in a vacuum. Reference should be made to publications that document best practice. These references can then be listed as documents consulted during the course of the risk management study.

29.5.7 Determine the Timing – the When

The final section of the ToR should describe the *when*: when should the risk management study be undertaken? The "when" refers to both the start date and the duration. Overall timing will be driven by a series of issues such as the budget for the study, the availability of the participants, the date by which the results are required, the availability of information to conduct the study (such as budget or programme), the availability of the analyst(s), the complexity of the task, the amount of information to be reviewed and the number of stakeholders involved. Commonly problems occur when the information supplied is incomplete, and when the time taken to "plug the gaps" exceeds expectations.

29.6 CLIENT RESPONSIBILITIES

For assignments to be a success, it is a common requirement for clients to undertake the following activities, to ensure that the participants engage in the process.

- Inform the participants that a consultant has been engaged and the objective(s) of the assignment.
- Inform the participants of the date, time and location of any meetings/workshops.

- Book the venue for the workshop and arrange for refreshments/food as appropriate.
- Participate in a "kick-off" meeting.
- Make available any background material for the purpose of a desktop study at the outset of the assignment.
- Make serviced workstations available as appropriate.
- Make time available to review the outputs of the assignment and comment.

29.7 REMUNERATION

The financial terms of a commission may vary widely. They may, for instance, be based on a fixed lump sum, day rates or an open book arrangement. The financial element of a proposal can lose a consultancy a commission or expose it to a loss. A fixed lump sum requires the task in hand to be precisely defined to provide certainty over the time required for completion. Day rates provide no risk to a consultancy, provided of course the day rate is sufficient to cover the consultant's salary, expenses, overhead and profit. An open-book arrangement is where the client is afforded complete disclosure of the consultant's costs to the consultancy (in terms of salary and benefits, national insurance and so on) and a day rate is agreed composed of the disclosed costs (staff and overhead) and an agreed profit percentage.

29.8 SUMMARY

This chapter has examined the preparation of a risk management proposal by a consultant. It walks through the steps of preparation, writing the proposal and determining its content. A structured term of reference is described. The client responsibilities are explored – no assignment can be carried out in a vacuum and the client organisation itself is a significant determinant in whether or not as assignment is a success.

29.9 REFERENCES

Chapman, R. (2003) Risky business. *Project*, October.

Office of Government Commerce (2007) *Management of Risk: Guidance for Practitioners*, 2nd edition. The Stationery Office, London.

30
Implementation

The previous chapter examined the preparation of an enterprise risk management proposal by a consultant. This chapter explores the implementation of an enterprise risk management assignment for the benefit of consultants and clients alike. Each assignment naturally has a start, a middle and an end. The start entails ensuring that the assignment heads in the right direction towards satisfying the goals of the assignment, the middle involves management of the assignment as if it were a project in its own right, and the end entails completion of the deliverables, commonly describing the findings in a report and presenting the headline results. The key for enterprise risk management is ensuring that: it has added value to overall business performance; there are demonstrable benefits to spending money on consultancy fees and tying senior staff's time up in dialogue with consultants; and/or options have been proposed which have the potential to save the business a considerable sum of money. Following this chapter are a series of appendices which provide support to these chapters.

30.1 WRITTEN STATEMENT OF PROJECT IMPLEMENTATION

Prior to the appointment there will have been an offer, which will have been in the form of a terms of reference as described in Chapter 29 or in some other format. This document will be the baseline from which the enterprise risk management service will be delivered. It will be a reference point for the consultant to ensure that he/she delivers the agreed outputs on completion of the assignment or provides an explanation why the outputs have changed. The key will have been to leave the client with an output that provides a lasting enhancement to share value or provides support to securing a more favourable outcome for some business venture, whether it be organisational change, an acquisition, an investment or a new product launch.

30.2 MANAGEMENT

Any assignment should be treated as if it were a project in its own right, with objectives, a timeframe, resources, a budget and end deliverables. Applying project management principles in this way will provide greater certainty of delivering the required outcome. The greater the number of team members and/or client representatives participating in the assignment, the greater the importance of the application of management principles. It comes down to who is doing what, when, how and why. The task must be broken down into manageable "lumps" which can all be assigned objectives, a timeframe, budget and resources.

30.2.1 Objectives

The objectives of the assignment must be translated into the activities and tasks to be undertaken. Any ambiguity or uncertainty about the assignment objectives must be ironed out at the beginning, otherwise this will lead to disappointment downstream. Hence, it is important to understand the question that the sponsor is attempting to answer. To put it another way, what is the decision that the sponsor wishes to make as a result of the assignment? Understanding the client's question will inform you of the information to be gathered, the individuals to be consulted, the findings that will have to be conveyed, and possibly the tools and techniques that will have to be used. Sponsor questions might be as follows:

- How likely is my project to be completed on time?
- What is the risk profile of my business proposal?
- Of the options under examination, taking account of the estimated risks, which one is likely to give me the best return on capital employed?
- How will embarking on this activity affect the rest of my business?
- What financial contingency should I allow for my capital project at, say, an 80% confidence level?
- What are the risks to my organisational change programme?
- How will the market risk profile change for product A over the next five years?

It is also important to know how the information provided to the client on completion of the assignment will be used, as this will be a vital test as to its appropriateness. While assumptions may be made about how the information will be used, it is worth asking the direct question.

30.2.2 Planning the Project

The premise is that the future can be envisaged and delivered. Once the objectives of the assignment have been clearly defined it is essential to plan the execution of the work in a logical, structured manner. A project plan (Gantt chart or programme) converts a project brief into a form that everyone understands. It will provide a structured basis for carrying out the assignment, for its control and completion by the due date. Preparation of the plan will be an iterative process, in that as additional information becomes available, the sequencing of activities and task durations will have to be modified. If the assignment entails participation by different specialists from the client team and/or the assignment delivery team, all the participants should be consulted during the preparation of the plan to gain their buy-in. The primary activities need to be identified and then subdivided into the individual tasks that make up the activities. A task is completed by a single individual, whereas an activity may be completed by a group of people. Activities (and tasks) may be carried out in parallel and series. Activities that are carried out in parallel are independent and can be carried out concurrently. Activities that are carried out in series are dependent and carried out sequentially.

The value of the plan will be greater when a large team is involved, the assignment spans a long period of time, client participation is required for a large number of the activities or external parties have to be consulted. The plan should include any interim milestones such as reviews, data gathering tasks, decision gates or approvals. Plans can be prepared by placing sticky notes on a flip chart to establish the correct order in which to carry out tasks, as they can be readily moved around until the preferred sequence is established. An alternative method is to use a software tool and construct a plan on the screen, working it up from scratch and then forwarding it on to other participants for their comments. The use of software enables a plan to be adjusted quickly and provides the opportunity to carry out "what-if" analysis to explore what effect changes in time taken for or sequence of activities will have on the overall duration. The ability to carry out certain tasks concurrently will be entirely dependent on the size of the assignment team.

30.2.3 Consultant Team Composition

If the assignment warrants more than one individual working on it, a team may be formed. Tasks may be assigned to an individual based on their individual capabilities, depth of knowledge, previous experience, problem-solving skills, speed of working or time management ability. Commonly a mixture of these abilities is sought. Each task in the plan needs to be owned by someone in the risk management team. It will be their responsibility to ensure that their allocated tasks are completed in accordance with the plan. In addition, it will be their role to confirm at the outset that all the activities have been identified, the dependencies have been mapped, the estimates of durations are appropriate, and the work is completed in accordance with the dates included in the plan by monitoring progress and taking corrective steps to fix any shortfall. Whenever work is split between many participants it must be made explicit what the interfaces are, who has to supply what information to whom and when, and how all the parts will be brought together to form a cohesive answer at the end. Whether team members have worked together before will influence the ease with which they communicate, share information, understand each other's working methods, anticipate each other's likely behaviour and performance and gel as a team.

30.2.4 Interface with Stakeholders

In the instance where a consultant is appointed, it is very rare for members of the client organisation not to be involved. These individuals have the power to influence, simply support, actively encourage, positively enhance or curtail an assignment. Their support is commonly critical to the successful completion of an assignment in terms of the value of the outputs. They can operate in both an overt and covert manner, more often than not driven by their personal values and goals, as opposed to those of the business. For some, protecting their position, impressing superiors and/or maintaining reputation may be prime drivers behind behaviours. Their behaviour will also be influenced by the change that is being proposed to which the risks are initially being identified and assessed. Cope (2003) identifies four generic stakeholders, whose behaviours resonate with the author's personal experience. Cope amusingly labels them "key person", "loose cannon", "little interest" and "desperate Dan"; however, their behaviours can make the management of any assignment difficult and more complex than it needs to be. "Key person" is a critical player in the assignment who has the power to affect change, but whose actions are unpredictable. "Loose cannon" is the individual who, while being involved in the assignment, lacks the appropriate knowledge and does not have a real appreciation of the need for the change to which the risks are being assessed. "Little interest" is on the fringe but has the ability to affect the change, their interests lie elsewhere and they have no desire to get involved, initially at least. They may decide to get involved late in the day, disagree with some aspect of the assignment and withdraw funding, or change the direction of the proposed change. "Desperate Dan" has a strong desire to get involved in the assignment, but has little capability, power or understanding of the need for change. The stakeholders will have to be monitored through an assignment and, where possible, influenced to secure successful completion of the assignment.

30.2.5 Data Gathering

Data gathering is a critical process. It must be established what data needs to be obtained, where it is held and how it will be accessed. To fully understand the information obtained, it will be necessary to discuss its content with members of the client organisation. This can be accomplished through face-to-face interviews, focus groups or attending meetings where the subject matter is reviewed and discussed as part of routine business activities. When planning how to gather the data, it can help to take a slightly different approach in setting the boundaries to what is sought. Data gathering conventionally pushes for hard, tangible and established facts, whereas the potential problems and risks may lie in the soft, intangible areas of the business that the hard facts will never tell you about. If a clinical view is taken of the problem, the resulting data will provide a firm foundation for the risk study, but that is all; it will have no heart. Businesses, projects and change programmes are run by people. So risks will not solely emanate from, say, interest rates, outsourcing or competitor behaviour, but from how individuals within the business chase opportunities or respond to emerging risks.

30.2.6 Budget

The budget of the assignment will need to be managed from a number of standpoints. If a fixed lump sum has been offered, the time spent will have to be carefully managed to achieve the dual goals of achieving the assignment within the pre-planned period to ensure a profit is achieved and delivering the promised outputs. If the assignment has been accepted on a time-charge basis due to uncertainty over the exact scope of the assignment, the time spend should be recorded against specific tasks to provide an audit trail and afford client review. In either situation, progress should be monitored on a regular cycle to ensure that planned expenditure against completion of the tasks matches expectations. Where it does not, the causes need to be addressed. If the scope of the assignment changes radically through its life and a mismatch between resource requirements and budget occurs, this will have to be addressed with the sponsor of the assignment.

30.2.7 Assessment of Risk

When assessing risk, the culture of the organisation needs to be understood. Business culture will influence how activities are undertaken, methods of working, quality control, decision making, how individuals are rewarded, how poor performance is addressed and what is considered important. Members of a project team may deliberately suppress risk exposure, if they want a project to proceed due to personal commitment to date or if their jobs are dependent on the project proceeding. The reverse may be true in the situation where risks are assessed as part of an acquisition evaluation where risks are deliberately inflated, again as a result of uncertainty over job retention. Risks assessed in relation to a new IT project may be artificially suppressed as it is felt the new system will significantly reduce administrative tasks but the return on investment envisaged by the board will be achieved.

30.2.8 Deliverables

The deliverables must reflect the original terms of reference or explain any deviation. Ideally, they should exceed what was commissioned. The deliverables may take a number of forms, but their sole purpose is communication. Hence, the most appropriate forms of communication should be selected, whether this be a histogram, radar chart, risk map, cumulative frequency curve, pie chart, table, scatter diagram, programme, PERT chart, flow chart, influence diagram, decision tree or other technique. In many instances the output will inform a decision. Any assumptions made in settling on the figures to use in, say, a financial model must be made explicit so that the weight that can be attached to the findings is representative of the quality of the input.

30.2.9 Presentation of the Findings

Findings are commonly presented in a report. Before commencing writing the report, it is important to identify the readers, understand what they already know and what they need to know, find out how much knowledge they are likely to have and what their involvement with the assignment has been to date. Collect the material together, have full access to the information needed to write the report, formulate the objectives (what the piece of writing must convey) and then write. Consider whether the readers will have the same level of knowledge of the subject matter and how the report should be modified to suit. Think through whether it will be appropriate to include a glossary of technical terms. Keep confidential information confidential. It may be unlikely that all client personnel are as well briefed as the consultant. The findings must be thought through and provide a rational argument.

What should the response to the findings be? Do nothing? Is it reasonable to assume that the risk will disappear over time because of other factors, or should the planned business activity be abandoned as the risks far outweigh any potential benefit? Deeper diagnosis: are the results inconclusive or is there sufficient uncertainty around the aspects not studied in depth to warrant further investigation? The premise being that gathering further information on the issue will afford greater clarity, leading to greater risk reduction. Should an alternative strategy be sought? Is the risk sufficiently large that an alternative approach is warranted, one that bypasses the problem and does not attempt to tackle it? Would it be appropriate to ignore and plough through? Is the momentum behind the planned activity so great that it alters the business's risk appetite and the business is prepared to take on more risk in the firm belief that the activity will achieve the anticipated contribution to bottom-line performance? Once the "riskiness" of a particular course of action has been assessed, this is not an end in itself. What specific response actions should be considered? It may be necessary to implement a series of interrelated response actions to address the risk identified.

30.2.10 Key Factors for Successful Implementation

The development and implementation of risk management systems to identify, assess, evaluate, plan and manage enterprise-wide risks requires a methodical, structured approach. There are a series of key factors that can increase the probability of success:

- clarifying the objectives of the assignment so they are clearly understood by both the sponsor and those carrying out the assignment;
- developing a clear plan with objectives, deliverables, responsibilities and timeframe;
- assigning an experienced and knowledgeable risk management professional to lead the assignment;
- using consistent terminology throughout any reports produced, and providing a glossary of terms where appropriate;
- providing an audit trail that can be followed on completion of the assignment in terms of the documents that were referred to during the course of the assignment, any initial work which formed the foundation of the assignment, the individuals involved and, where appropriate,

Table 30.1 Customer delight questionnaire

For each characteristic, rate the extent to which the statement is true of your organisation today using the following scale:
1 – Not true at all
2 – True to a small extent
3 – Moderately true
4 – True to a great extent
5 – True all the time without any reservation

Vision and commitment		
1	Our organisation is totally committed to the idea of creating delighted customers at the end of every transaction.	☐
2	We seek to do things right first time, every time.	☐
3	Executives always demonstrate by their actions their personal commitment to customer satisfaction.	☐
4	Our driving intention is always to exceed customer expectations in those things that matter most to them.	☐
5	We promote and reward employees on the basis of their demonstrated commitment to customer care.	☐
6	Everybody in our organisation has confirmed their personal commitment to total quality.	☐
7	Satisfying customer needs always takes precedence over satisfying our own internal needs.	☐
8	We reward with praise or tangible benefits every example of exceptional customer service.	☐
9	When mistakes are made we focus on problem solving and not the apportionment of blame.	☐
10	We communicate fully to customers our intention to give them superior service.	☐
	Your score	☐
Client/customer relationships		
1	When it comes to selling we play a consultative role with our customers.	☐
2	In advertising, selling and promotion we avoid promising more than we can deliver.	☐
3	We know the attributes of our products which customers value most.	☐
4	Information from customers is fully utilised in designing our service and product offering.	☐
5	We strive to be the leader in our industry in terms of customer retention.	☐
	Your score	☐
Client/customer problems		
1	We monitor all customer complaints.	☐
2	We regularly ask customers to give us feedback on our performance.	☐
3	Customer complaints are analysed to identify quality or service problems.	☐
4	We identify and eliminate internal procedures which cause customer problems.	☐
5	We refuse to live with convenient internal policies or procedures which fail to give added value to our customers.	☐
	Your score	☐
Client/customer understanding		
1	We know how our customers define "quality".	☐
2	We provide opportunities for all employees, whatever their function, at some time to meet with customers.	☐
3	We clearly understand what our customers expect of us.	☐
4	Our key managers clearly understand our customers' requirements.	☐
5	Our top team has frequent contact with customers.	☐
	Your score	☐

Table 30.1 *(Continued)*

Making it easy for clients and customers to do business with us		
1	We make it as convenient as possible for our customers to do business with us.	☐
2	Employees are encouraged to go "above and beyond" to serve customers well.	☐
3	Employees are told, as clearly as we know how, what they are free to do on their own authority to satisfy customers.	☐
4	We make it easy for customers to complain to us if they believe they have cause.	☐
5	We do everything reasonable to resolve customer complaints quickly.	☐
	Your score	☐
Empowerment		
1	We treat all employees with respect at all times.	☐
2	Employees at all levels have a good understanding of our products and services.	☐
3	Employees who work for customers are supported with resources to enable them to do their job well.	☐
4	At all levels of the organisation employees are empowered to act on their own judgement to make things right for a customer.	☐
5	Employees feel that they are part of an exciting enterprise.	☐
	Your score	☐
Training and development		
1	Decisions are pushed down to the lowest levels in the organisation capable and qualified to make them.	☐
2	No lower-level employee is expected to make a decision for which they lack the skills, knowledge, experience or confidence to ensure a good outcome.	☐
3	Managers are trained in the complexities of developing the autonomy of the workforce.	☐
4	Employees at all levels make at least some significant decisions about their own work.	☐
5	Employees are cross-trained so that they can support and fill in for each other when necessary.	☐
	Your score	☐
Business and organisational growth		
1	Instead of competing with each other, functional groups cooperate to achieve shared goals.	☐
2	We study the best practices of other companies to see how we may do things better.	☐
3	We work continuously to improve our products and processes.	☐
4	When a new product or service would meet a known customer need, we bust a gut to make it available to them.	☐
5	We have a comprehensive quality policy throughout our organisation.	☐
6	We recognise and respect the needs of the internal customer at every level.	☐
7	Our employees understand that quality means consistently meeting customer need at lowest cost, and they strive to reduce costs without damaging customer service.	☐
8	Our employees value and use creativity to provide exceptional service and build profitability.	☐
9	The key values of the organisation are known to and owned by all.	☐
10	We invest in the development of innovative ideas.	☐
	Your score	☐
Employee attitude to quality		
1	Every employee fully understands that total quality requires them consistently to meet internal and external customer needs at the lowest possible cost.	☐
2	Every employee recognises that they have a personal role to play in the marketing of our products and services and actively seek to create, identify and satisfy customer needs at a profit.	☐
	Your score	☐

Source: Reproduced with permission from *High Income Consulting: How to Build and Market your Professional Practice 2E* (1997) Nicholas Brealey Publishing.

the roles that they undertook, the software employed so that the results can be replicated, and any assumptions that were made to place the results in context;
- ensuring the key business representatives participated in the assignment.

30.3 CUSTOMER DELIGHT

Lambert (1997) provides a customer delight questionnaire, which could be completed by employees of a consultancy providing enterprise risk management. It provides a tool with which consultants may challenge themselves in terms of their current practices to see where improvement in the provision of services could be accomplished. It provides a unified focus on those things that matter. The complete questionnaire is included in its entirety in Table 30.1. It could be argued that it is too long and repetitive in places.

30.4 SUMMARY

This chapter has looked at implementing an enterprise risk management assignment, focusing on the management of delivery. Significant to delivery are a series of issues such as clarity of the objectives, programming and resources. In addition, common to all assignments are a number of critical success factors which influence both delivery and customer satisfaction. A customer delight survey for completion by employees of consultancies engaging in enterprise risk management is included, based on the questionnaire proposed by Lambert for all consultancies.

30.5 REFERENCES

Cope, M. (2003) *The Seven Cs of Consulting: The Definitive Guide to the Consulting Process*. Pearson Education, Harlow.

Lambert, T. (1997) *High Income Consulting: How to Build and Market your Professional Practice*, 2nd edition. Nicholas Brealey Publishing, London.

Appendix 1
Successful IT: Modernising Government in Action

This appendix should be read in conjunction with Section 7.3.

In terms of tools and techniques, the government report entitled *Successful IT: Modernising Government in Action* recommends the use of a project profile model and a summary risk profile.

PROJECT PROFILE MODEL

The report recommended the use of a *project profile model* (borrowed from the US government) to provide a standard set of high-level criteria against which senior project representatives can assess the characteristics and degree of difficulty of a proposed project, in order to establish the appropriate project controls, including the risk profile and corresponding risk strategy. The example of a model included in the review is included in Table A1.1. The intention was that the model would be piloted by OGC on a range of projects with a view to placing an updated version on the government intranet as a diagnostic tool. It was not intended to be an exhaustive project risk analysis model, but an aid to a fuller project risk analysis. The model requires the senior project representative to assess the project against a number of criteria, each of which is weighted to provide an overall score for the project. The review proposed that:

- a total score less than 20 suggests the project is relatively low risk – peer reviews and other project controls are likely to be managed from within the sponsoring department;
- a total score in the range 21–40 suggests the project is higher risk – peer reviews should involve other departments or agencies and may require support from OGC and/or CITU; and
- a total score of 41 or more suggests the project is high risk and will require OGC and/or CITU involvement.

It was thought the project profile model would be used as a starting point in determining the risk profile and corresponding risk strategy of a project, but recognised there will be other project-specific factors that would need to be taken into account. Other factors the report thought should be considered during the assessment of risk include:

- the effect of government priorities on the allocation of resources to the project;
- externally imposed time delays, such as waiting for requirements from other departments;
- capability of the supplier in terms of technology, expertise, skills, etc.;
- inexperience of the government department in projects of a particular size or complexity; and
- inadequate reliable estimates, feasibility studies, user trial programmes or other similar data upon which to base a risk assessment.

Table A1.1 Project profile model proposed by the Cabinet Office to measure project complexity

Criteria	Comments	Value	Score
Business Impact			
Total value of the business benefits in £	Total (as opposed to annual) value, calculated in line with HM Treasury guidance	Up to £10 million	2
		£10–100 million	4
		More than £100 million	1
Total value of the business costs in £	Total (as opposed to annual) costs, calculated in line with HM Treasury guidance. Excludes IT costs which are covered later	Up to £5 million	1
		£5–50 million	2
		More than £50 million	4
		Less than 1000	1
Number of individuals affected	Refers to internal personnel within government – i.e. includes technical and business staff and users, but excludes citizens, suppliers, etc.	1000–10 000	4
		More than 10 000	6
		No significant change to organisation.	1
Impact on business processes (includes changed processes)	Refers to the impact that the project will have on the organisation (both during development and after implementation). Allocate a score between 1 and 6	Major new legislation or significant new processes requiring new skills, new organisation and major new procedures	6
Impact on government services at implementation	Refers to the impact that the project will have outside the organisation, for example on the public and businesses (both during development and after implementation). Allocate a score between 1 and 6	Impact contained internally within the organisation	1
		Impact potentially disruptive to large sectors of the public and business	6
Impact on other projects and changes	The degree to which the project is dependent on and connected to other projects and changes. Allocate a score between 1 and 8	Standalone project	1
		Supporting wider departmental change initiative	3
		Supporting cross-cutting change initiative	6
		Supporting EU or 3rd country initiative.	8

Technical Impact			
Total IT costs	Total (as opposed to annual) IT costs, calculated in line with HM Treasury guidance. For commercial contracts this will be the total charge to department rather than cost to supplier	Up to £10 million £10–100 million More than £100 million	2 3 1
Number of IT practitioners (including internal and outsourced suppliers)		Up to 50 50 to 100 More than 100	1 2 3
Degree of innovation	The extent to which the project involves innovative solutions, and the level of familiarity and experience available. Allocate a score between 1 and 4	Stable, proven technology, widely implemented, familiar to organisation and suppliers Technology or scale of its planned use unproven, and organisation and some suppliers inexperienced in its application	1 4
Impact on legacy systems and data	The degree to which the project will need to develop interfaces to existing systems and data stores. Allocate a score between 1 and 4	Greenfield development Extensive data conversion, migration and integration issues, and bespoke interfaces to existing applications and platforms needed	1 4
Scope of IT supply (note: for this criterion score for each element, i.e. may be cumulative)	The range of activities that will be undertaken by the IT supplier, and the extent to which these will impact on the business processes of the organisation	Deliver infrastructure Deliver packaged software Deliver bespoke application Deliver new business processes Deliver package with significant bespoke elements Transfer of IT staff	1 1 3 3 4 4
Client/Supplier Arrangements			
Client-side organisation	The complexity of the client-side arrangements. Allocate a score between 1 and 4	Single business stream within department Cross-cutting involving multiple departments	1 4
Supply-side organisation	The complexity of the supply-side arrangements	Single internal Single external Multiple with prime contractor Multiple without prime contractor	1 2 3 4

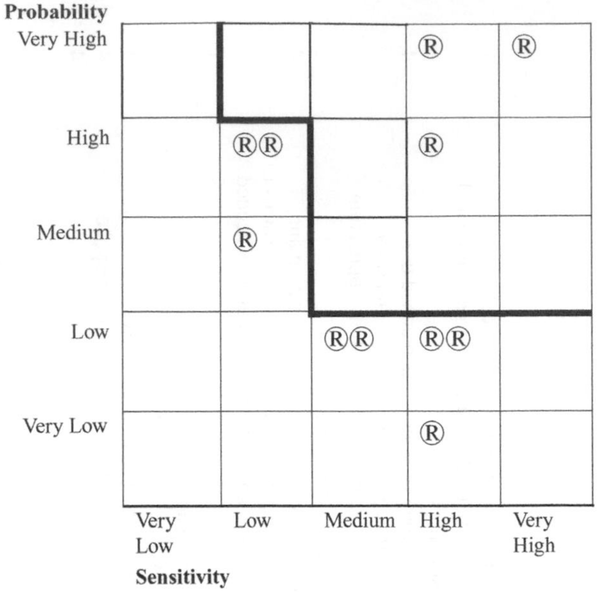

Figure A1.1 Summary risk profile: each ® represents a risk, and the thick black line is the risk tolerance line

This tool is not unique to the Cabinet Office and has been used by a variety of organisations. This author has used a similar model within training modules for GCHQ, for instance. Such tables must be appropriate for the project under examination and the results obtained are an aid to decision making rather than an end in themselves.

SUMMARY RISK PROFILE

The report also recommends the use of a *summary risk profile*, illustrated in Figure A1.1, as a simple mechanism to increase the visibility of risks. The term used for this tool by this author and other practitioners is *risk map*. Both are a graphical representation of information normally found on existing risk registers. The report describes the project manager or risk manager updating the diagram in line with the risk register on a regular basis and providing it to the person with overall responsibility for a project. Figure A1.1 shows an example of the summary risk profile which illustrates a project's risks in terms of probability and severity, with the effects of mitigating action taken into account. The line represents a set level of tolerance, below which the risks are regarded as being effectively managed. Attention is drawn to those above the line by the risk manager, which require immediate action, enabling the senior project representative to target his/her actions.

Appendix 2
Sources of Risk

This appendix should be read in conjunction with Sections 7.5.1 and 9.7.2. The following tables record different authors' views of the sources of risk which can be used as a prompt when carrying out risk identification at the time of investment decision making and during ongoing business activity.

Tables A2.1–A2.3 list categories of risk referred to by Cooper (2004), Day (2001) and Holliwell (1998), respectively.

Table A2.1 Cooper categories

• Market	• Credit
• Liquidity	• Technological
• Legal	• Health
• Safety	• Environmental
• Reputation	• Business probity

Table A2.2 Day categories

• Commercial and administration	• Management competence and drive
• Market shifts, especially with new technology	• Competitive responses leading to reduced demand
• Knowledge and information dissemination	• Financial – liquidity, profitability and financial structure
• Currency issues both on the supply and sales sides	• Legal issues
• Political events in home and overseas markets	• Partners, suppliers and subcontractors
• Quality issues leading to reduced sales	• Economic cycles and the effect on demand prices
• Technical ability of the company	• Resource availability leading to lower production
	• Innovation, copyright and availability of new technologies

Table A2.3 Holliwell categories

• Competitors	• Country
• Criminal/fraud	• Economic
• Environmental	• Legal
• Information	• Operational
• Market	• Political
• Personal	• Public relations
• Product/industry	• Technological
• Resources	• Financial (counterparty, funding, currency, interest rate)
• War/terrorism	

BSI publication PD 6668 (BSI 2001) lists the possible threats to a business which can damage an organisation's reputation, with catastrophic effects in the short term and long-lasting consequences (Table A2.4). Friend and Zehle (2004) categorise business risks as shown in Table A2.5.

Table A2.4 BSI categories

• Fraud	• Unethical dealings
• Product and/or service failure	• Public perception
• Lack of business focus	• Exploitation of workers and/or suppliers
• Environmental mismanagement	• Occupational health and safety mismanagement and/or liability
• Regulatory action	• Civil action
• Failure to respond to market changes	• Failure to control industrial espionage
• Failure to take account of widespread disease or illness among the workforce	• Failure to complete
• Failure to take on new technology	• Failure to invest
• Failure to control IT effectively	• Failure to establish a positive culture
• Vulnerability of resources (material and human)	• Failure to establish effective contingency arrangements in the event of a product and/or service failure
• Failure to establish effective continuity arrangements in the event of a disaster	• Inadequate insurance provision

Table A2.5 Friend categories

• Operational	• Financial
• Industry	• Political

The *Orange Book* (HM Treasury 2001) provides a table of the sources of risk (Table A2.6), which naturally has a government department focus and subdivides the sources into four main categories labelled "external", "financial", "activity" and "human resources".

Table A2.6 HM Treasury's (2001) most common categories of risk ("risk self-assessment" tool)

External	
Infrastructure	Relating to infrastructures such as computer networks, transport systems for staff, power supply systems
Economic	Relating to economic factors such as interest rates, exchange rates, inflation
Legal and Regulatory	Relating to the laws and regulations which if complied with should reduce hazards (e.g. Health and Safety at Work Act)
Environmental	Relating to issues such as fuel consumption, pollution
Political	Relating to possible political constraints such as a change of government
Market	Relating to issues such as competition and supply of goods
Act of God	Relating to issues such as fire, flood, earthquake

Table A2.6 (*Continued*)

External	
Financial	
Budgetary	Relating to the availability of resources or the allocation of resources
Fraud or Theft	Relating to the unproductive loss of resources
Insurable	Relating to the potential areas of loss which can be insured against
Capital Investment	Relating to the making of appropriate investment decisions
Liability	Relating to the right to sue or be sued in certain circumstances
Activity	
Policy	Relating to the appropriateness and quality of policy decisions
Operational	Relating to the procedures employed to achieve particular objectives
Information	Relating to the adequacy of information which is used for decision making
Reputational	Relating to the public reputation of the organisation and consequent effects
Transferable	Relating to risks which can be transferred or the transfer of risks at inappropriate cost
Technological	Relating to the use of technology to achieve objectives
Project	Relating to project planning and management procedures
Innovation	Relating to the exploitation of opportunities to make gains
Human Resources	
Personnel	Relating to the availability and retention of staff
Health and Safety	Relating to the well-being of people

The revised edition of the *Orange Book* (HM Treasury 2004) provides an updated expanded table of the sources of risk (Table A2.7), which again has a government department focus but reflects the developments in enterprise risk management in its new categories of risk, now labelled "external", "operational" and "change".

Table A2.7 HM Treasury (2004) summary of the most common categories or groupings of risk

Category of risk	Examples/explanation
1. External (arising from the external environment, not wholly within the organisation's control, but where action can be taken to mitigate the risk) [This analysis is based on the "PESTLE" model – see the Strategy Survival Guide at www.strategy.gov.uk]	
1.1 **Political**	Change of government, cross-cutting policy decisions (e.g. – the euro); machinery of government changes
1.2 **Economic**	Ability to attract and retain staff in the labour market, exchange rates affect costs of international transactions; effect of the global economy on UK economy
1.3 **Socio-cultural**	Demographic change affects demand for services; stakeholder expectations change
1.4 **Technological**	Obsolescence of current systems; cost of procuring best technology available, opportunity arising from technological development
1.5 **Legal**	EU requirements/laws which impose requirements (such as health and safety or employment legislation)
1.6 **Environmental**	Buildings need to comply with changing standards; disposal of rubbish and surplus equipment needs to comply with changing standards

Table A2.7 (*Continued*)

Category of risk	Examples/explanation
2. Operational (relating to existing operations – both current delivery and building and maintaining capacity and capability)	
2.1 Delivery	
2.1.1 Service/product failure	Fail to deliver the service to the user within agreed/set terms
2.1.2 Project delivery	Fail to deliver on time/budget/specification
2.2 Capacity and capability	
2.2.1 Resources	Financial (insufficient funding, poor budget management, fraud) HR (staff capacity/skills/recruitment and retention) Information (adequacy for decision making; protection of privacy) Physical assets (loss/damage/theft)
2.2.2 Relationships	Delivery partners (threats to commitment to relationship/clarity of roles), Customers/Service users (satisfaction with delivery), Accountability (particularly to Parliament)
2.2.3 Operations	Overall capacity and capability to deliver
2.2.4 Reputation	Confidence and trust which stakeholders have in the organisation
2.3 Risk management performance and capability	
2.3.1 Governance	Regulatory and propriety/compliance with relevant requirements/ethical considerations
2.3.2 Scanning	Failure to identify threats and opportunities
2.3.3 Resilience	Capacity of systems/accommodation/IT to withstand adverse impacts and crises (including war and terrorist attack) Disaster recovery/contingency planning
2.3.4 Security	Of physical assets and of information
3. Change (risks created by decisions to pursue new endeavours beyond current capability)	
3.1 PSA targets	New PSA targets challenge the organisation's capacity to deliver/ability to equip the organisation to deliver
3.2 Change programmes	Programmes for organisational or cultural change threaten current capacity to deliver as well as providing opportunity to enhance capacity
3.3 New projects	Making optimal investment decisions/prioritising between projects which are competing for resources
3.4 New policies	Policy decisions create expectations where the organisation has uncertainty about delivery

REFERENCES

BSI (2001) *Managing Risk for Corporate Governance*, BSI PD 6668. British Standards Institution, London.

Cooper, B. (2004) *The ICSA Handbook of Good Boardroom Practice*. ICSA Publishing, London.

Day, A.L. (2001) *Mastering Financial Modelling: A Practitioner's Guide to Applied Corporate Finance*, Pearson Education, London.

Friend, G. and Zehle, S. (2004) *Guide to Business Planning*. Profile Books, London.

HM Treasury (2001) *Management of Risk: A Strategic Overview*. London.

HM Treasury (2004) *Management of Risk – Principles and Concepts*. The Stationery Office, Norwich.

Holliwell, J. (1998) *The Financial Risk Manaual: A Systematic Guide to Identifying and Managing Financial Risk*. Pearson Education, UK.

Appendix 3
DEFRA Risk Management Strategy

This appendix should be read in conjunction with Section 7.9.

The DEFRA risk strategy is divided into six sections: (1) introduction and purpose; (2) aim, principles and implementation; (3) identifying risks; (4) assessing risks; (5) addressing risks; and (6) reviewing and reporting risks. These sections are described in turn below. A detailed description of the strategy is given here to avoid reinventing the wheel and learning from others' experience and methods.

INTRODUCTION

Summarising the *introduction*, the strategy states that the Department's aims are to:

- create an environment of "no surprises";
- be in a stronger position to deliver its business objectives;
- manage opportunities to be in a better position to provide both improved services and better value for money; and
- avoid risk management failure (which can be significant and high profile).

AIM, PRINCIPLES AND IMPLEMENTATION

In terms of the *aim, principles and implementation*, the strategy declares that the drivers for change include:

- the government's desire to promote effective risk management as part of its response to the report on the BSE inquiry;
- the need to respond to the recommendations of the Turnbull Report; and
- the obligation on the Permanent Secretary to sign and publish statements of internal control with the annual accounts.

The *aim* of the Department is described as follows: to become one of the leading departments in risk management; to be an exemplar of good practice; to strike a balance between risk and opportunity; to keep the strategy up to date; and to deliver the aim through the four linked elements of the strategy – identify, assess, address and review, and reporting. The *principles* are described as being transparency, coordination, credibility (in the eyes of the public) and effectiveness. *Implementation* is described as being accomplished through: establishing on the DEFRA intranet the detailed risk management guidance and register; developing the DEFRA website to make the stakeholders and public aware of the risk management approach; conducting training to promote awareness; and providing good practice guidance for staff.

IDENTIFYING RISKS

The strategy states that the risk management approach will: be "objective-driven"; be used right across the business; strive to be better at identifying longer-term risks or risks that are currently "over the horizon"; be alive to the general type of external or foresight risk (explained through examples such as worldwide events – economic shocks or political upheaval); and be dependent on good stakeholder involvement and good systems of gathering intelligence. In addition, DEFRA stated that its intention was to carry out "horizon scanning" to seek to identify new issues that may pose future risks to its objectives (or that may provide a new means of meeting the objectives) and "surveillance" to identify important changes in the health of particular populations, often as part of a disease control programme. A key building block of the strategy was declared to be the creation (and subsequent maintenance) of a central risk register of its important risks, built up from each of the Directorates and Agencies. Importantly, the strategy stated that "this should be a living process, not a tick-box approach and must not become bureaucratic". It was thought that a culture that systematically identified risks would be well placed to assess risks and opportunities. To this end DEFRA planned to be in that position – to bring improved performance through the calculated taking of opportunities.

ASSESSING RISKS

The strategy considers that to assess risks adequately DEFRA will identify the consequences of a risk materialising and give it a risk rating. Consequence categories identified by DEFRA are listed in Box A3.1.

Box A3.1 Consequence categories

Political	e.g. Ministerial embarrassment
Financial	e.g. high government expenditure to fight an animal health disease outbreak
Societal	e.g. vCJD cases[1]
Operational	e.g. targets missed
Legal	e.g. claims against the Department
Environmental	e.g. habitat damage
Reputational	e.g. loss of public confidence in the Department's advice

As a way of having a means of comparing risks and addressing the most serious, DEFRA have defined a four by four probability impact grid called a risk rating matrix which uses the scales VL (very low), L (low), M (medium) and H (high). In terms of likelihood, these scales are defined as in Box A3.2. In terms of impact, thesey are defined as in Box A3.3.

[1] See Chapter 7, note 8.

Box A3.2 Likelihood rating definitions

VL	rare – the risk may occur in exceptional circumstances
L	possible – the risk may occur in the next three years
M	likely – the risk is likely to occur more than once in the next three years
H	almost certain – the risk is likely to occur this year or at frequent intervals

Box A3.3 Impact category definitions

VL	low financial losses; no public health effects; service delivery unaffected; no legal implications; unlikely to affect the environment; unlikely to damage reputation
L	medium financial losses; minor or reversible health effects; reprioritising of delivery required; minor legal concerns raised; minor impact on the environment; short-term reputation damage
M	major financial loss; significant public health effects; deadlines renegotiated with customers; potentially serious legal implications (e.g. risk of successful legal challenge); significant environmental impact; longer-term damage to reputation
H	huge financial loss; key deadlines missed; very serious legal concerns (e.g. high risk of successful legal challenge with substantial implications for the Department); major environmental impact; loss of public confidence

ADDRESSING RISKS

The aim of the strategy is to prevent the realisation of risks; to this end, DEFRA has built on HM Treasury's guidance contained in the *Orange Book*. Having assessed each risk, a response is defined by selecting one of the "four Ts" described in the *Orange Book* – transfer, tolerate, treat or terminate. In addressing the risks, DEFRA seeks to adopt an appropriate response to reduce the risk to "as low as reasonably practicable" to suit the circumstances (this is known as the ALARP approach). DEFRA states that it will have contingency plans for all risks which have been addressed as having a potentially high impact, irrespective of the likelihood and plans would be rehearsed. Additionally business continuity plans will be prepared to keep the business running during times of change or disruption, and disaster recovery plans for serious events such as terrorist activity, contamination of sites or a national catastrophe.

REVIEWING AND REPORTING RISKS

DEFRA stresses that risk management is a dynamic process (and not static), and that risk assessments, countermeasures and contingencies will have to be kept up to date. Additionally, the management board will keep the main risks under regular review. It is intended to develop a baseline risk register and monitor progress against it. Evidence is required by DEFRA that its interventions are having the desired effect. Directorates and Agencies will be required to regularly review their registers and report significant changes. DEFRA will assess this updating process, and examples of its lines of enquiry are given in Box A3.4.

Box A3.4 Measures for determining the "currency" of the risk management process

1. How regularly [are registers reviewed and change reported] in each Directorate and Agency?
2. What level of input is the Director or Chief Executive having to this review?
3. How static is the register? Do the changes appear cosmetic and presentational or are they substantial, involving the identification of new countermeasures or additional actions? Are the additional actions being tackled?
4. Is there an audit trail, outlining the actions that have been taken and indicating their relative success?
5. Is the departmental register being made available to staff and are they amending their individual registers in the light of risks or actions in related Directorates or Agencies?
6. What is the turnover in terms of new risks being identified and existing risks being terminated? What is the general trend in the rating of the risks that have been registered? Is the number of high and very high risks decreasing?

ROLES AND RESPONSIBILITIES

The driver behind clarifying "who does what" within DEFRA in terms of risk management is based on the perception that the Department will be exposed to risks being unmanaged, causing damage or loss that could otherwise have been influenced, controlled or avoided. This section of the strategy identifies 17 affected parties (some being individuals, such as the Permanent Secretary, and others being within a group, such as the DEFRA Management Board), who have a role in the implementation of risk management in its widest terms. They are too numerous to report on in full here. The role of risk owners is described as ensuring the risk countermeasures are in place, reviewing contingencies and developing additional actions as required. Risk managers are described as the individuals responsible for the day-to-day implementation of countermeasures, monitoring their impact and reporting on their effectiveness.

Appendix 4
Risk: Improving Government's Capability to Handle Risk and Uncertainty

This appendix should be read in conjunction with Section 7.10.

The Cabinet Office report *Risk: Improving Government's Capacity to Handle Risk and Uncertainty* is structured into five areas: (1) the government's role and responsibilities; (2) improving government's handling of risk; (3) improving capacity; (4) handling the communication of risk; and (5) the role of leadership and culture change. Each of the areas is examined below.

THE GOVERNMENT'S ROLE AND RESPONSIBILITIES

The report provides a framework for understanding the roles government plays in handling risk and the responsibilities that are expected of it in each of its roles. The report defines three roles, a regulatory role, a stewardship role and a management role.

Regulatory roles. Governments generally have a regulatory role in providing the legal framework where the activities of business and individuals give rise to risk to others. The report advises that governments will not normally intervene where individuals take risks voluntarily and where they alone are affected. However, where these risks (taken voluntarily) have direct or indirect consequences on others, such as on the environment, government may intervene through regulation to limit or control that activity. Legislative proposals that have an impact on business, charities or the other voluntary sector have a regulatory impact assessment, which includes a risk assessment of the problem being addressed and of the proposal itself. In addition, governments generally seek to ensure that those who impose risks on others bear the cost of the consequences of the risk – an example being the "polluter pays principle" which transfers to the polluter the cost of cleaning up the environmental damage.

Stewardship role. Again governments generally have a stewardship role to protect individuals, businesses and the environment from risks imposed on them from outside such as major flooding, risks to public health, external threats to security or risks to economic stability. Government intervention in response to this type of risk can take two main forms: action to reduce the likelihood of the risk occurring (e.g. through the provision of flood defence); and action to mitigate the consequences (e.g. through the provision of health care). However, in either case government cannot take the risk entirely away from members of the public. Where there is a risk that an activity may cause serious harm to others and that those taking part may not be able to cover their liabilities, government may require them to pool their risks by taking out insurance first (the requirement for third-party motor insurance is cited as an example). Another concern for government is the provision of critical services provided by business such as energy, water and telecommunications, where the effects of service failure on the wider public would be severe. Government's role is seen to be to monitor and, when necessary, take action to ensure that critical networks continue to function.

Management role. The report describes government as having a management role in relation to its own business, including the delivery of public services and the performance of the regulatory and stewardship functions. In relation to its own business, government has a responsibility to identify and manage risks. Typical risks may include IT failure, delay or unbudgeted expenditure, and the risk of taking on too many high-risk projects simultaneously. The report provides an example where government will not provide a service direct, but will enter into a contract with another party to deliver it on its behalf – privately run prisons. While governments normally retain responsibility for the outcome of the service (in this case protection of the public), they may transfer responsibility for achieving specific objectives (and the associated risks) to another body. In addition, the government finds that when essential services go wrong, the public still look to it to put them right, regardless of whether these services are provided privately. The failure of Railtrack is cited as an example where government retains ultimate responsibility for the continuity of such services through its role as regulator, when (as the report describes) those tasked with providing the service are no longer capable of doing so. The report claims that where responsibility for a risk lies with government departments and agencies, there are well understood procedures for ensuring that it is adequately handled.

IMPROVING GOVERNMENT'S HANDLING OF RISK

This particular chapter sets out the challenge to government (as the report describes it) and considers that government needs to handle risk at three main levels: strategic, programme and operational. At a strategic level, decisions involve: the formulation of strategic objectives, resource allocation and choosing between options. At programme level policies and delivery plans are selected for the benefit of society, necessitating decisions on procurement, acquisition, funding, organisation, projects, service quality and business continuity. At project and operational level decisions are made on technical issues, managing resources, schedules, providers, partners and infrastructure. The report states that the "handling of risk at all three levels has been found wanting in recent crises and policy failures" and that reports by the National Audit Office (NAO) and the Public Accounts Committee (PAC) have discovered systematic weaknesses. The causes of these weaknesses are explored, examining the inherent complexity and riskiness of government business and reviewing the social context within which government works, which is considered to be more demanding. Examples cited where risk management was considered inadequate are given in Box A4.1.

Box A4.1 Inadequate risk management: unsuccessful government practices or projects

- The management of risk on the Individual Learning Accounts programme was criticised. By the end of January 2002, the total number of complaints received had reached nearly 18 300, with unsettled claims reaching nearly £15 million.
- The NAO report on the passport delays of the summer of 1999 stated that the cost of failure was £12.6 million. One of the general lessons learnt was that formal risk assessments should be carried out for all new computer systems (NAO 1999).

- The Phillips Inquiry report on BSE highlighted that several aspects of the government's handling of risk and uncertainty were unsatisfactory, notably the timing, implementation and enforcement of mitigation measures, the use of independent scientific experts and failure to communicate with the public on the risks to humans.
- The Cullen Report on the Ladbroke Grove rail crash identified that there was a "persistent failure to carry out risk assessments by whatever method was available" (Cullen 2001). There was (and possibly remains) a public perception that the railways were the responsibility of government.
- The NAO report on the handling of foot and mouth disease estimated that its mismanagement cost the public and the private sector in excess of £8 billion. According to the report, the organisation of government into departments makes it harder for government to deal with cross-cutting risks.

The NAO (2000) report *Supporting Innovation* and the PAC (2001) report *Managing Risk in Government Departments* both highlighted weaknesses in risk management practice and made recommendations for improvement. The surveys conducted by the Strategy Unit confirmed the findings of the two earlier reports. The cause behind these shortcomings was seen to be a mixture of poor communication across functional government departments, and the robustness of the IT and other systems that support the delivery of services. In addition, it is thought the social context within which government functions is more demanding in that while the traditional risks to life, health, economic well-being and housing have greatly reduced in modern societies, awareness of risk has risen and expectations on risk management in particular have risen. This is thought to be the product of the growing sense that risks can be controlled or are the product of human activity, rather than being effects of fate or random chance. In addition:

- there is declining trust in institutions borne out by studies by MORI, and the OECD, declining trust in government leadership (e.g. over the measles, mumps and rubella vaccination campaign) and scientific advice (recently rocked by the BSE crisis);
- the public expect that as the standard of service has improved in the private sector, it should be reflected in the public sector, particularly in terms of access, faster services and greater simplicity;
- there has been a rise in the number of groups that have been willing to become activists and an influential media has been found to amplify concerns significantly through their need to provide 24-hour coverage;
- the public have greater access to information through the media and websites – the government is not the sole source of information on risks and finds itself the subject of challenge;
- advances in science and technology have created novel and highly uncertain risks (e.g. radiation from mobile phones, computer fraud and genetically modified crops), and the government is increasingly being asked to assess, communicate and mitigate these risks (e.g. through regulation) with relatively little historical experience to draw on;
- there are greater expectations in terms of corporate governance and the handling of risk, reflecting developments in the private sector such as the Combined Code and the Turnbull Report.

IMPROVING CAPACITY

The government perceives that there is an expectation that it should manage risk well to cut waste and inefficiency and reduce unanticipated problems and crises that undermine the public's trust. The government considers that it should improve capacity to handle risk in five areas:

- ensuring that decisions take account of risk by embedding risk handling in the decision processes;
- firmly establishing risk management techniques;
- organising to manage risk by making sure that risk is placed with those best placed to manage it;
- developing skills and making sure that those involved in decision making are equipped to give due weight to risk issues and that they are supported by professional expertise;
- ensuring quality through the application of standards and benchmarking.

Common approach. The report states that while decisions are made at three commonly recognised levels – strategy, programme and operation/project – which have their own distinct characteristics, common approaches are necessary for all three. Risks have to be identified and assessed, responsibilities assigned, judgement made as to their importance, mitigation and consultancy plans considered, success of mitigation reported and details/decisions effectively communicated. It considers that there are particular weaknesses in risk analysis in the policy phase of the process of policy development and delivery. The report recommends an explicit systematic approach in order to improve the quality of decisions to provide an audit trail of risk judgements and to combat silo thinking by joining up risk management actions across departments, a clear parallel with enterprise risk management.

Embedding risk. It is considered that risk needs to be more clearly an integral part of the way government's business is done. Risk practice is uneven and, critically, not well integrated into the initial development of policy options. The lack of explicitness about risk issues and their management is a key concern. This is thought to undermine accountability and often lacks an audit trail of judgements about risks, making it impossible to regularly review risk judgements.

Barriers to overcome. The barriers to the implementation of risk management in government recorded in Box A4.2 are clearly echoed in business.

Box A4.2 Barriers to implementation of risk management

- A lack of planning – decisions often need to be made quickly and risk management will be compromised if information is not readily available
- Pressure on resources – encouraging planning on optimal assumptions
- Short planning horizons – traditionally ministers have been more focused on announcements than on longer-term implementation and delivery when risks might be realised
- Lack of good-quality relevant information
- Limited in-house skills, experience and tools
- The real difficulty of assessing and balancing risks and opportunities and weighing, for example, financial versus other risks
- Fear of failure acting as a disincentive to innovation
- In some cases political anxiety about explicit acknowledgement of risk

Practical approaches to the barriers. The report considers that the culture must support well-judged risk taking, embed risk thinking clearly into existing planning and operational decisions, implement an explicit appraisal of risks (as well as the benefits and costs) in all the main business processes, and ensure strategic risks are regularly considered by department boards and the Civil Service Management Board.

Policy making. As policy making is the process by which governments translate their political vision and priorities into actions to deliver outcomes, the report considers that failure to explicitly consider risk management in policy making and decision making can lead to serious problems, with the fallout being borne by the public. In addition, opportunities for high risk/reward options are passed over through a lack of confidence. At present the report advises there is no structured and enforced requirement to consider risks. Some very high-priority policies have been implemented without adequate attention to the attendant risks, often leading to very costly corrective action. The report recommends that a wider consideration of risk be included in policy making to provide an adequate review before proposals move into full development.

Business planning. The report recommends business planners refer to the Cabinet Office guide *Your Delivery Strategy: A Practical Look at Business Planning and Risk* (Cabinet Office 2001) which provides guidance and incorporates other sources information, including the *Orange Book*. In addition, business plans need to include better quality risk management plans.

Project and programme management. The report is guarded and states that "perhaps" risk management is best developed in the area of projects and programmes. In addition, the best-managed projects have moved well beyond the passive logging of risks and have very active approaches to identifying, assessing and managing risks. The report recommends that departments follow the OGC guidance on managing risk in projects and programmes and apply this guidance to their gateway reviews, where risks are required to be assessed and plans to manage them signed off before moving to the next stage of the project.

Investment appraisal. The report states that decision making needs to be underpinned by a series of issues such as the:

- focus of investment appraisals on benefits, costs and risks;
- explicit identification and assessment of risks;
- development of risk mitigation plans for priority risks from conception to appraisal and into execution;
- inclusion of risk identification and assessment within all key submissions at all levels;
- use of proformas or templates;
- use of post-project evaluations as a way of formally reviewing risk outcomes at the operational level;
- development of cost–benefit analysis incorporating risk assessment for inclusion within option appraisals; and
- development of HM Treasury's guide to investment appraisal (the *Green Book*), to deal with these issues.

Operational management. A Treasury-sponsored study found that adherence to risk management guidelines, identification and assessment, pooling of risks, data capture and financial incentives to improve risk management were all inadequate.

HANDLING THE COMMUNICATION OF RISK

The report considers that one of the main challenges of risk management for government is to win public trust. This trust, the government perceives, has been harder to secure more recently, due to a rise in public expectations and the extent and depth of media coverage given to government activities. Hence, it is concluded that there is a need for:

- building confidence in government decisions involving risk;
- more transparency about decisions that have been reached and the basis on which this is done;
- a refocus of decisions to better reflect public concerns and considerations about what is important;
- providing sufficient information so that the public can make balanced judgements; and
- a greater involvement of stakeholders at an early stage in the decision process.

THE ROLE OF LEADERSHIP AND CULTURAL CHANGE

The report considers that a sharper focus on risk management needs to be led from the "top" by ministers and permanent secretaries who should actively engage in providing the lead in:

- driving implementation of improvements in risk management identified in the report;
- making key judgements and providing a clear focus and direction;
- ensuring that managers are equipped with the skills, background knowledge and tools;
- supporting innovation; and
- ensuring clear accountability for managing risks.

In addition, the report considers that the following are important to embedding and effectively implementing risk management:

- getting the culture right;
- being aware of the risks;
- eliminating factors which hamper well-judged risk taking – mismatches between accountability, responsibility and authority to act; and
- overcoming the deterrent to risk taking from past criticism by the media and the PAC.

REFERENCES

Cabinet Office (2001) *Your Delivery Strategy: A Practical Look at Business Planning and Risk*. Cabinet Office, and HM Treasury, HM Government, London, September.

Cullen, Lord (2001) *Ladbroke Grove Rail Inquiry, Part: 1 Report*. HSE Books, London.

National Audit Office (1999) *United Kingdom Passport Agency: The Passport Delays of Summer 1999*, HC 812, Session 1998–99. Report by the Comptroller and Auditor General, 27 October. The Stationary Office, London.

National Audit Office (2000) *Supporting Innovation: Managing Risk in Government Departments*. Report by the Comptroller and Auditor General, 17 August. The Stationery Office, London.

Public Accounts Committee (2001) *First Report Session 2001–2002, Managing Risk in Government Departments*, November. The Stationery Office, London.

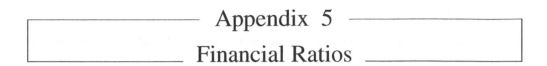

Appendix 5
Financial Ratios

This appendix should be read in conjunction with Section 8.7.1.

PROFITABILITY

The following ratios may be used to evaluate the profitability of a business.

Return on Ordinary Shareholders' Funds

The return on ordinary shareholders' funds (ROSF) compares the amount of profit for the period available to the ordinary shareholders with the ordinary shareholders' stake in the business. The ratio, which is normally expressed in percentage terms, is given by

$$\text{ROSF} = \frac{\text{net profit after taxation and preference dividend (if any)}}{\text{ordinary share capital plus reserves}} \times 100.$$

The net profit after taxation and after any preference dividend is used in calculating the ratio as this figure represents the amount of profit available to the ordinary shareholders.

Return on Capital Employed

The return on capital employed (ROCE) is a fundamental measure of business performance. The ratio expresses the relationship between the net profit generated by the business and the long-term capital invested in the business. The ratio is expressed in percentage terms as

$$\text{ROCE} = \frac{\text{net profit before interest and taxation}}{\text{share capital} + \text{reserves} + \text{long-term loans}} \times 100.$$

It should be noted that the profit figure used in the ratio is the net profit before interest and taxation. This figure is used because the ratio attempts to measure the returns to all suppliers of long-term finance before any deductions for interest payable to lenders or payments of dividends to shareholders are made. ROCE is considered by many to be a primary measure of profitability as it compares inputs (capital invested) with outputs (profit). This comparison is of vital importance in assessing the effectiveness with which funds have been deployed.

Net Profit Margin

The net profit margin relates the net profit for the period to the sales during that period. The ratio is expressed as

$$\text{net profit margin} = \frac{\text{net profit before interest and taxation}}{\text{sales}} \times 100.$$

Gross Profit Margin

The gross profit margin relates the gross profit for the period to the sales generated for the same period. It is expressed as

$$\text{gross profit margin} = \frac{\text{gross profit}}{\text{sales}} \times 100.$$

EFFICIENCY

Efficiency ratios examine the ways in which various resources of the business are managed. The following ratios consider some of the more important aspects of resource management.

Average Stock Turnover Period

Stocks often represent a significant investment for a business. For some types of business (e.g. manufacturers), stocks may account for a substantial proportion of the total assets held. The average stock turnover period measures the average number of days for which stocks are being held. The ratio is calculated as follows:

$$\text{stock turnover period} = \frac{\text{average stock held}}{\text{cost of sales}} \times 365 \text{ days (calculated to the nearest day).}$$

Average Settlement Period for Debtors

Businesses are commonly concerned with how long it takes for customers to pay the amounts owing. The speed of payment can have a significant effect on the cash flows of a business. The average settlement period for debtors calculates how long, on average, credit customers take to pay the amounts that they owe to the business. The ratio is as follows:

$$\text{average settlement period} = \frac{\text{trade debtors}}{\text{credit sales}} \times 365 \text{ days (calculated to the nearest day).}$$

Average Settlement Period for Creditors

The average settlement period for creditors tells a reviewer how long, on average, the business takes to pay its trade creditors. The ratio is calculated as follows:

$$\text{average settlement period} = \frac{\text{trade creditors}}{\text{credit purchases}} \times 365 \text{ days (calculated to the nearest day).}$$

Sales to Capital Employed Ratio

The sales to capital employed ratio examines how effectively the long-term capital employed of the business has been generating sales revenue. The ratio is calculated as follows:

$$\text{sales to capital employed ratio} = \frac{\text{sales}}{\text{long-term capital employed}}.$$

Sales Per Employee

The sales per employee ratio relates sales generated to a particular business resource. It provides a measure of the productivity of the workforce. The ratio is

$$\text{sales per employee} = \frac{\text{sales}}{\text{number of employees}}.$$

The Relationship between Profitability and Efficiency

The overall return on funds employed within a business will be determined by both the profitability of the sales and by the efficiency in the use of capital. Hence the ROCE ratio can be divided into two main elements: net profit to sales and sales to long-term capital employed:

$$\text{ROCE} = \frac{\text{net profit before interest and taxation}}{\text{sales}} \times \frac{\text{sales}}{\text{long-term capital employed}}.$$

LIQUIDITY

Current Ratio

The current ratio compares the "liquid" assets (cash and those assets held which will soon be turned into cash) of the business with the current liabilities (creditors due within one year). The ratio is calculated as follows:

$$\text{current ratio} = \frac{\text{current assets}}{\text{current liabilities}}.$$

Acid Test Ratio

The acid test ratio represents a more stringent test of liquidity. It can be argued that for many businesses the stock in hand cannot be converted into cash quickly. The acid test ratio is calculated as follows:

$$\text{acid test ratio} = \frac{\text{current assets (excluding stock)}}{\text{current liabilities}}.$$

Operating Cash Flows to Maturing Obligations

The operating cash flows to maturing obligations ratio compares the operating cash flows to the current liabilities of the business. It provides a further indication of the ability of the business to meet its maturing obligations. The ratio is expressed as

$$\text{operating cash flows to maturing obligations} = \frac{\text{operating cash flows}}{\text{current liabilites}}.$$

Gearing

Gearing occurs when a business is financed, at least in part, by contributions from outside parties. The level of gearing associated with a business is often an important factor when assessing risk. When a business borrows heavily, it takes on a commitment to pay interest charges and capital repayments. This can be a real financial burden and can increase the risk of a business becoming insolvent. It is common for businesses to borrow as they have insufficient funds to grow the business. A reason for borrowing may be that the loan interest is an allowable charge against tax and this can reduce the costs of financing the business:

$$\text{gearing ratio} = \frac{\text{long-term liabilities}}{\text{share capital} + \text{reserves} + \text{long-term liabilities}} \times 100.$$

Interest Cover Ratio

The interest cover ratio measures the amount of profit available to cover interest payable. The ratio may be calculated as follows:

$$\text{interest cover ratio} = \frac{\text{profit before interest and taxation}}{\text{interest payable}}.$$

INVESTMENT RATIOS

Dividend Yield Ratio

The dividend yield ratio relates the cash return from a share to its current market value. This can help investors to assess the cash return on their investment in the company. The ratio is given by

$$\text{dividend} = \frac{\text{dividend per share}/(1-t)}{\text{market value per share}} \times 100,$$

where t denotes tax credit.

Earnings Per Share (EPS)

The earnings per share of a company relates the earnings generated by the company during a period and available to shareholders to the number of shares in issue. For ordinary shareholders, the amount available will be represented by the net profit after tax (less any preference dividend where applicable). The ratio for ordinary shareholders is calculated as follows:

$$\text{earnings per share} = \frac{\text{earnings available to ordinary shareholders}}{\text{number of ordinary shares in issue}}.$$

Operating Cash Flow Per Share

It is commonly contested that, in the short term at least, operating cash flow (OFC) per share provides a better guide to the ability of a company to pay dividends and to undertake planned expenditures than the earnings per share figure. The OFC per ordinary share is calculated as follows:

$$\text{OFC per ordinary share} = \frac{\text{operating cash flows} - \text{preference dividends (if any)}}{\text{number of ordinary shares in issue}}.$$

Price/Earnings Ratio

The price/earnings (P/E) ratio relates the market value of a share to the earnings per share. The formula is

$$\text{P/E ratio} = \frac{\text{market value per share}}{\text{earnings per share}}.$$

Appendix 6
Risk Maturity Models

This appendix should be read in conjunction with Section 8.7.2.

As discussed in Section 8.7.2, risk maturity models are useful tools in understanding the degree of sophistication of a business risk management process, its reliability and effectiveness in identifying, assessing and managing risks and opportunities. Hillson (1997) proposes a risk maturity model and provides guidance to organisations wishing to develop or improve their approach to risk management, allowing them to assess their current level of maturity, identify realistic targets for improvement and develop action plans for increasing their risk capability. The model is composed of four levels, which are described in ascending order as “naïve”, “novice”, “normalised” and “natural”. The levels are defined as shown in Box A6.1.

Box A6.1 Hillson (1997) maturity model

Level 1 Naïve

The naïve risk organisation is unaware of the need for risk management and has no structured approach for dealing with uncertainty. Management processes are repetitive and reactive with little or no attempt to learn from the past or to prepare for future threats or uncertainties.

Level 2 Novice

The novice risk organisation is experimenting with [the] application of risk management, usually through a small number of nominated individuals, but has no formal or structured generic process in place. Although aware of the potential benefits of managing risk, the novice organisation has not effectively implemented risk processes and is not gaining the full benefits.

Level 3 Normalised

The normalised risk organisation has built management of risk into routine business processes and implements risk management on most or all projects. Generic risk processes are formalised and widespread, and the benefits are understood at all levels of the organisation, although they may not be consistently achieved in all cases.

Level 4 Natural

The natural risk organisation has a risk-aware culture, with a proactive approach to risk management in all aspects of the business. Risk information is actively used to improve business processes and gain competitive advantage. Risk processes are used to manage opportunities as well as potential negative impacts.

An alternative description of levels of maturity is proposed by the Central Computer and Telecommunications Agency (Government Centre for Information Systems 1993), again distinguishing between the levels of maturity by describing where in the organisation risk management is carried out and who is responsible for implementation (Box A6.2).

Box A6.2 Central Computer and Telecommunications Agency maturity levels

First level of maturity
The first type of organisation structure is the "virtual organisation", in which the management of risk is everyone's responsibility. In this situation, it is up to an interested individual manager to pursue good practice in respect to the management of risk.

Second level of maturity
The second level is where there is a separate management of risk group consisting of specialists who conduct analyses for operations, projects, and programmes and senior managers. Usually these groups operate on a task-by-task basis, examining a single high-risk project, for example. The usefulness of these groups depends greatly on the talents of the specialists involved and the individual managers' willingness to accept advice.

Third level of maturity
The third type of management of risk organisation exists when the specialist risk group is integrated within existing management groups at each organisational level. More formal mechanisms are needed to communicate risk information among these different groups. Although still mainly task oriented, more structured or formal management of risk approaches are put in place.

Fourth level of maturity
The fourth type of organisational structure is the fully integrated management of risk organisation. In this structure, the management of risk is everyone's responsibility, but formal mechanisms exist to help bring this about. A management of risk infrastructure that incorporates a standard analysis and management process exists.

Within the description of his model, Hillson describes four evaluation criteria – culture, process, experience and application – against which the four maturity levels are assessed. Each criterion using attributes of the typical organisation at each risk maturity model level. Hopkinson (2000) describes two Microsoft Access-based risk maturity models produced by a consultancy, one for use at the company (or business) level and one that is specifically applicable to the project environment. Both models adopt the four levels of maturity described by Hillson. The models determine the maturity of a risk management system (assumed here to be synonymous with process) by evaluating it against six criteria (called perspectives). For the company model these perspectives are management, risk identification, risk analysis, risk control, risk review and culture. For each perspective a series of questions are asked. The questions are weighted in accordance with the model's view of the significance of that question to the overall effectiveness of a risk management system. The overall assessment is considered to be only as high as the weakest score among the six assessments. Hopkinson explains that the rationale for this scheme of assessment is that the overall system for risk management is only as strong as its weakest area. The example he provides is "there is little point in having state of the art risk analysis, if the risk identification processes are so ineffective that many of the important risks are ignored". Hopkinson describes the characteristics of organisations operating at what he defines as level 4 (the most mature level) – see Box A6.3.

Box A6.3 Hopkinson risk maturity model for businesses, level 4

Management
- Board's risk management (RM) policy reported to shareholders
- Management leads RM by example. Practical definition of "significant risks"
- Practical definition of the risks to be borne
- Clear RM channels of communication

Risk Identification
- All sources of risk considered, including strategic, financial, technological, resource, disaster, projects, operational and external
- New risks identified in a timely manner
- Unusual events investigated for risk
- All employees can identify risks

Risk Analysis
- Consistent definition of probability
- Consistent definitions of impact
- Prioritisation influences agendas and promotes cost effectiveness
- Widespread availability of RM expertise
- Analysis traces risk source and secondary effects
- Risk records retained on state of the art tools

Risk Control
- Risk control actions based on cost–benefit analysis after considering all strategies
- Well-focused actions on individuals
- Actions are consistently completed
- Business continuity planning as appropriate

Risk Review
- Annual formal board review of RM effectiveness
- Strategy for review of all risks maximises cost effectiveness
- New information on significant risks is reported immediately
- Board regularly review significant risks
- Risk reports optimised for effectiveness

Culture
- Board's policy translated into management instructions understood by all employees
- Atmosphere of mutual trust
- Proactive risk management rewarded. Key managers have good RM skills and relevant experience in the core business

Table A6.1 describes a business risk maturity model developed by the author for assessing business risk management processes. It has four maturity levels – initial, basic, standard and advanced. Each level is assessed against five criteria – culture, system, experience, training and management.

Table A6.1 Business risk maturity model

	LEVEL 1 - INITIAL	LEVEL 2 - BASIC	LEVEL 3 - STANDARD	LEVEL 4 - ADVANCED
OVERVIEW	• Compliance only approach • Risk appetite not defined • No framework developed • Risk profile not defined • No senior management buy-in to RM as a decision tool	• RM established for business improvement • Risk appetite defined • Framework established • Risk system established • Risk profile defined	• RM built into routine business processes covering end-to-end production or delivery of services • Benefits recognised at all levels of the organisation	• RM considered critical to achievement of the business goals • Approach communicated to the organisation as a whole • Risk appetite transparent • Business seeks continuous improvement • Proactive upside (opportunity) RM • Sophisticated modelling techniques
CULTURE	• RM established to meet the Combined Code, the Listing Rules and annual reporting • Specific risk management roles not defined	• Risk exposure defined • Roles and responsibilities defined • Meeting structure defined • Decision-making mechanisms established	• Pro-active approach to RM to improve business performance • Central risk management function created • High level risks and responses debated at the board on a regular cycle	• RM culture lead by the chief executive • RM information used in decision making • RM roles and responsibilities included in the induction process, job descriptions and performance appraisals • Proactive enforcement of RM through employment contracts

SYSTEM	• Risk strategy unclear • Risk framework (and its constituent parts) embryonic	• RM framework under development • OR or business continuity management not addressed • Poor data collection and trend analysis	• RM strategy defined, relevant and practical • RM framework developed • OR and business continuity management frameworks being implemented	• RM strategy defined and kept under review • RM framework developed and benchmarked against best practice
EXPERIENCE	• Very limited understanding of systems, terminology or software	• Limited to small number of the audit committee and company secretary	• In-house core of experienced individuals in systems, modelling and response planning	• Risk awareness throughout the organisation plus external support
TRAINING	• No training provided in-house or from external support	• Training undertaken by audit committee members	• Risk manager appointed • Risk committee established	• Training and education programme provided to all business unit heads
MANAGEMENT	• Management practices focussed on satisfying the Combined Code and the Listing Rules	• Economic capital allocated to Operational Risk • Operational Risk management reactive • Risks reviewed on a yearly basis	• Guidance on risk–reward balance provided to line management • Early warning indicators established for OR • Economic capital allocated to risk	• Guidance on risk–reward balance provided to line management • Early warning indicators established for both OR and business context • Reputational risk addressed

REFERENCES

Government Centre for Information Systems (1993) *Introduction to the Management of Risk*. HMSO, Norwich.

Hillson, D. (1997) Towards a risk maturity model. *International Journal of Project and Business Risk Management*, 1(Spring), 35–45.

Hopkinson, M. (2000) Risk maturity models in practice. *Risk Management Bulletin*, 5(4).

Appendix 7
SWOT Analysis

This appendix should be read in conjunction with Section 8.7.3.

UNDERTAKING SWOT ANALYSIS

There are 17 commonly recognised steps in the implementation of a SWOT analysis as listed below. The term "factor" used below is a descriptive label to describe an issue, subject, or area of influence that will determine a business's ability to compete in the market place.

1. Establish the individuals who should be involved in the process. It should include employees from all key areas of the business.
2. Consider involving (if appropriate) key customers, suppliers or other sympathetic outsiders who know the market within which the business operates and can provide an objective independent view.
3. Arrange a workshop to identify the business's strengths and weaknesses and the opportunities and threats facing it.
4. Ask participants to collect and review information on internal management and external factors affecting the market within which the business operates, prior to the workshop.
5. Decide whether there is a suitable individual in-house who would have the skills and objectivity to manage the workshop. Otherwise appoint an external facilitator.
6. Prepare and issue a briefing pack to the participants including some basic details about the structure of the market and the business's performance within that market, so that discussion within the workshop is less subjective.
7. Decide on how the factors will be measured/quantified.
8. Hold the workshop. Brainstorm the factors.
9. List the strengths, weaknesses, opportunities and threats. Only important factors should be included, but some factors will invariably be more important than others. Each factor should be a short bullet point so that the SWOT analysis fits on one page.
10. Strive to make factor descriptions as specific as possible.
11. Where possible, quantify the factors.
12. Quantify in a readily comprehensible way – an amplification of a statement made about "broad distribution" may be "our products are distributed through 800 outlets compared with our nearest rival's 300 outlets".
13. When there are no further suggestions, score each factor and list the factors in order of importance.
14. Provide some explanation of the factors in the form of supporting paragraphs on a separate sheet.
15. Assess the significance of the SWOT analysis completed.
16. Create and execute an action plan to tackle weaknesses, capitalise on strengths and opportunities and deal with threats.
17. Use the analysis and action plan as a review tool before important decisions, so that decisions fit with what the analysis suggests.

RANKING STRENGTHS AND WEAKNESSES

Strengths matter only if a business can use them to exploit an *opportunity* or counter a threat. Similarly, a *weakness* is problematic if it relates to a *threat*. Therefore an external factor can be an *opportunity* or a *threat*. For example, if new technology is becoming available and a power tools company serving the construction industry has an excellent product development department that can take advantage of the new technology, this is an *opportunity*. In contrast, if a business cannot make use of the new technology, there is a *threat* from substitution[1] if rivals make use of the technology. The analysis should be undertaken bearing in mind the objective of strategic planning – to gain sustainable competitive advantage. A *strength* is a potential source of competitive advantage, such as core competencies or financial strength. As competitive advantage can only be sustained if customer needs are addressed, the market analysis is an important input into the SWOT analysis. To derive real advantage from a *strength*, it must be useful in satisfying the needs of customers. Similarly, if a *weakness* relates to specific customers' needs, this should be addressed as a matter of priority.

Table A7.1 Factors relevant to a SWOT analysis (based on Lynch 2000)

	Internal
Strengths	Weaknesses
Market dominance	Low market share
Core competencies	Few core competencies
Economies of scale	Old plant
Low-cost position	High-cost base
Leadership and management skills	Weak balance sheet and cash flow
Financial resources	Low R&R capability
Manufacturing skills and technology	Undifferentiated product
Research and development	Weak positioning
Brand and reputation	Quality problems
Differentiated products	Lack of distribution
	External
Opportunities	Threats
Technology innovation	New market entrants
New demand	Competitive price pressure
Diversification opportunity	Higher input prices
Market growth	Changing customer needs
Demographic and social change	Consolidation among buyers
Favourable political support	Threat from substitutes
Economic upswing	Capacity growth outstrips demand growth
Acquisition and partnerships	Cyclical downturn
Cheap funds	Demographic change
Trade liberalisation	Regulation and legislation
	Threat from imports

[1] Substitution: substitute products that perform the same function or satisfy the same need as an existing product. The threat from substitute products is particularly severe if the substitute product is cheaper or more cost effective. Whole industries have been wiped out by substitutes.

Table A7.1 provides a non-exhaustive checklist of factors that may be relevant to a SWOT analysis. However, each SWOT analysis will have to be tailored and made specific to the business under examination.

REFERENCE

Lynch, R. (2000) *Corporate Strategy*. Financial Times Management, London.

Appendix 8
PEST Analysis

This appendix should be read in conjunction with Section 8.7.4.

UNDERTAKING PEST ANALYSIS

Political

Local, national and international political changes can affect both costs and demand. Issues to consider are:

- Direct and indirect taxes, such as income tax and VAT, influence consumer spending and market demand.
- Corporate taxation has an impact on the profitability of a business.
- Public spending by central and local government has a direct impact on the level of demand within the economy.
- Regional and industrial policy can affect businesses at a micro level, and the availability of regional grants or other forms of assistance will boost local economies.
- Monetary policy and the level of interest rates will affect demand and a business's ability to service its debts.
- Exchange rate policy can have a critical effect on importers and exporters.
- Changes in international trade can create new markets.
- Competition law lays down rules on what a business can and cannot do and may be a crucial factor in the case of a merger or acquisition.
- Regulation and deregulation can have a dramatic impact on the business environment and individual business sectors.
- Education and training have a long-term impact on a business's ability to recruit suitably qualified staff and to compete in international markets.

Economic

Local, national and international economic factors to consider are as follows:

- *Business cycle*. Developed economies often follow a pattern known as the business cycle where periods of faster growth are followed by years of slower growth or even recession. Some sectors (e.g. construction, advertising, leisure and restaurants) are more susceptible to the impact of the business cycle than others (e.g. foodstuffs) where demand is less critical and tends to be more constant.
- *Employment levels*. These are closely related to the economy's position within the business cycle but also the state of the local economy. High levels of unemployment in a region will reduce demand but will also mean that labour is easier and cheaper to hire.

- *Inflation*. This can affect a business in many ways. For example, if the rate of increase in the price of raw materials is greater than the rate of inflation for the business's products, then the business will experience a fall in profitability over time.
- *House prices and stock market prices*. The growth or fall of house prices and the movement in stock market levels affect consumer confidence and hence consumer spending.

Interest rates and exchange rates, as mentioned under political factors, can critically affect a business's profitability. Exchange rates can make goods more or less expensive to overseas customers.

Social

Shifts in a country's demography and social cultural values usually occur over many years. However, with improvements in communication and increased employee mobility between countries, the speed of social and demographic change can be expected to increase.

- *Population growth*. The rate of growth of the population will have a direct impact on the size of the potential addressable market for a product or service. Population growth is typically higher in developing countries.
- *Age structure*. In the developed Western world economies are experiencing a significant increase in the average age of their populations. Differences in age structure of the population have implications for the overall level of saving compared with consumer spending, and for the relative sizes of the working and dependent sections of the population.
- *Social and cultural shifts*. Norms and values can change as a result of the composition of communities and the presence of ethnic groups. Changes may occur in religious, education and political preferences, impacting marriage and birth rates and buying preferences.

Technological

Changes in technology can have a rapid and dramatic impact on the economy. Issues to consider include:

- *Level of research and development by competitors*. This will provide an indication of whether any changes in technology-driven production processes or new products should be anticipated.
- *New markets*. Does the introduction of new technology create a new market for a particular technology-based product or service?
- *Rate of adoption of new technology*. It is often a considerable time before new technology gains mass-market appeal. The business plan must examine how long it will take the new product to penetrate the market (clearly some producer's products have remarkable appeal, such as the Apple iPad).
- *Production methods*. How might technology be utilised to improve production methods within the business, and how might competitors utilise technology to gain competitive advantage?

OTHER TYPES OF ANALYSIS

Later versions of PEST include both legal issues (making SLEPT), environmental issues (making PEEST), or legal and environmental issues (making SLEEPT). Possible definitions for legal and environmental factors are as follows:

- *Legal factors*. Changes in the law that might affect a firm (usually included under "political").
- *Environmental factors*. Green factors are becoming increasingly important to businesses; a business's environmental credentials will be linked to its reputation and corporate image.

Appendix 9
VRIO Analysis

This appendix should be read in conjunction with Section 8.8.6.

To conduct a resource-based analysis of a business, Barney (1991) proposes a structured approach based on analysing whether a resource is valuable, rare and imitable and whether the organisation is taking advantage of the resource.

- *Valuable*. A resource is valuable if it can be used, for example, to increase market share, achieve a cost advantage or charge a premium price (these features of a resource are not mutually exclusive, and hence a resource may have multiple attributes). Barney suggests that this question has to be answered first because a resource that is not valuable or is irrelevant cannot be a source of competitive advantage.
- *Rare*. If a valuable resource is not available to all competitors it is "rare" and therefore a potential source of competitive advantage. Rarity is important because if competitors possess the same resources, there is no inherent advantage in those resources. Of course different businesses can configure the same resources differently to achieve competitive advantage, but this is not the focus of the resource-based view of the firm.
- *Imitable*. If a resource is not readily copied or imitable, then the resource is a potential source of competitive advantage. To be advantageous the resource must be difficult or expensive for competitors to imitate or acquire, such as brand recognition/perception. If a resource is easy to imitate it offers only a temporary advantage, not a sustainable one.
- *Organisation*. A business must be capable of taking advantage of the resources at its disposal. If a resource is available, rare and difficult to imitate, a business must be able to exploit it, otherwise it is of little use. This may require reorganising the business.

The VRIO analysis framework illustrated in Table A9.1, which is based on the resources analysis proposed by Barney, links the VRIO resource analysis with strategic implications such as competitive advantage, the likely economic impact on the business and what this means in terms of SWOT analysis (see Appendix 7).

Table A9.1 Resource-based analysis of a business

Resource characteristics				Strategic implications		
Valuable	Rare	Costly to imitate	Organisation exploits it	Competitive implication	Impact on economic performance	SWOT category
No	–	–	No	Competitive disadvantage	Below normal	Weakness
Yes	No	–	↑	Competitive parity	Normal	Weakness or strength
Yes	Yes	No	↓	Temporary competitive advantage	Above normal	Strength and core competence
Yes	Yes	No	Yes	Sustainable competitive advantage	Above normal	Strength and long-term core competence

REFERENCE

Barney, J.B. (1991) Firm resources and sustained competitive advantage. *Journal of Management*, 17(1), 99–120.

Appendix 10
Value Chain Analysis

This appendix should be read in conjunction with Section 8.8.6.

CONFIGURATION OF RESOURCES

Competitive advantage is derived in part from the configuration of resources rather than simply the uniqueness of those resources. Therefore the analysis of a business should examine the link between resources and how they form part of a system, with the objective of adding value. This is an important concept as it can be applied right across a business and affects all major processes. Value added is the amount by which the selling prices exceed input costs. Input costs include bought-in products and services, salaries and the cost of capital equipment. Businesses exist to add value to a product or service. The ability to add value is closely linked to profitability. The identification of value added will allow a business to focus on improvements or strategic change on areas where little value is added. If any primary activity is making a loss for instance, it may be appropriate to outsource it, provided this does not introduce extensive risk, which is difficult to manage.

Michael Porter's concept of a value chain is a useful tool to examine value added across a business (Porter 1991). Examining, modifying and managing the value chain (the primary sequential activities of a business) can produce sustainable competitive advantage. Porter states that "competitive advantage results from the business's ability to perform the required activities at a collectively lower cost than rivals, or perform some activities in unique ways that create buyer value and hence allow the business to command a premium price".

The value chain identifies five primary and four support activities, as illustrated in Figure A10.1. A principal use of value chain analysis is to identify a strategy mismatch between different elements of the value chain. If a company competes on the basis of low costs, then every part of the value chain should be directed towards low costs. Friend and Zehle (2004) provide a topical example, low cost airlines which have looked at every aspect of the value chain and taken out costs at all stages of their business delivery: "Bookings are taken only via the internet rather than through travel agents; seats cannot be reserved; there are no paper tickets, free meals or drinks or lounges; and flights depart from secondary airports with lower landing fees."

The five primary activities are the sequential logistics, production and marketing processes, as described below. The primary activities can also be thought of as the main vertical functions of a business:

- Inbound logistics is the activity of receiving goods or services from suppliers and moving them onto the operations activity.
- Operations are where the production of the product or services takes place. Production may be broken down into further steps, for example producing intermediate goods from raw materials and then turning intermediate goods into the final product. Make or buy decisions can be made at every stage.

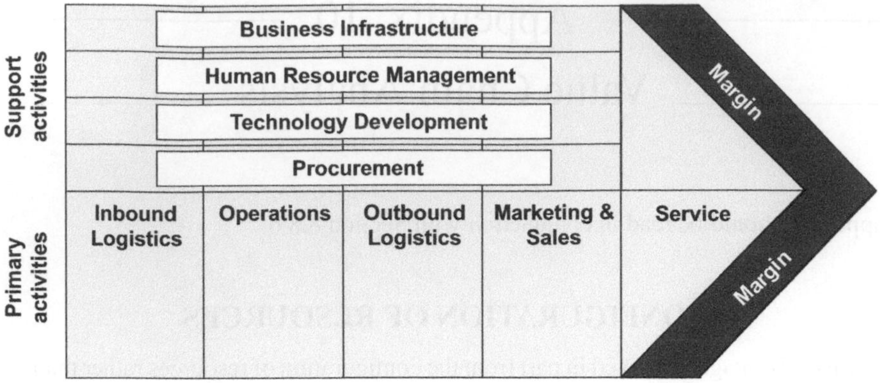

Figure A10.1 The value chain (Porter 1991)

- Outbound logistics covers order fulfilment, which is the warehousing of the finished goods and the distribution of the products or service to the customer. This is commonly outsourced.
- Marketing and sales include pricing, packaging and advertising as well as market research.
- Service refers mainly to after-sales service.

Support activities are horizontal in that they contribute to the different primary activities. For instance, the procurement department will deal with purchasing across all of the functions. The value chain has four support activities.

- Business infrastructure includes activities such as accounting, facilities, planning and general administration.
- Human resources management covers recruitment, training, labour relations and salaries.
- Technology development includes the development of new products or services or the enhancement of existing products and services.
- Procurement includes the purchasing of raw materials or intermediate goods as well as vehicles, office supplies and energy supplies.

REFERENCES

Friend, G. and Zehle, S. (2004) *Guide to Business Planning*. Profile Books Limited, London.
Porter, M.E. (1991) Towards a dynamic theory of strategy. *Strategic Management Journal*, 12, 95–117.

Appendix 11
Resource Audit

This appendix should be read in conjunction with Section 8.8.6.

The objective of a resource audit is to identify the key resources and ascertain how effectively they are utilised. The three primary resource attributes of a business are commonly recognised to be operational, human and financial. These are discussed in turn below.

- *Operational resources*. Operational assets, which consist of a variety of assets, not all of which may be recognised in the balance sheet, include: tangible assets such as buildings, plant and equipment; intangible assets such as brand names, rights and patents; and operational methods and systems, for example a just-in-time manufacturing system or a flexible manufacturing system. The identification of resources and whether they are put to good use may reveal extensive waste. For example, in recent years businesses have looked critically at the land and buildings they own. Lease-back deals and sell-offs have freed up substantial amounts of capital, and in some cases additional uses have been found for the land and buildings. An important task is to identify all of the resources that are not listed as assets on the balance sheet. There may be unused patents or rights, which could be used to economic advantage. Organisational resources comprise the overall organisational structure, departmental structures and the reporting systems of a business. They include the staffing of the departments and the management control assumptions. Analysis of the organisation may reveal duplication of activities, redundant activities or solely communication activities that are not adding value.
- *Human resources*. Leverage of human capital is seen as an untapped opportunity in improving competitiveness. As businesses become more knowledge-based and knowledge-dependent, human resources are increasingly viewed as a strategic asset. Annual reports of larger companies include information on human resources in an increasingly formalised manner. In the UK the Chartered Institute of Personnel and Development (CIPD)[1] has published a reporting framework on the value of human resources, to aid annual reporting of resources.
- *Financial resources*. Financial resources include all forms of funding – capital, debt, loans, vendor finance and creditors. Return on capital employed is perhaps the final measure of how successful a business is. Financial resources are crucial to the development and survival of a business. Most business plans have some kind of growth objective. Any new business activity, even within an existing business, requires funding.

Box A11.1, which includes an extract from the CIPD bulletin *Evaluating Human Capital* (Elias and Scarborough 2004), emphasises the merit of evaluation of how the management of people contributes to business performance.

[1] http://www.cipd.co.uk

Box A11.1 Human capital

Since 1997, the CIPD has been investigating the link between business performance and the way in which people are managed. CIPD considers that a wealth of evidence has been produced demonstrating that the key to competitiveness lies with the people of organisations, and particularly with the ability of the organisation to leverage the knowledge and skills of those people. Yet, it believes, most organisations still find it difficult to produce coherent measures of the worth and contribution of their people, that will be of use to their various stakeholders. CIPD says it is making strenuous efforts to develop tools and processes to enable organisations to better evaluate their human capital and communicate this to those who make the most important decisions about the long-term performance and viability of organisations.

Study

CIPD, in pursuit of the desire to develop some common principles upon which they could build practical guidance for practitioners, commissioned a six-month study, investigating the ways in which ten major UK-based firms from a variety of sectors evaluate their human capital. The findings are summarised in their bulletin entitled *Evaluating Human Capital* (Elias and Scarborough 2004).

Aims

The specific aim of the study was to investigate the contribution that human capital makes to business performance and the existing means of identifying and valuing that contribution. Additionally it aimed to relate the theoretical aspects of human capital to actual practice. The researchers were Professor Harry Scarborough of Warwick Business School, and Dr Juanita Elias at Cardiff University. The researchers' specific aims were: (1) to create a framework of shared understanding about the nature of human capital; (2) to identify some common principles for the analysis of human capital; and (3) to identify a possible process for the development of metrics for measuring the impact of human capital.

Findings

The study found a number of barriers and reasons why managers do not make more effort to evaluate human capital. Additionally the report found there is no single measure that can adequately reflect the richness of the employee contribution to corporate performance. As declared in the bulletin, the analysis suggests that measures are less important than the activity of measuring – of continuously developing and refining understanding of the productive role of human capital within particular settings. Of significance, the study considers that embedding such activities in management practices and linking them to the business strategy of the firm may enable firms to develop "a more coherent and ultimately strategic approach to one of the most powerful, if elusive, drivers of competitiveness".

Evaluation

This is not to suggest that metrics and other forms of information are irrelevant to the task of managing human capital. Rather, the report suggests that such information flows need to be embedded in wider processes of dialogue and exchange which, over time, enhance the knowledge of managers, employees and investors as to the value of human capital. The study concludes that the increasingly critical effect of human capital has not been matched by advances in management and accounting practices that would allow resources to be properly reflected in management decision making and the operation of capital markets.

The study considers that without advances in the internal measurement and reporting of human capital, management are unable to fully recognise the value of their employees' competencies and commitment for business performance. The bulletin states that without advances in the external reporting of human capital, capital markets are unable to allocate capital efficiently to firms whose principal assets are not reflected in their balance sheets.

REFERENCE

Elias, J. and Scarborough, H. (2004) *Evaluating Human Capital*, CIPD bulletin. Chartered Institute of Personnel and Development, London. First published 2002, updated 1 September 2004.

Appendix 12
Change Management

This appendix should be read in conjunction with Section 8.8.7.

There are a series of issues that are commonly recognised as ingredients of a successful change programme. The absence of any of these ingredients can lead to project failure and hence must be considered as risks. These issues include the following:

- *Organisational effectiveness*. Understanding the elements of organisations that lead to organisational effectiveness, such as strategy, structure, systems, staff, style, shared values and skills.
- *Information gathering*. Gaining clarification that a problem exists and that further action is warranted and required. Understanding specifically what is not working well and what corrective action is required.
- *Objectives*. Correctly translating the drivers for change into clear unambiguous objectives that are readily understood by the organisation. Comprehension is aided if they are short, sharp, succinct statements rather than long, wordy "monologues".
- *Translation of objectives*. Ensuring that the change project deliverables are a direct translation of the objectives and that deliverables are defined for all of the objectives. Furthermore, senior management must be readily able to understand the deliverable descriptions – they must not be "woolly" statements but hard outputs like restructuring, a new finance process or the appointment of a new staff member.
- *Language*. Ensuring a consistent language is used in all documents to describe both elements of the organisation, its structure and the change project.
- *Speed of change*. Adopting an appropriate rate of change to suit the scale and complexity of the change – that is, not being overoptimistic in change planning. Management is often too consumed at the strategic end and underestimates the scale of the challenge in executing programmes. Or the perceived urgency for change overrides appropriate levels of internal consultation.
- *Management team*. Establishing a change management team composed of members with the right blend of skills (reflecting the primary areas of change) and individuals who were responsible for the original assessment of the need for the change, to ensure continuity. Additionally, establishing subgroups responsible for discrete elements of the change project so that the project is broken down into manageable parts. Where subgroups are involved they must have the ability to seamlessly integrate with each other. Otherwise individuals will be working at cross-purposes with each other, resulting in the different pieces of the jigsaw not fitting together.
- *Change champion*. Having an appropriate change "champion" who has sufficient authority to ensure the change is implemented and overrides individuals who try to block changes that threaten their own interests and standing.
- *Change plan*. Establishing an activity plan or road map for the change effort that is realistic, effective and clear. The activity plan is the schedule for the change project and must include the key activities that reflect the objectives, the logical chronological sequence in which they

will be carried out, and the interrelationship between the activities. The activities reflect the change project deliverables.

- *Support*. Commitment planning – identifying the key people and groups whose commitment is needed to accomplish the change. For any large change process there is a critical mass of individuals or groups whose active commitment and support is necessary for the change to occur.
- *Managing personal transitions*. Recognising the common reaction cycle of staff to planned changes and how the cycle is accommodated in the process. John Hayes and Peter Hyde propose a model, which includes shock, denial, depression, letting go, testing, consolidation and internalisation/reflection.
- *Progress*. Monitoring progress to see that the deliverables are being produced to schedule and that they meet the objectives.
- *Training*. Ensuring staff are provided with the right training and development to enable them to acquire new skills and competencies to implement the changes, where these entail the introduction of new disciplines and/or processes.
- *Attitudes*. Modifying the attitudes and behaviours of individuals and groups to adopt new working practices which in the past have been resisted through self-interests. Many of these "softer" elements are quintessential to achieving lasting success but are not given the same level of attention as the hard measures of cost and time.
- *Receptivity to change*. There are three aspects to this:
  - Developing a willingness and a readiness for change by not merely stressing the positive aspects of the proposed change but by destabilising the status quo, so that the forces for change exceed the forces for stability.
  - Drawing staff into the process rather than treating them as objects of or obstacles to change. Making them responsible for it. Communicating the change to draw staff into the discussions and debates about the need for and the form of the change and allowing them the freedom to discuss the issues involved openly, to get them to convince themselves of the need for the change. In the words of Dale Carnegie: "People support a world that they help create."
  - Communicating to staff the pressures for change on a repetitive basis, highlighting the organisation's shortcomings to pave the way for change.
- *Performance feedback*. Providing feedback on the performance of processes or services. This provides an organisation with the opportunity to draw attention to any discrepancy between actual performance and desired present and future performance. The feedback has to be in a form that staff can relate to and act on. Feedback has to be timely. If it is very old news it has little potency.
- *Uncertainty*. Establishing a regular and effective communication process to significantly reduce people's levels of uncertainty. One of the major mistakes organisations make when introducing change is to fail to recognise and deal with the real and legitimate fears and concerns of managers and staff. Organisations need to recognise that change does create uncertainty and that some individuals may become so unsettled and anxious that they seek alternative employment before finding out what the changes entail. Close attention also needs to be paid to an organisation's past history of change and the extent to which this reduces or enhances people's fears and concerns.
- *Involvement*. Identifying and enrolling those whose assistance is necessary and those who are essential to make change happen. Where it is possible, ensuring that all those closely affected are involved in some, if not all, aspects.

- *Reinforcing desired behaviour*. Within organisations people generally do things that bring rewards or avoid criticism. Consequently, one of the most effective ways of sustaining the momentum for change is to reinforce the kinds of behaviour required to make it successful.
- *Resources*. Where staff and managers have to work long hours merely to get their normal work done, additional resources are provided, whether these are financial, human or both. Case studies show that where additional resources have not been provided at the outset, the project has suffered and the resources had to be provided eventually anyway to "catch up" on uncompleted tasks.

Appendix 13
Industry Breakpoints

This appendix should be read in conjunction with Section 8.8.8.

The following description of breakpoints is predominantly based on the description provided by Paul Strebel (1997).

An industry breakpoint occurs when the market is presented with a new offering so superior in terms of customer value (arising from delivered cost) that it completely changes the rules of the competitive game. More and more frequently, industries are being shaken by dramatic shifts in competitive behaviour that make the current strategies obsolete. Newcomers emerge out of nowhere to dislodge the established industry leaders. A new offering can cause a sharp shift in the industry's growth rate while the competitive response to the new business system results in a dramatic realignment of market shares. The breakpoints that everyone will be familiar with are those that occurred in the personal computer industry:

- Introduction by Apple offering individuals the possibility of decentralised computing power (enhanced customer value arising from a convenience and cost that was on a completely different level from that provided by centralised mainframes).
- Introduction of a personal computer by IBM that became an industry standard (enhanced customer value in terms of predominantly price).
- Introduction by Apple of the Macintosh, with a completely new level of user-friendliness (enhanced customer value in the form of hard disks, better graphics, greater speed and particularly new operating software).
- Introduction by such companies as Compaq of quality with portability at a very competitive price, driven by pronounced competition from decreasing demand as a result of a recession in the US in the late 1980s (enhanced customer value in terms of lower cost and greater functionality).
- Move towards laptops, workstations and integrated networks, triggered by more powerful chips supplied by Intel and software supplied by Microsoft (enhanced customer value in terms of lower cost and greater functionality).

There is a danger that the description of the causes of breakpoints can be oversimplified. For instance, it could be assumed that all the breakpoints for improvements in personal computers were driven by improvements in technology. However, this is not the case, as can be seen from the previous list, as the US recession was a trigger or breakpoint in the overall development of the computer industry. Breakpoints can be triggered by many factors other than technology, such as the economic cycle, government policy and shifting consumer preferences.

Strebel (1992) makes the important distinction between divergent and convergent breakpoints. He describes the two forms of breakpoints as follows:

- Divergent breakpoints are associated with sharply increasing variety in competitive offerings, resulting in more value for the customer.
- Divergent breakpoints arise when a competitor discovers a new business opportunity and seeks to explore its boundaries, making new offerings.

- Convergent breakpoints are associated with sharp improvements in the systems and processes used to deliver offerings, resulting in lowered delivered cost.
- Convergent breakpoints arise when (1) imitation of innovation by competitors has reached a point where it is impossible to differentiate offerings and offerings have converged; (2) returns made on the original innovation have declined; (3) businesses have exhausted improvements in total quality management, continual improvement and restructuring of the business system, in an attempt to cut costs and maintain market share; (4) cost cutting and consolidation have run their course and it is now extremely difficult to squeeze further costs out; and (5) businesses now seek new businesses opportunities.

REFERENCES

Strebel, P. (1992) *Breakpoints: How Managers Exploit Radical Business Change*. Harvard Business School Press, Boston.

Strebel, P. (1997) Breakpoints: How to stay in the game. In T. Dickson and G. Bickerstaffe (eds), *Financial Times Mastering Management*. Financial Times/Prentice Hall, London.

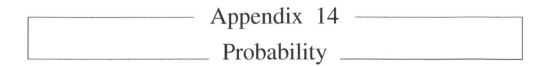

Appendix 14
Probability

This appendix should be read in conjunction with Section 10.7.1.

LOGIC PROBABILITIES

When the range of possible outcomes can be foreseen, assigning a probability to an event is a matter of simple arithmetic. Using a coin for the purposes of illustration, suppose you are going to toss it three times. What is the probability of only two tails? The set of all possible outcomes from tossing three coins is as follows (where, for example, one outcome from three tosses of the coin is three tails, or TTT): TTT, TTH, THH, THT, HTH, HTT, HHT and HHH. Of the eight possible outcomes, only three involve two tails. Hence there is a three in eight chance of two tails. The probability is $^3/_8$ or 0.375. Looking at it another way, each outcome has a $^1/_8 = 0.125$ chance of occurring, so the probability of two heads can be found by adding $0.125 + 0.125 + 0.125 = 0.375$. The likelihood of not getting two tails can be determined in one of two ways. The chance of not getting two tails can be determined by adding together the non-occurrences and dividing by the total number of possible outcomes, that is, $^5/_8$ or 0.625. A simpler method is to remember that as probabilities must add to 1, failure to achieve two heads must be $1 - 0.375 = 0.625$. This has highlighted two important rules. (1) If there are a series of outcomes where A occurs during n outcomes in total, the probability of event A is calculated by a/n. The abbreviated way of writing this relationship is $P(A) = a/n$. (2) The probability of an event not occurring is equal to the probability of it happening subtracted from one. Abbreviated this means $P(\text{not } A) = 1 - P(A)$.

OBJECTIVE PROBABILITIES

Objective probabilities are based on information usually gathered from past experience. So, for example, the manager of a railway operating company leasing rolling stock may be able to provide information concerning the possible life of a newly purchased train based on the records of similar trains purchased in the past.

SUBJECTIVE PROBABILITIES

On many occasions, especially with business problems, probabilities cannot be found from pure logic, or past data is neither appropriate nor available. In these circumstances, they have to be allocated subjectively. Subjective probabilities are based on opinion, experience or intuition. After considering all the available information, a probability value that expresses our degree of belief on the likely outcome is specified. You might say, for example, "considering current circumstances and my knowledge of the past behaviour of our competitors, I think that there is a 15% chance (i.e. a 0.15 probability) that they will imitate our new software product within one year". Such judgements are acceptable as the best that you can do, when hard facts are not available. However, it should be borne in mind that as subjective probability expresses a

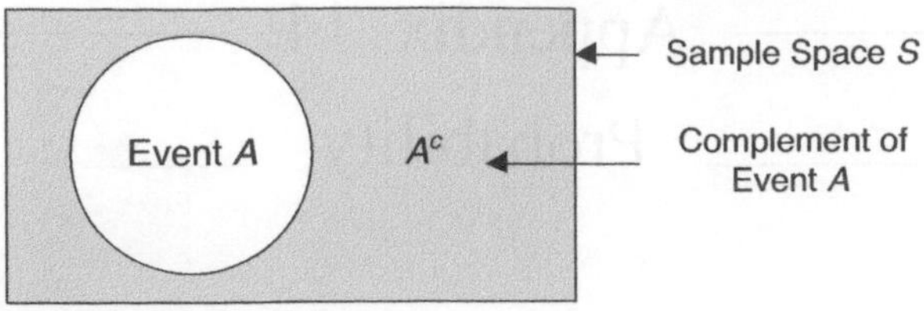

Figure A14.1 Complement of event *A*

person's degree of belief, it is personal. Using the subjective method, different people can be expected to assign different probabilities to the same risk event.

RELATIONSHIPS OF PROBABILITY

To be able to apply conditional probability, including Bayes' theorem, it is necessary to understand some basic concepts. The first is the complement of an event.

The *complement* of an event A is defined as the event consisting of all sample points that are not in A. A simple way of explaining this concept is with the aid of a Venn diagram (Figure A14.1). The rectangular area represents the sample space for the evaluation and as such contains all the sample points within the sample under examination. The circle represents event A and contains only the sample points that relate to A. The shaded region contains all the sample points not in event A and, by definition, is the complement of A. So as an example, the sample space could represent all members of a business unit within a pharmaceutical company and the circle (event A) could contain all those that had a PhD in chemistry.

In any probability application either event A or its complement A^c must occur. Hence

$$P(A) + P(A^c) = 1,$$

which we can also write as $P(A) = 1 - P(A^c)$. Consider the case of a telecommunications sales manager, who after reviewing sales reports states that 85% of new customer contacts result in a "no sale". By permitting A to denote the event of a sale and A^c to represent the event of no sale, the manager is stating that $P(A^c) = 0.85$. Using the equation $P(A) = 1 - P(A^c)$ we see that in this example

$$P(A) = 1 - P(A^c) = 1 - 0.85 = 0.15.$$

Hence we conclude that a new customer contact has a 0.15% probability of resulting in a sale.

The second concept to appreciate is the *addition rule for intersecting events*. The addition rule is used to find the probability of two events occurring, that is, the probability of event A occurring, event B occurring, or both. Given two events A and B, the *union* of A and B is defined as follows: the union of A and B is the event containing all sample points belonging to A or B or both. The union is denoted by $A \cup B$. The Venn diagram illustrating the union of two events is shown in Figure A14.2. The fact that the circles overlap indicates that some of the sample points are contained in both A and B. Given two events A and B, the event containing the sample points belonging to both A and B is called the *intersection* of A and B. The intersection is denoted by $A \cap B$.

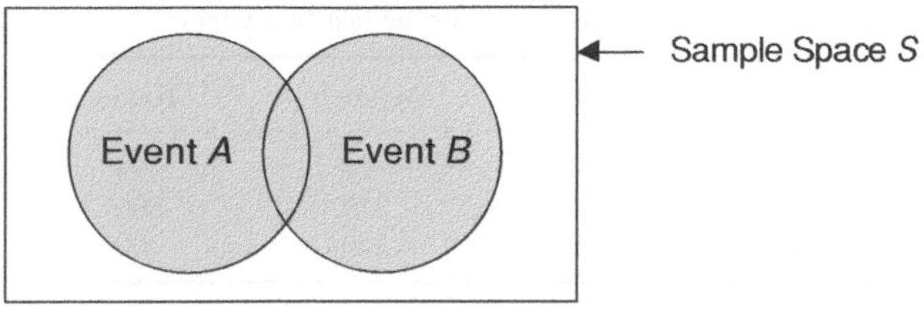

Figure A14.2 Union of events A and B

The addition rule is used to calculate the probability of the union of two events. The addition law is written as follows:

$$P(A \cup B) = P(A) + P(B) - P(A \cap B).$$

To understand this addition law intuitively, observe that the first two terms $P(A) + P(B)$ cover all of the sample points in the union of events A and B ($A \cup B$). However, as the sample points in $A \cap B$ are in both A and B, when the calculation $P(A) + P(B)$ is completed it is effectively counting the sample points in $A \cap B$ twice. Hence this duplication is corrected by subtracting $P(A \cap B)$.

The third concept to appreciate is the *addition rule for mutually exclusive events*. The addition rule is used to find the probability of two events occurring, that is, the probability of event A and B occurring together. Two events are said to be mutually exclusive if the events have no sample points in common. Hence, a requirement for events A and B to be mutually exclusive must be non-intersection between the sample points contained in the two events. The Venn diagram illustrating the union of two mutually exclusive events is shown in Figure A14.3. In this case $P(A \cap B) = 0$. The addition rule for mutually exclusive events can be written as follows:

$$P(A \cup B) = P(A) + P(B).$$

CONDITIONAL PROBABILITY

Frequently the probability of one event is influenced by whether a related event has already occurred. For instance, suppose we have an event A with probability $P(A)$. If we obtain new information and learn that a related event denoted by B has already occurred, it would be prudent to take advantage of this information by calculating a new probability for event A. This new probability of event A is called a *conditional* probability and is written $P(A \mid B)$. The notation "|" is used to indicate that we are considering the probability of an event A given the

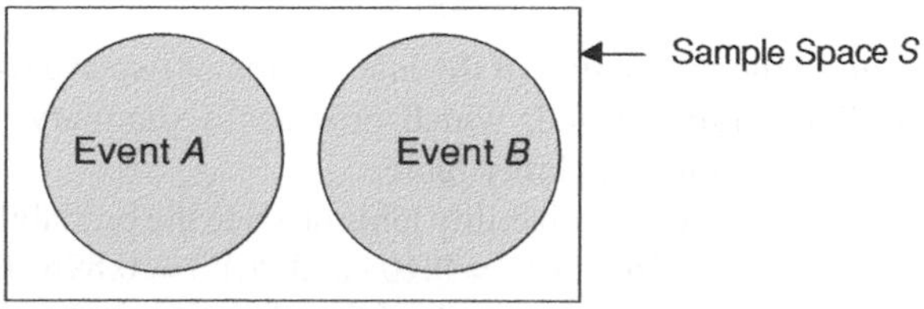

Figure A14.3 Mutually exclusive events A and B

Table A14.1 Promotions over the last two years

	Men	Women	Totals
Bonus	73	21	94
No bonus	587	219	806
Totals	660	240	900

condition event B has occurred. Therefore the notation $P(A \mid B)$ reads "the probability of A given B".

As an illustration of the application of conditional probability, consider the situation of the promotion status of male and female employees of a satellite sales office of a major software house in the South of England. There are 900 employees, 660 men and 240 women, located in this satellite office. Over the past two years 94 employees have received bonuses. The specific breakdown of bonuses for male and female employees is given in Table A14.1.

After reviewing annual bonuses, a group of female employees raised a discrimination case on the basis that 73 male employees had received bonuses but only 21 female employees had received bonuses. The board argued that the relatively low number of bonuses awarded to female employees was not due to discrimination, but due to the fact that relatively few females are employed. Conditional probability can be used to analyse the discrimination charge. Let M denote the event that an employee is male, F the event that the employee is female, A the event that the employee receives a bonus, and its complement A^c the event that the employee does not receive a bonus. Dividing the data values in Table A14.1 by the total of 900 employees enables us to summarise the available information with the following probability values: the probability that a randomly selected employee is a man *and* receives a bonus is

$$P(M \cap A^c) = \frac{73}{900} = 0.081;$$

the probability that a randomly selected employee is a man *and* does not receive a bonus is

$$P(M \cap A^c) = \frac{587}{900} = 0.652;$$

the probability that a randomly selected employee is a female *and* receives a bonus is

$$P(F \cap A) = \frac{21}{900} = 0.023;$$

and the probability that a randomly selected employee is a female *and* does not receive a bonus is

$$P(F \cap A^c) = \frac{219}{900} = 0.243.$$

As each of these values gives the probability of the intersection of two events, the probabilities are called *joint probabilities*. Table A14.2, which provides a summary of the probability information, is referred to as a joint probability table.

The values in the margins of the joint probability table provide the probabilities of each event separately: $P(M) = 0.733$, $P(F) = 0.267$, $P(A) = 0.104$ and $P(A^c) = 0.896$. These probabilities are referred to as *marginal probabilities* as a result of their location in the margins in the joint probability table. From the marginal probabilities it is possible to deduce that 73.3% of the

Table A14.2 Joint probability table

	Men (M)	Women (F)	Totals
Bonus (A)	0.081	0.023	0.104*
No bonus (A^c)	0.652	0.243	0.896*
Totals	0.733*	0.267*	1.000

*These figures are referred to as marginal probabilities.

employees are male, 26.7% are female, 10.4% received a bonus and 89.6% did not receive a bonus.

The probability that an employee receives a bonus and is a man, using the previously defined notation, is described as $P(A \mid M)$. To calculate $P(A \mid M)$, we must recognise that this notation simply means that we are attempting to establish the probability of event A, the award of a bonus, and that (in this instance) the award will be made to a male M. Hence $P(A \mid M)$ tells us that we are concerned with the probability of a bonus being awarded to one of 660 men. As 73 of the 660 male employees received a bonus, the probability of a male employee receiving a bonus is $73/660 = 0.11$. In other words, taking a very simplistic view, male employees had an 11% chance of receiving a bonus over the past two years. Let us demonstrate how conditional probabilities such as $P(A \mid M)$ can be calculated directly from related event probabilities rather than from the frequency data of Table A14.2.

The conditional probability (the probability of A given B) that a bonus will be awarded to an individual given that the individual is a male can be calculated as

$$P(A|M) = \frac{73}{660} = \frac{73/900}{660/900} = \frac{0.081}{0.733} = 0.11.$$

We now see that the conditional probability $P(A \mid M)$ can be calculated as 0.081/0.733. These figures can be found in Table A14.2. Observe that 0.081 is the joint probability of events A and M, where $P(A \cap M) = 0.081$. Observe also that 0.733 is the marginal probability that a randomly selected employee is male, that is, $P(M) = 0.733$. So the conditional probability $P(A \mid M)$ can be calculated as the ratio of the joint probability $P(A \cap M)$ to the marginal probability $P(M)$:

$$P(A|M) = \frac{P(A \cap M)}{P(M)} = \frac{0.081}{0.733} = 0.11.$$

The fact that conditional probabilities can be calculated as a ratio of a joint probability to a marginal probability provides the following general formula for conditional probability calculations for two events A and B:

$$P(A|B) = \frac{P(A \cap B)}{P(B)} \quad \text{or} \quad P(B|A) = \frac{P(A \cap B)}{P(A)}.$$

These principles can now be applied to the issue of discrimination against female staff in the award of annual bonuses. The critical issue involves the two conditional probabilities $P(A \mid M)$ and $P(A \mid F)$. That is, what is the probability of a bonus, given that the employee is male and what is the probability of a bonus, given that the employee is female? Obviously if these two probabilities are equal then the same proportions of men and women are receiving bonuses. It has already been established that $P(A \mid M) = 0.11$. Now using the values in the joint

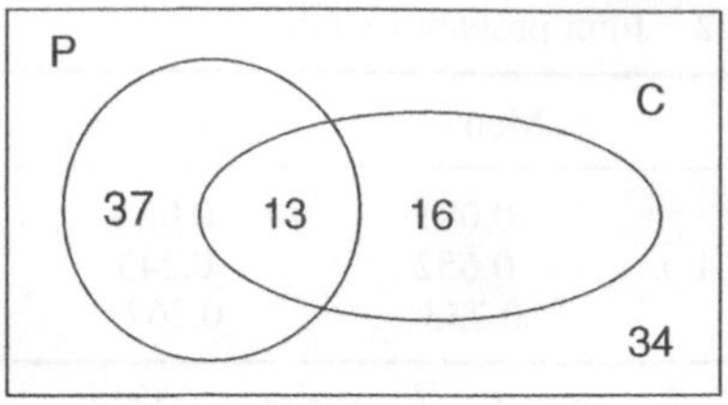

Figure A14.4 Venn diagram illustrating types of degree held by employees

probability table and the general formula for conditional probability, it is possible to calculate the probability that an employee is awarded a bonus, given that the employee is female. Using the previous formula but replacing B with F, we obtain

$$P(A|F) = \frac{P(A \cap F)}{P(F)} = \frac{0.023}{0.267} = 0.09.$$

The conclusion we draw is that there is an 11% chance of receiving a bonus if you are male and a 9% chance of receiving a bonus if you are female. Although the use of conditional probability in isolation will never prove that discrimination exists, the results can add weight to the argument. However, in this instance the difference is marginal and there may have been external influences which brought about the difference.

MULTIPLICATION LAW

Consider the situation where a pharmaceutical company department has 100 employees. Fifty of these employees studied physics at university, 29 studied chemistry and 13 studied both physics and chemistry (see Figure A14.4). The chance that an individual picked at random is found to have studied physics, having already established that they studied chemistry, may be represented by $P(P \mid C) = 13/29$ as there are 13 of the 29 employees who studied both physics and chemistry:

$$P(P|C) = \frac{P(P \cap C)}{P(C)} = \frac{13}{29}.$$

If we rearrange this equation, we have

$$P(P \cap C) = P(C|P) \times P(P).$$

For events A and B, the multiplication rule thus states that

$$P(A \cap B) = P(B|A) \times P(A).$$

INDEPENDENT EVENTS

When one outcome is known to have no effect on another outcome, then the events are said to be *independent*. For example, if the probability of a machine breaking down is 1/12 and the probability of stoppage of raw materials is 1/7, then it is possible to find the probability of the two events happening together by multiplying the two probabilities, because the occurrence of one of these events does not affect the probability of the other. So

$$\text{P(breakdown and stoppage of suppliers)} = 1/12 \times 1/7 = 1/84.$$

Table A14.3 Historical quality data of two suppliers

	Good parts (%)	Bad parts (%)
Supplier 1	97	3
Supplier 2	94	6

BAYES' THEOREM

In the previous description of conditional probability, it was indicated that revising probabilities when new information is obtained is an important phase of probability analysis. Often we commence an analysis with the initial or prior probability estimates for specific events. Then, having obtained additional information, we update the prior probability values by calculating revised probabilities, referred to as *posterior* probabilities. Bayes' theorem provides a means for making these probability calculations. As an application of Bayes' theorem, consider an aircraft manufacturing firm that receives shipments of parts from two different suppliers. Let A_1 denote the event that a part is from supplier 1 and A_2 denote that a part is from supplier 2. Currently 72% of the parts purchased by the company are from supplier 1 and the remaining 28% are from supplier 2. Hence, if a part is selected at random, we would assign the prior probabilities $P(A_1) = 0.72$ and $P(A_2) = 0.28$.

The quality of the purchased parts varies with the supplier. Historical data suggests that the quality ratings of the two suppliers are shown in Table A14.3. If we let G denote the event that the part is good and the letter B denote that the event is bad, the information in Table A14.3 provides the following conditional probability values:

$$P(G|A_1) = 0.97, P(B|A_1) = 0.03;$$

$$P(G|A_2) = 0.94, P(B|A_2) = 0.06.$$

The tree diagram in Figure A14.5 describes the process of the business receiving a part from one of the two suppliers and then discovering that the part is good or bad. There are a total of

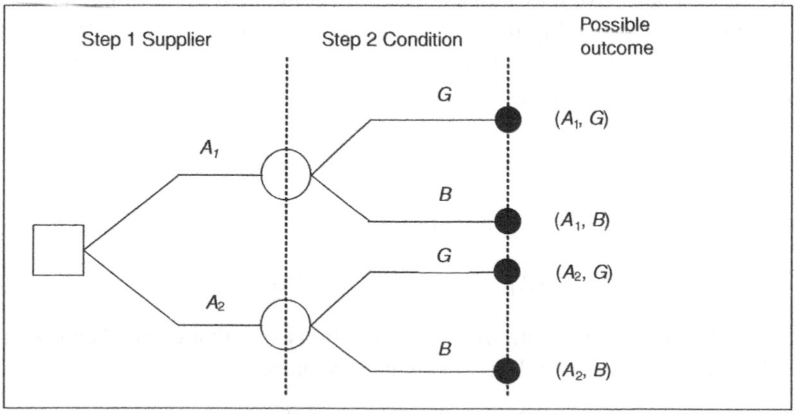

Figure A14.5 Tree diagram for two suppliers

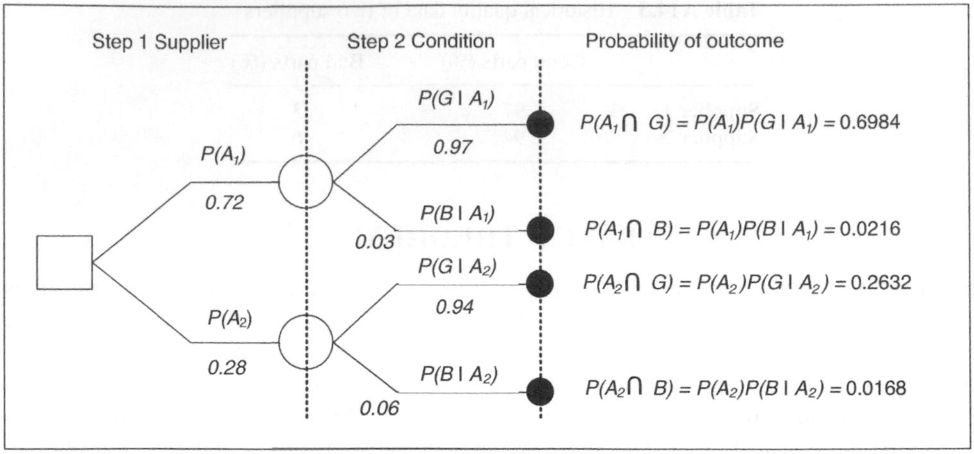

Figure A14.6 Probability tree diagram for two suppliers

four potential outcomes: two correspond to the part being bad and two correspond to the part being good.

Each of the experimental outcomes is the intersection of two events, so the multiplication rule can be used to calculate the probabilities. For instance:

$$\mathrm{P}(A_1, G) = \mathrm{P}(A_1 \cap G) = \mathrm{P}(A_1)\mathrm{P}(G|A_1).$$

The process of calculating these combined probabilities can be illustrated by a probability tree (see Figure A14.6). From left to right through the tree, the probabilities for each branch at step 1 are prior probabilities and the probabilities for each branch at step 2 are conditional probabilities. To find the probabilities of each outcome, we simply multiply the probabilities on all the branches leading to the outcome. Each of these joint probabilities is included in Figure A14.6, along with the known probabilities of each branch.

Suppose that the parts from the two suppliers are used in the business's manufacturing process (on the basis that faulty parts are not detected on arrival) and that a machine breaks down as a result of a defective part. Bayes' theorem, combined with Figure A14.6, can now be used to establish the probability that the part came from supplier 1 or 2. With B denoting, as before, the event that the part is bad, the objective now is to find the posterior probabilities $\mathrm{P}(A_1 \mid B)$ and $\mathrm{P}(A_2 \mid B)$. From the law of conditional probability, we know that

$$\mathrm{P}(A_1|B) = \frac{\mathrm{P}(A_1 \cap B)}{\mathrm{P}(B)}$$

and in addition, referring to the probability tree,

$$\mathrm{P}(A_1 \cap B) = \mathrm{P}(A_1)\mathrm{P}(B|A_1).$$

To find the probability $\mathrm{P}(B)$, we note from the probability tree that event B can occur in only two ways: $(A_1 \cap B)$ and $(A_2 \cap B)$. In consequence, we have

$$\begin{aligned}\mathrm{P}(B) &= \mathrm{P}(A_1 \cap B) + \mathrm{P}(A_2 \cap B)\\ &\ \mathrm{P}(A_1)\mathrm{P}(B|A_1) + \mathrm{P}(A_2)\mathrm{P}(B|A_2).\end{aligned}$$

Adopting the law of conditional probability and substituting $P(A_1 \cap B)$ with $P(A_1)P(B \mid A_1)$ and $P(B)$ with $P(A_1)P(B \mid A_1) + P(A_2)P(B \mid A_2)$ (from the probability tree) and writing a similar result for $P(A_2 \mid B)$, we obtain Bayes' theorem for the two events under examination: for event A_1 we have

$$P(A_1|B) = \frac{P(A_1)P(B|A_1)}{P(A_1)P(B|A_1) + P(A_2)P(B|A_2)},$$

and for event A_2 we have

$$P(A_2|B) = \frac{P(A_2)P(B|A_2)}{P(A_1)P(B|A_1) + P(A_2)P(B|A_2)}.$$

Using the equation above for the probability that the bad product came from supplier 1, we have

$$\begin{aligned} P(A_1|B) &= \frac{P(A_1)P(B|A_1)}{P(A_1)P(B|A_1) + P(A_2)P(B|A_2)} \\ &= \frac{0.72 \times 0.03}{0.72 \times 0.03 + 0.28 \times 0.06} = \frac{0.0216}{0.0216 + 0.0168} = 0.5625. \end{aligned}$$

Using the equation above for the probability that the bad product came from supplier 2, we have

$$\begin{aligned} P(A_2|B) &= \frac{P(A_2)P(B|A_2)}{P(A_1)P(B|A_1) + P(A_2)P(B|A_2)} \\ &= \frac{0.28 \times 0.06}{0.72 \times 0.03 + 0.28 \times 0.06} = \frac{0.0168}{0.0216 + 0.0168} = 0.4375. \end{aligned}$$

It can therefore be concluded that if a bad part is encountered during manufacture there is likely to be more than a 50/50 chance that it came from supplier 1. Note that the two aggregated results (0.5625 + 0.4375) equal 1.

Bayes' theorem is applicable when the events for which we wish to calculate posterior probabilities are mutually exclusive and their union occupies the entire sample space.

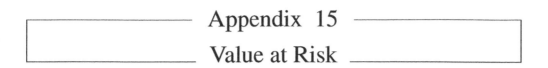

Appendix 15
Value at Risk

The following is a list of recommended reading on the subject of value at risk:

Alexander, C. (2009) *Market Risk Analysis: Volume IV: Value at Risk Models*. John Wiley & Sons, Ltd, Chichester.

Best, P. (1998) *Implementing Value at Risk*. John Wiley & Sons, Ltd, Chichester.

Butler, C. (1999) *Mastering Value at Risk: A Step-by-Step Guide to Understanding and Applying VAR*. Financial Times Pitman, London.

Choudhry, M. and Ketul, T. (2006) *An Introduction to Value-at-Risk*, 4th edition. John Wiley & Sons, Ltd, Chichester.

Dowd, K. (1998) *Beyond Value at Risk: The New Science of Risk Management*. John Wiley & Sons, Ltd, Chichester.

Duc, F. and Schorderet, Y. (2008) *Market Risk Management for Hedge Funds: Foundations of the Style and Implicit Value-at-Risk*. John Wiley & Sons, Inc., Hoboken, NJ.

Holton, G. (2003) *Value at Risk: Theory and Practice*. Academic Press, Amsterdam.

Jorion, P. (2006) *Value at Risk:The New Benchmark for Managing Financial Risk*, 3rd edition. McGraw-Hill, New York.

Pearson, N.D. (2002) *Risk Budgeting: Portfolio Problem Solving with Value-at-Risk*. John Wiley & Sons, Inc., New York.

Saunders, A. (1999) *Credit Risk Measurement: New Approaches to Value-at-Risk and Other Paradigms*. John Wiley & Sons, Inc., New York.

Saunders, A. and Allen, L. (2010) *Credit Risk Management in and out of the Financial Crisis: New Approaches to Value at Risk and Other Paradigms*, 3rd edition. John Wiley & Sons, Inc., Hoboken, NJ.

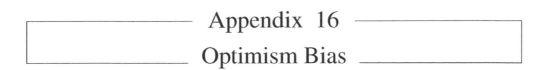

Appendix 16
Optimism Bias

METHOD ADOPTED IN CALCULATING OPTIMISM BIAS

The method proposed for calculating optimism bias is based predominantly on the recommendations of the UK Department for Transport (DfT 2004, 2007a, 2007b, 2010), the *Green Book* (HM Treasury 2003, 2003b) and the Mott MacDonald Review of Large Public Procurement in the UK (Mott MacDonald 2002).

The method adopted (for the adjustment for investment costs for optimism bias) follows a ten-step approach as follows and as illustrated in Figure A16.1:

1. Determine the reference class and the type of project.
2. Decide on the stage of the project.
3. Decide on the applicable upper bound value for optimism bias provided for the project type.
4. Decide on the contributory factors to the upper bound optimism bias for the project type.
5. Carry out a risk assessment of the contributory factors, discerning the "mitigation factor" (between 0 and 1) and the "cost of risk management".
6. Add together the effectively managed contributions to the optimism bias for each project risk area (as a percentage).
7. Calculate the revised optimism bias percentage.
8. Calculate the capital impact of optimism bias.
9. Calculate the cost of risk mitigation.
10. Calculate the adjusted capital estimate.

METHOD FOR CALCULATING OPTIMISM BIAS FOR COST

This section is based on an example Metro project called XYZ – which has reached life cycle stage 4, called single option development.

Step 1: Determine the reference class and the type of project

The first step involves determining the reference class and the type of project, according to the typology given in Table A16.1. Transport schemes are divided into a number of reference classes which are treated as statistically different, but where the projects within each of the reference classes can be treated as statistically similar. Bent Flyvbjerg (DfT 2004) concluded that within each of the reference classes identified in Table A16.1, the risk of investment cost overruns can be treated as statistically similar.

Step 2: Decide on the stage of the project

Optimism bias assumes that as a project progresses through its life cycle, the requirements, scope definition, schedule, cost estimate, change management and risk analysis become more developed and mature and hence the possible underestimation of project costs and duration diminishes. Hence, the suggested percentage levels of optimism bias reduce as a project progresses through the life cycle. To be able to select the appropriate/recommended level of

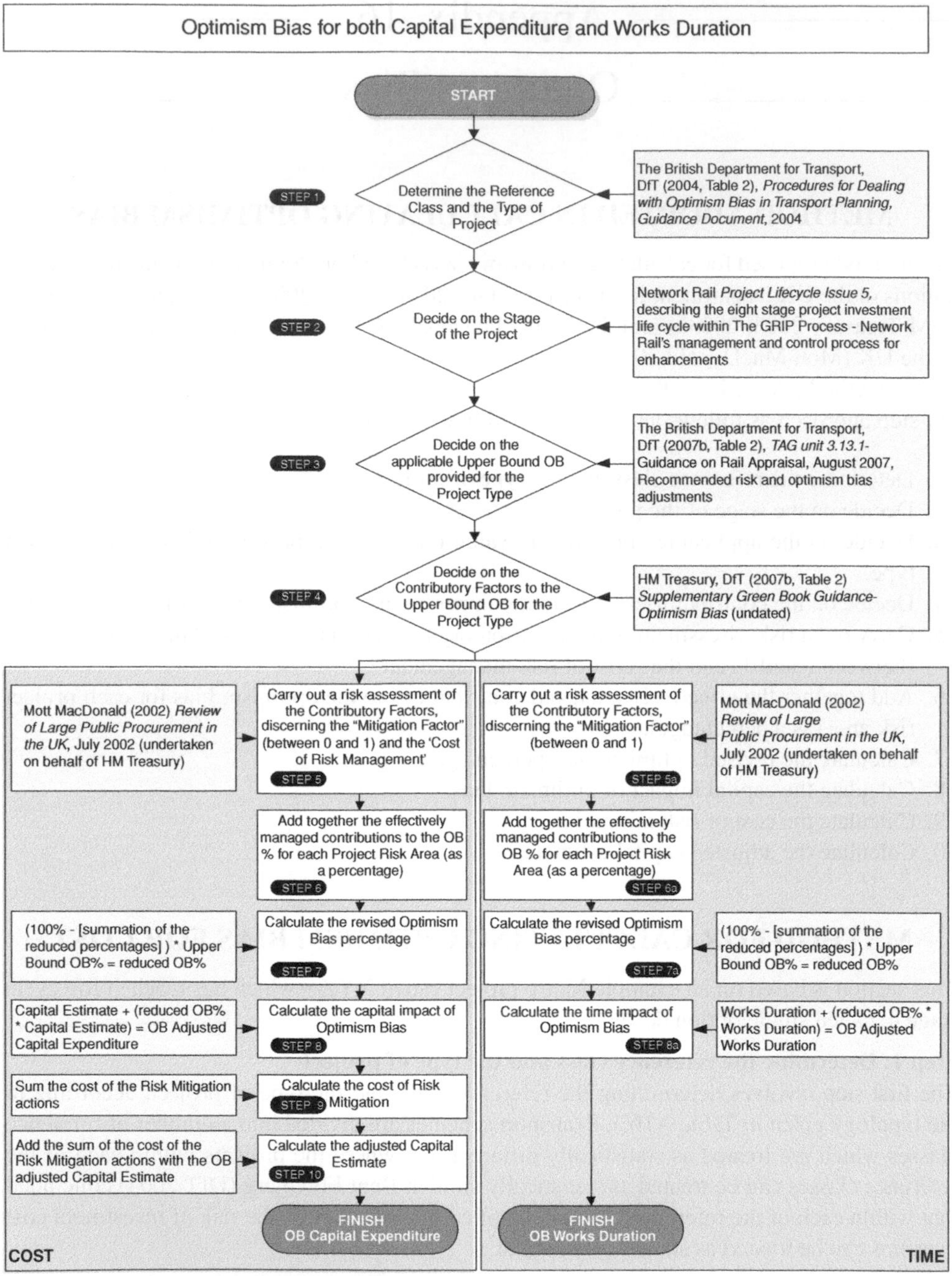

Figure A16.1 Optimism bias for capital expenditure

Table A16.1 Project reference classes and types of project

Reference class categorisation	Type of project
Roads	Motorway Trunk roads Local roads Bicycle facilities Pedestrian facilities Park and ride Bus lane schemes Guided buses on wheels
Rail	Metro Light rail Guided buses on tracks Conventional rail High speed rail
Fixed links	Bridges and tunnels
Building projects	Stations and terminal buildings
IT projects	IT system development

Source: DfT (2004).

optimism bias pertinent to a particular project stage it is necessary to decide on the stage that has been reached for the project under examination. The Network Rail Guide to Rail Investment Projects (GRIP) investment life cycle stage descriptions have been adopted here as illustrated in Table A16.2, and the stage identified for the development of the business case for the Metro is "single option development".

Step 3: Decide on the applicable upper bound value for optimism bias provided for the project type

Once the project classification, type and stage have been selected, then the upper bound optimism bias figure is to be selected from the UK DfT guidance. (Step two of the approach described in the Supplementary Green Book Guidance (HM Treasury 2003b, p. 3), states that projects should use the appropriate upper bound value for optimism bias as the starting value for calculating the optimism bias level.)

Table A16.3 shows the recommended uplifts for optimism bias in rail projects drawn from DfT Transport Analysis Guidance (TAG) unit 3.13.1 entitled "Guidance on Rail Appraisal" (DfT 2007b). This guidance is based on Strategic Rail Authority Appraisal Criteria (Strategic Rail Authority, April 2003) updated in 2007 by the DfT to bring it in line with current DfT appraisal guidance.

The upper bound for optimism bias for this project currently at the single option development stage, based on Table A16,3, is 18%.

Step 4: Decide on the Contributory factors to the upper bound optimism bias for the project type

The approach to calculating optimism bias described in the Supplementary Green Book Guidance (HM Treasury 2003b, p. 3) states that projects should reduce the upper bound value (for optimism bias) by the extent to which the contributing factors can be managed.

Before the contributing factors can be managed they need to be identified. In this example methodology the contributing factors identified by the Mott MacDonald research will be

Table A16.2 Eight life cycle stages for rail projects based on UK Network Rail GRIP model

Network Rail Investment Life Cycle Stages							
1	2	3	4	5	6	7	8
Outline Definition	Pre-Feasibility	Option Selection	Single Option Development	Detailed Design	Construction Test and Commission	Scheme Handover	Project Close Out
		→		←			
		Develops options for addressing the identified constraints and delivering the required incremental network capability. Assesses the options and selects the most appropriate one, together with confirmation that the outputs can be economically delivered.	Develops the selected single option to the point of engineering scope freeze and in sufficient detail to allow finalisation of the business case and scheduling of implementation resources.				

Source: Network Rail Project Lifecycle Issue 5.
(The investment stages within the life cycle reflect the significant business and technical milestones in the project's development and delivery.)

Table A16.3 Recommended risk and optimism bias adjustments for rail projects

Project Development Level*	Level 1	Level 2	Level 3	Level 4	Level 5
Activity	Pre-Feasibility	Project Defini-tion	Option Selection	Single Option Refinement	Design Development
	Capital expenditure				
QRA	No	No	No	QRA at mean estimate	QRA at mean estimate
Optimism Bias	66%	50%	40%	18%	6%
	Operational expenditure				
QRA	No	No	No	QRA at mean estimate	QRA at mean estimate
Optimism Bias	41% of present value Opex	1.6% per annum	1% per annum	Evidence based	Evidence based

Source: DfT (2007b, Table 2).
*Definition of project development levels is consistent with Network Rail's project development definition in GRIP.

adopted, specifically those described in Table 3 (optimism bias upper bound guidance for civil engineering projects) contained within the Supplementary Green Book Guidance. Table 3 is subdivided into "non-standard civil engineering" and "standard civil engineering".

The Supplementary Guidance defines these types of civil engineering as follows:

> *Standard civil engineering projects*: are those that involve the construction of facilities, in addition to buildings, not requiring special design considerations e.g. most new roads.
>
> *Non-standard civil engineering projects*: are those that involve the construction of facilities, in addition to buildings, requiring special design considerations due to space constraints or unusual output specifications e.g. innovative rail, road, utility projects, or upgrade and extension projects.

Due to the nature of the Metro in this example, which it is envisaged will include rail which is elevated, at grade and within tunnels (cut-and-cover and bored) and the latest technology to operate the rolling stock, this project has been classed as a non-standard civil engineering project.

Table A16.4 is an extract from Table 3 in the Supplementary Green Book Guidance which describes the risk area contributions which lead to the incidence of optimism bias, under the headings of procurement, project specific, client specific, environment and external influences.

Step 5: Carry out a risk assessment of the contributory factors, discerning the "mitigation factor" (between 0 and 1) and the "cost of risk management"

When calculating optimism bias, the extent to which the risks are mitigated is measured by a mitigation factor. The mitigation factor has a value between 0.0 and 1.0, where 0.0 means that risks in a project area are not mitigated, and 1.0 means all risks in a project area are fully mitigated. A value between 0.0 and 1.0 represents partial mitigation of the risks in the project area. Clear and tangible evidence must be observed and independently verified for the mitigation of risks in project risk areas before reductions in optimism bias should be made.

Table A16.4 Risk area contributions

Optimism bias upper bound guidance for non standard civil engineering

Risk area contributions to recorded optimism bias %		Capital Expenditure	Works Duration
Procurement	Complexity of contract structure		4
	Late contractor involvement in the design		< 1 (0.5 assumed)
	Poor contractor capabilities		2
	Government guidelines		
	Dispute and claims occurred		16
	Information management		
	Other	2	1
Project specific	Design complexity	8	5
	Degree of innovation	9	13
	Environmental impact	5	
	Other		3
Client specific	Inadequacy of the business case	35	3
	Large number of stakeholders		
	Funding availability	5	
	Project management team	2	
	Poor project intelligence	9	3
	Other		
Environment	Public relations		
	Site characteristics	5	
	Permits/consents/approvals		
	Other		
External influences	Political		19
	Economic	3	24
	Legislation/regulations	8	
	Technology	8	6
	Other	1	< 1 (0.5 assumed)
	TOTAL	**100**	**100**

Source: HM Treasury (2003b).

The proposed strategies for the mitigation of project risks and management of project risk areas form part of the business case.

Step 6: Add together the effectively managed contributions to the optimism bias for each project risk area (as a percentage).

Example calculation (capital expenditure)
Project: Rail Project XYZ

Project level:	4
Project activity:	Project definition
Upper bound optimism bias:	18% (see Table A16.3)
Project capital estimate:	AED 100million (for simplicity)

Table A16.5 Rail project XYZ details

Contributory Factor	% Contribution to Optimism Bias	Mitigation Factor	Cost of Risk Management
Design complexity	5	0.9	AED 2 million
Inadequacy of the business case	25	0.3	AED 1 million
Poor project intelligence	10	0.6	AED 3 million
Site characteristics	5	0.4	AED 5 million

Table A16.5 would be the output of a risk register recording the individual risks which give rise to the contributory factors, their respective response actions, an assessment of how successful the response actions would be (i.e. whether the risk would be eradicated in its entirety or whether there would be residual risk), and the cost of response actions where appropriate (such as specific studies which would remove assumptions and uncertainty from the design process).

$$\text{Final capital expenditure: AED 100m} + (18\% \times \text{AED 100m}) = \text{AED 150m}$$

(taking account of optimism bias)

$$\text{Managed optimism bias} = \text{reduction in optimism bias} = (5 \times 0.9) + (25 \times 0.3) + (10 \times 0.6) + (5 \times 0.4) = 20\%.$$

Step 7: Calculate the revised optimism bias percentage
Developing the example outlined in step 6:

$$\text{Resultant capital expenditure optimism bias} = (100\% - 20\%) \times 18\% = 14.4\%.$$

Step 8: Calculate the capital impact of optimism bias
Developing the example outlined in step 6, the forecast capital expenditure for this example (excluding the cost of risk management) is AED 114.4 million, calculated as follows:

$$\text{AED100m} + (14.4\% \times \text{AED 100m}) = 114.4\text{m}.$$

Step 9: Calculate the cost of risk mitigation
Developing the example outlined in step 6, the cost of risk mitigation/risk response planning is simply the summation of the costs of the individual response costs, which in this case is

$$\text{AED}\,(2 + 1 + 3 + 5) = \text{AED 11 million}.$$

Step 10: Calculate the adjusted capital estimate
Developing the example outlined in step 6, the forecast capital expenditure for this example including optimism bias and the cost of risk management is AED 125.4 million, calculated as follows:

$$\text{AED 114.4m} + \text{AED}\,(2 + 1 + 3 + 5) = \text{AED 114.4m} + \text{AED 11m} = \text{AED 125.4m}.$$

This figure may change if the risk response actions are not as effective as envisaged or the anticipated costs of risk response actions exceed expectations.

REFERENCES

Department for Transport (2004) *Procedures for Dealing with Optimism Bias in Transport Planning*, guidance document prepared by Bent Flyvberg in association with COWI on behalf of the Department for Transport, June. http://flyvbjerg.plan.aau.dk/0406DfT-UK%20OptBiasASPUBL.pdf

Department for Transport (2007a) *Transport Analysis Guidance*, Unit 3.5.9: The Estimation and Treatment of Scheme Costs (http://www.dft.gov.uk/webtag).

Department for Transport (2007b) *Transport Analysis Guidance*, Unit 3.13.1: Guidance on Rail Appraisal (http://www.dft.gov.uk/webtag).

Department for Transport (2010) *Transport Analysis Guidance*, Unit 2.7.1: Transport Appraisal and the Treasury Green Book (http://www.dft.gov.uk/webtag).

Department of Communities and Local Government (2007) *Adjusting for Optimism Bias in Regeneration Projects and Programmes*, Guidance Note. DCLG, London. http://www.communities.gov.uk/publications/corporate/adjustingoptimism

Federal Transit Administration (2003) *Project and Construction Guidelines*, Chapter 3: General Management Principles for Transit Capital Projects, Section 3.5.5. http://www.fta.dot.gov/publications/reports/other_reports/planning_environment_1338.html

HM Treasury (2003a) *Appraisal and Evaluation in Central Government*. The Stationery Office, Norwich.

HM Treasury (2003b) Supplementary Green Book guidance on optimism bias. http://www.hm-treasury.gov.uk/media/885/68/GreenBook_optimism_bias.pdf

Mott MacDonald (2002) *Review of Large Public Procurement in the UK*. Mott MacDonald, Croydon.

Index

Printed and bound by CPI Group (UK) Ltd, Croydon, CR0 4YY

27/10/2024

14580391-0002